Engineering Biotechnology

Engineering Biotechnology

V Sree Krishna
Vice Principal and Head, Department of Biotechnology
JB Institute of Engineering and Technology (JBIET)
(Moinabad, Ranga Reddy District, Hyderabad, Andhra Pradesh)

S Vijaya Sree
MBBS and DGO from Osmania Medical College
Hyderabad, Andhra Pradesh

Publishing Globally

NEW AGE INTERNATIONAL (P) LIMITED, PUBLISHERS
New Delhi • Bangalore • Chennai • Cochin • Guwahati
Hyderabad • Kolkata • Lucknow • Mumbai
Visit us at **www.newagepublishers.com**

Published by New Age International (P) Ltd., Publishers
First Edition: 2014

BRANCHES

- **Bangalore** 37/10, 8th Cross (Near Hanuman Temple), Azad Nagar, Chamarajpet, Bangalore-560 018
Tel.: (080) 26756823, Telefax: 26756820, **E-mail: bangalore@newagepublishers.com**
- **Chennai** 26, Damodaran Street, T. Nagar, Chennai-600 017
Tel.: (044) 24353401, Telefax: 24351463, **E-mail: chennai@newagepublishers.com**
- **Cochin** CC-39/1016, Carrier Station Road, Ernakulam South, Cochin-682 016
Tel.: (0484) 2377004, Telefax: 4051303, **E-mail: cochin@newagepublishers.com**
- **Guwahati** Hemsen Complex, Mohd. Shah Road, Paltan Bazar, Near Starline Hotel, Guwahati-781 008
Tel.: (0361) 2513881, Telefax: 2543669, **E-mail: guwahati@newagepublishers.com**
- **Hyderabad** 105, 1st Floor, Madhiray Kaveri Tower, 3-2-19, Azam Jahi Road, Nimboliadda, Hyderabad-500 027
Tel.: (040) 24652456, Telefax: 24652457, **E-mail: hyderabad@newagepublishers.com**
- **Kolkata** RDB Chambers (Formerly Lotus Cinema) 106A, 1st Floor, S.N. Banerjee Road, Kolkata-700 014
Tel.: (033) 22273773, Telefax: 22275247, **E-mail: kolkata@newagepublishers.com**
- **Lucknow** 16-A, Jopling Road, Lucknow-226 001
Tel.: (0522) 2209578, 4045297, Telefax: 2204098, **E-mail: lucknow@newagepublishers.com**
- **Mumbai** 142C, Victor House, Ground Floor, N.M. Joshi Marg, Lower Parel, Mumbai-400 013
Tel.: (022) 24927869, Telefax: 24915415, **E-mail: mumbai@newagepublishers.com**
- **New Delhi** 22, Golden House, Daryaganj, New Delhi-110 002
Tel.: (011) 23262368, 23262370, Telefax: 43551305, **E-mail: sales@newagepublishers.com**

ISBN: 978-81-224-3615-0

₹ 250.00

C-13-06-6954

Printed in India at Nisha Enterprises, Delhi.
Typeset at Pagitek Graphics, Delhi.

PUBLISHING FOR ONE WORLD
NEW AGE INTERNATIONAL (P) LIMITED, PUBLISHERS
7/30 A, Daryaganj, New Delhi-110002
Visit us at **www.newagepublishers.com**

Dedicated
to
our son
Engineer, S. Gopala Krishna, B.E. (CSE)
for
his cooperation and isolation he has experienced during our total concentration while authoring this book

Acknowledgement

We acknowledge our sincere thanks to all our well-wishers who have encouraged us to author this book. I extend my sincere thanks to Prof. G. Venkateshwarh Head, Department of Biotechnology, who has been teaching, guiding me since my M.Sc., Biological Sciences to till date.

We extend our sincere gratitudes to Hon'ble Sri Joginpala Bhaskar Rao Garu Chairman, J.B. Group of Educational Institutions, and respected Sri J. Venkata Krishna Rao Garu, Secretary J.B.I.E.T. for their strong support and encouragements.

We extend our sincere thanks to our sisters and brothers Smt. Anasuya, Smt. Sumathi, Dr. V. Rama Krishna, Sri V. Radha Krishna and Sri V. Bala Krishna for their endeavourance and support in our life.

We extend our thanks to Sri Saumya Gupta (Managing Director), New Age International (P) Ltd., Publishers for accepting our proposal to author this book. The authors extend their profound thanks to the editorial team of M/s New Age International (P) Ltd., Publishers for showing keen interest in editing this book to this presentable level.

We also extend our sincere gratitudes to Mr. Sudershan (Branch Manager) and all other staff of Bangalore branch for their support in bringing out this book.

Last but not least, we are specially thankful to our son Mr. S. Gopala Krishna for having spared his valuable time and patience in providing privacy, thus enabling us to author this book with devoted dedication.

V. Sree Krishna
S. Vijaya Sree

Preface

Engineering Biotechnology has initiated with reference to study about human anatomy, human physiology, biochemistry for understanding various biochemical reactions taking place from cellular level to various systems. Apart from this Engineering Biotechnology encompasses microbiology, to understand several useful microbes and pathogenic microorganisms, which are causative factors for different diseases.

The book Engineering Biotechnology provides the readers with an introduction to current state and future prospects of this break through technology. It offers an entiring look into a future engrossed in Engineering Biotechnology—its applications with the help of overviews on the enabling technologies being used currently in various biotechnology related industries.

Potential of Engineering Biotechnology to articulate and programme matter with human anatomy and physiology has invited scientists to explore innumerable applications of Biotechnology to mankind.

The increasing demand and importance of Engineering Biotechnology is an inspiration for the authors by contributing their ideas, observations and findings for the benefit of students of Medical Sciences, M.Sc. (Biotechnology), B.E., B.Tech., Biotechnology, Pharmacy, research scholars, academicians, technologists and industrialists associated with it or its allied field.

V. Sree Krishna
S. Vijaya Sree

Syllabus of III Semester B.E., Biotechnology as per Visveshvariah Technological University Belgaum, Karnataka

Part A: Human Physiology

Sub Code	:	06BT-33	IA Marks	:	25
Hrs/Week	:	04	Exam Hours	:	03
Total Hrs.	:	52	Exam Marks	:	100

Unit 1: Skeletal & Muscular System

Cartilage and bone; Comparison between cartilage and bone; Functions of skeletal system; Joints; Muscles of limb movement. Principal types of muscles; General properties of muscles; Mechanism of muscle contraction and relaxation, Red and white muscle fibers.

7 Hours

Unit 2: Circulatory System

Structure, Composition and functions of blood. Blood groups and Rh factor. Immunity and antibody formation. Heart and blood vessels, Arterial and venous system. Properties of heart muscle. Action of Heart and Heart Beat. Blood pressure. Lymph and lymphatic system.

6 Hours

Unit 3: Digestive System

Overview of digestive system, functional anatomy of digestive system: mouth, phaynx, esophagus, the stomach the small and large intestine. Digestive glands, Enzymes; Physiology of Digestion and Absorption. Energy requirements of the body.

6 Hours

Unit 4: Respiratory System

Introduction; Structure of respiratory organs; Mechanism of breathing; Pulmonary air volumes, Gas exchange in the lungs; Respiratory adjustments in exercise, Artificial respiration; Kinds of respiration; Transport of respiratory gases in the blood; Cellular respiration; Respiratory quotient; Some respiratory disorders; Control of respiration.

7 Hours

Unit 5: Exretory System

Methods of exretion; Physiological processes involved in excretion; Kidneys; Anatomy and physiology, Nephron and its structure. Functions of nephron; Nephron physiology and mechanism of urine formation; Regulation of urine formation; Osmoregulation by kidney.

6 Hours

Unit 6: Nervous System

Introduction; Role of nervous system; Generalized neuron; Morphological types of neuron; Physiological or functional types of neurons; Main properties of nervous tissue; Stimulus; Mode of action of nerves;

Conduction of nerve impulses; Reflex action; Central nervous system; The brain; The spinal cord; Peripheral nervous system and reflux activity. Special senses; tongue, smell, eye, hearing and balance.

10 Hours

Unit 7: Endocrine System

Introduction; Endocrine systems of vertebrates; Pituitary gland; Thyroid gland; Parathyroid gland; Pancreas; Adrenal or suprarenal glands; Sex glands; Gastrointestinal mucosa; Thymus gland; Pineal gland; Summary of different endocrine glands; Their hormones and influence; Summary of the effect of hyper secretion and hyposecretion of some important endocrine glands.

6 Hours

Unit 8: Reproductive System

Physiology of male and female reproduction systems, invitro fertilization, fertility in males and females, factors influencing fertility, test tube baby, perm count preservation of sperms.

4 Hours

Question Paper Pattern

For very 6–7 hours of teachning one question to be set. Eight questions to be set (four from each part) out of which five full questions (considering at least two from each part) to be answered.

Part B: Biochemistry

Sub Code	:	06BT-33	IA Marks	:	25
Hrs/Week	:	04	Exam Hours	:	03
Total Hrs.	:	52	Exam Marks	:	100

Unit 1: Basic Concepts

Types of Chemical reactions, pH, buffers and their properties, concentration of solution. Stereochemistry of carbon compounds.

4 Hours

Unit 2: Introduction to Biomolecules

Carbohydrates, fats and lipids, structure and properties of phospholipids, glycolipids, steroids, amino acids and proteins, biologically important peptides, purines, pyrimidines, nucleic Acids—DNA and RNA.

6 Hours

Unit 3: Principles of Bioenergetics

Energy, energy flow cycle, energy conversion. Structure and properties of ATP. High energy compounds, Thermodynamic considerations, Coupling reactions of ATP. High energy compounds, Thermodynamic considerations, Coupling reactions of ATP and NDP (Nucleotide di phosphate); photosynthesis ancillary Pigments, Photosystems PS I & II;

8 Hours

Unit 4: Transport Mechanism

Biological membranes: structure, permeability, properties, passive transport and active transport, facilitated transport, energy requirement, mechanism of Na^+/K^+, glucose and amino acid transport.

Organization of transport activity in cell. Action Potentials. Role of transport in signal transduction processes.

8 Hours

Unit 5: Carbohydrate Metabolism

Glycolysis—metabolism. Aerobic and anaerobic pathway and regulation, TCA cycle, NADPH Cycle, Calvin Cycle, Glyoxylate cycle, Pentose Phosphate Pathway. Electron transport chain and oxidative phosphorylation, energetics, energy balance sheet, oxidative stress, Gluconeogenesis regulation of gluconeogenesis. Biosynthesis of polysaccharides. Disorders of carbohydrate metabolism.

10 Hours

Unit 6: Lipid Metabolism

Biosynthesis of fatty acids, cholesterol, phospholipids, glycolipids, biodegradation of fatty acids. Disorders of lipid metabolism.

5 Hours

Unit 7: Amino Acid Metabolism

Biosynthesis of amino acids starting from acetyl CoA (with reference to oxaloacetate family). Biodegradation of amino acids, deamination, transformation and urea cycle. Disorders of amino acid metabolism.

6 Hours

Unit 8: Nucleic Acid Metabolism

Biosynthesis, biodegradation and regulation of Purines, pyrimidines and nucleic acids. Disorders of nucleic acid metabolism.

5 Hours

Question Paper Pattern

For every 6–7 Hours of teaching one question to be set. Eight questions to be set (four from each part) out of which five full questions (considering at least two from each part) to be answered.

Part C: Microbiology

Sub Code	:	06BT-33	IA Marks	:	25
Hrs/Week	:	04	Exam Hours	:	03
Total Hrs.	:	52	Exam Marks	:	100

Unit 1: Introduction

The Scope of Microbiology, History of Microbiology, Microbial diversity and Taxonomy, Prokaryotes and Eukaryotes, Types of Microorganisms, Bacteria, Viruses, Fungi, Protozoa.

4 Hours

Unit 2: Microscopic Examination

Bright-Field Microscopy, Dark-Field Microscopy, Phase-Contrast Microscopy, Fluorescence Microscopy, Electron Microscopy.

4 Hours

Unit 3: Microorganisms

The morphology and fine structure of bacteria, Cultivation of bacteria, Media and types of media nutritional requirements, Reproduction and growth, Factors affecting growth, Measurement of growth, Pure culture and cultural characteristics.

4 Hours

Structure, Classification, Modes of reproduction and growth of algae, Fungi, Protozoa and Viruses.

4 Hours

Unit 4: Microbial Techniques

Pure culture techniques (streak-plate, spread plane, pour plate), Staining of microorganisms (Simple staining, differential staining), Types of stains theories of staining, Enumeration of microorganisms (Direct microscopic count, by pour plate technique, Streak plate, membrane filtration, Electronic methods, etc.), Characterization: Biochemical tests and 16-s RNA homology studies.

10 Hours

Unit 5: Microbiology: Control of Microorganisms

Control of Microorganisms by physical methods (heat, filtration, radiation), Control by chemical ethods (phenol and phenolic compounds, Alcohols, Halogens, Dyes, Detergents, Aldehydes, Heavy metals, etc.), Amibiotics and other chemotherapeutic agents.

5 Hours

Unit 6: Microbial Metabolism

Aerobic and anaerobic growth, Metabolic pathways, Primary and Secondary metabolism.

5 Hours

Unit 7: Medical Microbiology

Common diseases caused by Microbes such as Tuberculosis, Leprosy, Typhoid, Darrhea, Pneumonia, AIDS, Polio, Rabies, Ameobiosis, Malaria.

8 Hours

Unit 8: Environmental and Industrial Microbiology

Microbiology of soil, Biofertilizers, Aquatic microbiology, Microbiology of domestic water and waste water, Microbes in bioremediation.

4 Hours

Traditional microbial processes using yeasts and bacteria, Microbes as source of protein, Microbial Insectricides, Enzymes from microbes, rDNA technology and useful products from microbes.

4 Hours

Contents

Part C: Microbiology

<u>Part A</u>
Human Physiology

UNIT 1 Skeletal and Muscular System

CARTILAGE AND BONE

Cartilage consists of a dense network of collagen fibres and elastic fibres firmly embedded in chondroitin sulfate, a gel like component of the ground substance. Cartilage can endure considerably more stress than loose and dense connective tissues. Whereas the strength of cartilage is due to its collagen fibres, its resilience (ability to assume its original shape after deformation) is due to chondroitin sulfate.

The cells of mature cartilage called chondrocytes occur singly or in groups within spaces called lacunae in the matrix. A membrane of dense irregular connective tissue called the perichondrium covers the surface of most cartilage. Unlike other connective tissues, cartilage has no blood vessels or nerves, except in the perichondrium. Three kinds of cartilage are recognized: hyaline cartilage, fibrocartilage and elastic cartilage.

Bone

Cartilage, joints and bones makeup the skeletal system. The skeletal system supports soft tissues, protects delicate structures, and works with skeletal muscles to generate movement. Bones store calcium and phosphorus; house red bone marrow, which produces blood cells; and contain yellow bone marrow, a storage site for triglycerides. Bones are organs composed of several different connective tissues, including bone or osseous tissue, the periosteum, red and yellow bone marrow, and the endosteum (a membrane that lines a space within bone that stores yellow bone marrow). Bone tissue is classified as either compact or spongy, depending on how its matrix and cells are organized.

The basic unit of compact bone is an Osteon or Haversian system. Each Osteon has four parts:

1. The lamellae are concentric rings of matrix that consists of mineral salts; mostly calcium and phosphates due to which the bone attains its hardness, and collagen fibres, which give bone its strength. The lamellae are responsible for the compact nature of this type of bone tissue.
2. Lacunae are small spaces between lamellae that contain mature bone cells called osteocytes.
3. Projecting from the lacunae are canaliculi, networks of minute canals containing the processes of osteocytes. Canaliculi provide routes for nutrients to reach osteocytes and for wastes to leave them.
4. A central (Haversian) canal contains blood vessels and nerves.

FUNCTIONS OF SKELETAL SYSTEM

Bone tissue and the skeletal system perform several basic functions:

1. *Support:* The skeleton serves as the structural frame work for the body by supporting soft tissues and providing attachment points for the tendons of most skeletal muscles.
2. *Protection:* The skeleton protects many internal organs from injury. For example, cranial bones protect the brain, vertebrae protect the spinal cord, and the rib cage protects the heart and lungs.
3. *Assistance in movement:* Because skeletal muscles attach to bones, when muscles contract, they pull on bones. Together bones and muscles produce movement.
4. *Mineral homeostasis:* Bone tissue stores several minerals, especially calcium and phosphorus. On demand, bone releases minerals into the blood to maintain critical mineral balances (homeostasis) and to distribute the minerals to other parts of the body.
5. *Blood cell production:* Within certain bones, a connective tissue called red bone marrow produces red blood cells, white blood cells and platelets, a process called hemopoiesis (hemo = blood; poiesis = making). Red bone marrow consists of developing blood cells, adipocytes, fibroblasts, and macrophages within a network of reticular fibres. It is present in developing bones of the fetus and in some adult bones, such as the pelvis, ribs, breast bone, backbones, skull, and ends of the arm bones and thigh bones.
6. *Triglyceride storage:* Triglycerides stored in the adipose cells of yellow bone marrow are an important chemical energy reserve. Yellow bone marrow consists mainly of adipose cells, which store triglycerides and a few blood cells. In the new born; all bone marrow is red and is involved in hemopoiesis. With increasing age, much of the bone marrow changes from red to yellow.

Joints

A joint, also called an articulation or arthrosis, is a point of contact between two bones, between bone and cartilage, or between bone and teeth. When we say one bone articulates with another bone. It means that the bones form a joint.

Joints are classified structurally, based on their anatomical characteristics, and functionally based on the type of movement they permit.

The structural classification of joints is based on two criteria:

1. The presence or absence of a space between the articulating bones, called a synovial cavity, and
2. The type of connective tissue that binds the bones together.

Structurally, joints are classified as one of the following types:

1. *Fibrous joints:* The bones are held together by fibrous connective tissue that is rich in collagen fibres, they lack a synovial cavity.
2. *Cartilaginous joints:* The bones are held together by cartilage; they lack a synovial cavity.
3. *Synovial joints:* The bones forming the joint have a synovial cavity and are united by the dense irregular connective tissue of an articular capsule, and often by accessory ligaments.

The functional classification of joints relates to the degree of movement they permit. Functionally, joints are classified as follows:

1. *Synarthrosis:* An immovable joint. The plural is termed synarthroses.
2. *Amphiarthrosis:* A slightly movable joint on (Amphi) both sides. The plural is termed amphiarthroses.

3. *Diarthrosis:* A freely movable joint. All diarthroses are synovial joints. They have a variety of shapes and permit several different types of movements.

As one examines the structure of each type of joint, they can also explore its functional attributes as detailed below.

Gomphoses

A gomphosis or dento-alveolar joint is a type of fibrous joint in which a cone shaped peg fits into a socket. Examples of this is articulations of the roots of the teeth with the sockets of the alveolar processes of the maxillae and mandible. The dense fibrous connective tissue between a tooth and its socket is the periodontal ligament (Fig. 1.1).

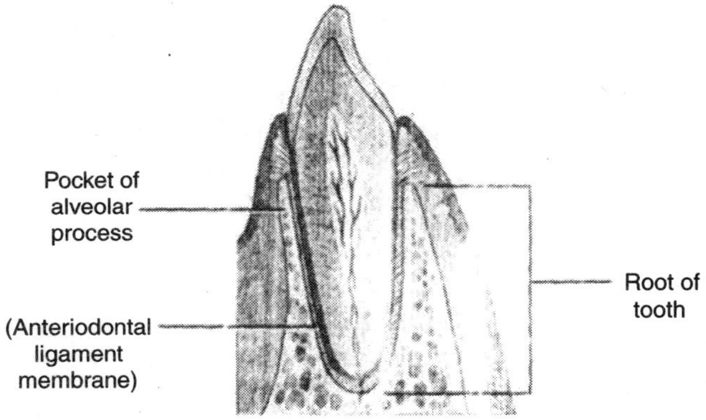

Fig. 1.1 Gomphosis between tooth and socket of alveolar process

CARTILAGINOUS JOINTS

Similar to a fibrous joint, a cartilaginous joint lacks a synovial cavity and allows little or no movement. The two types of cartilaginous joints are synchondroses and symphyses.

Synchondroses

A synchondroses is a cartilaginous joint in which the connecting material is hyaline cartilage. An example of a synchondrosis is the epiphyseal plate that connects the epiphysis and diaphysis of a growing bone.

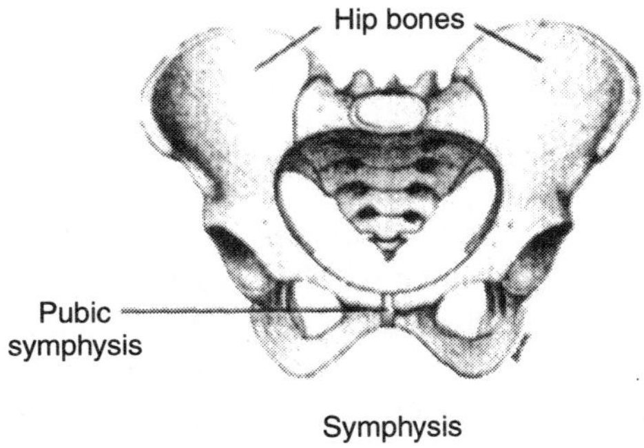

Fig. 1.2 Cartilaginous joints

Symphyses

A symphysis is a cartilaginous joint in which the ends of the articulating bones are covered with hyaline cartilage, but a broad, flat disc of fibrocartilage connects the bones. All symphyses occur in the midline of the body. The pubic symphysis between the anterior surfaces of the hip bones is one example of the symphysis (Fig. 1.2).

SYNOVIAL JOINT

Synovial joints have certain characteristics that distinguish them from other joints. The unique characteristics of a synovial joint is the presence of a space called a synovial cavity between the articulating bones. This cavity allows a joint to be freely movable. Therefore, all synovial joints are classified as diarthroses. The bones at synovial joint are covered by articular cartilage, which is of hyaline in nature. The cartilage covers the articulating surface of the bones with a smooth, slippery surface but does not bind them together. Articular cartilage reduces friction between bones in the joint during movement and helps to absorb shock. (Fig. 1.3).

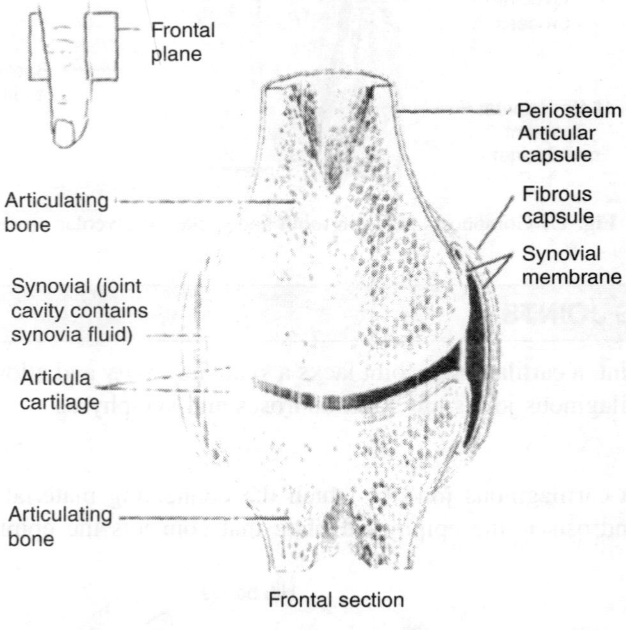

Fig. 1.3 Structure of a typical synovial joint

Types of synovial joints

Although all synovial joints have a generally similar structure, the shapes of the articulating surfaces vary enabling various types of movements are possible. In total there are six subtypes of synovial joints viz., planar, hinge, pivot, condyloid, saddle, and ball-and-socket joint.

Planar joints: The articulating surfaces of bones in a planar joint are flat or slightly curved (Fig. 1.4*a*). These joints permit side to side and back-and-forth gliding movements. These joints are said to be nonaxial because the motion they allow does not occur around an axis.

Examples of planar joints are the intercarpal joints between carpal bones and the wrist, intertarsal joints between tarsal bones at the ankle, sternoclavicular joints between the manubrium of the sternum and the clavicle sternocostal joints between the sternum and ends of the costal cartilages at the tips of the second through seventh pair of ribs and vertebrocostal joints between the heads and tubercles of ribs and transverse processes of thoracic vertebrae.

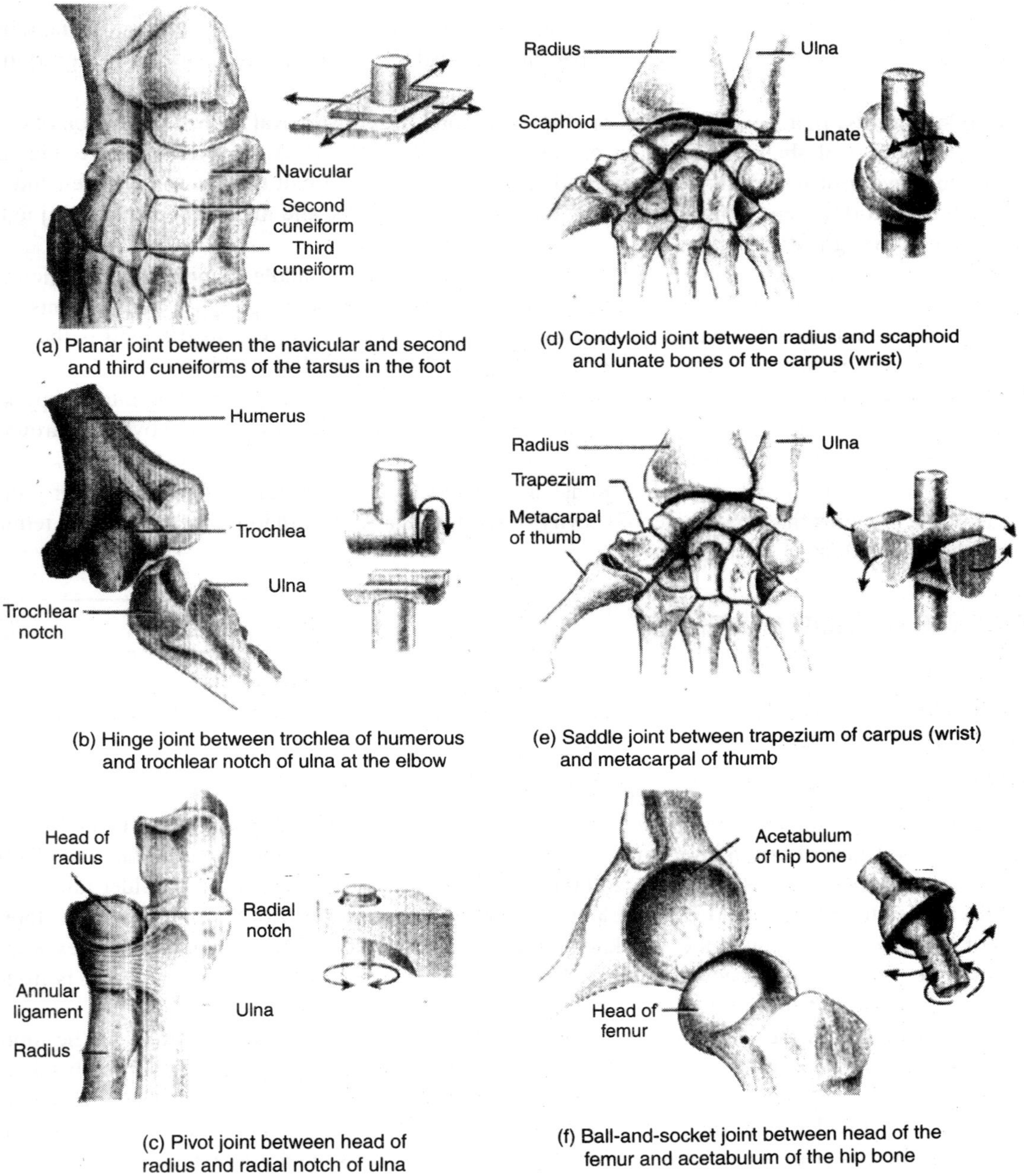

(a) Planar joint between the navicular and second and third cuneiforms of the tarsus in the foot

(d) Condyloid joint between radius and scaphoid and lunate bones of the carpus (wrist)

(b) Hinge joint between trochlea of humerous and trochlear notch of ulna at the elbow

(e) Saddle joint between trapezium of carpus (wrist) and metacarpal of thumb

(c) Pivot joint between head of radius and radial notch of ulna

(f) Ball-and-socket joint between head of the femur and acetabulum of the hip bone

Fig. 1.4 Subtypes of synovial joints

Hinge joints: In a hinge joint, the convex surface of one bone fits into the concave surface of another bone (Fig. 1.4*b*). As the name implies, hinge joints produce an angular, opening-and-closing motion like that of a hinged door. Hinge joints are uniaxial because they allow motion around a single axis. Examples of hinge joints are the knee, elbow, ankle and interphalangeal joints.

Pivot joints: In a pivot joint, the rounded or pointed surface of one bone articulates with a ring formed partly by another bone and partly by a ligament (Fig. 1.4*c*). A pivot joint is uniaxial because it allows rotation around its own longitudinal axis only. Examples of pivot joints are the otlanto-axial joint in which the atlas rotates around the axis and permits the head to turn from side to side, as in signifying 'no'.

Condyloid joints: In a condyloid joint or ellipsoidal joint the convex oval shaped projection of one bone fits into the oval shaped depression of another bone (Fig. 1.4*d*). A condyloid joint is biaxial because the movement it permits is around two axes. One can notice that they can move their index finger both up-and-down and from side-to-side. Examples are the wrist and metacarpophalangeal joints for the second through fifth digits.

Saddle joints: In a saddle joint, the articular surface of one bone is saddle shaped, and the articular surface of the other bone fits into the 'saddle' as a sitting rider would sit (Fig. 1.4*e*). Saddle joints are biaxial producing side-to-side and up-and-down movements. An example of a saddle joint is the carpometacarpal joint between the trapezium of the carpus and metacarpal of the thumb.

Ball-and-socket joints: This joint consists of the ball-like surface of one bone fitting into a cup like depression of another bone (Fig. 1.4*f*). Such joints are multiaxial because they permit movement around three axes plus all directions in between.

Examples of functional ball-and-socket joints are the shoulder and hip joints. At the shoulder joint, the head of the humerus fits into the glenoid cavity of the scapula. At the hip joint, the head of the femur fits into the acetabulum of the hip bone.

MUSCLES OF LIMB MOVEMENT

As it is a well known fact that human beings possess one pair each of hands and legs as limbs, muscles that move the wrist, hand, thumb and fingers as well as muscles that act on thigh bones and leg bones including of the foot are discussed here under.

Muscles that move the forearm (radius and ulna)

The biceps brachii is the large muscle located on the anterior surface of the arm. As indicated by its name, it has two heads of origin both from the scapula. The muscle spans both the shoulder and elbow joint. It also supinates the forearm at the radioulnar joints and flexes the arm at the shoulder joint. The brachialis is deep to the biceps brachii muscle. It is the most powerful flexor of the forearm at the elbow joint. For this reason it is called the 'workhorse' of the elbow flexors. The brachioradialis flexes the forearm at the elbow joint, especially when a quick movement is required or when a weight is lifted slowly during flexion of the forearm.

The triceps brachii is the large muscle located on the posterior surface of the arm. It is the more powerful of the extensors of the forearm at the elbow joint. As its name implies, it has three heads of origin, one from the scapula (long head) and two from the humerus (lateral and medial heads). The long head crosses the shoulder joint; the other heads do not. The anconeus is a small muscle located on the lateral part of the posterior aspect of the elbow that assists the triceps brachii in extending the forearm at the elbow joint.

Some muscles that move the radius and ulna are involved in pronation and supination at the radio-ulnar joints. The pronators as suggested by their names are the pronator teres, and pronator quadratus muscles. The supinator of the forearm is the aptly named the supinator muscle.

Generally we use the powerful action of the supinator when we twist a corkscrew or turn a screw with screw driver.

In the limbs functionally related skeletal muscles and their associated blood vessels and nerves are grouped together by fascia into regions called compartments. In the arm, the biceps brachii, brachialis, and coracobrachialis muscles comprise the anterior (flexor) compartment. The triceps brachii muscle forms the posterior (extensor) compartment. Fig. 1.5.

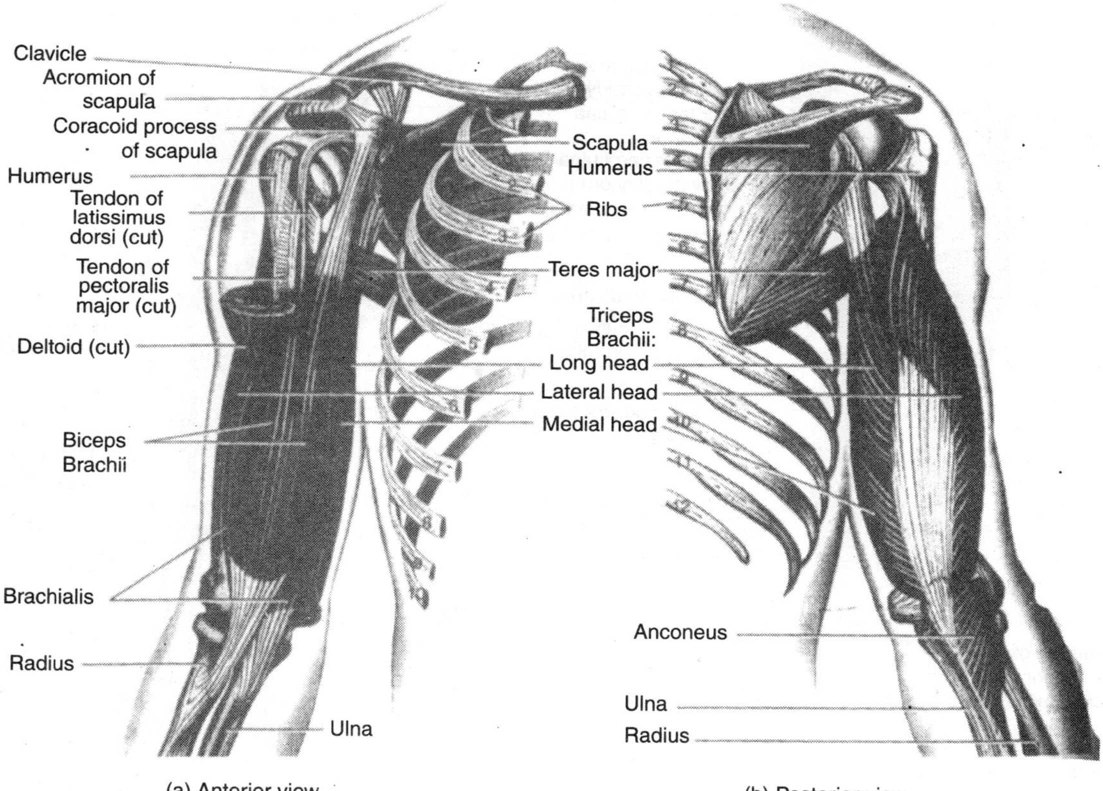

Fig. 1.5 Showing anterior and posterior muscles of forearm

Muscles which move the wrist, hand, thumb and fingers:

Muscles of the forearm that move the wrist, hand, thumb and fingers are many and varied. The muscles of the forearm are divided into two groups:

1. Anterior compartment muscles
2. Posterior compartment muscles.

The anterior compartment muscles of the forearm originate on the humerus, typically insert on the carpals, metacarpals and phalanges and function as flexors. The bellies of these muscles form the bulk of the forearm. The posterior compartment muscles of the forearm originate on the humerus, insert on the metacarpals and phalanges, and function as extensors within each compartment, the muscles are grouped as superficial or deep.

The superficial, anterior compartment muscles are arranged in the following order from lateral to medial; flexor carpi radials, palnaris longus and flexor carpiulnaris. The flexor digitorum superficialis muscle is deep to the other three muscles and is the largest superficial muscle in the forearm.

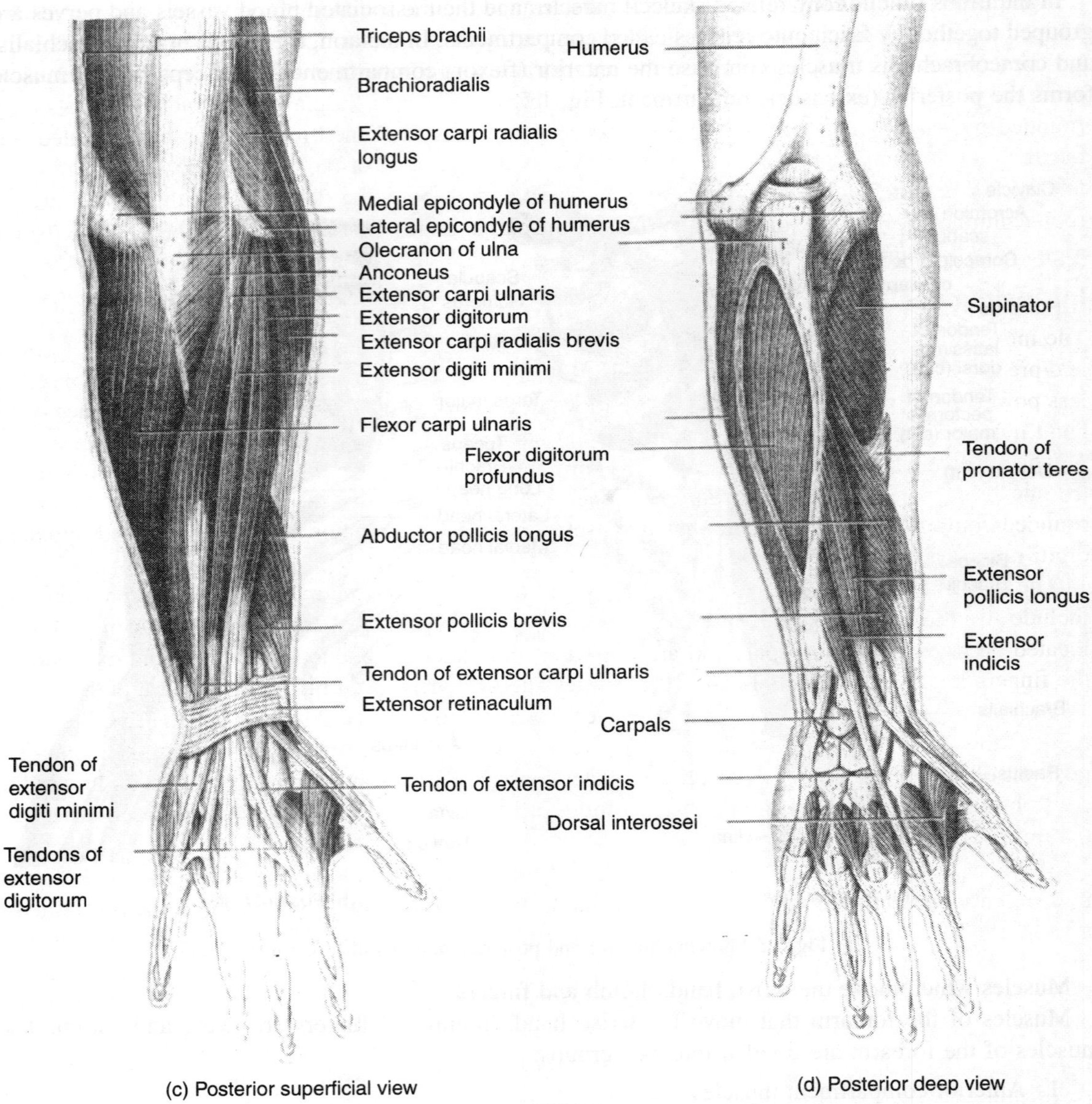

Fig. 1.6 Posterior superficial and deep view of forearm

The deep anterior compartment muscles are arranged in the following order from lateral to medial; flexor pollicis longus and flexor digitorum profundus ends in four tendons that insert into the phalanges of the fingers, whereas the flexor pollicis longus ends in the distal phalanx of the thumb.

The superficial posterior compartment muscles are arranged in the following order from lateral to medial; extensor carpi radials longus, extensor carpi radials brevis, extensor digitorum which occupies most of the posterior surface of the forearm and divides into four tendons that insert into the middle and distal phalanges of the fingers. Extensor digitiminimi, a slender muscle generally connected to the extensor digitorum and the extensor carpiulnaris.

The deep posterior compartment muscles are arranged in the following order from lateral to medial: abductor pollicis longus, extensor pollicis brevis, extensor pollicis longus and extensor indicis.

The tendons of the muscles of the forearm that attach to the wrist or continue into the hand, along with blood vessels and nerves are held close to bones by strong fasciae. The tendons are also surrounded by tendon sheaths. At the wrist, the deep fascia is thickened into fibrous bands called retinacula. The flexor retinaculum is located over the palmar surface of the carpal bones. The long flexor tendons of the digits and wrist and the median nerve pass through the flexor retinaculum. The extensor retinaculum is located over the dorsal surface of the carpal bones. The extensor tendons of the wrist and digits pass through it (Fig. 1.6).

Intrinsic muscles of the hand

The intrinsic hand muscles are located in the palm. They produce the less powerful but more intricate and precise movements of the digits. The intrinsic hand muscles are located in the palm. They produce less powerful but more intricate and precise movements of the digits. The functional importance of the hand muscles is readily apparent because certain hand injuries can cause permanent disability.

The intrinsic hand muscles are divided into three groups viz., (1) thenar, (2) hypothenar and (3) intermediate. The four thenar muscles act on the thumb and form the thenar eminence, the lateral rounded contour on the palm. The thenar muscles include the abductor pollicis brevis, and opponens digiti minimi.

The eleven intermediate muscles act on the digits except the thumb. The intermediate muscles include the lumbricals, palmar interossei, and dorsal interossei. Both sets of interossei muscles are located between the metacarpals and are important in abduction, adduction, flexion and extension of the fingers and in movements in skilled activities such as writing, typing and playing a piano.

Besides free motion, the hand performs several special movements. Power grip is the forcible movement of the fingers and thumb against the palm, as in hand shaking. Precision handling is a change in position of a handled object that requires exact control of finger and thumb positions, as in winding a watch or threading a needle. Pinching is compression between the thumb and index finger or between the thumb and first two fingers.

Movements of the thumb are very important in the precise activities of the hand. They are defined in different planes from comparable movements of the fingers because the thumb can be positioned at a right angle to the fingers. Figure 1.7 shows the five main movements of the thumb. They include flexion, extension, abduction, adduction and opposition. Opposition is the most distinctive digital movement. It gives human beings and other primates the ability to precisely grasp and manipulate objects.

Flexion – Movement of thumb medially across the palm.
Extension – Movement of the thumb laterally away from the palm.
Abduction – Movement of the thumb in an anteroposterior plane away from the palm.
Adduction – Movement of the thumb in an anteroposterior plane toward the palm.
Opposition – Movement of the thumb across the palm so that the tip of the thumb meets the tip of a finger.

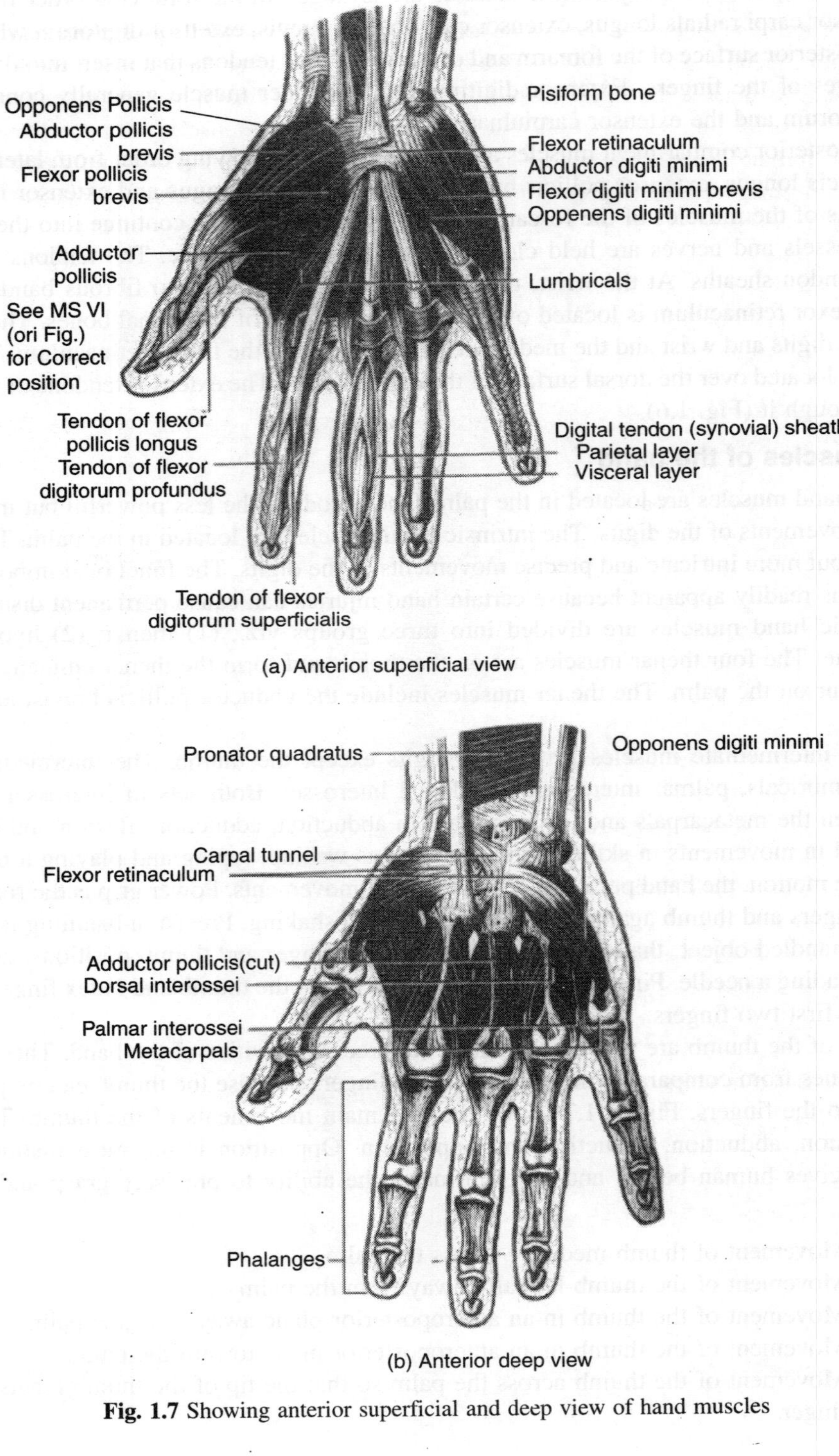

Fig. 1.7 Showing anterior superficial and deep view of hand muscles

Muscles that move the Femur (Thigh bone)

The majority of muscles that move the femur originate on the pelvic girdle and insert on the femur. The psoas major and iliacus muscles have a common insertion. Together, they are known as the iliopsoas muscle because they share a common insertion. There are three gluteal muscles: gluteus maximus is the largest and heaviest of the three muscles and is one of the largest muscles in the body. It is the chief extensor of the femur. The gluteus medius is mostly deep to the gluteus maximus and is a powerful abduction of the femur at the hip joint. It is a common site for an intramuscular injection. The gluteus minimus is the smallest of the gluteal muscles and lies deep to the gluteus medius.

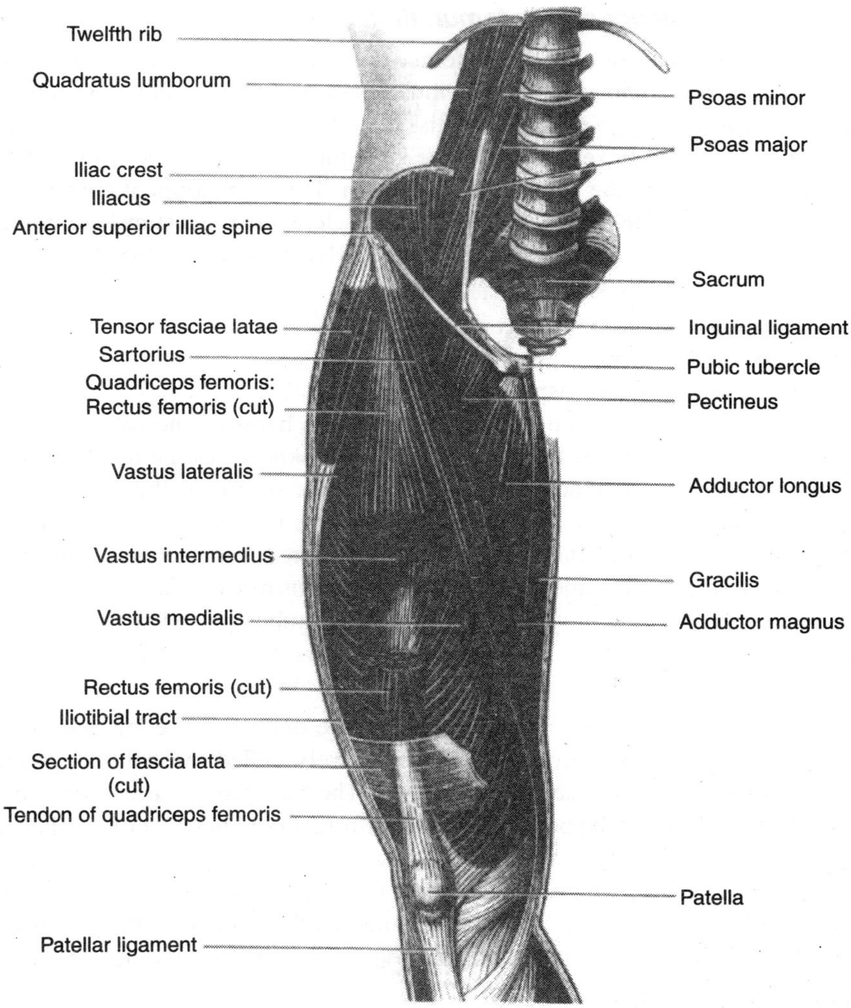

Fig. 1.8 Showing anterior view of femur muscles

The tensor fascialata muscle is located on the lateral surface of the thigh. The fascialata is a layer if deep fascia, composed of dense connective tissue, that encircles the entire thigh. It is well developed laterally where, together with the tendons of the tensor fasciae and gluteus maximus muscles, it forms a structure called the iliotibial tract. The tract inserts into the lateral condyle of the tibia.

The piriformis, obturator internus, obturator externus, superior gemellus, inferior gemellus, and quadratus femoris muscles are all deep to the gluteus maximus muscle and function as lateral rotators of the femur at the hip joint.

Three muscles on the medial aspect of the thigh are the adductor longus, adductor brevis and adductor magnus. They originate on the pubic bone and insert on the femur. All three muscles adduct, flex and medially rotate the femur at the hip joint. The pectineus muscle also adducts and flexes the femur at the hip joint. (Fig. 1.8).

Action of the muscles that act on the femur, tibia, and fibula

Deep fascia separate the muscles that act on the femur (thigh bone) and tibia and fibula (leg bones) into medial anterior and posterior compartments. The muscles of the medial (adductor) compartment of the thigh adduct the femur at the hip joint. The gracilis, the other muscle in the medial compartment, not only adducts the thigh, but also flexes the leg at the knee joint. The muscles of the anterior (extensor) compartment of the thigh extend the leg and flex the thigh. This compartment contains the quadriceps femoris and sartoris muscles. The quadriceps femoris muscle is the biggest muscle in the body, covering most of the anterior surface and sides of the thigh. The muscle is actually a composite muscle usually described as four separate muscles:

1. Rectus femoris, on the anterior aspect of the thigh,
2. Vastus lateralis, on the lateral aspect of the thigh,
3. Vastus medialis, on the medial aspect of the thigh, and
4. Vastus intermedius, located deep to the rectus femoris between the vastus lateralis and vastus medialis. The common tendon for the four muscles is known as the quadriceps tendon, which inserts into the patella. The tendon continues below the patella as the patellar ligament, which attaches to the tibial tuberosity. The quadriceps femoris muscle is the great extensor muscle of the leg. The sartorius is a long narrow muscle which forms a band across the thigh from the ilium of the hip bone to the medial side of the tibia. The various movements it produces help effect the cross-legged sitting position in which the heel of one limb is placed on the knee of the opposite limb. It is known as the tailors muscle because tailors often assume this cross-legged sitting position.

The muscles of the posterior (flexor) compartment of the thigh flex the leg and extends the thigh. This compartment is composed of three muscles collectively called the hamstrings viz., (1) biceps femoris, (2) semitendinosus and (3) semimembranous. The hamstrings, are so-named because their tendons are long and string like in the popliteal area and from an old practice of butchers in which they hung hams for smoking by these long tendons. Because the hamstrings span two joints, *i.e.*, hip and knee they are both extensors of the thigh and flexors of the leg. The popliteal fossa is a diamond-shaped space on the posterior aspect of the knee bordered laterally by the tendons of the biceps femoris muscle and medially by the tendons of the semitendinosus and semimembranosus muscles (Fig. 1.8).

Intrinsic muscles of the foot

The muscles in this exhibit are termed intrinsic muscles because they originate and insert within the foot. The deep fascia of the foot forms the plantar aponeurosis that extends from the calcaneous bone to the phalanges of the toes. The aponeurosis supports the longitudinal arch of the foot and encloses the flexor tendons of the foot.

The intrinsic muscles of the foot are divided into two groups; dorsal and plantar. There is only one dorsal muscle, the extensor digitorum brevis, a four-part muscle deep to the tendons of the extensor digitorum tongus muscle, which extends toes 2-5 at the metatarsophalangeal joints.

The planar muscles are arranged in four layers, the first superficial layer being termed the first layer. Three muscles are in the first layer. The abductor hallucis, which lies along the medial border of the sole, is comparable to the abductor pollicis brevis in the hand, and abducts the great toe at the metatarsophalangeal joint. The flexor digitorum brevis, which lies in the middle of the sole, flexes toes 2-5 at the interphalangeal and metatarsophalangeal joints. The abductor digiti minimi, which lies along the lateral border of the sole, is comparable to the same muscle in the hand, and abducts the little toe.

The second layer consists of the quadratus plantae, a rectangular muscle that arises by two heads and flexes toes 2-5 at the metatarsophalangeal joints, and the lumbricals, four small muscles that are similar to the lumbricals in the hands. They flex the proximal phalanges and extends the distal phalanges of toes 2-5.

Three muscles comprise the third layer. The flexor hallucis brevis, which lies adjacent to the plantar surface of the metatarsal of the great toe, is comparable to the same muscle in the hand, and flexes the great toe. The abductor hallucis, which has an oblique and transverse head like the adductor pollicis in the hand, adducts of the great toe. The flexor minimi brevis, which lies superficial to the metatarsal of the little toe, is comparable to the same muscle in the hand, and flexes the little toe.

The fourth layer is the deepest and consists of two muscle groups. The dorsal interossei are four muscles that abduct toes 2-4, flex the proximal phalanges and extend the distal phalanges. The three planar interossei adduct toes 3-5, flex the proximal phalanges, and extend the distal phalanges. The interossei of the feet are similar to those of the hand; however, their actions are relative to the midline of the second digit rather than the third digit as in the hand.

Muscles that move the foot and toes

Muscles that move the foot and toes are located in the leg. The muscles of the leg, like those of the thigh, are divided by deep fascia into three compartments: (*i*) anterior, (*ii*) lateral and (*iii*) posterior. The anterior compartment of the leg consists of muscles that dorsiflex the foot. In a situation analogous to the wrist, the tendons of the muscles of the anterior compartment are held firmly to the ankle by thickenings of deep fascia called the superior extensor retinaculum and inferior extensor retinaculum.

Within the anterior compartment, the tibialis anterior is a long thick muscle against the lateral surface of the tibia where it is easy to palpate. The extensor hallucis longus is a thin muscle between and partly deep to the tibialis anterior and extensor digitorum longus muscles. This feather like muscle is lateral to the tibialis anterior muscle, where it can easily be palpated. The fibularis lertins muscle is part of the extensor digitorum longus, with which it shares a common origin.

The lateral fibular compartment of the leg contains two muscles that plantar flex and evert the foot; the fibularis longus and fibularis brevis.

The posterior compartment of the leg consists of muscles in superficial and deep groups. The superficial muscles share a common tendon of insertion, the calcaneal tendon, the strongest tendon of the body. It inserts into the calcaneal bone of the ankle. The superficial and most of the deep muscles plantar flex the foot at the ankle joint. The superficial muscles of the posterior compartment are the gastrocnemius, soleus and plantaris – the so called calf muscles. The large size of these muscles is directly related to the characteristic upright stance of human beings. The gastronemius is the most superficial muscle and forms the prominence of the calf. The soleus, which lies deep to the gastronemius, is broad and flat. It derives its name from its resemblance to a flat fish (sole). The plantaris is a small muscle that may be absent sometimes these are two of them in each leg. It runs obliquely between the gastronemius and soleus muscles.

The deep muscles of the posterior compartment are the popliteus, tibialis posterior, flexor digitorum longus and flexor hallucis longus. The popliteus is a triangular muscle that forms the floor of the popliteal fossa. The tibialis posterior is the deepest muscle in posterior compartment. It lies between the flexor digitorum longus and flexor hallucis longus muscles. The flexor digitorum longus is smaller than the flexor hallucis longus, even though the former flexes four toes, whereas the latter flexes only the great toe at the interphalangeal joint. (Fig. 1.9).

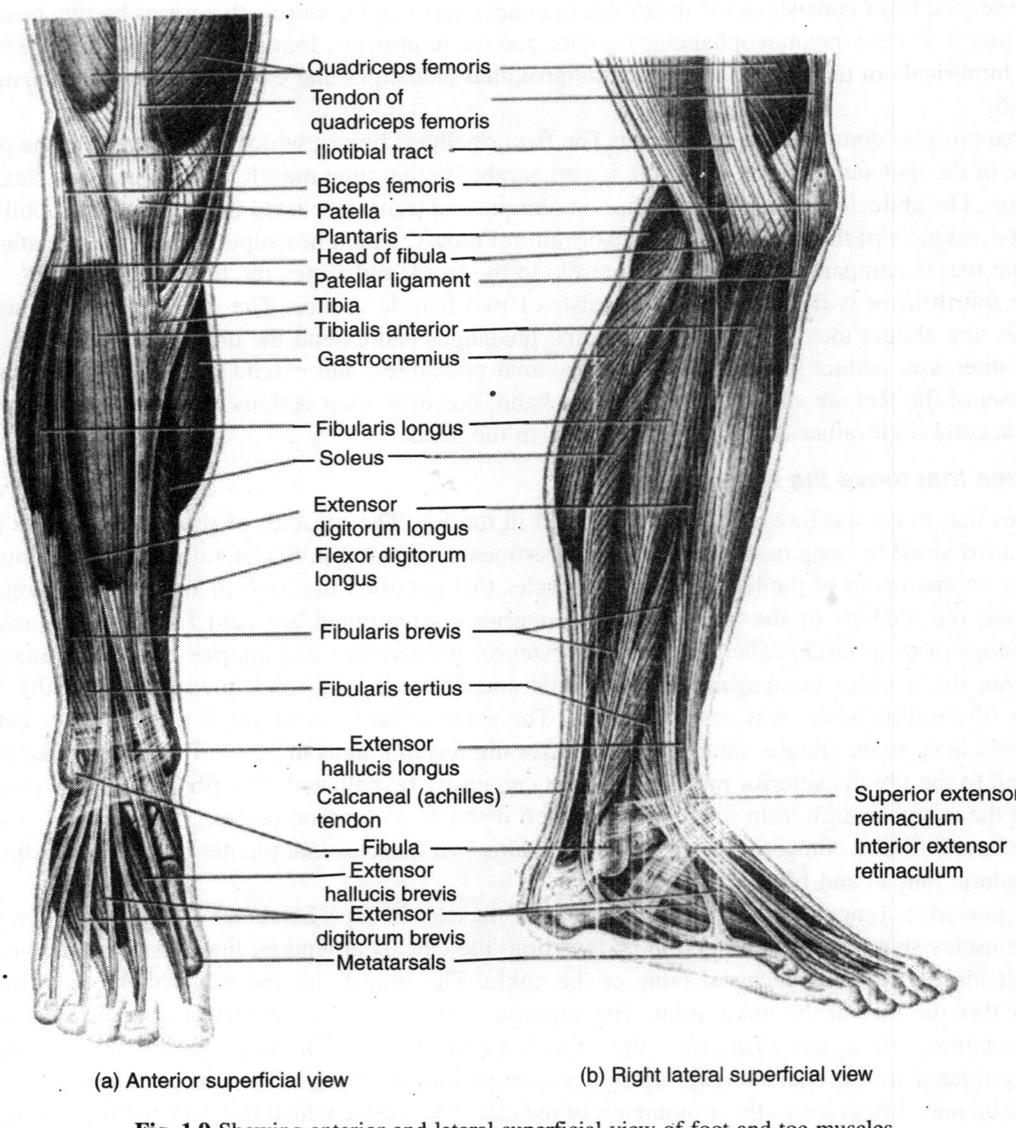

Fig. 1.9 Showing anterior and lateral superficial view of foot and toe muscles

PRINCIPLE OF MUSCLES

Human beings muscle tissues are of three types. They are skeletal, cardiac and smooth. Although the different types of muscle tissue share certain properties, they differ from one another in their microscopic anatomy location and control by the nervous and endocrine systems.

Skeletal muscle tissue: It is named so because most skeletal muscles function to move bones of the skeleton. Skeletal muscle tissue is striated. Alternating light and dark bands (striations) are seen when the tissue is examined with a microscope (Fig. 1.10). Skeletal muscle tissue works mainly in a voluntary

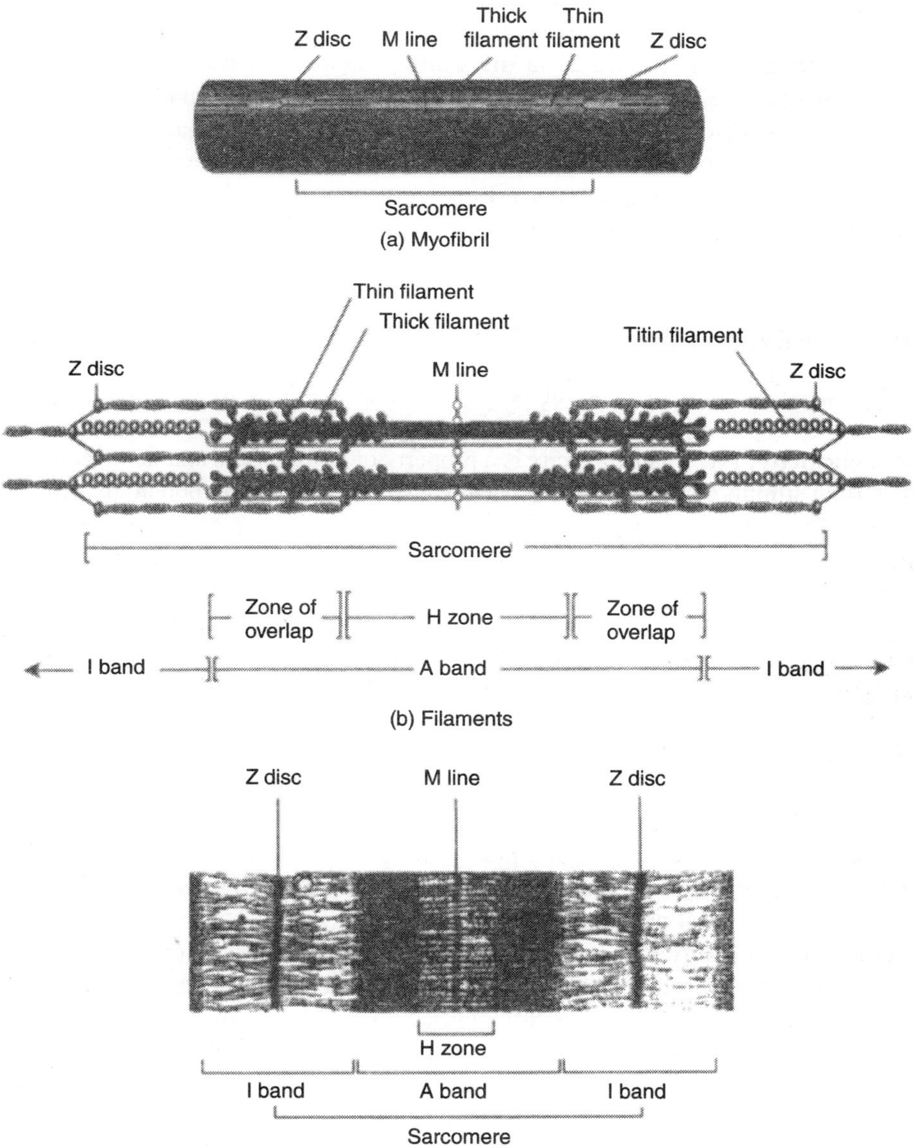

Fig. 1.10 Showing zones and bands of skeletal muscles

manner. Its activity can be consciously controlled by neurons *i.e.*, nerve cells that are part of the somatic division of the nervous system. Most skeletal muscles also are controlled subconsciously to some extent. For instance, the diaphragm continues to alternately contract and relax while we are asleep to power breathing. Also imagine about contracting the skeletal muscles that maintain posture or stabilize body positions.

Cardiac muscle tissue: Only the heart contains cardiac muscle tissue, which forms most of the heart wall. Cardiac muscle is also striated, but its action is involuntary. The alternating contraction and relaxation of the heart is not consciously controlled. Rather, the heart beats because it has a pacemaker that initiates each contraction. This built-in or intrinsic rhythm is termed autorhythmicity. Several hormones and neurotransmitters can adjust heart rate by speeding or slowing the pacemaker.

Smooth muscle tissue: This is located in the walls of hollow internal structures, such as blood vessels, airways, and most organs in the abdominopelvic cavity. It is also found in the skin, attached to hair follicles. Under a microscope, this tissue lacks the striations of skeletal and cardiac muscle tissue. For this reason, it looks non-striated or smooth. The action of smooth muscle is usually involuntary and some smooth muscle tissue has autorhythmicity. Both cardiac muscle and smooth muscle are regulated by neurons that are part of the autonomic division of the nervous system and by hormones released by endocrine glands.

GENERAL PROPERTIES OF MUSCLE

Human muscle tissue has four (4) special properties that enable it to function and contribute to homeostasis:

1. **Electrical elasticity:** Electrical elasticity is a property of both muscle and nerve cells, is the ability to respond to certain stimuli by producing electrical signals such as action potentials. An action potential (AP) or impulse is a sequence of rapidly occurring events that take place in two phases (Fig. 1.10).

During the depolarizing phase, the negative membrane potential decreases towards zero and eventually becomes positive. Then the repolarizing phase restores the membrane potential to the resting state of –70 mV. During an action potential two types of voltage-gated channels open and then close. These channels are present mainly in the plasma membrane of the ascon and ascon terminals. The first channels that open, the Na^+ channels, allow Na^+ to rush into the cell, which causes the depolarizing phase. Then K^+ channels open, allowing K^+ to flow out, which produces the repolarizing phase. Together, the depolarization and repolarization phases last about 1 m second (0.001 sec) in a typical neuron. Action potentials can travel along a cells plasma membrane due to the presence of specific voltage-gated channels. For muscle cells, two main types of stimuli trigger action potentials, one is autorhythmic electrical signals arising in the muscle tissue itself, such as occurs in the hearts pacemaker. The other is chemical stimuli, such as neurotransmitters released by neurons, hormones distributed by the blood, or even local changes in pH.

2. **Contractility:** Contractility is the ability of muscle tissue to contract forcefully when stimulated by an action potential. When muscle contracts, it generates tension *i.e.*, force of contraction while pulling on its attachment points. In an isometric contraction (iso = equalimetric = length or measure), the muscle develops tension but not shorten. An example is holding a brick in an outstretched hand. If the tension generated is great enough to overcome the resistance of the object to being moved, the muscle shortens and movement occurs. In an isotonic contraction (-tonic = tension), the tension developed by the muscle remains almost constant while the muscle shortens. An example is lifting a book off a table.

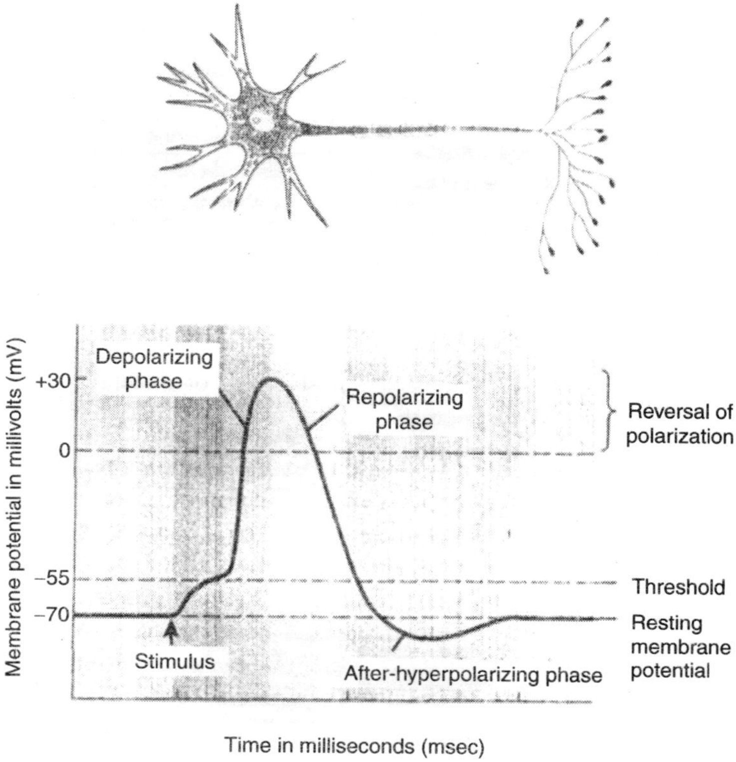

Fig. 1.11 Showing action potential to stimuli

3. **Extensibility:** Extensibility is the ability of muscle to stretch without being damaged. Extensibility allows a muscle to contract forcefully even if it is already stretched. Normally, smooth muscle is subject to the greatest amount of stretching. For example, each time the stomach fills with food, the muscle in its wall is stretched. Cardiac muscle also is stretched each time the heart filled with blood. However, during normal activities, the stretch on skeletal muscle remains almost constant.

4. **Elasticity:** Elasticity is the ability of muscle tissue to return to its original length and shape after contraction or extension.

Mechanism of Muscle Contraction and Relaxation

Human skeletal muscle shortens during contraction because the thick and thin filaments slide past one another.

Muscle contraction occurs because myosin heads attach to and 'walk' along the thin filaments towards the M line (Fig. 1.12). As a result, the thin filaments slide inward and meet at the centre of a sarcomere. They may even move so far inward that their ends overlap (Fig. 1.12). As the thin filaments slide inward, the Z discs come closer together, and the sarcomere shortens. However, the lengths of the individual thick and thin filaments do not change. Shortening of the sarcomeres causes shortening of the whole muscle fibre, which leads to shortening of the entire muscle.

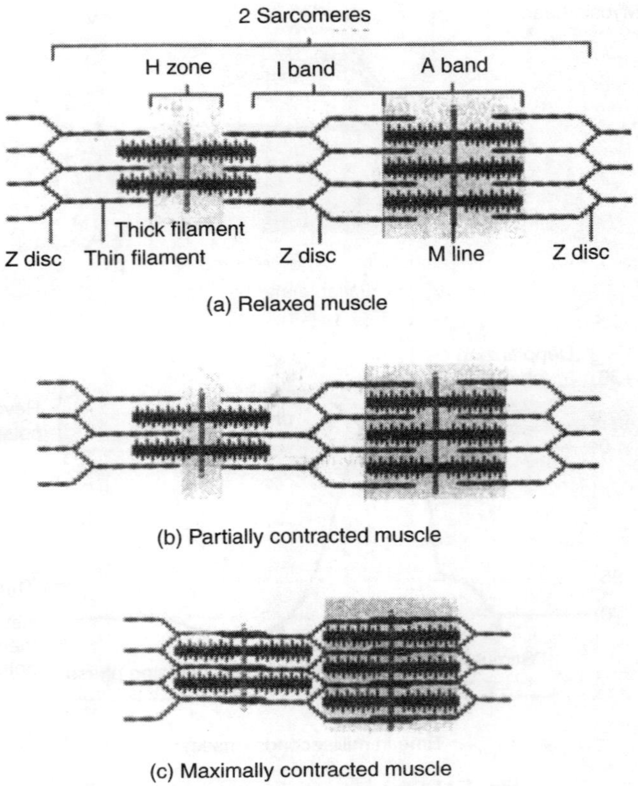

Fig. 1.12 Showing mechanism of muscle contraction

The Contraction Cycle

At the onset of contraction, the sercoplasm reticulum releases calcium ions (Ca^{2+}) into the cytosol. There they bind to troponin and cause the troponin tropomyosin complexes to move away from the myosin binding sites on actin. Once the binding sites are 'free', the contraction cycle–the repeating sequence of events that causes the filaments to slide-begins. The contraction cycle consists of four (4) steps as depicted in Fig. 1.13.

1. **ATP hydrolysis:** The myosin head includes an ATPase, on enzyme that hydrolyzes ATP into ADP (adenosine diphosphate) and a phosphate group. This hydrolysis reaction energizes the myosin head. One can notice that the products of ATP-hydrolysis–ADP and a phosphate group–are still attached to the myosin head.
2. **Attachment of myosin to actin to form crossbridges:** The energised myosin head attaches to the myosin binding site on actin and then releases the previously hydrolysed phosphate group.
3. **Power stroke:** The release of the phosphate group triggers the power stroke of contraction. During the power stroke, the pocket on the myosin head rotates and releases the ADP. The myosin head generates force as it rotates towards the centre of the sarcomere, sliding the thin filament past the thick filament toward the M line.

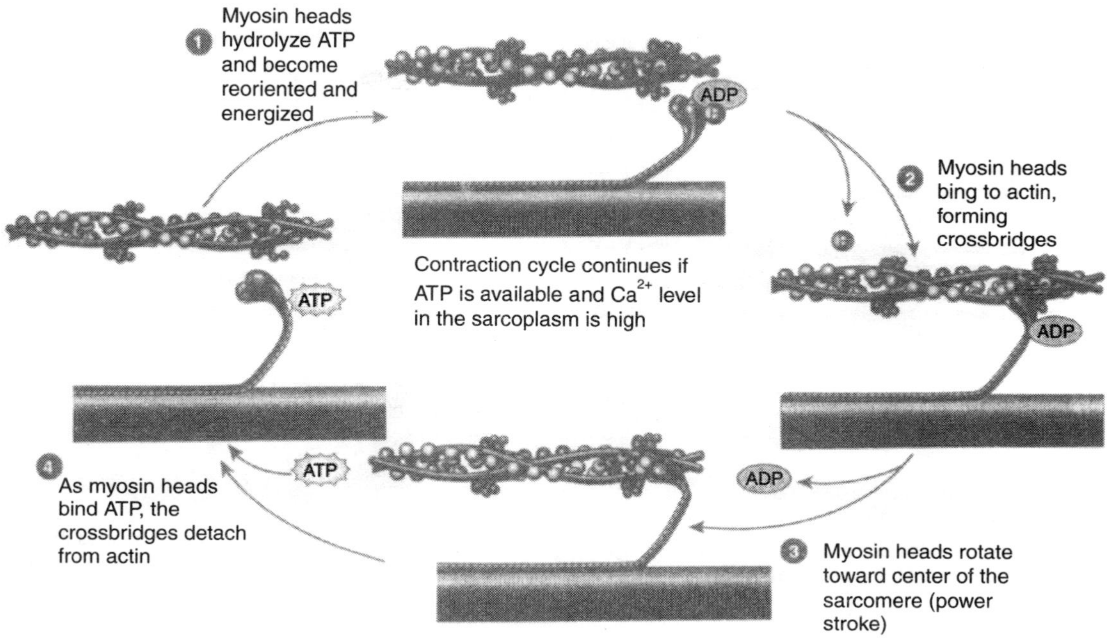

Fig. 1.13 Showing contraction cycle

4. **Detachment of myosin from actin:** At the end of the power stroke, the myosin head remains firmly attached to actin until it binds another molecule of ATP. As ATP binds to the ATP-binding pocket on the myosin head, the myosin head detaches from actin.

The contraction cycle repeats as the myosin ATPase again hydrolyses ATP. The reaction reorients the myosin head and transfers energy from ATP to the myosin head. Then the myosin head is again energized and ready to combine with the next myosin-binding site on the thin filament. The contraction cycle repeats for as long as ATP is available and the Ca^{2+} level near the thin filament is sufficiently high. The myosin heads keep rotating back and forthwith each power stroke, pulling the thin filaments toward the M line. Each of the 600 myosin heads in one thick filament attaches and detaches about five times per second. At any one instant, some of the myosin heads are attached to actin and generating force, whereas, others are detached and getting ready to bind again.

Excitation-contraction Coupling

An increase in Ca^{2+} concentration in the cytosol starts muscle contraction, whereas a decrease stops it. When a muscle fibre is relaxed, the concentration of Ca^{2+} in its cytosol is very low, only about 0.1 micromole per litre. A huge amount of Ca^{2+}, however, is stored inside the sarcoplasmic reticulum (Fig. 1.14). When these channels are open, Ca^{2+} flows out of the SR into the cytosol around the thick and thin filaments. As a result, the Ca^{2+} concentration in the cytosol rises ten fold or more. The released calcium ions combine with troponin, causing it to change shape. This conformational change moves the troponin-tropomyosin complex away from the myosin binding sites on actin. Once these binding sites are free, myosin heads bind to them, and the contraction cycle begins. The events just described constitute excitation-contraction coupling, the steps that connect excitation to contraction.

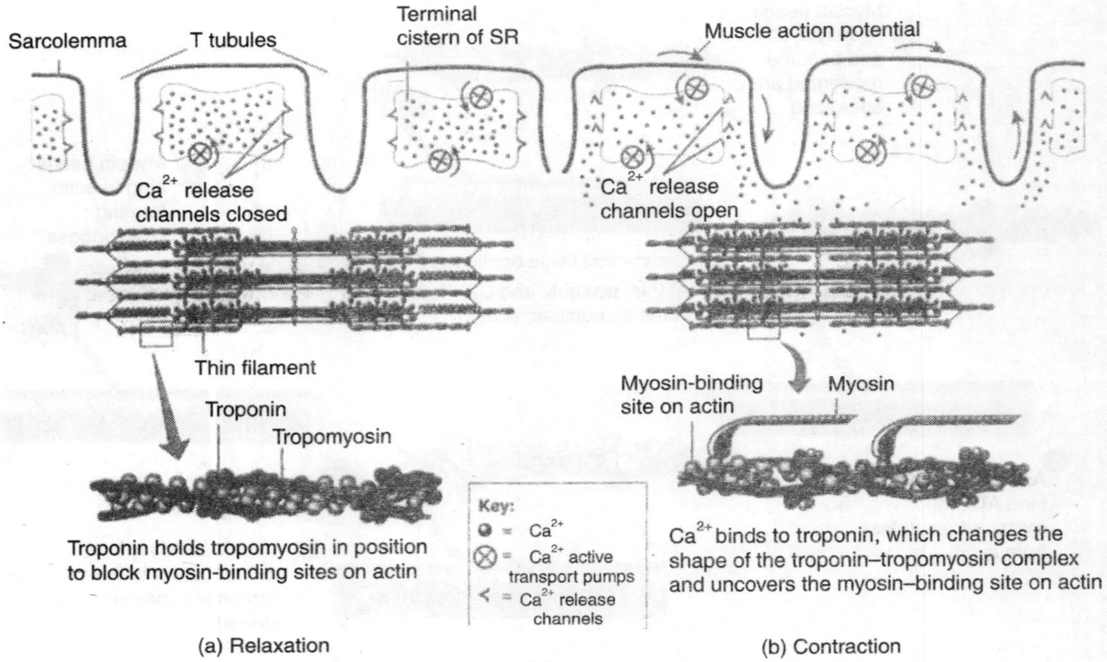

Fig. 1.14 Showing role of Ca in the regulation of contraction

The sarcoplasmic reticuluin membrane also contains Ca^{2+} active transport pumps that use ATP to constantly move Ca^{2+} from the cytosol into the SR. While muscle action potentials continue to propagate through the T tubules, the Ca^{2+} release channels are open. Calcium ions flow into the cytosol more rapidly than they are transported back by the pumps. After the last action potential has propagated throughout the T tubules, the Ca^{2+} release channels close. As the pumps move Ca^{2+} back into the SR, the concentration of calcium ions in the cytosol quickly decreases. Inside the SR, molecules of a calcium-binding protein, appropriately called calsequestrin, bind to the Ca^{2+} enabling even more Ca^{2+} to be sequestered within the SR. As a result, the concentration of Ca^{2+} inside the SR is 10,000 times higher than in the cytosol in a relaxed muscle fibre. As the Ca^{2+} level drops in the cytosol, the troponin-tropomyosin complexes inside back over and cover the myosin binding sites, and the muscle fibre relaxes.

RED AND WHITE MUSCLE FIBRES

Slow oxidative (SO) fibres are smallest in diameter and thus are the least powerful type of muscle fibres. They appear dark reddish in colour because they contain large amounts of myoglobin and many blood capillaries. Because they have many large mitochondria, so fibres generate ATP mainly by aerobic cellular respiration, which is why they are called oxidative fibres. These fibres are said to be 'slow' because the ATPase in the myosin heads hydrolyzes ATP relatively slowly and the contraction cycle proceeds at a slower pace than in 'fast' fibres. As a result, so fibres have a slow speed of contraction. Their twitch contractions last from 100 to 200 m sec, and they take longer to reach peak tension. However, slow fibres are very resistant to fatigue and are capable of prolonged, sustained contractions for many hours. These slow-twitch, fatigue-resistant fibres are adapted for maintaining posture and for aerobic, endurance-type activities such as running a marathon.

WHITE MUSCLE FIBRE

Fast glycolytic (FG) fibres are largest in diameter and contain the most myofibrils. Hence, they can generate the most powerful contractions. FG fibres have low myoglobin content, relatively few blood capillaries, few mitochondria, and appear white in colour. They contain large amounts of glycogen and generate ATP mainly by glycolysis. Due to their large size and their ability to hydrolyze ATP rapidly. FG fibres contract strongly and quickly. These fast-twitch fibres are adapted for intense anaerobic movements of short duration, such as weight lifting or throwing a ball, but they fatigue quickly, strength training programmes that engage a person in activities requiring great strength for short times produce increases in the size, strength and glycogen content of fast glycolytic fibres. The FG fibres of a weight lifter may be 50% larger than those of a sedentary person or endurance athlete. The increase in size is due to increased synthesis of muscle proteins. The overall result is muscle enlargement due to hypertrophy of the FG fibres.

UNIT 2

Circulatory System

STRUCTURE, COMPOSITION AND FUNCTIONS OF BLOOD

Blood is a liquid connective tissue. It has two major components – plasma and formed elements. Plasma constitutes 55% while the remaining 45% constitutes the formed elements. Blood is pumped by the heart and circulates through the vessels.

Plasma: Plasma is a straw coloured fluid. It contains about 90% water, three types of plasma proteins albumin, globulin, and fibrinogen. Inorganic salts like sodium bicarbonate, iron, calcium, potassium, magnesium, phosphorus etc. Plasma also contains organic nutrients like amino acids, glucose, fatty acid, glycerol etc. Hormones, nitrogenous wastes in the form of urea and uric acid are also carried in the plasma. Nitrogen, oxygen, carbondioxide are dissolved in the plasma.

Plasma – Its Composition and Functions

- **Water:** Maintains blood volume, pressure, forms lymph, supplies water to cells, carry other components.
- **Inorganic ions:** Maintain osmotic balance pH balance, carbon-di-oxide transport to lungs.
- **Plasma proteins, Fibrinogen, Prothrombin:** Help in osmotic and pH balance; participates in blood clotting mechanism.
- **Albumen, Globulin:** Act as antibodies.
- **Glucose, Amino acids and Fatty acids:** In the process of transit to and fro cells.
- **Urea, Carbondioxide, Oxygen, Hormone etc.:** In the process of being transported from one region of body to another.

Formed Elements

There are three types of formed elements, erythrocytes, leucocytes and thrombocytes.

Erythrocytes (Red blood corpuscles): They are enucleated, circular, disc like biconcave cells, red in colour due to presence of the iron pigment, haemoglobin. (Among mammals, camel is an exception with nucleated erythrocytes. Erythrocytes arise from the bone marrow. They have a life span of 100 to 120 days old and worn out red corpuscles are destroyed in the liver and spleen (Liver is considered graveyard of red blood cells. One cubic millimeter of blood contains about 4.5 to 5 million red corpuscles.

Leucocytes (white blood corpuscles) are large, amoeboid cells. They also arise from the bone marrow. One cubic millimeter of blood contains about 7000-8000 white cells. They are defensive in function.

Leucocytes are of two types, (*i*) granulocytes and agranulocytes. In granulocytes, the cytoplasm is granular. In total there are three types of granulocytes-neutrophils, acidophils and basophils (Fig. 2.1).

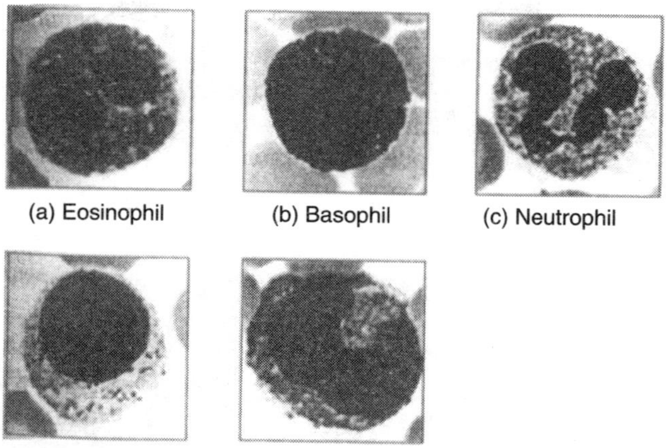

(a) Eosinophil (b) Basophil (c) Neutrophil

(d) Small lymphocyte (e) Monocyte

Fig. 2.1 Showing types of white blood cells

Neutrophils: Neutrophils reacts faintly with both acidic and basic stains and contain a multi-lobed nucleus, they are phagocytic in function (Fig. 2.1).

Acidophils: Acidophils are also called as eosinophils, which react positively with acidic stains like eosin; they possess a bilobed nucleus; they are antiallergic in function (Fig. 2.2).

Basophils: Basophils stain positively with basic stains; they possess S-shaped nucleus. They secrete heparin and histamine. Basophiles, cytoplasm is smooth without granules; thus they are also termed as agranulocytes. Agranulocytic basophils are two types: (*i*) Lymphocytes and (*ii*) Monocytes.

(*i*) Lymphocytes: Lymphocytes contain a large spherical nucleus; they produce antibodies.

(*ii*) Monocytes: Monocytes contain a large bean-shaped nucleus; they serve as scavengers (Fig. 2.1).

Thrombocytes are other component of blood which are also known as platelets. They arise as fragments of large cells known as megakaryocytes. They help in clotting of blood.

Certain important properties of leukocytes (white cells) are represented in the following table.

Table 2.1: Showing types of leucocytes in comparison

Character	*Neutrophil*	*Basophil*	*Acidophil*	*Monocyte*	*Lymphocyte*
Percentage of blood	60%	1%	3%	6%	30%
Nucleus	Multilobed	S-Shaped	Bilobed	Bean shaped	Large, round
Cytoplasm	Neutrophilic	Basophilic	Eosinophilic	Basophilic	Basophilic
Granules	Fine granular	Coarse, granular	Coarse, granular	Absent	Absent
Functions	Phagocytic	Secretes heparin and histamine	Anti-allergic and wound healing	Scavenging	Formation of antibodies

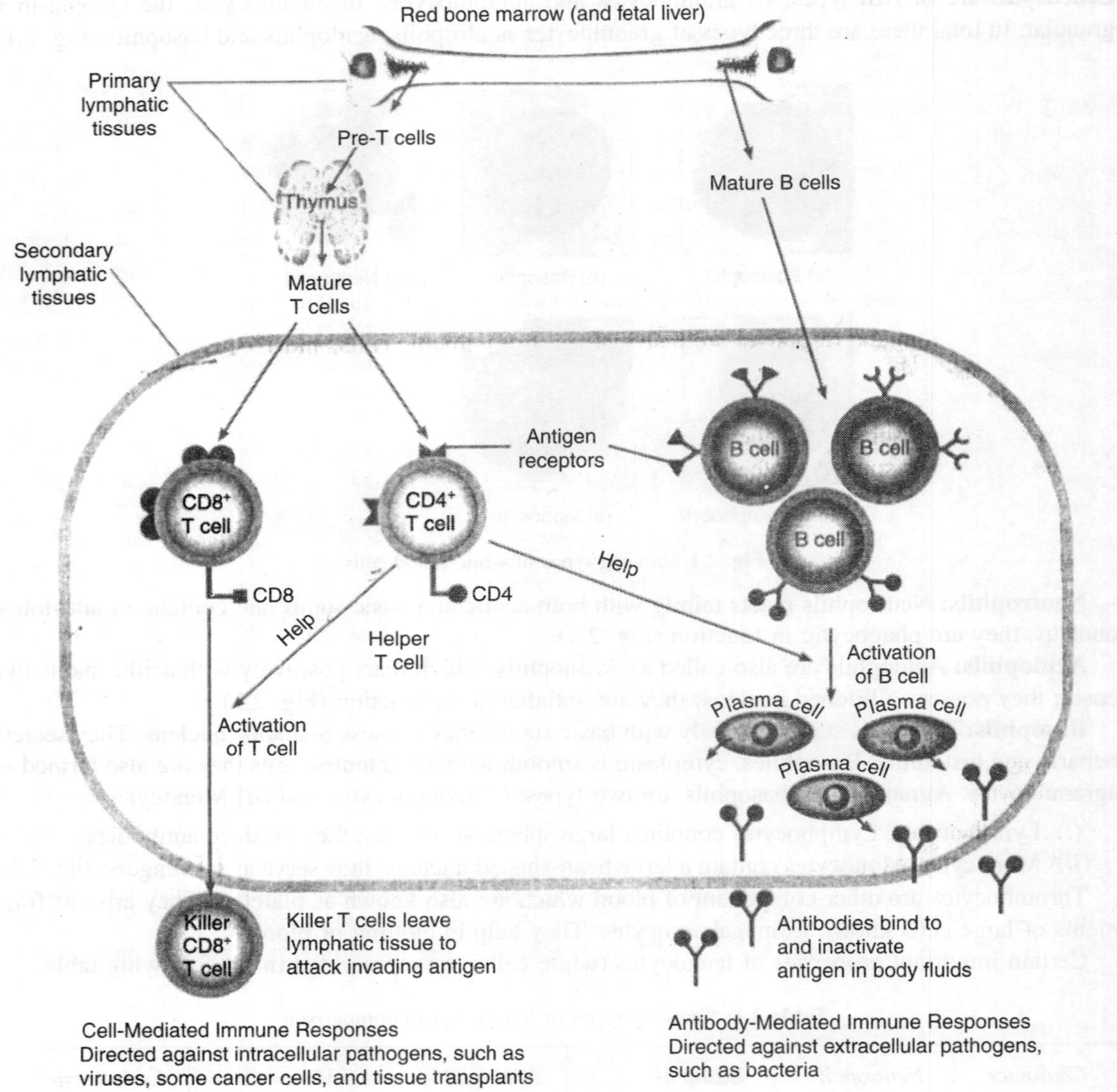

Fig. 2.2 Showing immune response

FUNCTIONS OF BLOOD

One of the important functions of blood is transportation of oxygen to tissues and carbon dioxide to the lungs. Blood also transports digested food to tissues. Besides these, it also transports nitrogeneous wastes from tissues to kidneys. Further bloods functions involve maintenance of fluid balance, maintenance of pH. Human body temperature is regulated by blood. Blood performs coagulation at the site of injury. And blood defends human body against germs.

Blood Groups and Rh Factor

Whenever there is loss of more blood due to an accident or during surgery, the doctors advise the relatives to arrange for blood for the patient to meet the emergency. This transfer of blood from a healthy individual (donor) to another individual (recipient), is known as blood transfusion.

When the practice of blood transfusion began, it was noticed that the blood of the donor may not always match the blood of the patient who receives it. As a result, the blood of the patient clumps or agglutinates, proving fatal to the life of the patients. Landsleiner discovered that mismatching of blood was due to a reaction between antigens present in the red blood cells of the donor, and antibodies present in the plasma of receiver or recipient. Two different antigens were discovered and named A and B. Their corresponding antibodies which react and clump the blood were termed '*a*' and '*b*'. In human beings, four different blood groups are identified *i.e.*, A, B, AB and O. Blood groups usually are inherited from parents. Successful transfusion is possible between the same group of or certain groups, as shown in the following table.

Table 2.2: Showing blood group details

Blood group	*Antigen present in RBC*	*Antibody present in plasma*	*Possibility of Blood donation*
A	A	b	Can donate blood to A and AB.
B	B	a	Can donate blood to B and AB.
AB	A, B	none	Can receive blood from all but donate only to AB.
O	none	a, b	Can donate blood to all but can receive only from O.

We can see from the table 2.2, that O group can donate blood to all the groups, and, hence, a person having this blood group is called universal donor. Interestingly, a person with AB group can receive blood of all other groups but cannot donate to other groups. Hence persons with AB group can be considered as the universal recipient.

Another factor in blood, called Rh factor, is also genetically determined. Most people are Rh positive (Rh^+). A few are Rh negative (Rh^-). An Rh negative mother may lose her baby if the baby has Rh^+ blood. However, these are medically available techniques and procedures to save the baby in such cases. Many other blood groups have also been identified.

Immunity and Antibody Formation

There are a number of injections agents that can use the human body as a sanctuary to increase their progeny. However, the human immune system provides a defence mechanism to establish immunity to infection.

Immunity is therefore, "All those physiological mechanisms that endow the man with the capacity to recognize materials as foreign to itself and to neutralize, eliminate or metabolize them with or without injury to its own tissues".

Antibody Formation

The specific reactivity induced in host by antigenic stimulus is called immune response (Fig. 2.2). Immune response may be beneficial, harmful or nonreactive (tolerance). Immune response is of two types:

(*a*) Humoral immunity
(*b*) Cell mediated immunity

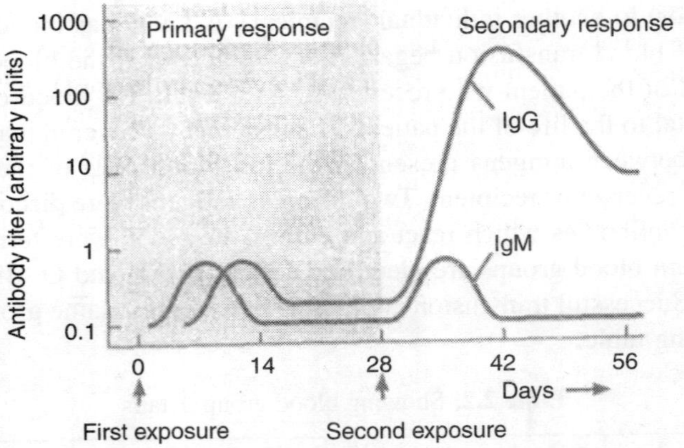

Fig. 2.3 Showing production of antibodies in primary and secondary responses

Humoral immunity provides primary defence against extracellular pathogens, helps in defences against virus, infections, immediate hypersensitivity and certain autoimmune diseases. Cell mediated immunity protects against fungi, viruses, intracellular pathogens, participates in graft reflection, immunity, against cells, delayed hypersensitivity and certain autoimmune diseases.

Humoral Immune Response

It consists of three steps:

(*i*) Entry and distribution of antigen in tissue and its contact with immunocompetent cells.
(*ii*) Processing of antigen and control of antibody forming process.
(*iii*) Secretion of antibody, its distribution in tissue, body fluid and manifestation of its effects.

Primary and Secondary Responses

Primary Response (Fig. 2.3)

Primary response to initial antigenic stimulus is slow, sluggish shortlived with a long log phase and low antibody titer that does not persist for long time. Antibodies formed are predominantly IgM.

Secondary response (Fig. 2.3)

Secondary response to subsequent stimuli is prompt, powerful prolonged, and with much higher level of antibody. It losts for long time. Antibodies predominantly formed are IgG.

When an antigen is injected into a human being carrying specific antibody in a temporary fall in the level of circulating antibody occurs due to combination of antigen and antibody. This is called negative phase. Then it is followed by increase in titer of the antibody exceeding initial levels.

Figure 2.3 Secondary immune response: Effect of repeated antigenic stimuli A, B, C: (1) Secondary immune response, (2) Negative immune response (3) Negative phase, (4) High level of antibody after booster dose.

Heart and Blood vessels

The circulatory system in human being is highly developed. It consists of a central pumping organ-the heart, a set of blood vessels which supply oxygenated blood to various tissues called arterial system, and another set of blood vessels which gather the deoxygenated blood from different tissues and return it to the heart, called venous system. Blood flows through these blood vessels. The circulatory system in human being is termed closed type as blood circulates in concealed blood vessels.

The human being heart is about the size of a fist and is placed obliquely in thoracic cavity within the mediastinum. It is bounded by the lungs laterally, the sternum anteriorly, and the thoracic vertebrae posteriorly. The wide superior portion is the base, and the inferior pointed end is the apex of the heart.

An average adult heart measures approximately 12 cms from base to apex. In males it weighs between 280-340 grams (average 300 g) while in females it weighs between 230-280 grams (average 250 grams). The pyramidal shape heart is a hollow, fibro-muscular organ (Fig. 2.4).

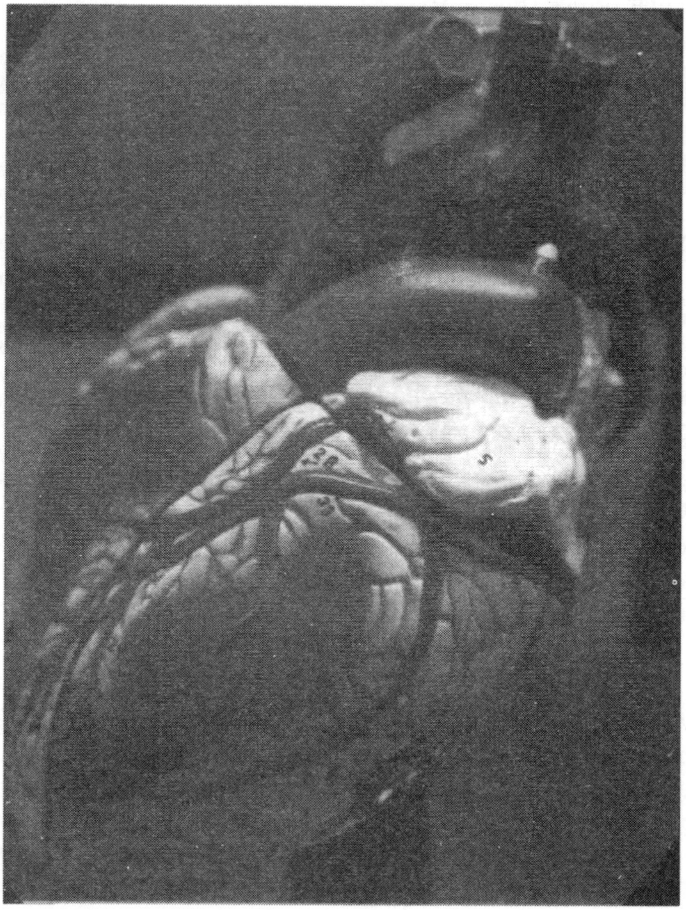

Fig. 2.4 Showing human heart structure

The human heart is covered and protected by the pericardium. It protects the heart. The cavity between pericardium and heart is filled with a serous fluid, called pericardial fluid which protects heart from external shocks as a shock-absorbant.

The human being heart is made up of three distinct layers:

(*a*) Superficial epicardium
(*b*) Middle myocardium and
(*c*) Inner endocardium.

These three layers are highly vascularised. The endocardium is a glistening white sheet of squamous epithelium and the middle myocardium is very thick and is composed of cardiac muscles.

Internal structure: In a vertical section, the heart is divisible into right and left sides. Each side has an upper chamber. Auricle (little ear) or Atrium (entry way) and a lower chamber called ventricle (venter-underside). A slim inter-atrial septum separates the two auricles. Similarly an inter-ventricular septum separates the two ventricles. This septum is much thicker. The right atrium opens into the right ventricle by means of a right auriculo-ventricular orifice. The left atrium opens into the left ventricle by a left auriculo-ventricular orifice (Fig. 2.5).

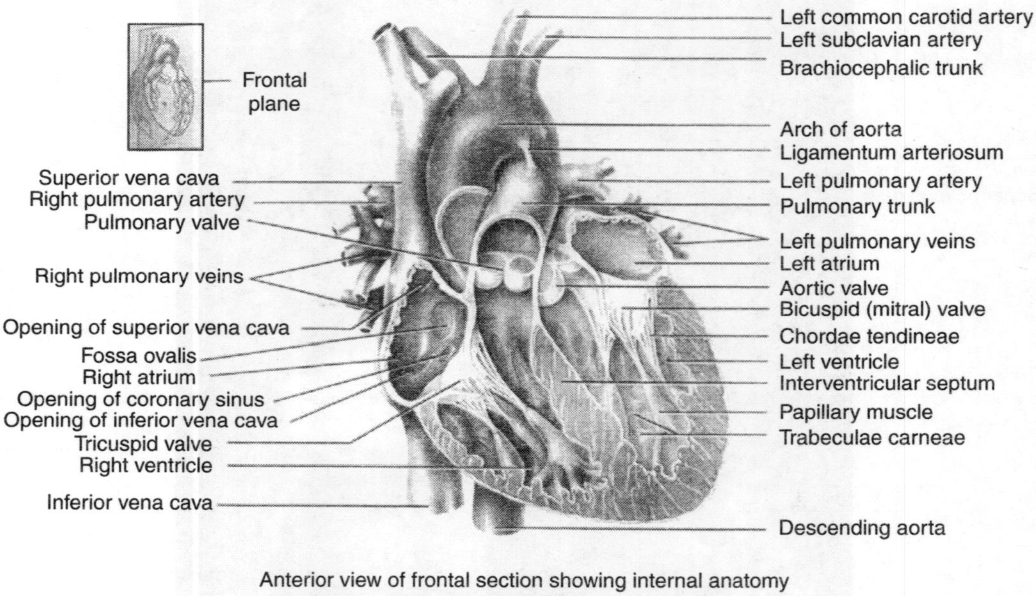

Fig. 2.5 Showing internal structure of human heart

The walls of the ventricles are very thick and highly muscular. Internally they show irregular ridges of muscles called trabeculae carneae (trabecula = little beam; carnea = flesh). The larger of these muscles which are cone like are called papillary muscles. These muscles have string-like cords, attached to them called the chordae tendinae (tendinous = strands). The other ends of these cords are attached to the flaps of tricuspid and bicuspid valves. Chordae tendinae are glistening white fibrous collagen cords, chordae tendinae hold the flaps of these valves in position when the ventricles contract. (Fig. 2.5) The left ventricle is the largest and strongest chambers of the heart.

Major Blood Vessels of the Four Chambers

1. **Right atrium:** Three prominent veins open into the right atrium. They are:

   Superior vena cava,
   Inferior vena cava, and
   Coronary sinus.

All these return oxygen - poor blood to the right atrium.

Superior vena cava returns impure blood from the parts of the body superior to the diaphragm.

Inferior vena cava brings impure blood from the parts of the body inferior to the diaphragm coronary sinus returns blood from the myocardium of the heart.

2. **Right ventricle:** From the right ventricle, a prominent artery called pulmonary trunk arises, it divides into the right and left pulmonary arteries. These branches carry impure blood to the lungs where it is oxygenated.

3. **Left atrium:** The left atrium receives two right and two left pulmonary veins from the lungs. Oxygen-rich blood is returned to the left atrium through these veins.

4. **Left ventricle:** From the left ventricle arises a largest blood vessel; Aorta: Aorta distributes blood to the systemic circulation. The aorta begins as a short ascending aorta; curves to the left to form the aortic arch, then, descends posteriorly, and continues as the descending aorta (Fig. 2.6).

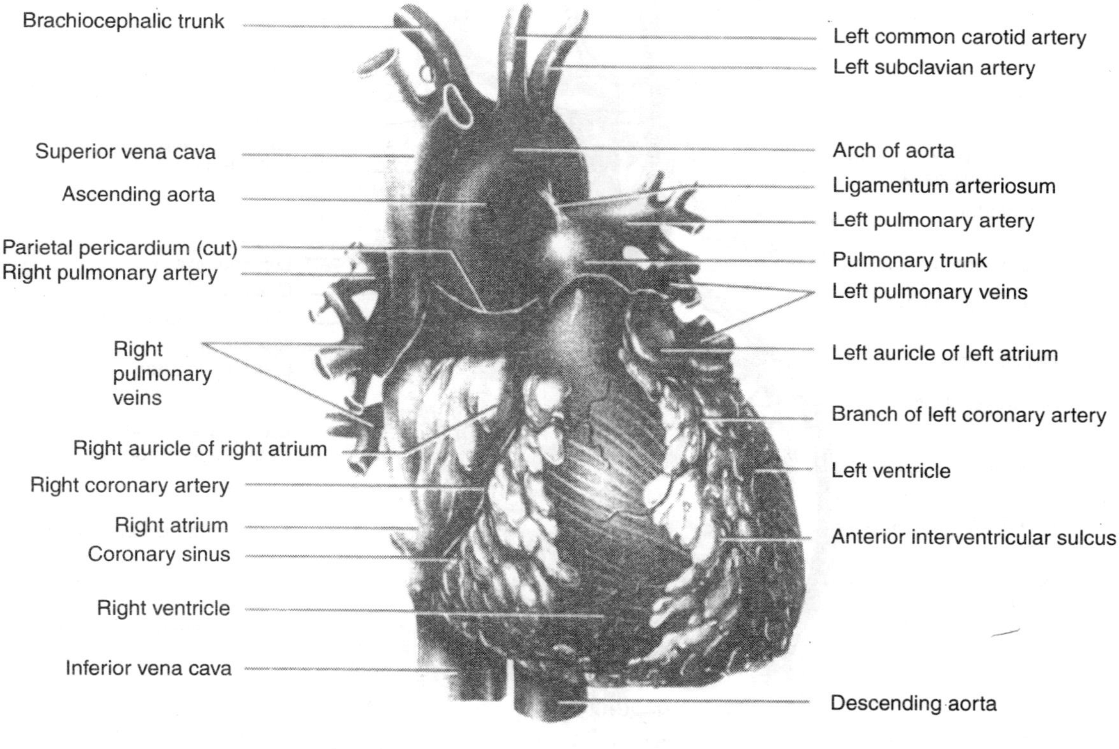

Anterior external view showing surface features

Fig. 2.6 Showing right, left atriums and right, left ventricles

Valves of Human Heart

The heart has four valves that control the one way flow of blood. Two atrio-ventricular valves and two semilunar valves (semi-help; lunar = moon). The right atrio-ventricular orifice is guarded by the right atrio-ventricular valve) or the tricuspid valve (tri = three; cusp = flap). The left atrio-ventricular orifice is guarded by the left atrioventricular valve, or the bicuspid valve or the mitral valve (miter = tall, head dress) because it resembles a bishop's headdress. The tricuspid and bicuspid valves are structurally similar, except the respective number of flaps they possess.

These valves prevent blood from flowing back into the atria. Blood in the right ventricle goes through the pulmonary valve (semilunar) to enter the pulmonary artery. The aortic (semilunar) valve is located between the left ventricle and the aorta. These two semilunar valves are identical with each having three pockets that fill with blood thus preventing blood from flowing backward into the ventricles (Fig. 2.7).

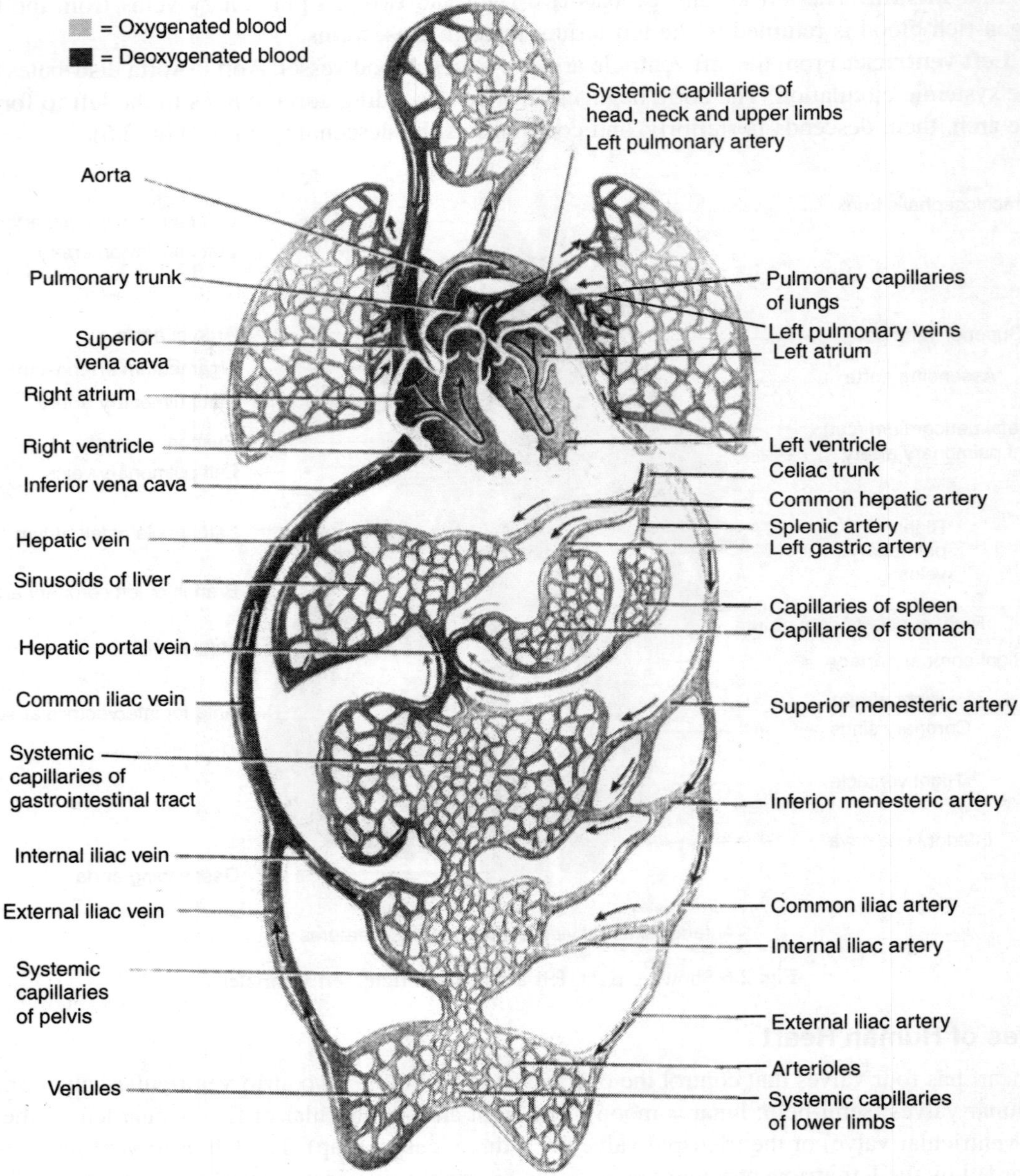

Fig. 2.7 Blood vessels are organized into various routes that deliver blood to tissues of the body

Blood vessels are organised into circulatory routes that carry blood to specific organs in human body (Fig. 2.7).

The routes are parallel. In most cases, a portion of the cardiac output flows separately to each tissue of the body. Every organ receives its own supply of freshly oxygenated blood. The two main circulatory routes are, Arterial system and Venous system.

Arterial system: The arterial system of circulation includes the arteries and arterioles that carry oxygenated blood from the left ventricle to systemic capillaries, including the veins and venules that return deoxygenated blood to the right atrium. Blood leaving the aorta and flowing through the systemic arteries, is a bright red colour. As blood flows through capillaries, it loses some of its oxygen, and picks up carbon dioxide, becoming a dark red colour. All systemic arteries branch from the aorta. completing the circuit, all the veins of the systemic circulation drain into the superior vena cava, the inferior vena cava or the coronary sinus, which in turn empty into the right atrium. The bronchial arteries, which carry nutrients to the lungs, also are part of the systemic circulation.

The aorta is the largest artery of the body with a diameter of 2-3 cm. Its four principal divisions are the ascending aorta, arch of the aorta, thoracic aorta, and abdominal aorta. The portion of the aorta that emerges from the left ventricle posterior to the pulmonary trunk is the ascending aorta. The beginning of the aorta contains the aortic valve. The ascending aorta gives off the coronary artery branches that supply blood to myocardium of the heart. Then it turns to the left to, forming the arch of the aorta, which descends and ends at the level the inter-vertebral disc between the fourth and fifth thoracic vertebrae. As the aorta continues to descend, it lies close to the vertebral bodies, passes through the diaphragm, and divides at the level of the fourth lumbar vertebra into two common iliac arteries, which carry blood to the lower limbs. The section of the aorta between the arch of the aorta and the diaphragm is called thoracic aorta; the section between the diaphragm and the common iliac arteries is the abdominal aorta. Each division of the aorta gives off arteries that branch into distributing arteries that lead to various organs, the arteries divide into arterioles and then into capillaries that serve the systemic tisues; *i.e.*, all tissues except the alveoli of the lungs.

The ascending aorta is about 5 cm in length, and begins at the aortic valve. It is directed superiorly, slightly anteriorly, and to the right. It ends at the level of the sternal angle, where it becomes the arch of the aorta. The beginning of the ascending aorta is posterior to the pulmonary trunk and right auricle; the right pulmonary artery is posterior to it. At its origin, the ascending aorta contains three dilations called aortic sinuses. Two of these, the right and left sinuses give rise to the right and left coronary arteries respectively.

The left and right coronary arteries arise from the ascending aorta. They form a crown like ring around the heart giving off branches to the artrial and ventricular myocardium. The posterior interventricular branch of the right coronary, artery supplies both ventricles and the right ventricle is supplied by the marginal branch. The circumplex branch supplies blood to left, atrium and left ventricle.

The arch of the aorta is 4-5 cm in length and is the continuation of the ascending aorta. It emerges from the pericardium posterior to the sternum at the level of the sternal angle. The arch is directed superiorly and posteriorly to the left and then inferiorly it ends at the intervertebral disc between the fourth and fifth thoracic vertebrae where it becomes the thoracic aorta. Three major arteries branch from the superior portion of the arch of the aorta the branchiocephalic trunk, the left common carotid and the left subclavian the first and largest branch from the arch of the aorta is the brachiocephalic trunk. It extends superiorly, bending slightly to the right, and divides at the right sternoclavicular joint to form the right subclavian, artery and right common carotid artery. The second branch from the arch of the aorta is the left common carotid artery, which divides into the same branches with the same names as the right common carotid artery. The third branch from the arch of the aorta is the left subclavian artery, which distributes blood to the left vertebral artery and vessels of the left upper limb. Arteries branching from the left subclavian artery are similar in distribution, and name to those branching from the right subclavian artery.

The thoracic aorta, is about 20 cm long and is a continuation of the arch of the aorta. It begins at the level of the intervertebral disc between the fourth and fifth thoracic vertebrae, where it lies to the midline and extends through an opening in the diaphragm, which is located anterior to the vertebral column at the level of the intervertebral disc between the twelfth thoracic and first lumbar vertebrae.

Along its course, the thoracic aorta sends off numerous small arteries, visceral branches to viscera and parietal branches to body wall structures.

The abdominal aorta is the continuation of the thoracic aorta. It begins at the aortic hiatus in the diaphragm and ends at about the level of the fourth lumbar vertebrae, where it divides into the right, and left common iliac, arteries. The abdominal aorta lies anterior to the vertebral column.

Similar to the thoracic aorta, the abdominal aorta gives off visceral and parietal branches. The unpaired visceral branches arise from the anterior surface of the aorta and include the celiac trunk and the superior mesenteric and inferior mesenteric arteries. The paired visceral branches arise from the lateral surfaces of the aorta and include the suprarenal, renal and gonadal arteries. The unpaired parietal branch is the median sacral artery. The paired parietal branches arise from the posterolateral surfaces of the aorta, and include the inferior phoenic and lumbar arteries.

The abdominal aorta ends by dividing into the right and left common iliac arteries. These in turn divide into the internal and external iliac arteries. In sequence the external iliacs become the femoral arteries in the thighs; the popliteal arteries posterior to the knee and the anterior and posterior tibial arteries in the legs.

VENOUS SYSTEM

While arteries distribute blood to different parts of the human body, veins drain blood away from them. Although only one systemic artery, the aorta, takes oxygenated blood away from the heart, three systemic veins the coronary sinus, superior vena cava and inferior vena cava return deoxygenated blood to the heart. The coronary sinus receives blood from the cardiac veins; the superior vena cava receives blood from other veins superior to the diaphragm, except the alveoli (air sacs) of the lungs; the inferior vena cava receives blood from veins inferior to the diaphragm.

Most of the blood draining from the head passes into three pairs of veins the internal jugular vein, external jugular vein and vertebral veins. Within the brain all veins drain into dural venous sinuses and then into the internal jugular veins. Dural venous sinuses are endothelial-lined venous channels between layers of the cranial dura matter.

Although the brachiocephalic veins drain some portions of the thorax most thoracic structures are drained by a network of veins, called the azygos system that runs on either side of the vertebral column. The system consists of three veins

- the azygos,
- the hemiazygos, and
- accessory hemizygos veins, all these veins ultimately empty into superior vena cava.

Blood from the abdominal and pelvic viscera and abdominal wall returns to the heart via the inferior vena cava.

The inferior vena cava does not receive veins directly from the gastrointestinal tract, spleen, pancreas and gall bladder. These organs pass their blood into a common vein, the hepatic portal vein, which delivers the blood to the liver. The superior mesenteric and splenic veins unite to form the hepatic portal vein. This special flow of venous blood is called hepatic portal circulation. After passing through the liver for processing, blood drains, into the hepatic veins which empty into the inferior vena cava.

The hepatic portal circulation carries venous blood from gastrointestinal organs and spleen to the liver. A vein that carries blood from one capillary network to another is called a portal vein. The hepatic portal vein (hepat = liver) receives blood from capillaries of gastrointestinal organs, and the spleen and delivers it to the sinusoids of the liver. After anneal, hepatic portal blood is rich in nutrients absorbed from the gastrointestinal tract. The liver stores some of them and modifies others before they pass into the general circulation. For example, the liver converts glucose into glycogen for storage thereby reducing blood glucose level shortly after a meal. The liver also detoxifies harmful substances such as alcohol, that have been absorbed from the gastrointestinal tract and destroys bacteria by phagocytosis, while the liver is receiving nutrient rich but deoxygenated blood via the hepatic portal vein, it also recieves oxygenated blood via the hepatic portal vein, it also receives oxygenated blood via the hepatic artery, a branch of the celiac trunk. The oxygenated blood mixes with the deoxygenated blood in sinusoids. Eventually blood leaves the sinusoids of the liver through the hepatic veins, which drain into the inferior vena cava.

The superior mesenteric and spleenic veins unite to form the hepatic portal vein.

The superior mesenteric vein drains blood from the small intestine and proportions of the large intestine, stomach and pancreas through the jejunal, ileal, ileocolic, right colic, middle colic pancreatic, duodenal, and right gastropiploic veins. The splenic vein drains blood from the stomach, pancreas and large intestine portions. The cystic vein, which also opens into the hepatic portal vein, drains the gall bladder.

The Pulmonary Circulation

The pulmonary circulation (pulmo = lung) carries deoxygenated blood from the right ventricle to the air sacs (alveoli) within the lungs and returns oxygenated blood from the air sacs to the left atrium. The pulmonary trunk emerges from the right ventricle and divides into right pulmonary artery left pulmonary artery.

The Fetal Circulation

The circulatory system of a fetus, called the fetal circulation, exists only in the fetus and contains special structures that allow the developing fetus to exchange materials with its mother. It differs from the postnatal (after birth) circulation because the lungs, kidneys and gastrointestinal organs do not begin to function until birth. The fetus obtains O_2 and nutrients and eliminates CO_2 and other wastes into the maternal blood.

The exchange of materials between fetal and maternal circulation occurs through the placenta, which forms inside the mother's uterus and attaches to the umbilicus (navel) of the fetus by the umbilical cord. The placenta communicates with the mother's cardiovascular system through many small blood vessels that emerge from the uterine wall. Wastes from the fetal blood diffuse out of the capillaries and mother's uterine veins.

Nutrients travel the opposite route from the maternal blood vessels to the intervillous spaces to the fetal capillaries. Normally, there is no direct mixing of maternal and fetal blood because all exchanges occur by diffusion through capillary walls.

Blood passes from the fetus to the placenta via two umbilical arteries.

PROPERTIES OF HEART MUSCLE

The properties present in other muscles are also shown by the cardiac muscle. Besides this, heart muscles show certain special features (Fig. 2.8):

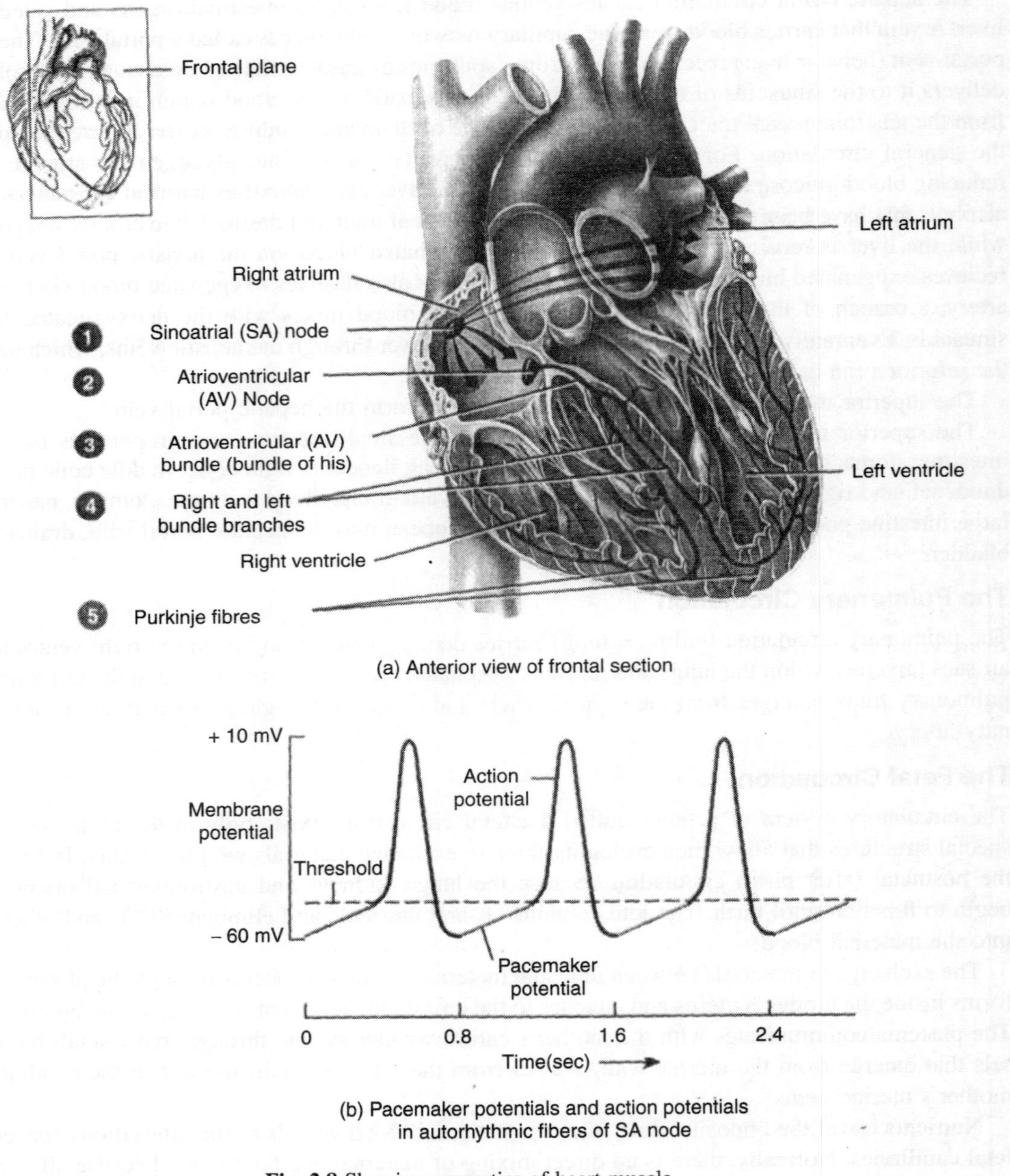

Fig. 2.8 Showing properties of heart muscle

1. **Rhythmicity:** One of the main characteristic features of the cardiac muscle is that it can initiate its own impulse rhythmically. Due to higher, rhythmical property of the S. A. node, it controls the rest of cardiac muscle and thus heart beats at the rhythm of the S.A. node. When the S.A. node fails, the A.V. node takes the charge and if it fails, the atrium and after wards ventricle take the charge of maintaining heart beat.

2. **Conductivity:** Conduction of impulse through different parts of the heart is established. The A.V. node transmits the impulse through the bundle of His and its branches to the ventricles. From the apex of the heart through the purlinge fibres the impulse is conducted to the base. Conduction in the bundle of his and the Purlinge fibre is 1 metre per second, still lies in the ventricular muscles 0.4 metre per second and least in the S.A. node 0.05 metre per second and A.V. node 0.1 metre per second.

3. **Excitability and contractibility:** Like other muscles, the cardiac muscle is excitable by adequate stimuli and responds by contraction. The fundamental contractile unit of the cardiac muscle is myofibril which contains the protein units, actin and myosin. During contraction these two units are associated in presence of ATP and thus the fibre is shortened, but during rest these are dissociated, again with the resynthesis of A.T.P. myosin itself is an enzyme. 'ATP-ase' capable of dephosphorylation of ATP. Ca^{++} ion activates the ATP-ase activity-favouring prompt association of actomyosin and ADP complex. Excess calcium always keeps the muscle unit in contracting state (calcium rigor) due to association of more contractile units. K^+ ions do not favour association of actin and myosin.

Therefore, if excess K^+ is added in the extracellular fluid them the heart muscle gradually stops in diastole.

4. **Allornone response:** If aquiscent heart muscle is stimulated at widely spaced electrical shocks of increasing strength then muscle contracts as a whole only when the threshold strength is reached.

5. **Staircase:** In a stannies preparation if the ventricular muscle is stimulated with inducted current, the first few contractions gradually increase in size and then it becomes steady. This is known as treppe or staircase phenomenon. This phenomenon is only observed in quiscent heart but not in active normal heart. From the experiments on staircase phenomenon, it can be suggested that it is caused by the decrease of internal potassium. After the analysis of potassium ion it is reported that the favourable condition is mostly related with the release of potassium from the cell which reenters during rest at a speed depending the condition of the membrane.

6. **Refractory period:** The refractory period of the heart is long and can be divided into three parts.

(*i*) Absolute refractory period,
(*ii*) Relative refractory period,
(*iii*) Supernormal period.

Absolute refractory period extends throughout the whole period of contraction. Any stimulus, however, strong will fail to elicit a response if it falls within this period. For this reason, heart muscle cannot be tetanised. This long refractory period ensures enough time for recovery of the cardiac muscle. This is the reason why cardiac muscle cannot be fatigued.

Relative refractory period—starts immediately after the absolute refractory period and involves the first part of relaxation. Only a very long stimulus will be effective. This period beings when the transmembrane potential during repolarisation phase has just reached the threshold potential and ends just before the repolarisation phase is ceased.

Supernormal period—is limited from the point of termination of repolarisation to the beginning of slow diastolic repolarisation phase.

7. **Tone:** Heart muscle possesses tone. This tone is independent of nerves and can be adjusted. In this way, it can maintain a fairly constant tension upon its varying contents.

It has been observed that the various properties of cardiac muscles are not developed in all the tissues of heart in the same order certain properties have developed specially in certain tissues while others are not. It is also observed that these functions are related to the size and chemical composition of the muscle cells. Thus cardiac muscles can be divided into four groups, as follows:

(*a*) The smallest fibres with least glycogen at the nodes.
(*b*) The broader fibres with more glycogen in the ventricles.
(*c*) The still broader fibres with more glycogen in the atria.
(*d*) The broadest fibres with abundant glycogen in the purlinge fibres, bundle of his and its branches. As the size of the fibre increases, the rate of conduction and the glycogen content also increase. But the duration of systole, the refractory period and the rhythmicity increase in the reverse.

Action of Heart and Heart Beat

It is well established that the atrial rhythm was higher than the ventricular rhythm. From sinuauricular node the heart beat first starts.

Cardiac impulse originated at the sinuauricular node is transmitted over both the atria like concentric waves and thus the P wave is produced in E.C.G. (Electrocardiogram). The spread of electrical impulse through the S.A. node is very slow (0.05 m per second) but the same through the functional tissues that connect the node to the atrial musculature or to the A.V. node through the internodal atrial bundle (Fig. 2.8).

Conduction over A.V. node: There is also a considerable delay of transmission of impulse in the A.V. node. This delay is observed maximally at the functional region between the atrium and atrioventricular node.

Conduction over bundle of His and the right and left bundle branches: The impulse from the bundle of His passes quickly through the right and left bundle branches and ultimately reaches the Purkinje fibres and ventricular muscle fibres as well.

Conduction through Purkinje systems. The impulse, after passing through the right and left bundle branches, passes into the Purkinje fibres and also its multiple ramification within the subendoranial surfaces of both ventricles. The impulse then travels from the endocardium to the epicardium of ventricular muscle perpendicularly conduction through ventricular muscles.

In human beings the mid portion of the interventricular septum is activated normally in a left to right direction. So the depolarisation of the ventricular muscle begins at the left side of the interventricular septum because the Purkinje fibres arises more proximally from the left bundle branch than from the right bundle branch and actinates the left side of the septum initially. After this, the impulse comes down the septum to the apex of the heart, and next portions of myocardium that is activated is the anteroseptal region of the myocardium. The impulse then proceeds along the right and left ventricular walls to the aterioventricular groove. The spread of excitation through the ventricle proceeds from the endocardium to the epicardium and thus the whole of the right and left ventricular walls are depolarised the posterobasal regions of the left ventricle. The portions of the ventricles that are excited lastly (Fig. 2.8) the pulmonary conus, and the uppermost portion of the interventricular septum.

Blood Pressure

Blood pressure is the lateral pressure exerted by blood on the vessel wall while flowing through it. Four terms are in common use:

(*i*) Systolic pressure (S.P.)—The maximum pressure during systole,
(*ii*) Diastolic pressure(D.P.)—The minimum pressure during diastole.

(*iii*) Pulse pressure (P.P)—The difference between systolic and accepted diastolic pressure.

(*iv*) Mean pressure (M.P.)—It is the arithmetic mean of the diastolic and the systolic pressures. But approximation to the mean pressure could be obtained by adding the diastolic pressure with one-third of the pulse pressure.

In an adult the relation, among the three presssures should be as follows:
S.P./D.P./P.P. = 3/2/1 *i.e.*, if systolic pressure is 120,
diastolic pressure should be 80, and pulse pressure 40 mm of Hg.

Significance of blood pressure. Systolic pressure undergoes considerable fluctuations. Excitements, exercise, obesity and heavy meals etc. increases it while sleep, during rest and relaxation it diminishes. The height of systolic pressure indicates:

(*a*) The extent of work done by heart,
(*b*) The force with which the heart is working,
(*c*) The degree of pressure which the arterial walls have to withstand.

Diastolic pressure undergoes much less fluctuations in health and remains within a limited range. Increase of diastolic pressure indicates that the heart is nearing towards failure. Diastolic pressure is the measure of peripheral resistance. It indicates the constant load against which heart has to function. Pulse pressure generally varies directly as the stroke volume. But this quantitative relation may not be true in all cases.

The normal function of blood pressure is

(*i*) To maintain a sufficient pressure head to keep the blood flowing.
(*ii*) To provide for the motive force of filtration at the capillary bed—thus assuring nutrition the tissue cells, formation of urine, lymph and so on.

LYMPH AND LYMPHATIC SYSTEM

Lymph is a modified tissue fluid. It is the clear watery appearing fluid found in lymphatic vessels and is formed by the passage of substance from blood capillaries into tissue spaces.

Composition of lymph. Detailed examination of lymph depicts that it possesses a large number of leucocytes ranging from 500 to 75,000 per cu. mm. No blood platelets are present. Noncellular composition of lymph is water 94% and solid is –6%.

(*a*) **Proteins:** Total protein content is roughly half that of plasma and varies from 2.0–4.5%. It varies according to the part of the body from which it is collected. Lymph from the liver contains three times (6%) as much protein as that coming from the limb (2%).

Three varieties of proteins are found—albumin, globulin and fibrinogen. Traces of prothrombin are also found.

(*b*) **Fats.** During fasting condition fat content is low but after a fatty diet it may be 5.0–15%.

(*c*) **Carbohydrates:** Sugar 132.2 mgm per 100 ml.

(*d*) **Other constituents** (Expressed in mgm per 100 ml, urea, 23.5 mgm; Non protein nitrogen substance, 34.8 mgm; Creatinine, 1.4 mgm, chlorides, 711 mgm. Total phosphorus, 11.8 mgm; Inorganic phosphorus, 5.9, mgm; calcium, 9.84 mgm. Enzymes and antibodies also are present.

Lymphatic system. The lymphatic vessels at the periphery are microscopic and vessels called lymphatic capillaries. These tiny vessels are situated in the intercellular spaces and their walls formed by endothelial cells supported by the fibrous connective tissue (Fig. 2.9). These capillaries repeatedly join together to form bigger lymphatic vessels, which pass through the lymph nodes receive more tributaries and gradually increase in size. The right lymphatic duct and the thoracic duct which open respectively, at the right and left subclavian veins. The thoracic duct emerges from the cisterna chyli. The cisterna chyli, being situated on the front of the body of the second lumbar vertebra, receives entire lymph coming from two hind limbs and elimentary canal. The lymphatic vessels are provided with values which help the lymph stream to flow in the direction of the chest. The primary lymphatic vessels of intestinal villi are known as lacteals; they are filled with milk white fluid, chyle during the course of digestion. Chyle's chemical composition is similar to lymph. In the central nervous system there are nolymphatics. Here, cerebrospinal fluid takes the place of lymph. Lymphatic capillaries, are not found in the cartilage, spleen, epidermis, internal ear and eyeball.

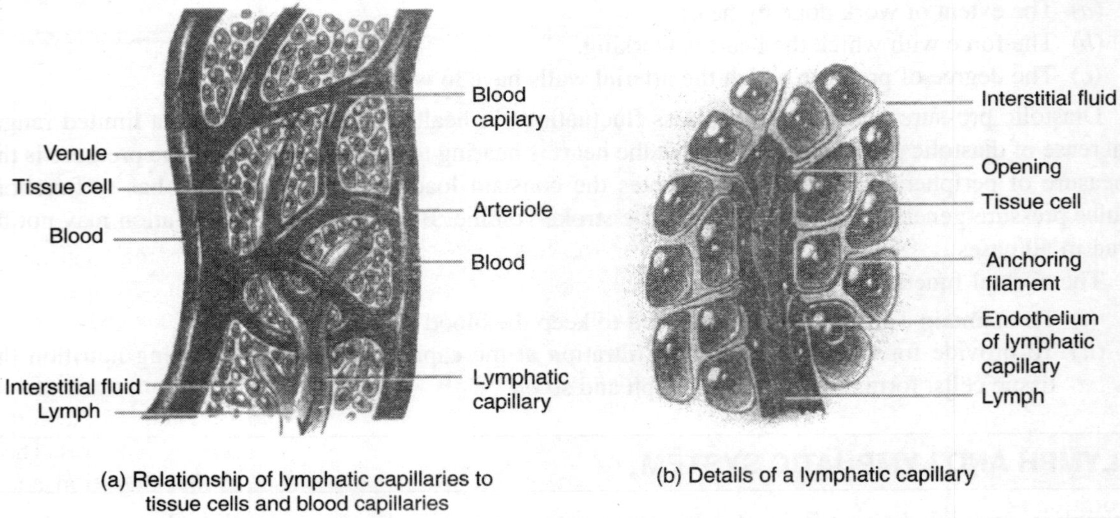

Fig. 2.9 Showing lymphatic vessels

The function of lymphatics is to carry tissue fluid from tissues to veins and the retun of water and protein from the interstitial fluid to blood from which they come. And the function of lacteals is to help in the absorption of digested food materials generally fats from the intestine.

UNIT

3

Digestive System

OVERVIEW OF DIGESTIVE SYSTEM

The human digestive system includes a long muscular canal consisting of the mouth (guarded by lips and teeth, tongue, pharynx, oesophagus, stomach, small intestine, large intestine, rectum and anal canal. The ducts of salivary glands open into the mouth. The junction of stomach with oesophagus is guarded by cardiac sphincter. Following the pyloric sphincter at distal and of stomach, the small intestine begins. Small intestine consists the following parts-duodenum, jejunum and ileum (Fig. 3.1).

The duodenum receives food from the stomach where the duodenum joins the jejunum, connective tissue and muscle fibres thicken the mesentery. The bile duct and pancreatic duct open in the mesentery through ampula of vater. The small intestine is very long and is roughly measures about 7.6 metres. The long length of this small intestine provides sufficient time and surface area, so that digestion and absorption of food stuff could be completed.

The small intestine opens into the next part of the alimentary canal—the large intestine. The opening between them is guarded by iliocolic sphincter. In this large intestine water is absorbed and the faeces become formed. The large intestine opens into the last part–rectum and anal canal.

The anal canal opens outside through the anal orifice.

Peritoneum is a serous membrane and lines the interior of the abdominal cavity. The parietal outer layer is in contact with the body wall and the inner visceral layer envelops the abdominal organs. The mesentery is the continuation of the peritoneum and extends to the small and large intestines from dorsal body wall. The lesser omentum of the stomach attaches to the liver and the greater omentum hangs from the greater curvature of the stomach over the intestine to the colon, as an apron. In this fold fat could accumulate.

Functions of the Digestive System

The human digestive system serves various functions like:

(*i*) Ingestion of food, (*ii*) Digestion of food, (*iii*) Secretion of various digestive juices, (*iv*) Absorption of water, salts, vitamins and end products of food digestion, (*v*) Excretion-heavy metals, toxins, certain alkaloids, etc. (*vi*) Movements certain types of movements are present in this whole of the gastrointestinal tract. The functions of these movements are to facilitate admixture of food with digestive juices, to propel food onwards, to help blood and lymphatic circulation althrough the intestinal wall. Defeacation

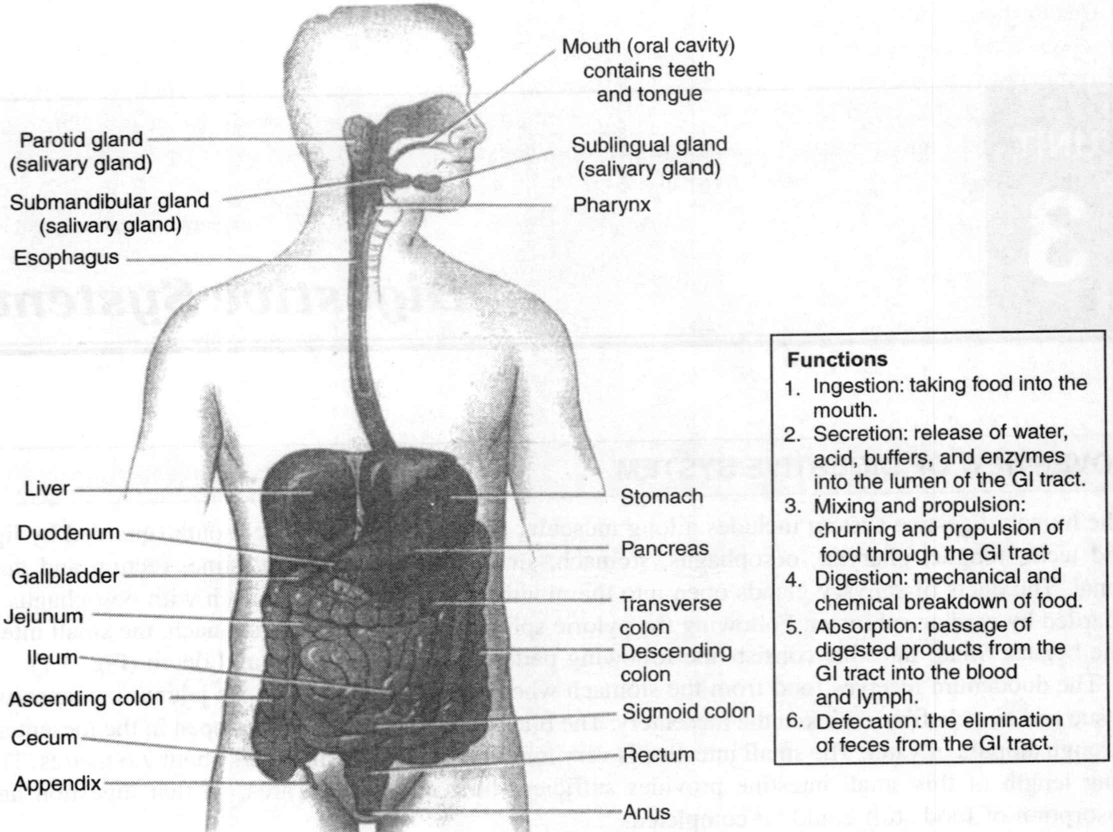

Fig. 3.1 Showing human digestive system

is also due to the movements of large intestine, (*vii*) Erythropoesis-stomach manufacturers a substance called the intrinsic factor. The extrinsic factor is vitamin B_{12}; the intrinsic factor interacts with it and helps in the absorption of the extrinsic factor. The extrinsic factor promotes the maturation of the erythroid cells. Pernicious anaemia is associated with gastric leison, (*viii*) Regulates blood reaction—the alimentary canal takes part in the regulation of blood reaction. (*ix*) Regulates blood sugar—It takes part in the regulation of blood sugar. (*x*) Maintains water balance—The phenomenon of thirst is an important function of the digestive tract by which the fluid balance of the body is maintained.

Functional Anatomy of Digestive System

Mouth: The mouth is guarded by lips and teeth. Mouth the opening and beginning of digestive system leads into oral cavity; also known as buccal cavity, which is bounded by the upper and lower jaws. On the roof of the mouth cavity, anteriorly is the hard palate which continues posteriorly as the soft palate on the floor of the buccal cavity, tongue is present. There are teeth both in the upper and lower jaws.

Teeth in human beings are not similar in type. As they are of different types, such teeth are called heterodont. There are two sets of teeth in human beings, one set is in early life-called milk dentition which is replaced in the adult by another set called-permanent dentition. Due to the presence of two sets of teeth the condition is called diphyodont.

Four types of teeth are present in human beings. These are: (1) Incisors (i), the cutting teeth in the front part of the mouth, (2) Canines (c), The tearing teeth with pointed edges found behind the incisors (3) Premolars (pm) and (4) Molars (m). The premolars and molars are usually called grinders which have flat edges and are found behind the canines. These teeth, assist in grinding the food.

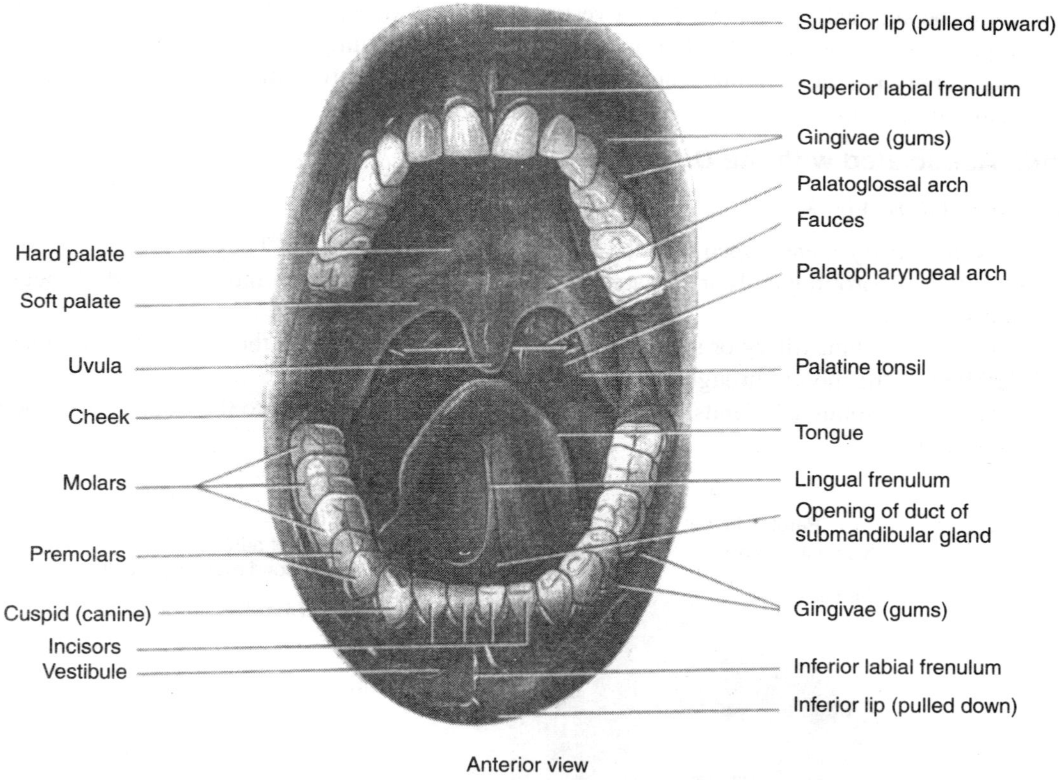

Fig. 3.2 Showing arrangement of teeth

The nature, type and arrangement of teeth in mammals is called dentition. The arrangement of human being teeth is expressed in the form of a dental formula which is as follows:

$$i\frac{2}{2}\,c\frac{1}{1}\,pm\frac{2}{2}\,m\frac{3}{3}=\frac{8}{8}=16\times 2=32$$

The buccal cavity leads into the pharynx, which further leads into oesophagus. The ten inches long oesophagus is muscular.

The oesophagus leads into the stomach which is a muscular chamber. The stomach is more or less 'J' shaped and internally lined with mucous membrane, in which there are present large number of gastric glands, secreting gastric juice. The narrow distal end of the stomach is called pylorus.

The stomach leads into the small intestine which is about 23 feet long and is divided into three regions, the duodenum, the jejunum and the ileum. The duodenum is about ten inches long and is provided with duodenal digestive glands. The bile duct from the liver and pancreatic duct from the pancreas open into duodenum.

In the jejunum and the ileum the mucous membrane internally protrude to produce microscopic finger like processes which project into the lumen of the intestine. These structures are called villi which increase the absorptive surface of the intestine. The intestinal glands secrete intestinal juice.

The ileum opens into the first part of the large intestine called caecum which is provided with a diverticulum called vermiform appendix, which is a blind sac. The appendix is frequently infected and then it has to be surgically removed. It is a vestigial organ. A degenerated organ without function.

The large intestine, also called colon, is divided into an ascending colon, a transverse, colon and a descending colon. The descending colon continues behind as the rectum. The rectum opens to the exterior through the anus.

Glands Associated with the Digestive System

1. Salivary glands (Fig. 3.3)

Three pairs of salivary glands open into the buccal cavity through ducts. They are:

(*a*) One pair of parotid glands are located in front of the ears and open into the mouth through two ducts of stensen.

(*b*) One pair of submaxillary or submandibular glands, located behind the lower jaw and open into the floor of the mouth through two ducts of wharton.

(*c*) One pair of sublingual glands, located below the tongue and open into the floor of the mouth by several ducts.

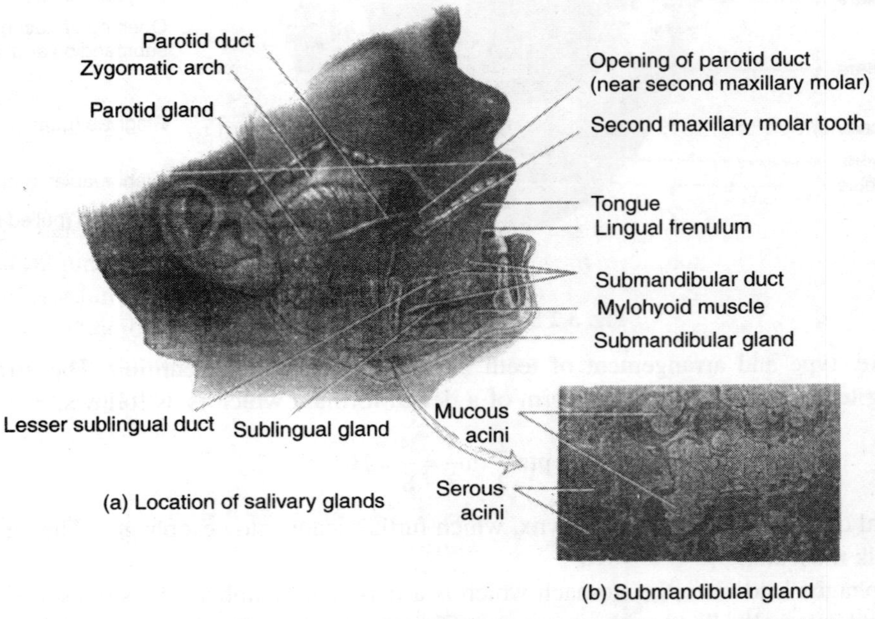

Fig. 3.3 Showing three pairs of salivary glands, open into the buccalcavity through ducts

The salivary glands are composed of lobules which are composed of serous and mucous cells. Saliva secretion by salivary glands occurs through the ducts. The saliva contains amylase enzyme (amylopsin or plyatin).

In addition to these 3 pairs of salivary glands, many small salivary glands called buccal glands are also present located in the mucous membrane of the buccal cavity.

2. Gastric Glands

Gastric glands are present in the stomach located beneath the interanl epithelial lining. These are simple tubular glands. It is estimated that 3,50,00,000 gastric glands are present in the stomach.

Gastric glands, by virtue of their location are different types:

(*i*) Cardiac glands—in the upper part,
(*ii*) glands or body glands—in the middle fundic
(*iii*) Pyloric glands—in the pylorus.

The gastric glands secrete gastric juice.

3. Intestinal Glands

Intestinal glands are present in the mucous membrane of the intestine. They are tubular shaped glands and secrete intestinal juice, which is also known as succus entericus, which is emptied into the intestine through vault like glandular structures between the villi and are called 'crypts of Lieberkuhn'. The mucosa is produced into circular folds called 'folds of kerchering'.

4. Liver or Hepar (Fig. 3.4)

Liver is the largest gland of the human body. It weighs about 1.5 kg. It is located on the right side and is in contact with the undersurface of the diaphragm. Human liver consists of 4 lobes. The right, left, quadrate and caudate lobes. Liver is an exocrine gland, as its secretions are taken into the destination through ducts.

The liver gland is made up of small lobules, in which the hepatic cells are arranged in the form of hepatic plates or hepatic cords. Hepatic plates branch and anastomose. The lobule has the central or intralobular vein. The liver cells, are polyhedral and remain in close contact with blood spaces, called sinusoids, which separate the hepatic plates. These sinusoids are useful to communicate with the intralobular vein or central vein and receive blood from the hepatic artery and hepatic vein. Liver cells possess specialised kupffer cells, and they are meant for phagocytosis.

Bile canaliculi, empty spaces are present between the hepatic plates and they open into small bile ducts; they subsequently join to form the hepatic duct which, opens into the duodenum. The portal vein, the bile duct and the hepatic artery together constitute the hepatic triad. The portal vein brings venous blood to the liver from stomach, intestine and spleen. Hepatic artery supplies blood to the liver lobules.

Bile: The secretion of digestion gland liver is called bile. It is viscous yellowish brown or greenish fluid with a pH of 7 to 8. Bile contains water, bile salts, bile pigments, inorganic salts, lecithin and cholesterol.

Functions of Bile

(*a*) Bile does not contain any enzyme; however, it is essential for digestion of fats by reducing them into small droplets, through a process 'emulsification'.
(*b*) Bile has activating influence on lipases.
(*c*) Bile being a good solvent, helps in absorption.
(*d*) Certain toxic substances are also excreted through bile.

Bilirubin, which is golden yellow and biliverdin which is green are bile pigments formed as a result of breakdown of haemoglobin of wornout red blood cells in the spleen. The liver uses these pigments for the synthesis of bile.

When intestine is empty without food, bile instead of flowing into the duodenum, stored in the gall bladder. Thus gall bladder is a reservoir for bile and its duct cystic duct, which joins the hepatic duct, which is joined by pancreatic duct before opening into the duodenum.

General Functions of Liver

(*i*) Secretion of bile

(*ii*) Serving in protein synthesis.

(*iii*) During 'ornithine cyle' urea is produced.

(*iv*) Poisons are disposed off by liver.

(*v*) Wornout red blood corpuscles are broken down by liver.

(*vi*) Liver helps in excretion through bile pigments.

(*vii*) Excess quantity of glucose is converted into glycogen by liver; this process is termed as 'Glycogenesis'.

(*viii*) Glycogen is converted into glucose by liver and released to the blood; this process is termed as 'Glycogenolysis'

(*ix*) Proteins and fats are converted into glucose by liver; this process is termed as 'Gluconeogenesis'.

(*x*) Excess glucose available is converted into fat by liver; this process is termed as 'Lipogenesis'.

(*xi*) Liver helps for the manufacture of erythrocytes in embryo, this process is termed 'Haemopoiesis'.

(*xii*) Liver acts as store house of blood.

(*xiii*) Vitamins are synthesized and stored in liver.

(*xiv*) Alcohol metabolism is accomplished in liver.

(*xv*) Phagocytosis is accomplished by Kupffer cells.

(*xvi*) Liver plays a vital role in the maintenance of homeostasis of human being.

5. Pancreas (sweet bread) Fig. 3.4

Pancreas is a soft pink-whitish coloured gland lying between the stomach and duodenum, it measures about six inches in length. The exocrine part of it is composed of cells called Acini or lobules with a central lumen. The cells are granular connective tissues with blood vessels are present between the acini. The exocrine part of pancreas secretes the pancreatic juice which is carried to the duodenum by the 'Ampulla of Vater'.

Endocrine part: The endocrine part of the pancreas (gland without duct) comprises groups of cells called 'islets of langerhans'. The islets of langerhans consist two types of cells – alpha cells and beta cells.

Pancreatic Gland Secretions

1. *Insulin:* Insulin is secreted by the beta cells, which regulates the sugar metabolism in blood. Deficiency of insulin leads to relatively high concentration of glucose in blood and this condition is termed 'Diabetes mellitus'.
2. *Glucagon:* Glucagon is secreted by the alpha cells; which converts liver glycogen into glucose. Insulin and glucagon are directly released into the blood stream and are called hormones.

Digestion

Digestion of food includes mechanical digestion and physiological (chemical) digestion.

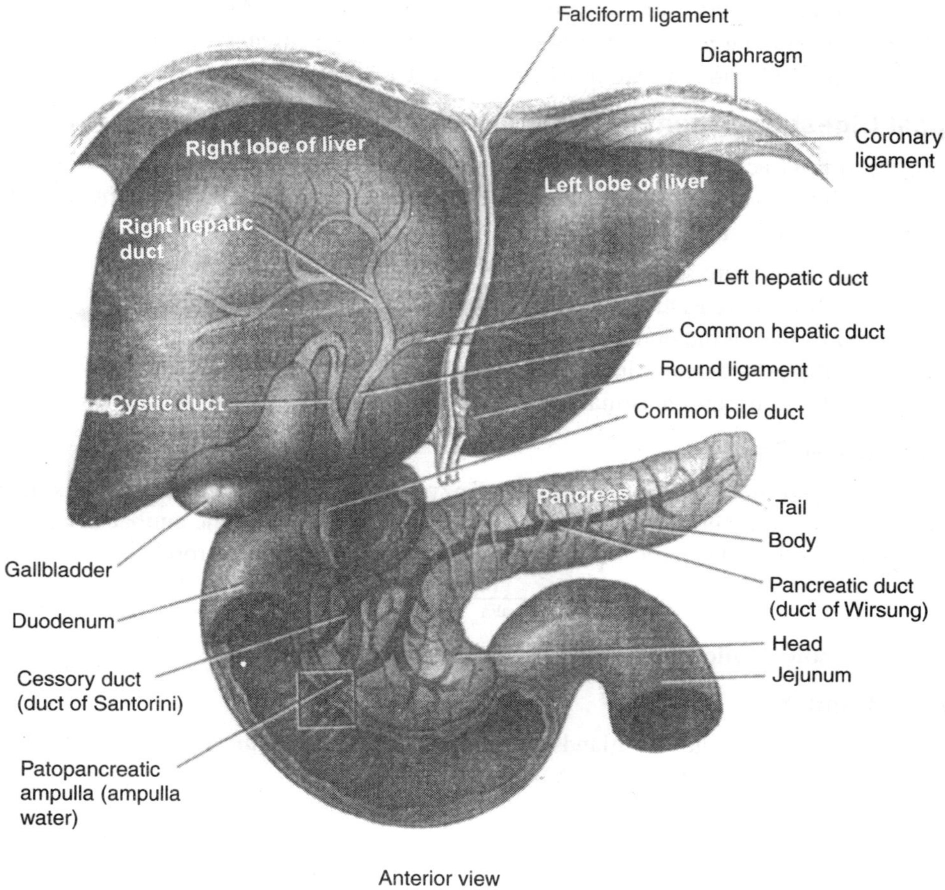

Fig. 3.4 Showing relation of the pancreas to the liver, gallbladder, and duodenum

Mechanical digestion involves the following processes:

(*a*) *Mastication:* In mastication food is cut and chewed with the help of teeth and mixed with saliva in the buccal cavity.

(*b*) *Deglutition:* Deglutition is also known as swallowing. The process of swallowing is accomplished by mechanical movements. Consequently the food bolus (or mass of masticated, chewed food) is taken to the stomach.

(*c*) *Movements of alimentary canal:* The mechanical movements of the gastrointestinal tract aids:

(*i*) to propel the food forwards.

(*ii*) to get the food mixed with digestive juices and enzymes.

(*iii*) to ensure proper circulation in the gastrointestinal tract.

The mechanical movements of the alimentary canal are of two types.

(*i*) Translatory movements and

(*ii*) Stationary movements.

Translatory movements: These movements propel the food forward. e.g., Peristalsis. This is the movement resulted due to alternate wave of contractions followed by expansions of the muscular walls.

These movements are called peristaltic movements. Peristaltic movements take place is oesophagus, stomach, small intestine and large intestine stationary movements. Stationary movements are localised and are useful in mixing the food. These movements occur in stomach and intestine.

Physiological Digestion (Chemical Digestion)

Physiological digestion is accomplished by the action of digestive juices and enzymes on the food particles in the gastrointestinal tract. Various digestive juices and enzymes secreted to the different parts of the alimentary canal are as follows:

1. Salivary glands contribute saliva in the mouth.
2. Gastric glands contribute gastric juice in the stomach.
3. Liver contributes bile in the duodenum.
4. Pancreas contributes pancreatic juice in the duodenum.
5. Intestinal glands contribute intestinal juice or succus entericus in small intestine.

Digestion in Buccal Cavity

Saliva is secreted by three pairs of salivary glands. Enzyme Amylase or Ptyalin, which is acidic (pH 6.02–7.05) splits starch into maltose *i.e.*, polysaccharides to disaccharides. The antibacterial enzyme of saliva called 'Lysozyme' kills bacteria. Further, saliva aids as lubrication of food.

$$\text{Starch} \xrightarrow[\text{Ptyalin}]{} \text{Maltose}$$

The food bolus is now swallowed into the stomach through oesophagus.

Digestion in Stomach

Gastric juice is secreted by the gastric glands of the stomach. Gastric juice is acidic (pH 0.9 to 1.5) composition of gastric juice is as follows:

(*i*) Water
(*ii*) Hydrochloric acid
(*iii*) Enzymes: Pepsin, Rennin and gastric lipase
(*iv*) Inorganic salts
(*v*) Intrinsic factor.

HLC in gastric juice performs the following functions:

(*a*) It acidifies the food and stops the action of ptyalin.
(*b*) Initially pepsin is secreted in an inactive form 'pepsinogen' which is converted into the active form pepsin by HLC.
(*c*) HLC kills bacteria present if any, in food.

Enzymes and Their Functions

(*i*) *Pepsin:* Pepsin digests proteins. It breaks down proteins into proteoses and peptones.

$$\text{Proteins} \xrightarrow[\text{Pepsin}]{} \text{Proteoses + Peptones}$$

(*ii*) *Rennin (or chymosin or Rennet):* Rennin converts caseinogen of milk into solid casein. This is called curding of milk. The curd like partly digested milk is further influenced by pepsin and is converted into peptone.

(*a*) $\text{Milk caseinogen} \xrightarrow[\text{Rennin}]{} \text{Solid casein (curd)}$

(*b*) Curd $\xrightarrow[\text{Pepsin}]{}$ Peptone

(*iii*) *Gastric lipase:* This enzyme splits down fats particularly butter and yolk of egg into fatty acid.

Fats $\xrightarrow[\text{Lipase}]{}$ Fatty acids

As a result of digestion the food is converted into a semifluid mass called chyme.

Intrinsic factor: The intrinsic factor present in the gastric juice is very essential for the absorption of vitamin B_{12} which is considered as the extrinsic factor.

The mechanical movements of the stomach exerted as a result of expansion and contraction of stomach muscles, help in mixing the food and also in passing the chyme into the duodenum in small quantities.

Digestion in the Small Intestine

(*a*) The Bile is secreted by the liver. Bile is useful for the total digestion of fats by reducing into small droplets. This process is called emulsification.

Fats $\xrightarrow[\text{Bile}]{}$ Emulsification

(*b*) Pancreatic juice which is alkaline, contains the following enzymes. (pH of pancreatic juice –8.0 to 8.3).

(*i*) **Trypsin:** Trypsin splits proteoses into tri-and dipeptides.

Proteoses and Peptones $\xrightarrow[\text{Trypsin}]{}$ tri and dipeptides.

(*ii*) **Chymotrypsin:** It also splits, proteoses and peptones into tri and dipeptides.

(*iii*) **Amylase or alpha amylase:** It splits starches and glycogen into maltose.

Starches and Glycogen $\xrightarrow[\text{Amylase}]{}$ Maltose.

(*iv*) **Pancreatic lipase or steapsin:** Emulsified fats are converted into fatty acids and glycerol by lipase.

Emulsified fats $\xrightarrow[\text{Steapsin}]{}$ Fatty acids with Glycerols.

(*v*) **Maltose:** It converts maltose into glucose.

Maltose $\xrightarrow[\text{Maltase}]{}$ Glucose

(*vi*) **Carboxpeptidase:** It converts proteins polypeptides into tri and dipeptides.

Polypeptides $\xrightarrow[\text{Carboxy-peptidase}]{}$ Tri and dipeptides

(*vii*) **Nuclease:** Nuclease converts nucleic acids into nucleotides.

Nucleic acids $\xrightarrow[\text{Nuclease}]{}$ Nucleotides

(*c*) Intestinal juice or succus entericus. This is secreted by the intestinal glands. This digestive juice is alkaline with a pH–8.3.

The important enzymes of intestinal juice are as follows:

(*i*) **Enterokinase:** It converts the inactive trypsinogen into active trypsin.

Trypsinogen $\xrightarrow[\text{Enterokinase}]{}$ Trypsin

(*ii*) **Erepsin:** Erepsin converts tri and dipepetides of proteins into amino acids.

$$\text{Tri and dipepetides} \xrightarrow[\text{Erepsin}]{} \text{Amino acids}$$

(*iii*) **Sucrase or Invertase:** It converts sucrose into glucose and fructose.

$$\text{Sucrose} \xrightarrow[\text{Sucrase}]{} \text{Glucose and fructose}$$

(*iv*) **Maltase:** Maltase converts maltose into glucose.

$$\text{Maltose} \xrightarrow[\text{Maltase}]{} \text{Glucose}$$

(*v*) **Lipase:** It converts emulsified fats into fatty acids and glycerols.

$$\text{Emulsified fats} \xrightarrow[\text{Lipase}]{} \text{Fatty acids and glycerols.}$$

(*vi*) **Phospholipase:** This converts phospholipids into fatty acids, glycerols and phosphoric acid.

$$\text{Phospholipids} \xrightarrow[\text{Phospholipase}]{} \text{Fatty acids, glycerols and phosphoric acid.}$$

(*vii*) **Phosphatase:** This enzyme converts organic phosphates into free phosphates.

$$\text{Organic phosphates} \xrightarrow[\text{Phosphatase}]{} \text{Free phosphates}$$

(*viii*) **Polynucleotidase:** It converts nucleic acids into nucleotides.

$$\text{Nucleic acids} \xrightarrow[\text{Polynucleotidase}]{} \text{Nucleotides}$$

(*ix*) **Nucleosidase:** This converts nucleosides into purine or pyrimidine bases and pentose sugars.

$$\text{Nucleoside} \xrightarrow[\text{Nucleosidase}]{} \text{Purine, Pyrimidine and Pentose sugar.}$$

(*x*) **Lactase:** It converts lactose into galactose and glucose.

$$\text{Lactose} \xrightarrow[\text{Lactase}]{} \text{Galactose and glucose.}$$

Absorption: About 90% of the absorption of digested food occurs in the small intestine.

Absorption of the digested food takes place in the stomach to a limited extent only, *e.g.* water and glucose. By the time chyme reaches the large intestine, absorption is almost complete.

Functions of Large Intestine (Colon)

(*a*) **Reabsorption:** Reabsorption of water takes place through the walls of large intestine so that the undigested matter becomes a semisolid mass called the stool (or faeces).

(*b*) **Secretion:** The large intestine secretes mucin which is helpful for lubricating the faeces so that the latter can easily pass through the rectum and anus.

(*c*) **Excretion:** It also excretes excess of calcium, iron and certain drugs.

(*d*) **Bacterial decomposition:** The decomposition of cellulose is accomplished by bacteria present in the large intestine.

Defaecation: Defaecation is the process of emptying the distal colon by ejecting out faeces, through rectum and anus. Both translatory and stationary movements of the walls of the colon help the process of defaecation.

(*e*) **Excretion:** It also excretes excess of calcium, iron and certain drugs.

(*f*) **Bacterial decomposition:** The decomposition of cellulose is accomplished by bacteria present in the large intestine.

UNIT

4

Respiratory System

Respiration can be defined as the process during which oxygen (O_2) of the fresh air, taken to the lungs during breathing is carried through blood, to the cells, where nutrients are broken down in the presence of oxygen by means of oxidative metabolism, for the recovery of energy from the substrates. Carbon-dioxide (CO_2) which is a waste gas, formed during cellular respiration or oxidative metabolism is taken to the lungs through blood and subsequently eliminated during breathing. Thus, respiration involves breathing and also oxidative metabolism of nutrients which occurs in cells in the presence of oxygen.

The human respiratory system consists of the nose, nasal cavity, pharynx, larynx, trachea, bronchi, bronchioles and the lungs. This respiratory system can be studied in upper and lower respiratory tracts. The upper one consists of the air passages of the nose, nasal cavities, pharynx, upper trachea and larynx. The lower respiratory tract consists of the lower trachea and the lungs.

The nose opens into nasal cavity. The nasal cavity is richly supplied with arteries, veins and capillaries. At the end of the nasal cavity is the pharynx. It is a common passage way for air and food. This is divided into three regions – nasopharynx, oropharynx and larynyopharynx.

The pharynx opens into the larynx or voice box. The thyroid cartilage of the larynx forms a median protrusion, called Adams Apple. Adam's Apple is more pronounced in males than in females. The larynx is located at the upper end of the trachea. Inside and stretched across the larynx are two elastic folds of tissue ligaments called the vocal cords.

Air rushes through the voice box and causes the vocal cords to vibrate producing sound waves. The walls of the trachea are strengthened with C-shaped rings of tough flexible cartilage. These cartilagenous rings protect the trachea, make it flexible and prevent it from collapsing or over expanding. Trachea enters the thoracic cavity and divides into two branches – the right and left bronchi. Each bronchus enters the lung of its respective side. The walls of the bronchi are supported by cartilage, thickening rings similar to that of trachea (Fig. 4.1).

The lungs are located inside the thoracic cavity. There are two lungs in human beings respiratory system. The right lung has three lobes, and it is larger of the two. The left lung has two lobes and is smaller, as the heart is placed slightly towards the left in the thoracic cavity (Fig. 4.2). The lungs are bounded and protected by the rib cage and diaphragm. Lining the entire cavity and encasing the lungs is the pleural membrane. It has an outer parietal layer, and an inner visceral layer. The gap between the two layers is filled with mucus that decreases friction and helps in the smooth movement of the lungs while breathing.

Rib cage: The sternum and the ribs constitute the thoracic cage prominently known as rib cage. The sternum is commonly called as breast bone. Sternum is placed in the anterior median part of the

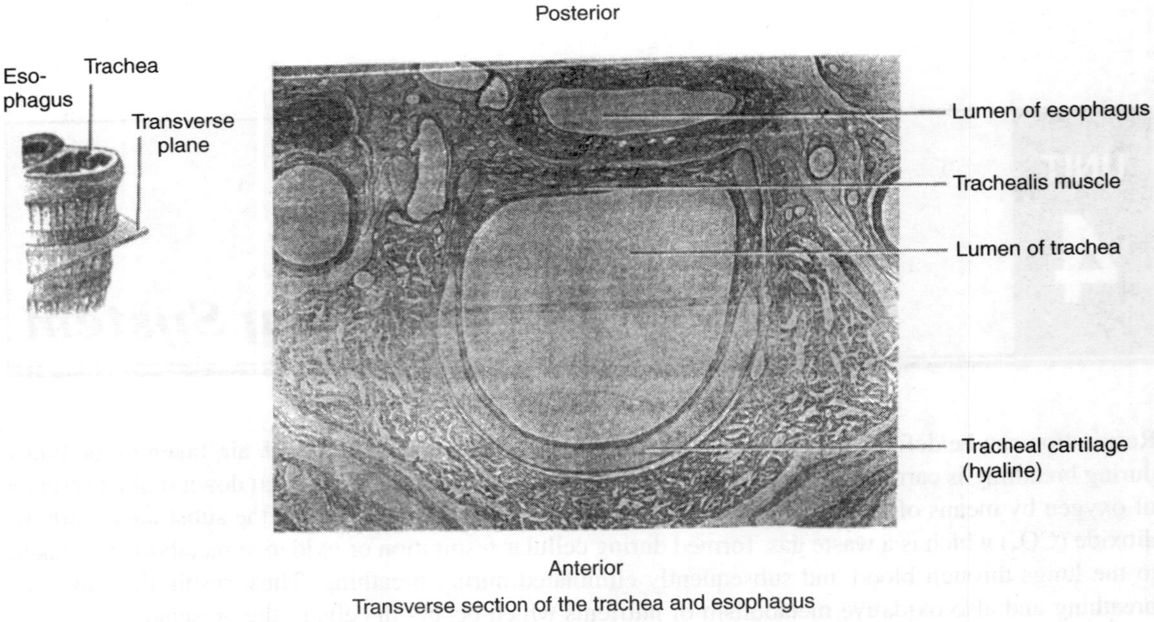

Fig. 4.1 Showing location of the trachea in relation to the esophagus

thoracic wall. There are 12 pairs of ribs or costae. Each rib is a curved bone. Each rib is attached to the thoracic vertebra posteriorly. It curves downwards and forwards. At its anterior end each rib has a costal cartilage. The first seven ribs have their own cartilages by which they are attached to the sternum. They are the true ribs. The 8th, 9th and 10th ribs are attached to the rib above by their cartilages; hence these ribs are considered as false ribs. The 11th and 12th ribs do not reach the sternum and are termed as floating ribs.

Mediastinum. Mediastinum is a tuft of connective tissue which separates thoracic cavity vertically into two distinct halves. It starts from the vertebral column at the back and moves anteriorly to the sternum at the front. It is divisible into a superior mediastinum and an inferior mediastinum (Fig. 4.3*b*).

The inferior mediastinum is further divisible into three parts. The area infront of the pericordium is the anterior mediastinum.

The area behind the pericardium is the posterior mediastinum. The pericardum with the heart inside forms the middle mediastinum. The mediastinum separates two lungs completely. Each lung is exclusively surrounded by the pleura. This arrangement prevents the clashing and interfering of the lungs with the heart during their functional performances. This arrangement also localizes any infection to a particular chamber, thus preventing it from spreading to other chambers.

Bronchial Tree: The bronchial tubes inside the lungs divide and subdivide into smaller and smaller branches. The primary bronchus that enters the lung divides into secondary bronchi, which in turn divide into tertiary bronchi. The bronchi are also supported by cartilages. Each tertiary bronchus divides into several primary bronchioles; the primary bronchioles divide into secondary bronchioles. They further divide into tertiary bronchioles. These branches resemble the branching of a tree and hence it is referred to as the bronchial tree or respiratory tree. Thus the trachea represents the trunk of an inverted tree with extensive branches that become smaller and smaller and ultimately end up in terminal bronchioles. Each terminal bronchiole leads into a smaller alveolar ducts. Every alveolar duct

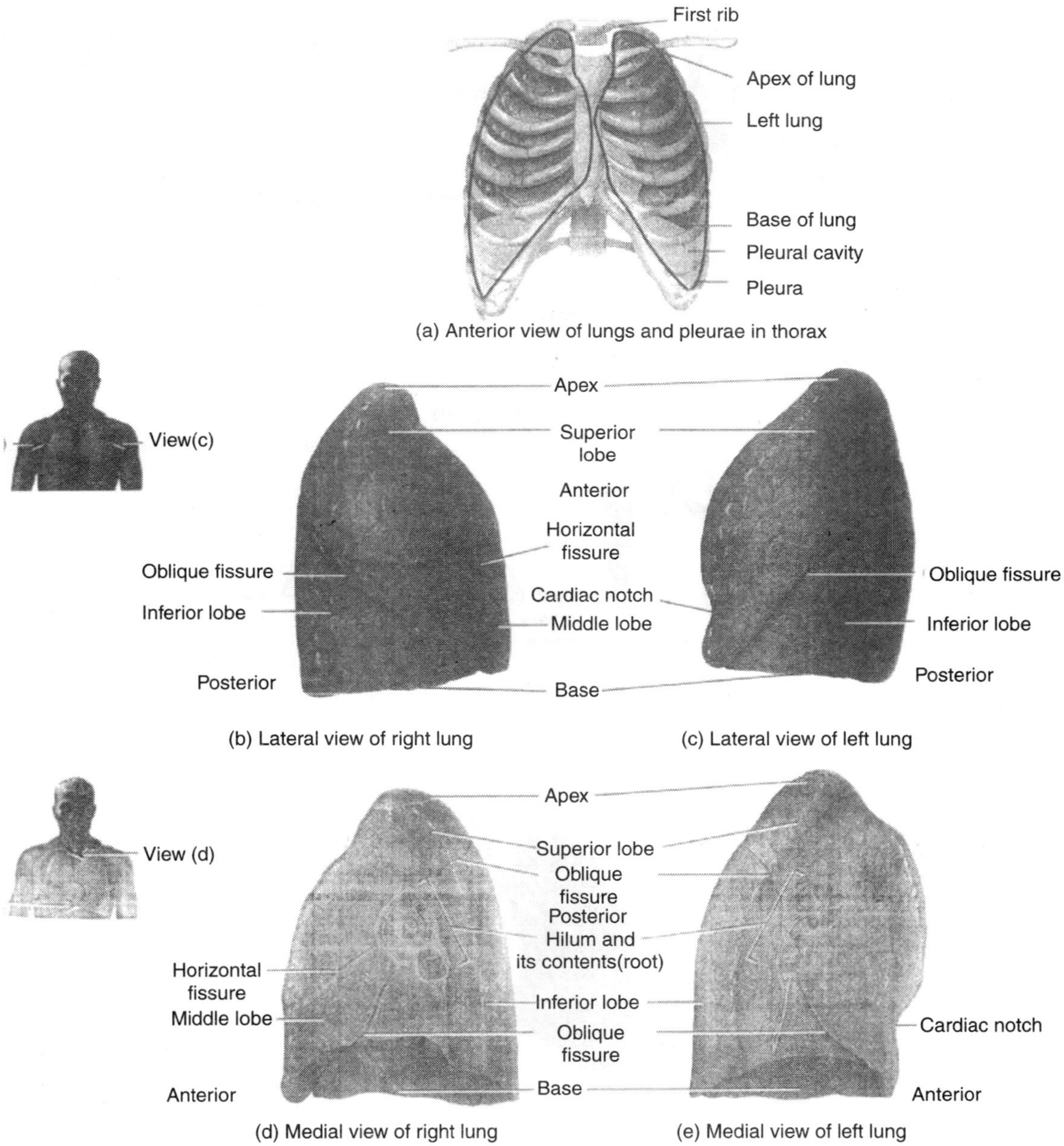

Fig. 4.2 Showing lungs are bounded and protected by the rib cage

becomes dilated to form the atrium. An atrium gives rise to number of air sacs. Each air sac gives out clusters of tiny hollow alveoli. The alveoli resemble like bunches of grapes. The exchange of gases take place in the alveoli. Thus the alveoli are the functional units of the lungs. A lung in total contains nearly 300 million, alveoli, and all of their surface are similar to that of a tennis court (Fig. 4.3b).

Functions of the different parts of the respiratory system:

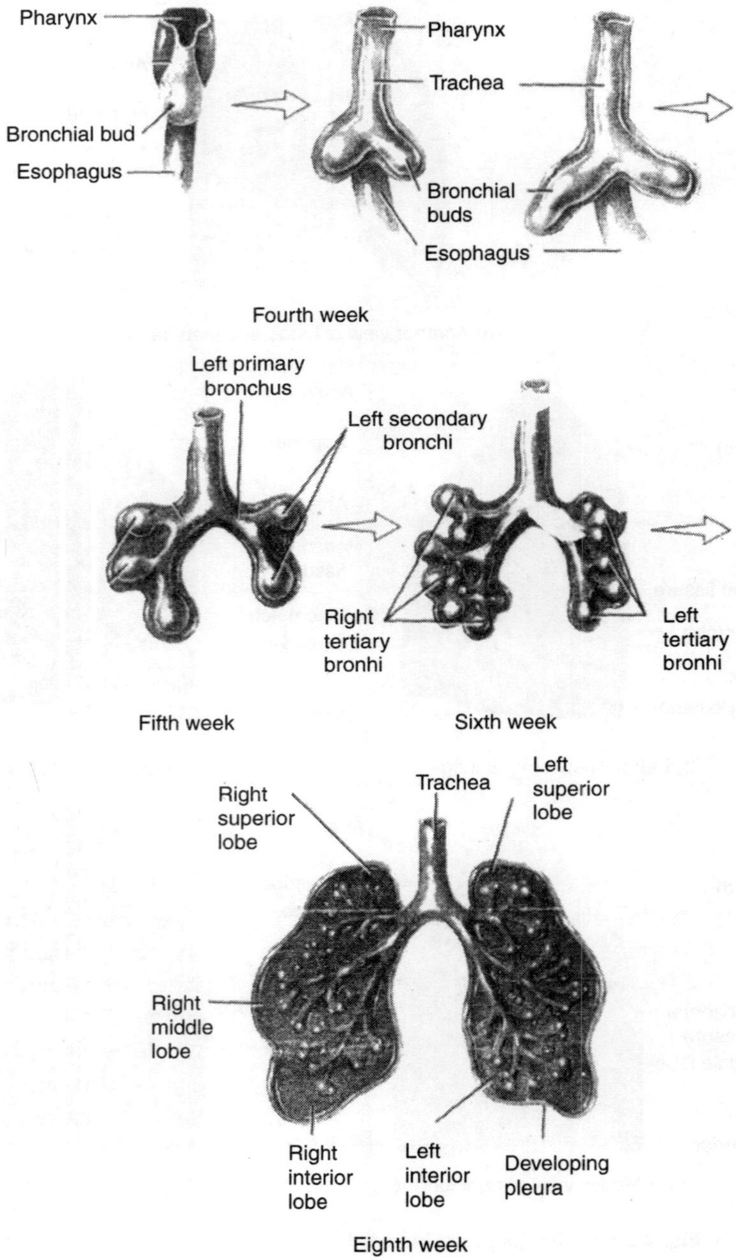

Fig. 4.3 (*a*) Showing human lungs containing several alveoli

A. Nose

(*i*) The nose is the gateway to the respiratory system.

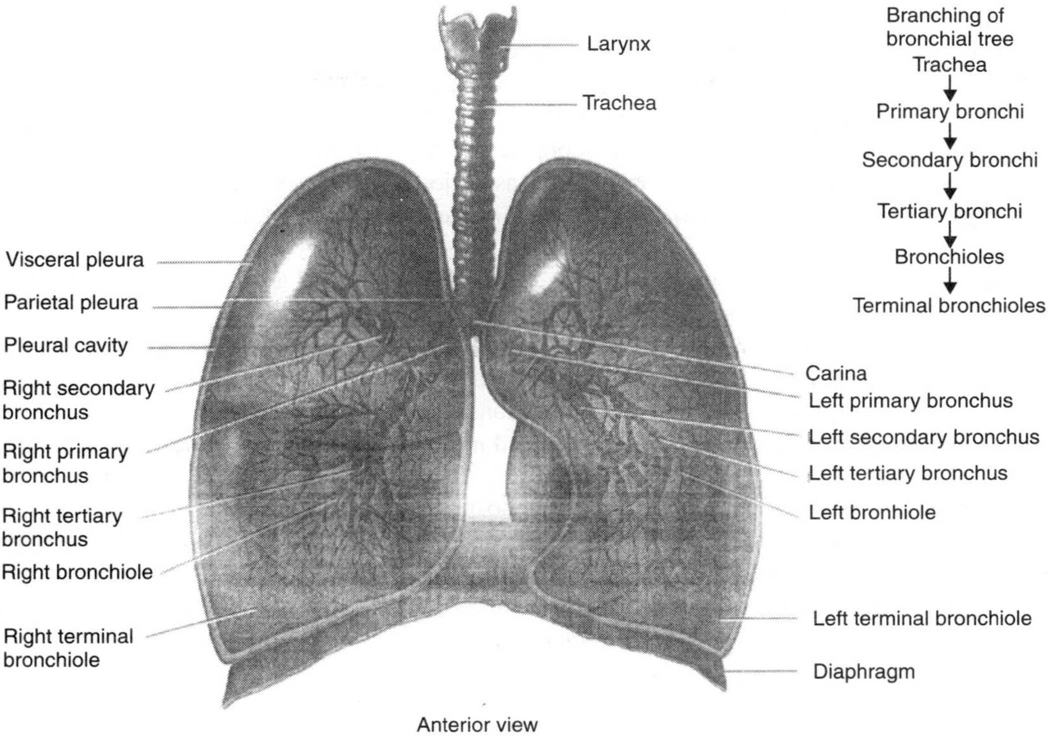

Fig. 4.3 (*b*) Showing branching of airways from the trachea: the bronchial tree

(*ii*) It filters the inspired air and cleanses it of foreign matter.
(*iii*) It moistens and warms the incoming air.
(*iv*) It houses of factory receptors.
(*v*) It serves as a resonating chamber for speech.

B. Nasal cavity: It contains hair follicles. The hair or vibrissae filter coarse particles such as dust, pollen etc., through the inspired air. The nasal cavity is lined by two types of mucous membranes:

(*i*) **Olfactory mucosa:** This type of membrane is present immediately behind the nostril; it possesses the smell receptors.

(*ii*) **Respiratory mucosa:** This type of membrane is present more interiorly; it is lined by ciliated cells. It contains goblet cells, mucous glands and serous glands. These glands secrete on an average, about half litre of sticky mucous per day. It contains lysozyme which is an antibacterial enzyme. Mucous traps dust, bacteria and other debris that is carried in the air. Lysozyme chemically attaches and destroys bacteria.

The cilia create a gentle current which carries the contaminated mucus towards the throat, where it is swallowed and digested by the stomach juices.

(*c*) **Pharynx:** This connects the nasal cavity to the larynx on one side and the oral cavity to the oesophagus on the other side. It acts as the passage way for both air and food.

(*d*) **Larynx:** It is placed behind the pharynx. This connects the pharynx to trachea. It helps to prevent the food from entering the lower respiratory tract. It possesses a pair of vocal cords which helps in producing the sound.

(*e*) **Trachea (wind pipe):** Trachea cleanses, warms and moistens the inhaling air. The tracheal lining cells produce mucus, which helps to trap dust and microorganisms and sweep them out of the air passage way with the help of tiny cilia into the digestive tract.

(*f*) **Bronchial tree:** It is made up of a series of air passage ways, which connect the trachea with alveoli. This passage also warms and moistens the incoming air.

(*g*) **Alveoli:** Alveoli are the ultimate microscopic respiratory structures. They are highly vascularised terminal chambers of bronchial tree. Alveoli serve as major areas of gas exchange.

(*h*) **Lungs:** Lungs are a pair of large sacs, placed in the thoracic cavity, literally hanging devoid of any support or any muscles.

(*i*) **Pleurae:** Pleurae are serous membranes divided into parietal and visceral layers. Besides compartmentalising lungs, they secrete a lubricating fluid.

Mechanism of Respiration. The main function of the respiratory system is to supply oxygen to and eliminate carbon dioxide from the human body. This occurs through four events *viz*., breathing, external respiration, internal respiration and cellular respiration.

The respiratory system, functionally can be divided into the (*i*) Conducting zone and, (*ii*) respiratory zone.

The conducting zone conducts the air from the atmosphere to the respiratory zone. The organs of the conducting zone cleanse, humidify and warm the air that enters the respiratory system. It includes the nostrils, nasal cavity, pharynx, larynx, trachea, bronchi, and bronchioles. The organs of the respiratory zone are placed deep inside the lungs. They include the respiratory bronchioles, alveolar ducts, atria air sacs and alveoli. All these parts are microscopic.

Breathing (inspiration and expiration)

Breathing is also considered pulmonary ventilation (L. ventilare-to fan). Every time we take a breath, or move air in and out of our lungs, two major activities occur:

(*a*) Inspiration or inhalation.
(*b*) Expiration or exhalation.

(*a*) **Inspiration:** Inspiration is a process during which the air is pulled into the lungs. Inspiration can occur when the lungs expand. The lungs can expand when the volume of the (chest) thoracic cavity increases. This process is accomplished by the diaphragm, and the inter costal muscles.

Diaphragm is a thick sheet of muscle that separates trunk into an upper thorax or chest and a lower abdomen. It is the major muscle of respiration. In a relaxed state, it is done-shaped, protruding into the chest cavity. There are two sets of intercostal muscles which are attached to ribs. These two sets of muscles operate antagonistic to each other. When externals contract, the internals relax and vice versa (Fig. 4.4).

These both events involve the application of certain physical and chemical principles as follows:

Boyle's law: According to this law when the temperature is constant, the pressure of any gas varies inversely with its volume. The product of pressure and volume is always constant. $P \times V = C$.

Thus, if the volume of a container is increased, the pressure falls, so that the product of pressure and volume always remains constant.

Pressure Equations

Atmospheric pressure: The atmospheric pressure of air at sea level is 760 mmHg. Intra-pulmonary (intra-alveolar) pressure. The pressure within the alveoli of the lungs is the intra-pulmonary pressure. When there is no air-movement, this pressure remains same as the atmospheric pressure. It rises and

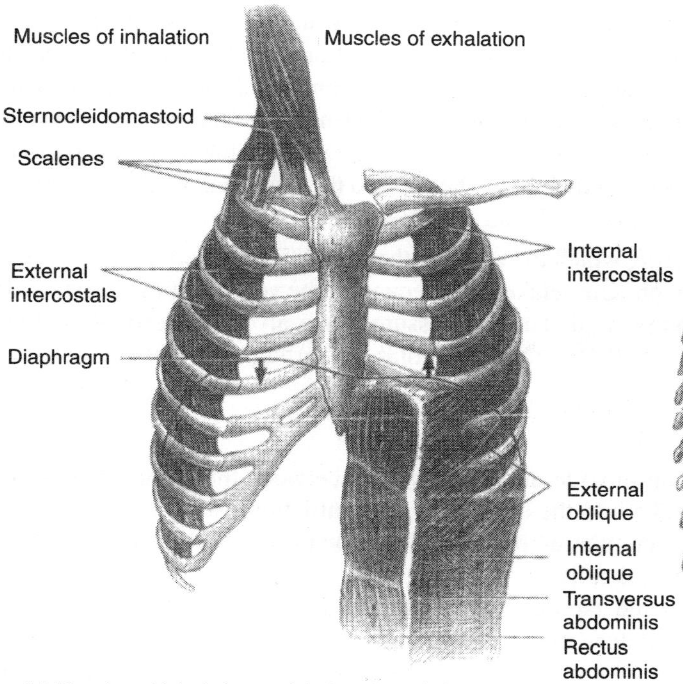

(a) Muscles of inhalation and their actions (left); muscles of exhalation and their actions (right)

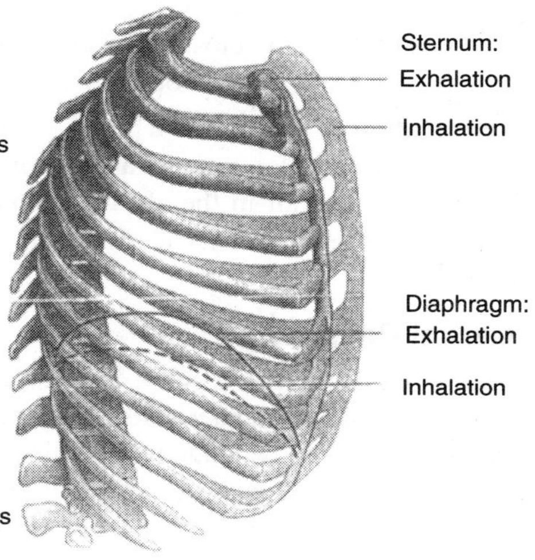

(b) Changes in size of thoracic cavity during inhalation and exhalation

(c) During inhalation, the ribs move upward and outward like the handle on a bucket

Fig. 4.4 Showing muscles of inhalation and exhalation and their actions

falls according to the phases of breathing. However, it always equalises itself with the atmospheric pressure outside the body.

Intra-thoracic (intra-pleural) pressure: The pressure within the thoracic cavity is the intra-pleural pressure. As it fluctuates with the phases of breathing, this pressure is not constant. This pressure remains always around 4 mm Hg; which is lower than the intra alveolar pressure (756 mm Hg). This pressure is said to be a negative pressure relative to atmospheric pressure and intra-pulmonary pressure.

The lungs has a natural tendency to recoil and assume the smallest size due to their great elasticity. The alveoli are coated with a fluid film, drawing the alveoli to their smallest possible size. Both these factors work together to establish a negative force in the thorax.

The rib cage-elasticity, however, is opposed to the above forces and tends to pull the thorax outward, thus enabling the lungs to enlarge.

During inspiration the air from the atmosphere enters the nasal cavity and reaches the pharynx. The pharnyx at its kind end has two passage ways one of them leads to the trachea while the other leads to the oesophagus. The air is conducted into the trachea while food is conducted into the oesophagus. These two routes criss-cross. Care should be taken to see that food does not enter the trachea. This task is performed by a flap like valve called epiglottis. When human being swallows the food the epiglottis protects the trachea by closing the air passage and deflects the food to the oesophagus. During breathing, however, the epiglottis opens widely, allowing the air into the lungs (Fig. 4.4).

(b) **Expiration:** Expiration is a process during which the air from the alveoli is expelled. In this process, the intercostal muscles contract and the diaphragm relaxes. Subsequently the volume of the chest cavity is minimised. The air in the alveoli is compressed. The air pressure in the alveoli exceeds that of the atmospheric pressure. This leads to the expulsion of the alveolar air through the passage ways to outside.

External Respiration

In this process, oxygen loading and carbon dioxide unloading occurs between the blood and alveoli. The air from the atmosphere has been drawn into the alveoli during ventilation. Nextly, the oxygen in the alveolar air is to be loaded to the blood circulating in capillaries, around the alveoli. The alveoli are lined by a single layer of squamous epithelial cells.

The basement membranes of the alveoli and capillary walls fuse and form the respiratory membrane (air-blood barrier). On one side of the respiratory membrane there is gas while on the other side there is blood. The movement of oxygen from the alveoli into the blood and the movement of carbondioxide from the blood into the alveoli are governed by two of the following principles:

1. **Dalton's law of partial pressures:** This law indicates that the total pressure exerted by a mixture of gases is the sum of the pressure exerted independently by each gas in the mixture. Further the partial pressure of each gas is directly proportional to its percentage in the total gas mixture.

2. **Henry's law:** When a mixture of gases is in contact with a liquid, each gas dissolves in the liquid in proportion to its partial pressure.

When the partial pressure of a gas phase is higher than in the liquid, there is a net movement of the gas from the gas phase into the liquid. Conversely, if the partial pressure of a dissolved gas has a higher partial pressure than in the gas phase, the gas molecules from the liquid move into the gas phase until partial pressures of the gas in the two phases are equalised. This is precisely what occurs between the alveolar air (gas phase) and the pulmonary capillary blood (liquid phase) (Fig. 4.5).

The barometric pressure of air is 760 mm mercury. It is made up of nitrogen roughly about 80% and oxygen approximately 20%. As per Dalton's law of partial pressure, the pressure exerted by each gas (*i.e.*, partial pressure) is directly proportional to its percentage in the total gas mixture. Accordingly, the partial pressure of oxygen in atmospheric air is about 152 mm of mercury. Similarly the percentage of carbon dioxide in atmospheric air is 0.04. Its partial pressure is 0.304 mm of mercury.

According to Henry's law, the partial pressure of oxygen in the gaseous phase (alveoli) is more than in the liquid phase (blood capillaries). Oxygen from the alveoli diffuses, into the blood in the capillaries down the concentration gradient.

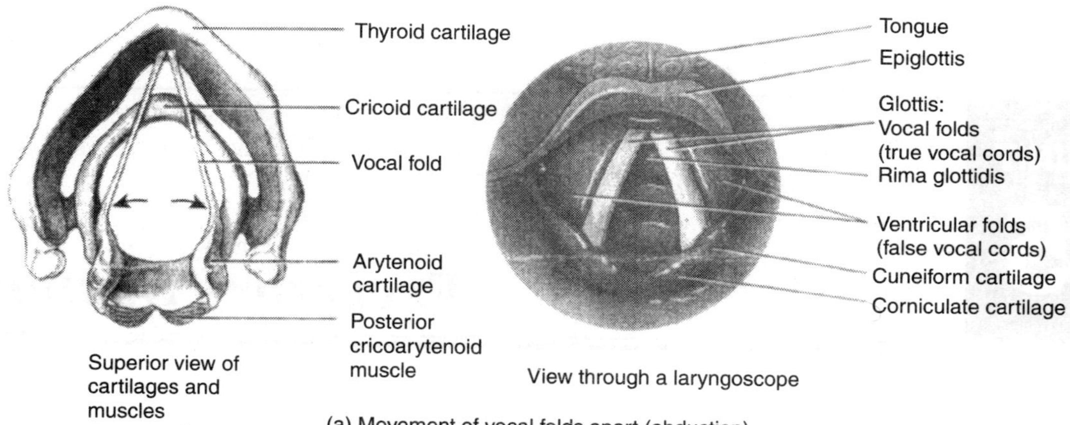

(a) Movement of vocal folds apart (abduction)

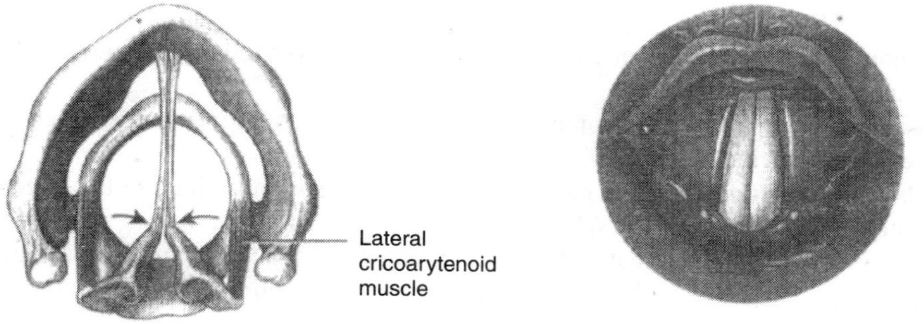

(b) Movement of vocal folds together (abduction)

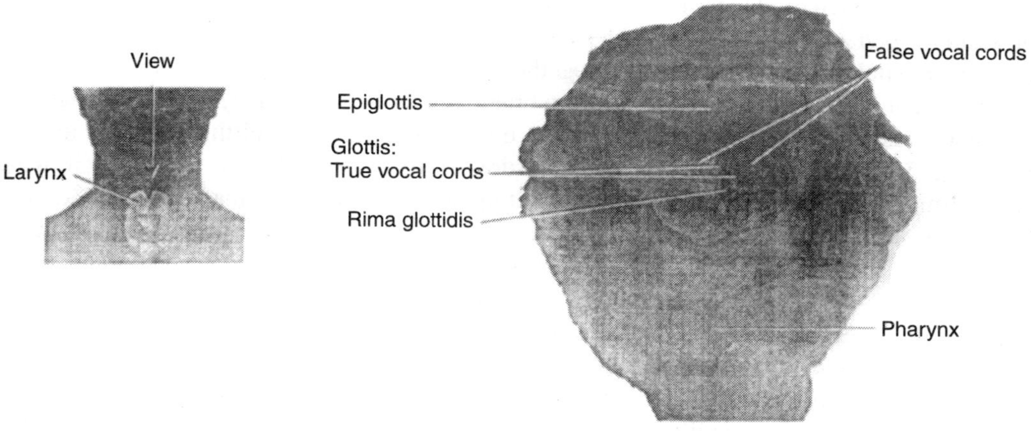

(c) Superior view

Fig. 4.5 Showing movement of vocal folds and epiglottis

UNIT 5
Excretory System

Kidneys, lungs, skin, gastrointestinal tract, salivary glands and liver are the main channels through which excretion takes place in human beings. There seems to be a general rule regarding the channels of excretion of different substances.

1. The soluble, non irritant, solid substances and water are mainly excreted through the kidneys and to a lesser extent through the skin.
2. Volatile substances, such as CO_2, ammonia, alcohol, ketone bodies, aromatic oils, water vapour etc. are excreted through the lungs.
3. Heavy metals are excreted through the gastrointestinal tract, especially through the large intestine and slightly through the liver and saliva.
4. Fats and fat derivatives are passed out of the body through the skin, as sebum, and through the liver along with bile. These are the broad principles on which excretion from human body takes place.

The kidneys perform excretory, homeostatic and endocrine functions. The urinary system consists of specialised organs which produce urine and eliminate it from the body. These specialised organs include pair of kidneys, pair of ureters, one bladder and one urethra. Urine secretion and its excretion from human body are vital functions because they together involve in maintaining bodily homeostasis kidneys in front of vertebral column from T_{12} to L_5 vertebrae. Human being's kidneys are bean shaped paired organs situated just behind the vertebral column in the abdomen, at the twelfth thoracic to third lumbar segment. (Fig. 5.1). An average-sized kidney measures 10 to 12 cm in length, 5 to 6 cm in width and 3 to 4 cm in thickness, each weighing 150 g in adult male and about 135 g in adult female. Usually, the right kidney is slightly smaller than the left one. A deep notch or concavity is present at the medial border – the hilus (hilum) and through this region the blood vessel, renal artery, vein, ureter and nerves pass.

In a longitudinal section, along its long axis of a kidney, on examining with naked eye, a dark reddish brown peripheral region the cortex, and the rest lighter area – the medulla are distinguishable. The medulla is again divisible into 10–15 conical areas, the renal pyramids with broad base towards the cortex and the vertex, apex or papilla projected into the lumen of the minor calyx. The lateral boundaries of the pyramid are formed by the projections of the darker cortex which are termed as renal columns of Bertin. The brownish lines which converge towards the apex – is due to the straight uriniferous tubules and the blood vessels parallel to them. The apex or papilla is known as the area cibrosa, which is perforated by a number of openings of the collecting tubules to the minor calyx.

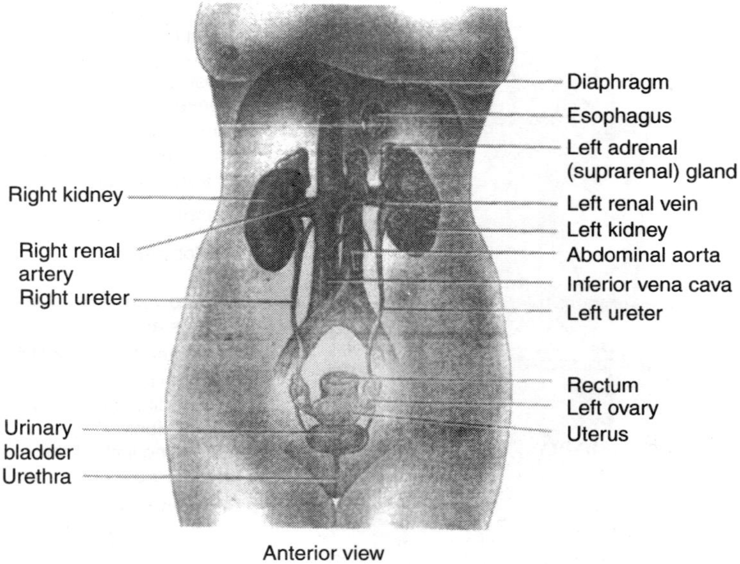

Fig. 5.1 Organs of the urinary system in a female

Functions of the Urinary System

1. The kidneys regulate blood volume and composition, help regulate blood pressure, synthesize glucose, release erythropoietin, participate in vitamin D synthesis, and excrete wastes in the urine.
2. The ureters transport urine from the kidneys to the urinary bladder.
3. The urinary bladder stores urine.
4. The urethra discharges urine from the body.

The renal medulla is prominent to be a site of origin for PGA_2, PGE_2 and $PGF_2 \propto$ phostaglandins concerned with blood pressure regulation.

The pair of kidneys are the chief excretory organs of human body. They are made up of numerous functional microscopic units called nephrons. They perform excretory and homeostatic functions. About one million nephrons are present in each human kidney. These nephrons ultimately drain into the pelvis of the ureter (Fig. 5.2). From here urine is collected in the urinary bladder passing through the ureters.

Urinferous tubules. These tubules consist two portions *i.e.*,

(*i*) secretory portion or nephron and
(*ii*) Nonsecretory portion or collecting portion.

Types of nephrons: According to their relative position in the cortex there are two types *i.e.*,

(*i*) The superficial nephrons – occupy the outer two thirds of the cortex and make up about 85% of the total number of nephrons. They function in normal conditions.
(*ii*) The juxtamedullary nephrons – occupy the inner third of the cortex and constitute about 15% of the total number of nephrons. They are large, act only in conditions of stress.

The nephrons consists of six successive parts:

(*a*) Renal malpighian corpuscle
(*b*) Proximal convoluted tubule
(*c*) Loop of Henle (a thin segment)
(*d*) Proximal straight tubule
(*e*) Distal straight tubule and
(*f*) Distal convoluted tubules.

Malpighian corpuscle: It is also termed as malpighian body or renal corpuscle.

This is found only in the cortex of the kidney. It measures about zoomicrons in diameter. This corpuscle is made up of two parts:

1. Glomerulus and
2. Bowmans's capsule.

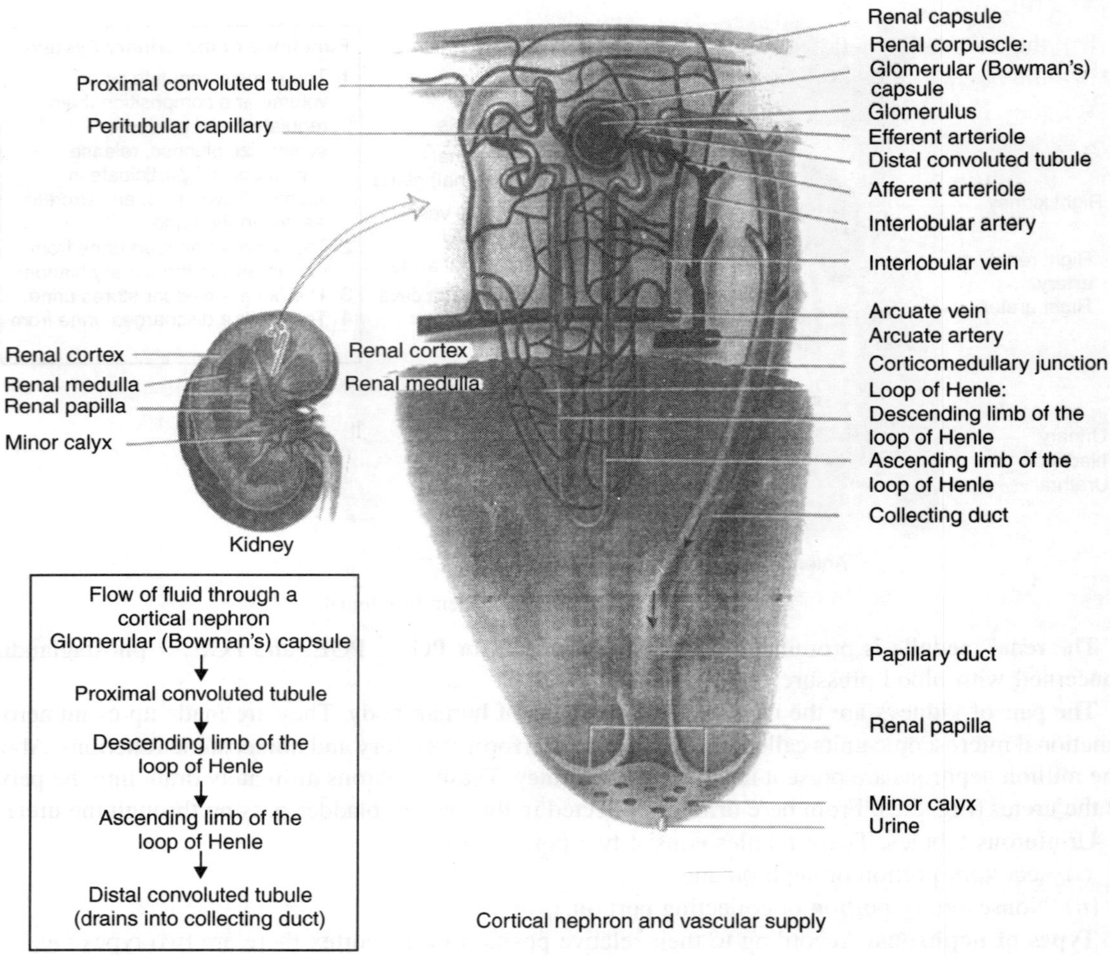

Fig. 5.2 The structure of nephrons and associated blood vessels

1. **Glomerulus:** This is a tuft of capillaries, invaginated in Bowman's capsule. The efferent arteriole breaks up into about fifty capillary loops and forms the glomicular tuft which lies within Bowman's capsule, a double walled epithelial sac. Each nephron starts with a tuft of renal blood capillaries invaginated into the end of a tubule. This structure is termed Glomerulus. There are about 20 to 50 glomerular capillaries which make a network. Just before entering the glomerulus, the media of the afferent arteriole is found to contain thick cuff of large modified muscle cells the juxtaglomerullar cells. The capillary tuft reunites and forms the efferent arteriole which passes out of the glomerulus. The afferent vessel is short and wide whereas the afferent vessel is narrow and longer.

This arrangement enables glomerular blood pressure much higher (70 mm Hg) than the capillary bed elsewhere and facilitates filtration. The total surface area of all the glomerular capillaries in two kidneys is about 1.73 square meters.

2. **Bowman's capsule:** This is the dilated blindend of the nephron, invaginated by the glomerular tuft—the vascular end. It consists of two layers—parietal and visceral. The parietal layer is made up of typical squamous epithelium of flat polygonal cells.

The filtering membrane of the malpighian capsule consists three layers, *viz.*

(*a*) the endothelial cell layer of the capillaries

(*b*) the basement membrane

(*c*) the epithelial cells of the visceral layer of Bowman's capsule. The cells of the visceral layer during development undergoes extensive modification and are known as podocytes. These give rise to a number of radiating tentacle-like cytoplasmic processes which in turn give up large number of small branches – the end feet or the pedicles. The pedicles are actually attached to the basal lamina.

Renal tubules: The glomerulus is described to have two poles. At one pole – the vascular, the blood vessels are attached. At the opposite pole – the tubular pole, the renal tubule begins. The human tubule is about 3.0 cm long, 20–60 microns wide, and since there are about two millions of nephrons the total length of the tubules is about 65 km or 40 miles. Just below the glomerulus, the tubule has a very short constricted portion—the neck after the neck, the tubules consists of the following parts.

1. First or Proximal convoluted tubule (PCT),
2. Henle's loop,
3. Distal or second convoluted tubule,
4. Collecting or straight tubule.

First or Proximal Convoluted Tubule (PCT)

It is about 14 mm long, external diameter 60 μm and internal diameter 15–25 μm. It is lined by a single layer of cubical cells, the free borders of which are striated or brush bordered. This brush border possesses microvilli which enhance the surface area of the cells for absorption in the lumen of the proximal tubule. There are several tubular invaginations, arising at the clefts between microvilli, and are known as apical canaliculi. These cells contain a large number of enzymes and these cells reabsorb about two-thirds of the water of the glomerular filtrate and all the glucose and part of sodium chloride and phosphate.

Henle's Loop

Henle's loop is a U-shaped loop, dipping for a variable length into the medulla. This can be divided into three parts, *viz.*,

(*i*) Descending limb

(*ii*) Thin-walled ascending limb and

(*iii*) Thick walled ascending limb.

The average length of the loop is about 20 mm.

In the thin-walled descending limb being devoid of obvious intra cellular structures isoosmotic fluid delivered from the PCT becomes increasingly hypertonic as it progresses from the renal cortex towards the renal papilla until it reaches the maximum concentration and the membrane of this portion is more permeable to H_2O than solute, urea, NaCl and in order to effect equilibrium, H_2O is removed. In the thin segment of the ascending limb the tubular fluid, becomes progressively less concentrated due to its backflow towards the renal cortex as a consequence of loss of NaCl from the renal tubule. The thick portion of the ascending loop is known as functional proximal tubule. Reabsorption of NaCl from the both cortical and medullary thick ascending limbs results in by active chloride reabsorption but that reabsorption of sodium occurs by means of a passive reabsorptive process in the maintenance of electrical neutrality.

Distal or Second Convoluted Tubule

This part of the nephron is situated in the cortex. It has a very limited number of microvilli.

The reabsorptive activity of the distal convoluted tubule is under the influence of ADH, Vasopressin, which is one determinant of the volume of urinary output. Distal tubular cells are capable of reabsorbing very little quantities of water and electrolytes due to few number of microvilli in addition to determining urinary acid base balance collecting or straight tubule. It is about 20 mm long and lined by cuboidal cells. Several collecting tubules from different nephrons join and form the duct of Bellini, which opens at the apex of the renal pyramid.

Juxtaglomerular Apparatus

Juxtaglomerular apparatus includes the

(*i*) granular juxtaglomerular cells in the afferent arteriole,

(*ii*) macula densa of the distal convoluted tubule

(*iii*) agranular Polkissen or 'Lacis' cells situated in the angle created by the entrance and exit of the afferent and efferent arterioles of each glomerulus.

Some prefer to term this as 'juxtaglomerular complex'.

The components of this apparatus have been considered to be related in some way to the control of blood pressure, renal blood flow, salt balance and erythropoiesis. Renal hypertension is claimed to be the cause of secretion of renin from the juxtaglomerular apparatus thus helping in the formation of angiotension. (Fig. 5.6).

During haemorrhage (Hypoxia) release of red blood cells from bone marrow in the circulation is increased.

The erythropoietin which is a hormone with a molecular weight of 25,000–40,000 enables certain stem cells. (erythropoietin sensitive stem cells) in the bone marrow to get converted into procrythroblast. It is also claimed that the action of erythropoietin is apparently mediated through the stimulation of mRNA synthesis.

The renal erythropoietic factor (REF) or erythrogenin is known to be the secretion of juxtaglomerular cells. It's production is enhanced by hypoxia, cobalt salts and androgens. Renin secretion is determined by the degree of stretch of the afferent glomerulus and also by the Na-concentration of the macula densa cells.

Macula densa is a modified epithelial cell in the portion of the distal convoluted tubule lying in contact with the afferent glomerular vessel of the same nephron.

'Lacis cell' is also known as Polkissen (polar cushion) cell. It is a cell mass formed mainly by the agranular or occasional granular cells. These cells lie in close contact with the macula densa cells and also within the vascular pole formed by the afferent and efferent glomerular vessels.

Functions of Kidney and Glomerulus.

Functions of Kidney

1. It excretes waste products, especially the nitrogenous and sulphur—containing end products of protein metabolism.
2. It maintains the normal hydrogen ion concentration of body fluids add electrolytes.
3. It maintains water balance of the body and thereby plasma volume.
4. It maintains the optimum concentration of certain constituents of blood by the process of selective reabsorption.

5. It eliminates drugs, and various toxic substances from human body.
6. It manufacturers certain new substances – ammonia, hippuric acid, and inorganic phosphates. Ammonia helps in preserving acid base equilibrium.
7. It helps in osmotic pressure maintenance in blood and tissues.
8. It maintains blood pressure during hypoxia in emergency condition by liberating renin from the juxtaglomerular apparatus.
9. It helps in the regulation of erythropoietin through the formation of erythropoietin by REF secreted from the glomerular cells.
10. Kidney plays an important role in vitamin D metabolism.

UNIT 6

Nervous System

INTRODUCTION

The human nervous system is one of the smallest and yet the most complex of the 11 body systems. Nervous system is an highly organised network of billions of neurons and even more neuroglia. The structures that make up the nervous system include the brain, cranial nerves and their branches, the spinal cord spinal nerves and their branches, ganglia, enteric plexuses and sensory receptors. The skull encloses the brain, which contains about 100 billion neurons. Twelve pairs of right and left cranial nerves, numbered I through XII, emerge from the base of the brain. A nerve is a bundle of thousands of axons plus associated connective tissue and blood vessels that lies outside the brain and spinal cord. Each nerve follows a defined path and serves a specific region of the body. For example, the right cranial nerve I carries signals for the sense of smell from the right side of the nose to the brain.

The spinal cord connects to the brain through the foramen magnum of the skull and is encircled by the bones of the vertebral column. It contains about 100 million neurons. Thirty one pairs of spinal nerves emerge from the spinal cord, each serving a specific region on the right or left side of the body. Ganglia (swelling) are small group of nervous tissue, containing primarily cell bodies of neurons, that are located outside the brain and spinal cord. Ganglia are closely associated with cranial and spinal nerves.

Role of nervous system: The nervous system carries out a complex array of tasks, such as sensing various smells, producing speech, remembering, providing signals that control body movements, and regulating the operation of internal organs. These diverse activities can be grouped into three basic functions: sensory, integrative and motor.

Sensory function: Sensory receptors detect internal stimuli, such as an increase in blood acidity, and external stimuli, such as a mosquito biting on your hand. The neurons that carry sensory information from cranial and spinal nerves into the brain and spinal cord or from a lower to a higher level in the spinal cord and brain are sensory or afferent neurons.

Integrative function: The nervous system integrates sensory information by analyzing and storing some of it and by making decisions for appropriate responses. Many of the neurons that participate in integration are interneurons, whose axons extend only for a short distance, and contact nearby neurons in the brain, spinal cord or a ganglion.Interneurons comprise the vast majority of neurons in the body.

Motor function: The human nervous system's motor function involves responding to integration decisions. The neurons that serve this function are motor or efferent neurons. Motor neurons carry information from the brain toward the spinal cord or out of the brain and spinal cord into cranial or spinal nerves. The cells and organs contacted by motor neurons in cranial and spinal nerves are termed effectors. Examples are muscle fibres and glandular cells.

Generalized neuron: Like muscle cells, neurons have the electrical excitability property, this enables the ability to produce action potentials or impulses in response to stimuli, Once they arise, action potentials propagate from one point to the next along the plasma membrane due to the presence of specific types of ion channels.

Parts of a neuron: Most neurons have three parts: (1) a cell body, (2) dendrites and (3) anaxon (Fig. 6.1). The cell body contains a nucleus surrounded by cytoplasm that includes typical organelles such as lysosomes, mitochondria and a golgi complex. Neuronal cell bodies also cotnain prominent clusters of rough endoplasmic reticulum, termed Nissl bodies, the sites of protein synthesis in neurons.

Two kinds of processes or extensions emerge from the cell body of a neuron: multiple dendrites and a single axon. Nerve fibre is a general term for any neural process like dendrite or axon. Dendrites are the receiving or input portions of a neuron. In many neurons the dendrites form a tree shaped array of processes extending from the cell body. Hence are called little trees. Their cytoplasm contains Nissl bodies, mitochondria, and other organelles. Nissl bodies are unique microscopic feature of neurons. Newly synthesized proteins produced by Nissl bodies are used to replace cellular components and to regenerate damaged axons in the peripheral nervous system.

A single axon of a neuron propagates nerve impulses toward another neuron, a muscle fiber or a gland cell. An axon is a long, thin, cylindrical projection that often joins the cell body at a cone shaped elevation called the axon hillock (small hill). In most neurons, impulses arise at the junction of the axon hillock and the initial segment, which is called the trigger zone, and then conduct along the axon. An axon contains mitochondria, microtubules, and neurofibrils. Because rough endoplasmic reticulum is not present, protein synthesis does not occur in the axon. The cytoplasm of an axon, called axoplasm, is surrounded by a plasma membrane known as the axolemma. Along the length of an axon, side branches called axon collaterals may branch off, typically at a right angle to the axon. The axon and its collaterals end by dividing into many fine processes called axon terminals.

The site of communication between two neurons or between a neuron and an effector cell is called a synapse. The tips of some axon terminals swell into bulb-shaped structures called synaptic end bulbs, whereas others exhibit a string of swollen bumps called varicosities. Both synaptic end bulbs and varicosities contain many tiny membrane enclosed sacs called synaptic vesicles that store a chemical neurotransmitter. Neurons were previously through to liberate just one type of neurotransmitter at all synaptic end bulbs. It is now established that many neurons contain two or even three neurotransmitters. When neurotransmitter molecules are released from synaptic vesicles, they excite or inhibit other neurons, muscle fibers, or gland cells.

The cell body is the portion of a neuron that synthesizes new cell products or recycles old ones. However, because some substances are needed in the axon or at the axon terminals, two types of transport systems carry materials from the cell body to the axon terminals and back. The slower system, which moves materials about 1–5 mm per day is called slow axonal transport. It conveys axoplasm in one direction only-from the cell body toward the axon terminals.

The faster system, which is capable of moving materials a distance of 200–400 mm per day, is called fast axonal transport. It uses proteins that function as 'motors' to move materials in both directions—away from and toward the cell body—along the surfaces of microtubules. Fast axonal transport moves various organelles and materials that form the membranes of the axolemma, synaptic end bulbs, and synaptic vesicles. Some materials transported back to the cell body are degraded or recycled; other influence neuronal growth.

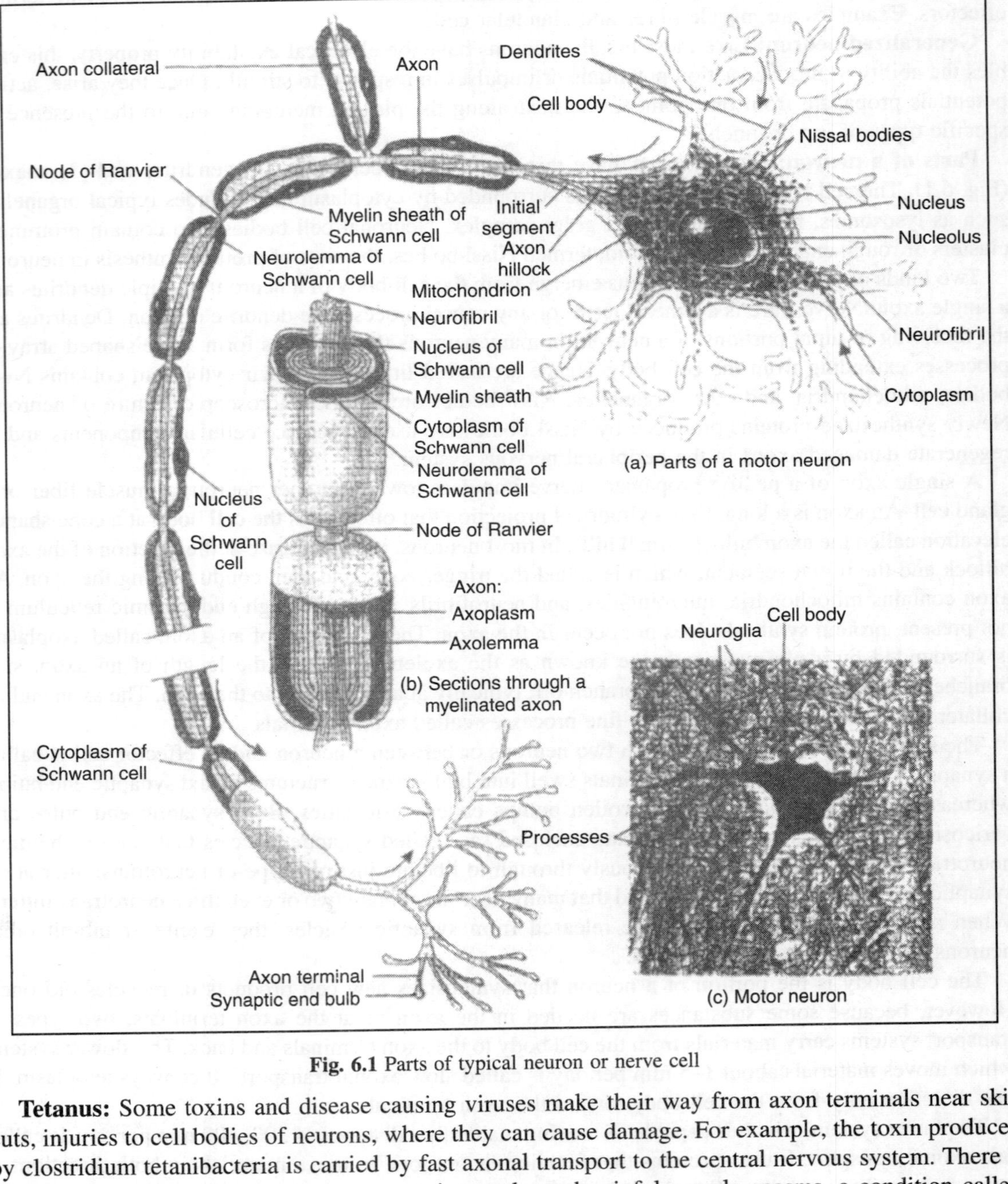

Fig. 6.1 Parts of typical neuron or nerve cell

Tetanus: Some toxins and disease causing viruses make their way from axon terminals near skin cuts, injuries to cell bodies of neurons, where they can cause damage. For example, the toxin produced by clostridium tetanibacteria is carried by fast axonal transport to the central nervous system. There it disrupts the actions of motor neurons, causing prolonged painful muscle spasms, a condition called tetanus. The closure the site of injury is to the brain, the shorter the transmit time. Therefore, treatment must begin quickly to prevent symptoms of tetanus.

MORPHOLOGICAL TYPES OF NEURONS

Neurons exhibit diversity in shape and size. Their cell bodies range in diameters from 5 micrometers (μm) *i.e.*, slightly smaller than a red blood cell, up to 135 μm. The pattern of dendritic branching is varied and distinctive for neurons in different parts of the nervous system. A few small neurons lack an axon, and many others have very short axons. The longest axon's however, are almost as long as a person's height, extending from the toes to the lowest part of the brain.

Based on the both structural and functional features, various neurons in the human body are classified. Neurons are classified structurally according to the number of processes extending from the cell body (Fig. 6.1)

1. Multipolar neurons usually have several dendrites and one axon. Most neurons present in the brain and spinal cord are of this type.
2. Bipolar neurons have one main dendrite and one axon. They are found in the retina of the eye in the inner ear, and in the olfactory area of the brain.
3. Unipolar neurons are sensory neurons that begins in the embryo as bipolar neurons. During development, the axon and dendrite fuse into a single process that divides into two branches at a short distance from the cell body. Both branches have the characteristic structure and function of an axon. They are long cylindrical processes that propagate action potentials. However, the axon branch that extends into the periphery has dendrites at its distal tip, whereas the axon branch the textends into the central nervous system ends in synaptic and bulbs. The trigger zone for nerve impulses in this neuron is at the function of the dendrites and axon. The impulses are propagated towards the synaptic end bulbs.

Certain neurons are nomenclatured after the histologist who first described them; for example, Purkinje cells in the cerebellum (Fig. 6.2(*a*)) and Renshaw cells in the spinal cord. Pyramidal cells found in the brain, have a cell body in the shape of pyramid (Fig. 6.2(*b*)).

PHYSIOLOGICAL OR FUNCTIONAL TYPES OF NEURONS

By virtue of their physiological functions also neurons are classified into neuroglia, unmyelinated and myelinated, axons, grey and white matter.

Neuroglia

Neuroglia orglia constitutes about 50% of the volume of central nervous system. They were mistaken by early histologists, as a 'glue' which holds nervous tissue together. It is now established that neuroglia are active participants in the operation of nervous tissue. Generally, neuroglia are smaller than neurons, and they are 5 to 50 times more numerous. In contrast to neurons, glia do not generate or propagate action potentials, and they can multiply and divide in the mature nervous system. Neuroglia multiply to fill in the spaces formerly occupied by neurons, and gaps formed due to injuries or diseases. Brain tumors derived from glia, is called gliomas, which is tend to be highly malignant and to grow-rapidly. Among the six types of neuroglia, four-astrocytes, oligo-dendrocytes, microglia, and ependymal cells are found only in the CNS. The remaining types – Schwann cells and satellite cells – are present in the peripheral nervous system.

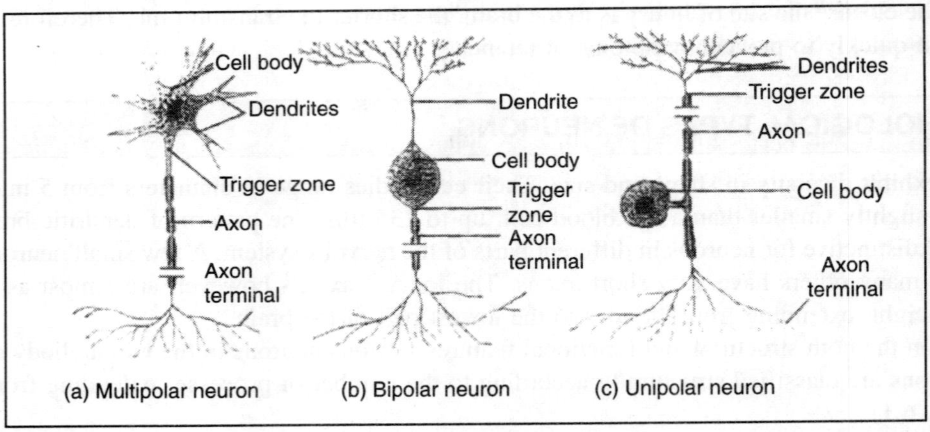

Fig. 6.1 Showing structurally classified neurons

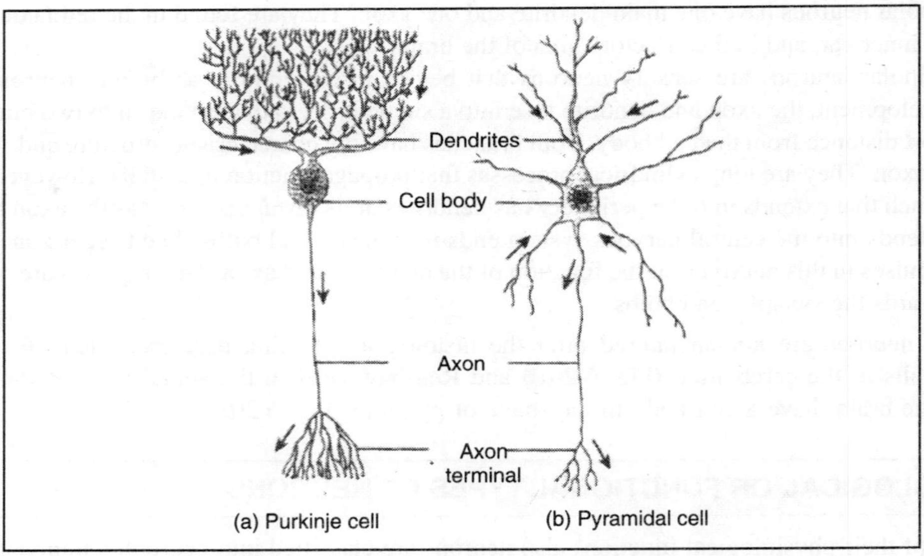

Fig. 6.2 Types of neurons named after the histologist

Myelination

The axons of neurons in human beings are surrounded by a multilayered lipid and protein covering, called the myelin sheath, that is produced by neuroglia. The sheath electrically insulates the axon of a neuron and enhances the speed of nerve impulse conduction. Axons without such a covering are unmyelinated.

Two types of neuroglia produce myelin sheaths: schwann cells in the PNS and oligo-dendrocytes in the CNS. Schwann cell membrane is the myelin sheath. The outer nucleated cytoplasmic layer of the Schwann cell which encloses the myelin sheath is the neurolemma. When an axon is injured, the neurolemma aids regeneration by forming a regeneration tube that guides and stimulates regrowth of the axon. Some gaps in the myelin sheath, called nodes of Ranvier appear at intervals along the axon.

In the CNS, oligodendrocytes are present. Every oligodendrocyte extends 15 broad, flat processes that spiral around CNS axons, forming a myelin sheath. A neurolemma is not present, however, because the oligodendrocyte cell body and nucleus do not envelop the axon. Only few nodes of Ranvier are present.

Axons in the CNS perform little regeneration in case of injury. This is partly due to the absence of a neuro-lemma are partly, due to an inhibitory influence exerted by the oligodendrocytes on axon regeneration.

The content of myelin enhances from birth to maturity, and its presence improves the speed of nerve impulse conduction. An infants responses to stimuli are neither as rapid nor as coordinated as those of an adult; because, myelination is still under progress during infancy. Diseases like multiple sclerosis and Tay sachs disease destroy myelin sheaths.

Gray and White Matter

In a longitudinal section of the human brain and transverse section of spinal cord, some regions look white and glistening, whereas others appear gray. The white matter is aggregation of myelinated and unmyelinated axons of many neurons. The whitish color of myelin endorses its name-white matter. The gray matter of the nervous system contains neuronal cell bodies, dendrites, unmyelinated axons, axon terminals and neuroglia. It looks grayish, because the Nissl bodies impart a gray colour. Blood vessels are present in both white matter and grey matter.

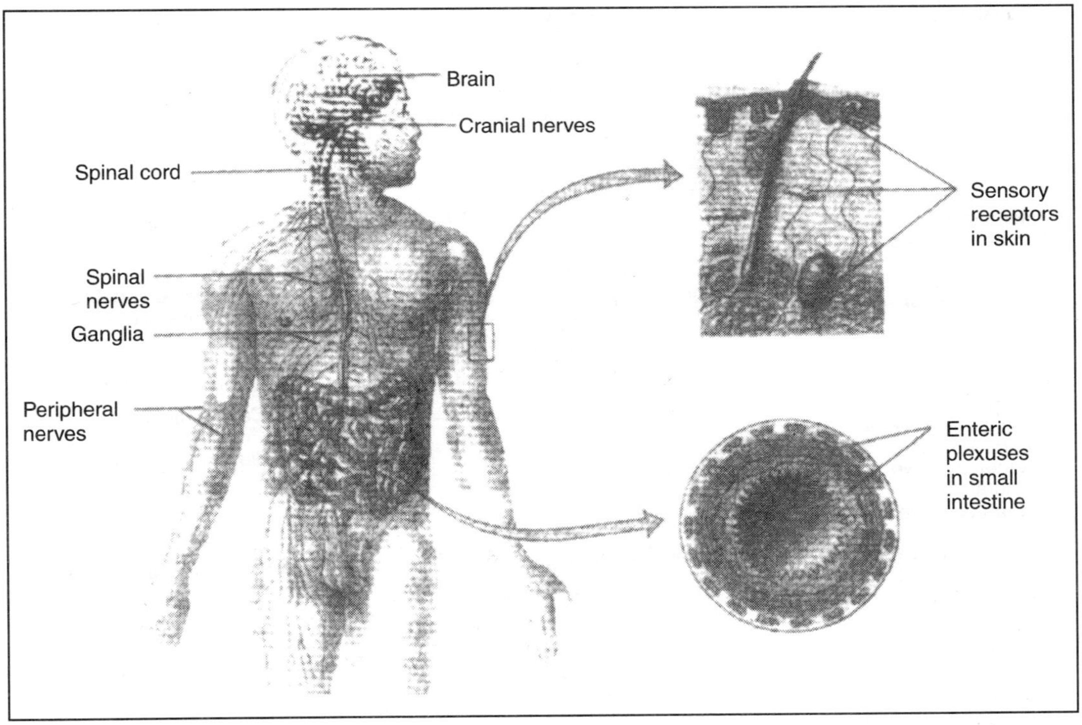

Fig. 6.3 Some of the components of nervous system

In the spinal cord, the white matter surrounds an inner core of gray matter shaped like a butterfly or the letter H. In the brain, a thin shell of gray matter covers the surface of the largest portions of the brain, the cerebrum and cerebellum (Fig. 6.3).

MAIN PROPERTIES OF NERVOUS TISSUE

Similar to muscle fibres, neurons are electrically excitable. They communicate with one another using two types of electrical signals;

1. Action potential which allow communication over short and long distances within the body.
2. Graded potential which are used for short-distance communication only.

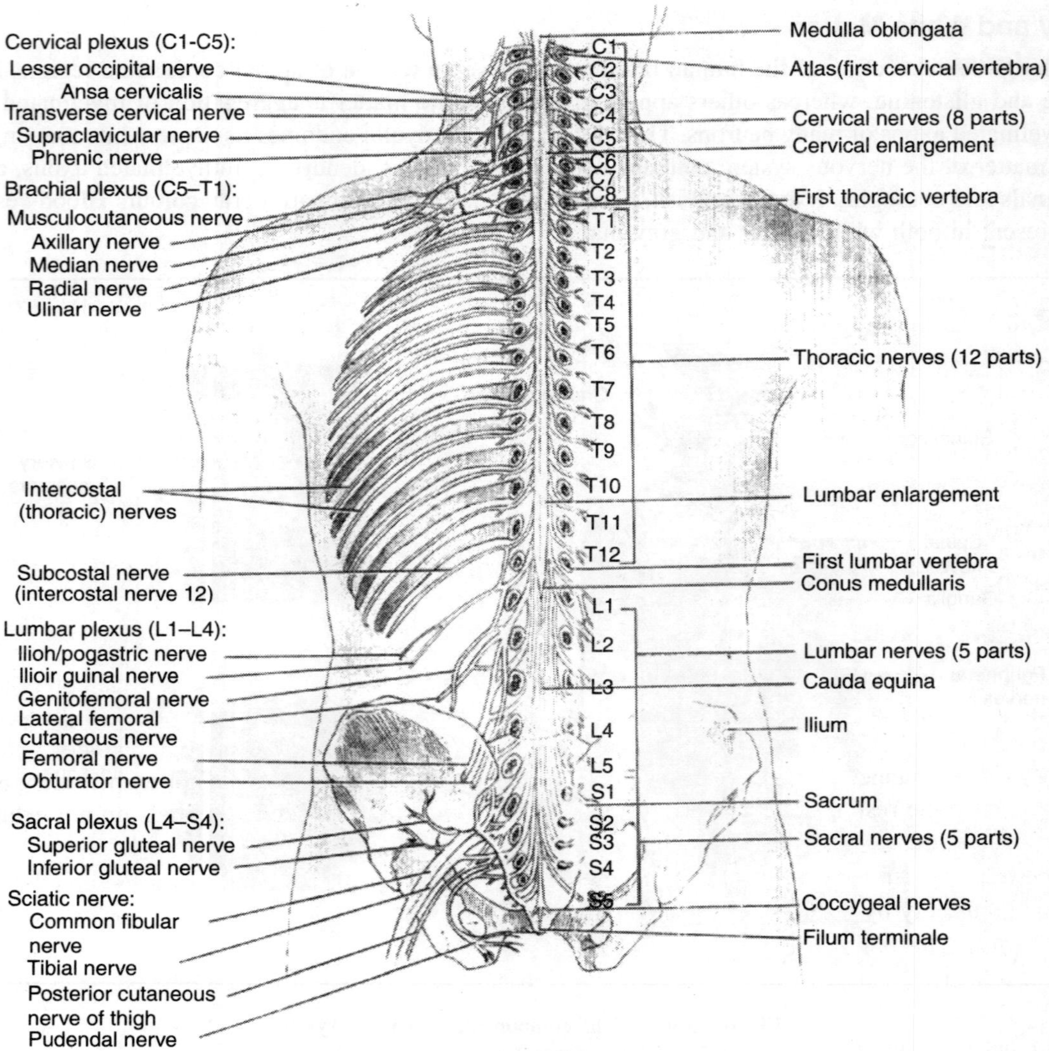

Posterior view of entire spinal cord and portions of spinal nerves

Fig. 6.4 Showing extenal anatomy of the spinal cord and the spinal nerves

Production of these two types of signals depends on two basic features of the plasma membrane of excitable cells; the existence of a resting membrane potential and the presence of specific types of ion channels.

Like many other cells in human body, the plasma membrane of excitable cells, this voltage is termed the resting membrane potential. The membrane potential is like voltage stored in a battery. If we connect the positive and negative terminals of a battery with a piece of wire, electrons flow along the wire. This flow of charged particles is called current. Among living cells the flow of ions constitutes the electrical current.

As the plasma membranes of neurons contain several kinds of ion channels that open or close in response to specific stimuli. To facilitate this graded potentials and action potentials occur. The lipid bilayer of the plasma membrane is a good insulator; the main paths for current to flow across the membrane are through the ion channels.

Depolarising Phase

When a depolarising graded potential or some other stimulus causes the membrane to depolarize to threshold, voltage gated Na^+ channels open rapidly. Both the electrical and the chemical gradients favour inward movement of Na^+ and the resulting in rush of Na^+ causes the depolarizing phase of the action potential (Fig. 6.4).

Repolarizing Phase

In addition to opening voltage gated Na^+ channels, a threshold level depolarization also opens voltage gated K^+ channels. The voltage gated K^+ channels open more slowly, however, so their opening occurs at about the same time the voltage gated Na^+ channels are closing. The slower opening of voltage gated K^+ channels and the closing of previously open Na^+ channels produce the repolarizing phase of the action potential. As the Na^+ channels are inactivated, Na^+ inflow slows. At the same time, the K^+ channels are opening, and K^+ outflow accelerates. Slowing of Na^+ inflow and acceleration of K^+ outflow causes the membrane potential to change from +30 mV to –70 mV. Repolarizaton also allows inactivated Na^+ channels to revert to the resting state. While the voltage gated K^+ channels are open, outflow of K^+ may be large enough to cause an after-hyperpolarizing phase of the action potential. During this phase the membrane is even more permeable K^+ than in the resting state.

Refractory Period

The length of time after an action potential begins during which an excitable cell cannot generate another action potential is called the refractory period. During the absolute refractory period, a second action potential cannot be initiated, even with a very strong stimulus. This period coincides with the period of Na^+ channel activation and inactivation. Inactivated Na^+ channels cannot reopen; they must, first return to the resting stage. In contrast to the action potentials, graded potentials do not exhibit a refractory period.

The relative refractory period is the length of time during which a second action potential can be initiated, but only by a larger-than-normal stimulus.

Stimulus: The nerve fibre can be stimulated by a suitable stimulus, which may be mechanical, thermal, electrical or chemical. In experiments electrical stimulation is usually employed, as its strength and frequency can be exactly controlled.

The below mentioned changes prove that a nerve has been excited:

(*a*) The gland or the muscle where the nerve ends will respond.

(*b*) The stimulated point on the nerve becomes electrically negative potential passes along the nerve and this wave of negative potential passes along the nerve; which could be detected by galvanometer.

(*c*) The action potentials – an electrical agitation always accompanies the travelling nerve impulse. In a resting cell the surface is positively charged and the interior is negatively charged. When the surface is stimulated and the permeability is increased consequently there is a reversal of polarisation. The surface at the stimulated spot becomes negative (cathode) creating catelectronic change, when this change rises to threshold level, impulse shall pass like self propagated disturbance by drawing positively charged particles from the neighbouring spots which in turn becomes cathodes. The depolarisation of the membrane, is the first step of the manifestation of an impulse. After an initial slow rise, depolarisation wave overshoots rapidly and reaches the isopotential (zero) line to approximately +35 mV. After that it reverses and begins to fall very rapidly towards the resting phase (–70 mV). Approximately, at two-thirds of repolarisation, the rate of fall is becomes abruptly slowed. This slower fall is known as negative after potential. The fast rise of depolarization wave and the rapid fall of repolarisation wave are considered as 'spike potential'. After reaching the bottom level the wave overshoots slightly but slowly in the hyperpolarisation way. This is considered as 'positive after-potential'. The entire sequence of potential changes in the nerve following excitation is considered as action potential.

MODE OF ACTION OF NERVES

For communicating information from one part of the body to another, nerve impulses should travel from where they arise at a trigger zone, often the axon hillock, to the axon terminals. This mode of travel is termed conduction. This depends on positive feedback.

Because sodium ions flow in, they cause depolarisation which opens voltage gated Na^+ channels in adjacent segments of the membrane. Further, because the membrane is refractory behind the leading edge of a nerve impulse, it normally moves in a single direction only – from where it is initiated at the trigger zone toward the axon terminals.

Conduction of Nerve Impulses

Nerve impulses are produced more rapidly along myelinated axons than along unmyelinated axons. If you compare parts a and b in figure (6.5) to find that the impulse propagates much farther along the myelinated axon in a few milliseconds. Saltatory conduction (saltat = leaping) is a special mode of impulse production that occurs along myelinated axons. It happens due to the uneven distribution of voltage-gated channels (Fig. 6.5). Hence, current carried by Na^+ and K^+ flows across the membrane mainly at the nodes.

During a nerve impulse propagation along a myelinated axon, an electric current carried by ions flows through the extracellular fluid surrounding the myelin sheath and through the cytosol from one node to the next. The nerve impulse at the first node generates ionic currents in the cytosol and extracellular fluid that depolarize the membrane to threshold, opening voltage-gated Na^+ channels at the second node. The resulting ionic flow through the opened channels constitutes a nerve impulse at the second node. Then, the nerve impulse at the second node generates an ionic current that opens voltage gated Na^+ channels at the third node, and so on. Each node repolarizes after it is depolarized.

Reflex action: A reflex action is a fast, involuntary, unplanned sequence of actions that occurs in response to a particular stimulus. Some reflex actions are in born; such as – pulling one's hand away

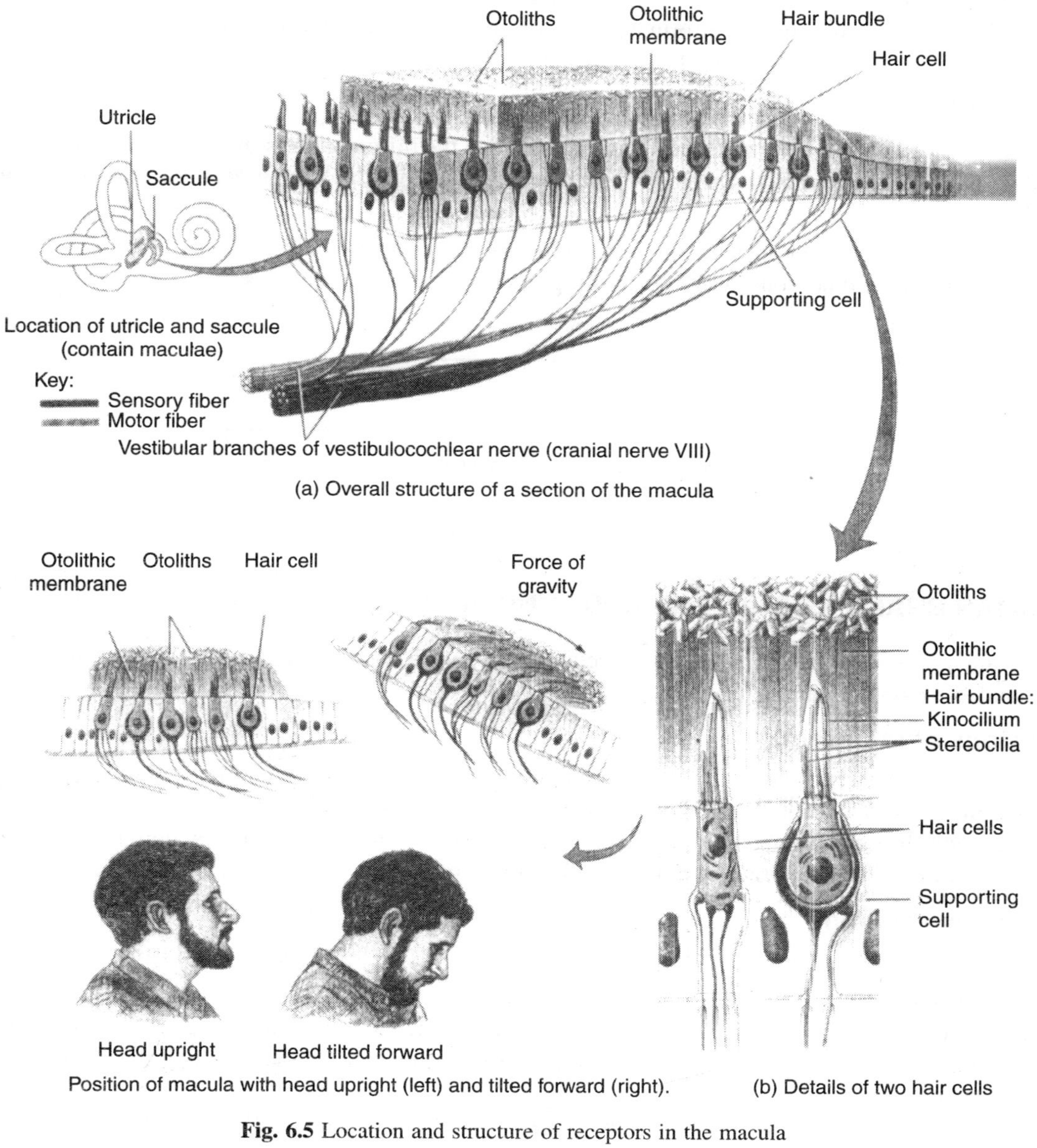

Fig. 6.5 Location and structure of receptors in the macula

from a hot source before one even feel that it is hot, other reflex actions are learnt or acquired. For example, one learns many reflexes during driving experiences. Slamming on the brakes in an emergency is an example. The integration occurred in the spinal cord gray matter the reflex is a spinal reflex. By contrast, if integration occurs in the brain stem rather than the spinal cord, the reflex is cranial reflex. An example is the tracking movements of one's eyes while reading. Somatic reflexes involve contraction of skeletal muscles. Automic (visceral) reflexes involve responses of smooth muscles, cardiac muscles and glands. Body functions such as digestion, urination, defeacation and heart beat are controlled by the autonomic nervous system through autonomic reflexes.

Nerve impulses contributing into, through and out of the central nervous system follow specific pathways, based on the kind of information, its origin and its destination. The pathway followed by nerve impulses which produce a reflex is a reflex circuit (reflex arc).

Nervous System

The nervous system is a communication system between the different parts of human body. It could be divided into several connected systems which function together. The human nervous system is broadly divided into

(*i*) The central nervous system consisting of the brain and spinal cord.

(*ii*) The peripheral nervous system consisting of cranial and spinal nerves.

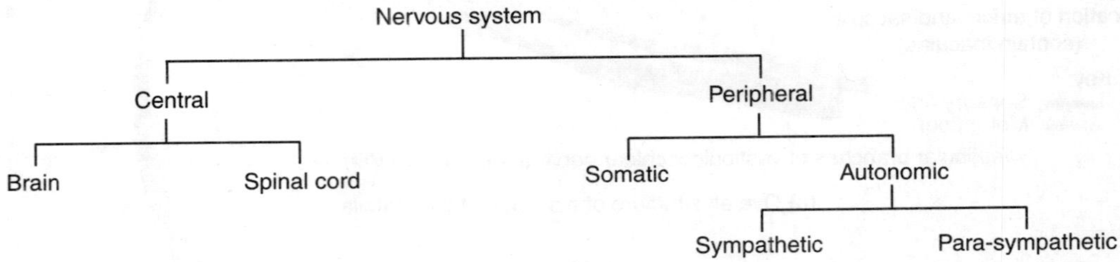

HUMAN BRAIN – STRUCTURE AND FUNCTIONS

The brain is also termed encephalon (encephalon = in the head). The brain is placed in a cavity surrounded by bony cranium.

Divisions of the Brain

The brain is divided into three regions – forebrain or prosencephalon, midbrain or mesencephalon and hind brain or rhombencephalon. Prosencephalon is divided into two parts:

Telencephalon and Diencephalon. Rhombencephalon is made of myelencephalon, metencephalon and cerebellum. The hind brain continues as the spinal cord.

1. Prosencephalon

The human fore brain is made up of:

- a pair of olfactory lobes
- a pair of large cerebral hemispheres, called the telencephalon. Due to the crossing over of the spinal tracts, the left hemisphere of the fore brain deals with the right side of the body and vice versa.
- a group of unpaired structures located deep within the cerebrum, called the diencephalon.

Olfactory lobes: Hidden beneath cerebral cortex are a pair of olfactory lobes. Each olfactory lobe has an olfactory tract which terminates in a bulbous olfactory bulb. Olfactory lobes are mainly concerned with the sense of smell.

The major part of forebrain is occupied by the cerebrum. It is made of two large ovoid parts called cerebral hemispheres. The two cerebral hemispheres are partially separated by a deep groove, which is called median longitudinal fissure. About 2-4 mm of the outer region of the cerebrum is known as cerebral cortex. The cerebral cortex of the left and right cerebral hemispheres are connected by a thick band of nerve fibres called the corpus collosum. The cerebral cortex is highly convoluted with several bumps and grooves. A bump/bulge on the cortex is termed a gyrus (plural: gyri). A groove is called a sulcus (plural: sulci). The cortex is made up of gray matter which is packed with millions of neurons with their cytons and dendrites. Beneath the cortex there is white matter.

Cerebral Hemispheres: The two cerebral hemispheres are divided into four lobes each by three deep fissures or sulci. The three fissures are:

(*i*) Fissure of rolando or central fissure

(*ii*) Parieto-occipital fissure

(*iii*) Sylvian fissure or lateral fissure.

The four lobes are:

- frontal lobe
- parietal lobe
- occipital lobe and
- temporal lobe.

The frontal lobes are the largest among the four. They are responsible for different functions like speech, intellectual performance and motor activities.

Speech: The second area of the frontal lobe controls the ability to form words. Damage in this area results in inability or difficulty in speaking; they understand speech and know what they want to say but cannot speak normally.

Intellectual performance: The frontal lobes also have a strong influence on personality and sharp intellectual functions; including motivation judgement, emotional control and abstract thinking. People with problems in the frontal lobes may not be able to continuously converse because they are distracted being attracted to any noise from another direction.

Motor activities: The back parts of the frontal lobes are the motor areas which control voluntary motor movements like moving an arm to reach an object.

Nerve cells in the motor area cross to the other side of the brain. Therefore, the right hemisphere controls movements on the left side of the body and the left hemisphere controls movement on the right side of the body. Therefore, any damage in the motor area on one side of the brain affects the opposite side of the body.

Parietal lobes: The anterior parts of the parietal lobes are concerned with the sense of touch. The posterior part of the parietal lobes and especially those on the right side of the brain control the ability to pay attention.

The right parietal lobe is concerned with perceiving the spatial relationship and location of objects. The left parietal lobe plays an important role in language and speech. It is also important for reading, writing, calculating numbers, and performing learned skilled acts such as repairing an electrical item or hair dressing.

The occipital lobes in the back part of the brain are important to one's ability to see (*i.e.*, vision).

The temporal lobes have many important functions. They help in recognizing objects and faces of people. They are also important for understanding speech and speaking. They also assist in understanding facial expressions and emotional tones of voice. The information that is seen, heard, smelt, felt or tasted gets into the temporal lobes, where it is stored briefly, and soon stored into long-term memory.

Diencephalon: It is the part of the forebrain that lies between the cerebral hemisphere and midbrain. It is made of the following regions. It is covered around by the cerebral hemispheres.

Thalamus (Inner chamber): A region composed of gray matter. Excepting olfaction, all sensory input passes through it on the way up to the somatic sensory regions of the cerebral cortex and then returns to it from there. Signals from the cerebellum pass through it on the way to the motor areas of the cerebral cortex.

Hypothalamus: It is the seat of autonomic nervous system. When the hypothalamus is damaged the normal homeostasis of body temperature, blood chemistry etc., goes out of control. It may become fatal.

Although it is of the size of only a pea, the hypothalamus is highest centre of autonomic nervous system. It is responsible for certain very important functions like body temperature control. For example, if one's body is too hot, the hypothalamus detects this, sends a signal to expand the capillaries of their skin. This leads to blood to be cooled faster. Pituitary gland is also controlled by hypothalamus.

Optic chiasma: In front of the hypothalamus two optic nerves arise. They cross each other and travel to the opposite sides. The right nerve innervates the left eye and left nerve innervates the right eye.

Infundibulum: It is a stalk-like body arising from the hypothalamus. The pituitary gland is borne at the tip of infundibulum.

Mesencephalon (meso = middle)

The mid brain or mesencephalon occupies only a small region in human beings on each side of the roof of the mid brain, there are two swellings. They are called superior and inferior colliculi. The two swellings of each side are referred to as corpora bigemina. The corpora bigemina of both sides together constitute corpora quadrigemina. They are the optic lobes.

Rhombencephelon (The hind brain) (Rhombus = diamond)

The main structures of the hind brain are the

- Pons varoli
- Cerebellum and
- Medulla oblongata.

Pons varoli (bridge): It is a thick bundle of fibres found on the ventral side of the brain, below the cerebellum. The fibres run transversely and connect the cerebellum and the fibres which run longitudinally connect medulla oblangata. The pons serves as a relay station carrying signals from various parts of the cerebral cortex to the cerebellum. Nerve impulses coming from the eyes, ears, and touch receptors are sent to the cerebellum. The pons also participates in the reflexes that regulate breathing cerebellum. The word 'cerebellum' is a latin word (little brain). It is located behind the brain stem. It is divided into two deeply convoluted hemispheres and has a cortex that surrounds these hemispheres. It is made of two cerebellar hemispheres. The two hemispheres are joined by a median strip called the vermis.

Function of the cerebellum is to coordinate body movements. It is also a centre for learning motor skills, *i.e.*, implicit memory. Medulla oblangata. The medulla looks like a swollen tip to the spinal cord; it is conical in shape. Nerve impulses arising here rhythmically stimulate the intercostal muscles and diaphragm – enabling breathing possible.

Brain stem: The brain stem is a general term for the area of the brain between the thalamus and spinal cord. Structures within the brain stem include the mid brain, pons and medulla oblangata some of these areas are responsible for the most basic functions of life such as breathing, heart rate, swallowing, heartbeat and blood pressure. The brain stem is also the part of the brain that connects the cerebral hemispheres and the cerebellum with the spinal cord.

Human Spinal Cord

The medulla oblangata continues behind as the spinal cord. Like the brain, the spinal cord is also wrapped in the three meninges—duramater, arachnoid process and piamater. The space between these meninges is filled with cerebrospinal fluid. The spinal cord is about 42–45 centimeters long and about 2 centimeters thick. It tapers at the end into a conical, narrow, sharp tip known as conus medullaries. From the conus medullaris, a fine connective tissue filament of about 20 cm in length arises and runs posteriorly to the coccygeal segment. It is known as 'filum terminale'.

The spinal cord does not extend to the end of the vertebral column. It terminates at the level of first lumbar vertebra; due to this the many caudal spinal nerve roots descend beyond the cord in reaching their destinations. By doing so, they form a divergent tuft of spinal roots, around the filum terminale. They are likened to the horse tail and therefore, are called cauda equina.

Peripheral Nervous System

Spinal nerves and the nerves branch from them are part of the peripheral nervous system (PNS). They connect the CNS to sensory receptors, muscles and glands in all parts of the body. The 31 pairs of spinal nerves are named and numbered according to the region and level of the vertebral column from which they emerge.

A typical spinal nerve has two connections to the cord; a posterior root and an anterior root. The posterior and anterior roots unite to form a spinal nerve at the intervertebral foramen. If the posterior root contains sensory ascons, a spinal anterior root contains motor axons, a spinal nerve is a mixed nerve. The posterior root contains a posterior root ganglion in which cell bodies of sensory neurons are located.

A short distance after passing through its intervertebral foramen, a (spinal) peripheral nerve divides into several branches. These branches are known as rami. The posterior (dorsal) ramus serves the deep muscles and skin of the dorsal surface of the trunk. The anterior (ventral ramus) serves the muscles and structures of the upper and lower limbs and the skin of the lateral and ventral surfaces of the trunk. In addition to posterior and anterior rami, spinal nerves also give off a meningeal branch. This branch reenters the vertebral canal through the intervertebral foramen and supplies the vertebrae, vertebral ligaments, blood vessels of the spinal cord, and meninges. Other branches of a peripheral nerve are the rami communicates, components of the autonomic nervous system.

Plexuses

The anterior rami of peripheral nerves except for thoracic nerves T2–T12 do not go directly to the body structures they supply. Instead, they form networks on both the left and right sides of the body by joining with various numbers of axons anterior rami of adjacent nerves. Such a network of axon is called a plexus. The principal plexus, are the cervical plexus, brachial plexus lumbar plexus and sacral plexus. A smaller coccygeal plexus is also present. Emerging from the plexures are nerves bearing names that are often descriptive of the general regions they serve or the course they take. Each of the nerves in turn may have several branches named for the specific structures they innervate.

Intercostal Nerves

The anterior rami of peripheral nerves T2–T12 do not enter into the formation of plexuses and are known as intercostal or thoracic nerves. These nerves directly innervate the structures they supply in the intercostal spaces. After leaving its intervertebral foramen, the anterior ramus of nerve T2 innervates the intercostal muscles of the second intercostal space and supplies the skin of the axilla and posteromedial aspect of the arm. Nerves T3–T6 extend along the costal groves of the ribs and then to the intercostal muscles and skin of the anterior and lateral chestwall. Nerves T7–T12 supply the deep track muscles and skin of the posterior aspect of the thorax.

Special Senses

Receptors for the special senses–taste, smell, vision, hearing and balance are housed in complex sensory organs such as the tongue eyes, nose and ears.

Both taste and smell are chemical senses; the sensation arise from the interaction of molecules with taste or smell receptors.

Sense of Taste (Tongue)

Taste is a chemical sense; to be detected by either sense, stimulating. Taste of gustation distinguishes five primary tastes *viz.*, sour, sweet, bitter, salty and umami. Because olfaction is much more sensitive than taste, a given concentration of a food substance may stimulate the olfactory system too. When persons with colds or allergies complain that they could not taste their food, they are reporting blockage of olfaction, not of taste.

The receptors for sensation of taste are located in the taste buds. Taste buds are found in elevations on the tongue called papillae which provide a rough texture to the upper surface of the tongue. Three types of papillae contain taste buds.

1. About one dozen very large, circular vallate papillae form an inverted V-shaped row at the back of the tongue. Each of thess papillae houses 100–300 taste buds.
2. Fungiform papillae are mushroom shaped elevations scattered over the entire surface of the tongue that contain about five taste buds each.
3. Foliate papillae are located in small trenches on the lateral margins of the tongue but most of their taste buds degenerate in early childhood.

In addition, the entire surface of the tongue has filiform papillae (Filiform = thread like). These pointed, thread like structures contain tactile receptors but no taste buds. They create friction between the tongue and food, making it easier for the tongue to move food in the oral cavity.

Chemicals that stimulate gustatory receptor cells, are known as testants. Once a testant is dissolved in saliva, it can make contact with the plasma membrane of the gustatory hairs, which are the sites of taste transduction. The result is receptor potential that stimulates exocytosis of synaptic vesicles from the gustatory receptor cell. In turn, the liberated neurotransmitter molecules trigger nerve impulses in the first order sensory neurons that synapse with gustatory receptor cells.

The sodium ions (Na^+) in a salty food enter gustatory receptor cells via Na^+ channels in the plasma membrane.

The accumulation of Na^+ inside causes depolarization, which opens Ca^{2+} channels. In turn inflow of Ca^{2+} triggers exocytosis of synaptic vesicles and liberation of neurotransmitter. The hydrogen ions in sour testants may flow into gustatory receptor cells via H^+ channel.

They also influence opening and closing of other types of ion channels; the result is a depolarization that leads to release of neurotransmitter.

However, other testants, responsible for stimulating sweet, bitter and umami tastes do not themselves enter gustatory receptor cells. Rather, they bind to receptors on the plasma membrane that are linked to G proteins. The G proteins then activate several different chemicals known as second messengers inside the gustatory receptor cells. Different second messengers cause depolarisation in different ways resulting – release of neurotransmitter.

If all testants cause release of neurotransmitter from many gustatory receptor cells, why does foods taste differently. Different tastes arise from activation of different groups of taste neurons. Further, although each individual gustatory receptor cell responds to more than one of the five primary tastes. Complete adaptation to a specific taste can occur in 1–5 minutes of continues stimulation. Taste adaptation is due to changes that occur in the taste receptors, in olfactory receptors and in neurons of the gustatory pathway in the central nervous system.

Three cranial nerves including axons of first-order gustatory neurons form taste buds. The facial nerve serves the anterior two thirds of the tongue, the glassopharyngeal nerve serves the posterior one-third of the tongue, and the vagus nerve serves the throat and epiglottis. From taste buds, impulses propagate along these cranial nerves to the medulla oblangata. From the medulla, some axons carrying taste signals project to the limbic system and the hypothalamus, whereas others project to the thalamus. Taste signals that project from the thalamus to the primary gustatory area in the parietal lobe of the cerebral cortex give rise to the conscious feeling of taste.

Smell

The nose contains 10–100 million receptors for the sense of smell or olfaction. The total area of the olfactory epithelium is 5 cm^2. It occupies the superior part of the nasal cavity. The olfactory epithelium consists of three types of cells. Olfactory receptors, supporting cells and basal stem cells.

Olfactory receptors are the primary neurons of the olfactory pathway. They are bipolar neurons whose exposed tip is a knob-shaped dendrite and whose axon ends in the olfactory bulb. The sites of olfactory transduction are the olfactory hairs, which are cilia that project from the dendrite. Chemicals with an odor can stimulate the olfactory hairs are termed odorants.

Supporting cells are columnar epithelial cells of the mucous membrane living the nose. They provide physical support, nourishment and electrical insulation for the olfactory receptors. Basal stem cells lie between the bases of the supporting cells and continually undergo cell division to produce new olfactory receptors, which survive for only a month and then or replaced.

Embedded in the connective tissue that supports the olfactory epithelium, are Bownman's glands of olfactory, which produce mucus that is carried to the epithelium by ducts. This secretion moistens the surface of the olfactory epithelium and dissolves odorants. Olfactory glands and nasal epithelium are innervated by branches of the facial nerve (cranial nerve VII), which can be stimulated by certain chemicals. Impulses in these nerves, in turn, stimulate the lacrinal glands in the eyes and nasal mucous glands. The resultants are tears and running nose after inhaling household, ammonia vapors, pepper and chilly powder.

Olfactory receptors react to odorant molecules when a depolarization develops and triggers one or more nerve impulses. Many different odorants bind to receptors that are linked to G proteins in the plasma membrane and activate the enzyme adenylate cyclase. The following chain of events result:

Production of cyclic adenosine monophosphate (AMP) → opening of sodium ion (Na^+) channels → inflow of Na^+ → depolarizing generator potential → Nerve impulses arise and propagate along axon of olfactory receptor.

Adaptation to odors occur rapidly. Olfactory receptors adopt by about 50% in the first second only but adapt very slowly thereafter.

On each side of the nose, bundles of unmyelinated axons of olfactory receptors extend through about 20 olfactory for amina. In total about 40 bundles of axons collectively form right and left olfactory nerves, which terminate in the brain in pairs called the olfactory bulbs; within which the axon terminals of olfactory receptors – the first-order neurons – form synapses in the olfactory pathway.

Axons of olfactory, bulb neurons extend posteriorly and from the olfactory tract. Because several olfactory tract axons end in the lateral area, it is the primary olfactory area, where conscious, awareness of smell begins.

Vision: (Eye) After considerable processing of visual signals in the retina at synapses among the various types of neurons, the axons of retinal ganglion cells provide output from the retina to the brain. They exit the eyeball as the optic nerve (cranial nerve II).

Photoreceptors and Photopigments

Rods and cones were named for the different appearance of each's outer segment, the distal end of the photoreceptor next to the pigmented layer. The outer segments of rods are cylindrical or rod-shaped, whereas those of cones are tapered or cone shaped (Fig. 6.1).

Transduction of light energy into a receptor potential occurs in the outer segment. The photopigments are integral proteins in the plasma membrane of the outer segment. In cones the plasma membrane folds back and forth in a pleated fashion, whereas in rods the pleats pinch off from the plasma membrae to form discs. The outer segment of each rod contains a stack of about 1000 discs, piled up like coins in a purse.

Outer segment of photoreceptors are renewd at a very rapid space. In rods one to three new discs are added to the base of the outer segment every hour while old discs slough off at the tip and are phagocytised by pigment epithelial cells. The inner segment contains the cell nucleus, golgi complex, and several mitochondria. At its proximal end, the photoreceptor expands into bulb like synaptic terminals filled with synaptic vesicles.

The first step in visual transduction is absorption of light by a photopigment, a coloured protein that undergoes structural changes when it absorbs light. Light absorption initiates the events that lead to the production of a receptor potential. The type of photopigment in rods is rhodopsin. Three different cone photopigments are present in the retina, one in each of three types of cones. Colour vision occurs from different colours of light selectively activating the different cone photopigments.

Photopigments associated with vision contain two parts: a glycoprotein termed as 'opsin' and a derivative of vitamin A. Vitamin A derivatives are formed from carotenoids, the plant pigments that give carrots their orange colour. Good vision depends on adequate dietary intake of carotenoid rich vegetables such as carrots, broccoli, spinach and yellow sanash, or foods that possess vitamin A, such as liver.

Retina is the light absorbing part of all photopigments in the eyes. In human retina, there are four different opsins, one for each cone photopigment and another for rhodopsin. Small variations in the amino acid sequences of different opsins permit the rods and cones to absorb different colours-wavelengths of incoming light. Rhodopsin absorbs blue to green light most effectively, whereas the three different cone photopigments most effectively absorb blue, green, or yellow-orange light.

The layer of pigment stores a large quantity of vitamin A and contributes to the regeneration process in rods.

Light and Dark Adaptation

When human beings emerge from dark surroundings into the light zone, light adoptation occurs – human visual system adjusts in seconds to the brighter environment by decreasing its sensitivity. In contrast when we enter a darkened room such as a theatre, our visual system undergoes dark adoptation – its sensitivity enhances slowly over several minutes.

When light level decreases abruptly, sensitivity enhances rapidly at first and then very slowly. In total darkness, full regeneration of cone photopigments occur during the first 8 minutes of the dark adaptation.

Rhodopsin regenerates very slowly, and human visual sensitivity increases. Release of Neurotransmitter by Photoreceptors.

The absorption of light and isomerization of retinal initiates chemical changes in the photoreceptor outer segment which lead to production of a receptor potential.

In darkness, sodium ions (Na^+) flow into photoreceptor outer segments through ligand gated Na^+ channels. The ligand that holds these channels open is cyclic GMP; (guanosine monophosphate) or cGMP. The inflow of Na^+, called the 'dark current', partially depolarizes the photoreceptor. As a result in darkness the membrane potential of a photoreceptor is about –30 mV; which is much closer to zero than a typical neurons resting membrane potential of –70 mV. The partial depolarization in darkness triggers continuous release of neurotransmitter at the synaptic terminals. The neurotransmitter in rods and perhaps in cones, is the glutamic acid-amino acid.

Dim lights course small and brief receptor potentials the partially turn off glutamic acid (Glutamate) release; brighter lights cause larger and longer receptor potentials which more completely shut down neurotransmitter release. In this way, light excites the bipolar cells that synapse with rods by turning of the release of an inhibitory neurotransmitter.

Hearing

The following events are involved in the process of hearing.

The auricle directs sound waves into the external auditory canal.

When sound waves strike the ear drum to vibrate back and forth. Depending on the intensity and frequency of the sound wave the distance, sound waves move is very small. The eardrum vibrates slowly in response to low-frequency sounds, and rapidly in response to high frequency sounds.

The mid area of the ear drum connects to the malleus, which also starts to vibrate. The vibration is transmitted from the malleus to the incus and then to the stapes.

As the stapes moves back and forth, it pushes the membrane of the oval window in and out. This vibrates about 20 times more vigorously than the eardrum because the ossicles efficiently transmit small vibrations spread over a large surface area into larger vibrations of a smaller surface.

The movement of the oval window-sets up fluid pressure waves in the perilymph of the cochlea. As the oval window bulges inward, it pushed the perilymph of the scalavestibuli.

Pressure waves are transmitted from the scala vestibuli to the scala tympani and eventually to the round window, causing it to bulge outward into the middle ear.

While the pressure waves deform the walls of the scalavestibuli and scalatympani, they also push the vestibular membrane back and forth, creating pressure waves in the endolymph inside the cochlear duct.

The pressure waves in the endolymph cause the basilar membrane to vibrate, which move the hair cells of the spiral organ against the tectorial membrane. Bending of the stereocilia produces receptor potentials which ultimately generates nerve impulses.

Balance (Equilibrium)

There are two kinds of equilibrium. One kind called static equilibrium which refers to the maintenance of the body position mainly the head relative to the force of gravity. The second kind, dynamic equilibrium, is the maintenance of body position mainly the head in response to sudden movements such as rotation, acceleration, and deceleration. Collectively the receptor organs for equilibrium are called the vestibular apparatus, which includes the saccule, utricle and semicircular ducts.

Sacule and utricle (otolithic organs): The walls of both the utricle and sacule contain a small thickened region called a macula (Fig. 6.13). The two maculae are perpendicular to one another, are the receptors for static equilibrium. They contribute to some extent dynamic equilibrium also. They also provide sensory information for static equilibrium on the position of head and essential for maintaining appropriate balance and posture.

Both the maculae consists of two kinds of cells; hair cells which are the sensory receptors and supporting cells. Hair cells appear as hair bundles which consists of 70 or more stereocilia and one kinocilium. A utricle gelotinous ottolthic membrane rests on the hair cells; this is of glycoprotein in nature. A layer of dense calcium carbonate crystals, called otoliths extends over the entire surface of the otolithic membrane.

As the otollithic membrane is situated on top of the macula, if anyone tilt their head forward, the otolithic membrance including the otoliths are pulled by gravity. It slides 'downhill' over the hair cells in the direction of the tilt, bending the hair bundles.

The hair cells depolarize and repolarize. Thus they release neurotransmitter at a faster or slower rate. The hair cells synapse with first order sensory neurons in the vestibular branch of the vestibulocochlear nerve *i.e.*, cranial nerve VIII.

Semicircular Ducts: The three semicircular ducts together with ultricle and sacule function in dynamic equilibrium. The ducts lie at right angles to one another in three planes. (Fig. 6.4). The two vertical ducts are the anterior and posterior semicircular ducts and the horizontal one is the lateral semicircular duct. This positioning permits detection of rotational acceleration or deceleration. When the head moves, the attached semicircular ducts and hair cells move with it. The endolymph within the ampulla, however, is not attached and lags behind, due to its inertia. As the moving hair cells drag along the stationary fluid, the hair bundles bend. Bending of the hair bundles produces receptor potentials. In turn, the receptor potentials lead to nerve impulses that pass along the vestibular branch of the vestibulocochlear nerve *i.e.*, cranial nerve VIII.

UNIT 7

Endocrine System

In human beings body glands can be divided into two categories. Glands with an internal secretion are considered endocrine glands and, those with an external secretion are considered exocrine glands. Endocrine glands do not possess any ducts or openings to the exterior, hence are known as ductless glands.

The endocrine (Greek, endon – within; crinein – to set apart) system consists of a number of ductless glands which manufacture certain chemical substances that take an essential part in the life processes of the human body. The word horma (Greek = Hormao-to excite or set in motion) was coined for such substances which circulate in the blood and bring about an effect on a distant organ.

Thus throughout the whole human life – from early stage of foetal development right up to the end of life – the endocrine glands exert a profound influence on various bodily activities including the mental and behavioural patterns.

Hormone: Hormone is a chemical agent which is released by endocrine glands and, travel via the blood stream to affect one or more systems. Hormones can also be defined as information transferring molecules.

It has also been observed that the nervous system and hormonal systems are interrelated playing a major role in homeostasis. They are interdependent and their function in homeostasis is integrated since there exists a reciprocal relationship between them. The following glands are grouped under endocrine system (Fig. 7.1):

1. Hypophysis (Pituitary)
2. Thyroid
3. Parathyroid
4. Adrenal gland or supra renal gland
5. Ovary
6. Testis
7. Islets of langerhans
8. Skin
9. Hypothalamus
10. Gastrointestinal mucosa
11. Thymus gland
12. Pineal gland

Examples of exocrine glands are:

1. Sweat gland
2. Lacrimal gland, and
3. Mammary gland etc.

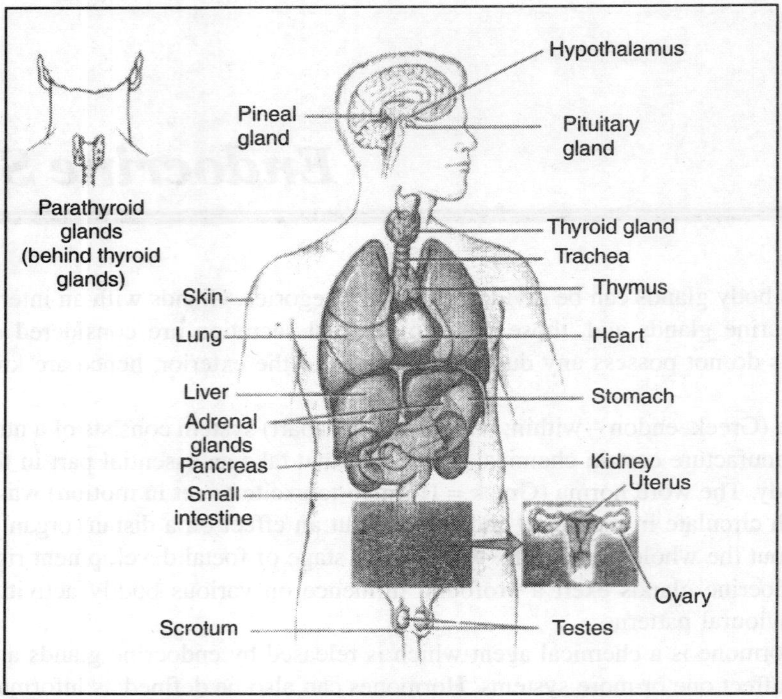

Fig. 7.1 Location of many endocrine glands

HYPOPHYSIS (PITUITARY GLAND)

In 16th century vesalius nomenclatured this gland, the pituitary (L. pituita = phlegm or mucus) collected wastes from the brain which flow through the nasal and pharyngeal passages. Whereas, when the true function of the gland has established the term hypophysis (hypo + Gr. Phyein = to grow) was substituted. But the gland is still often called the pituitary.

Pituitary gland is a reddish grey, small oval shaped structure situated, at the base of the brain; it weighs in adults with an average weight of 0.5 to 0.6 g; however, it varies with age, sex and physiologic state. It is slightly larger in females ranging from 0.6 to 0.7 g.

It consists of one anterior to be and another posterior to be. But microscopically it consists of six parts (Fig. 7.2).

(*i*) Pars anterior or Pars distalis
(*ii*) Pars tuberalis
(*iii*) Pars intermedia
(*iv*) Pars nervosa

(*v*) Median eminence of tuber cinereum and
(*vi*) Infundibulum (Pituitary stalk).

The major hormones secreted by the pituitary gland and their principal actions are enumerated in the following table:

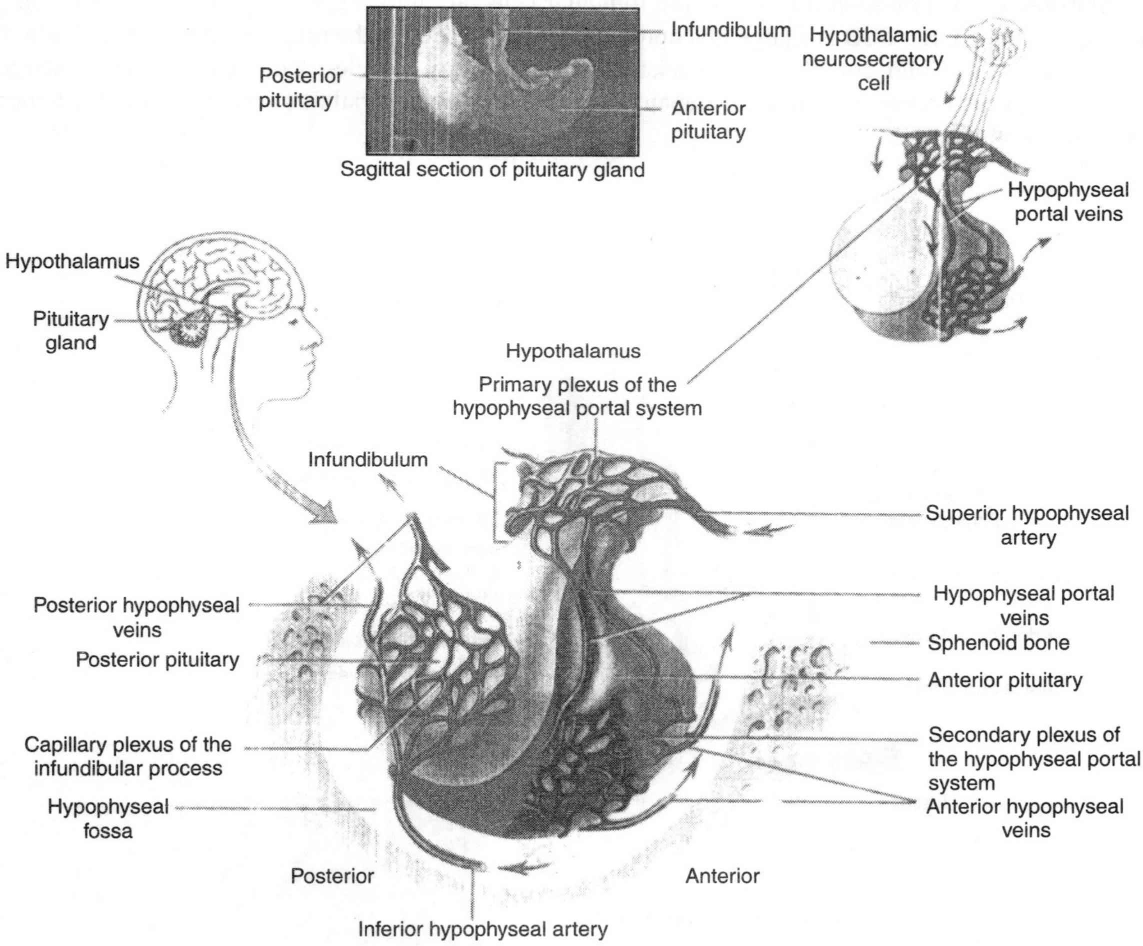

Fig. 7.2 Showing pituitary and hypothalamus glands

Thyroid: The thyroid gland is situated at the root of the throat, having two symmetrical lobes, one on either side of trachea joined by a thin portion of thyroid tissue called the isthmus. In adults, this gland varies from 20–25 g, however, it is influenced by age, sex, diet and reproductive state.

Blood flows form 3.5 to 6 ml per gram of thyroid per minute. This fast flow rate probably serves to supply the tissues with sufficient inorganic iodide.

Thyroid Hormones

The thyroid gland produces three hormones, thyroxine, triiodothyronine a calcitonin.

Thyroid gland synthesise andrelea iodinated thyronine molecules. These molecules influence metabolic processes and growth (Fig. 7.3).

Oxidation of iodide: Some of the amino acids in thyroglobulins are tyrosines that will become iodinated. However, negatively charged iodide ions cannot bind to tyrosin until they undergo oxidation (*i.e.*, removal of electrons) to iodine: $2\ I^- \rightarrow I_2$. The enzyme that catalyses this reaction is peroxidase.

Synthesis of thyroglobuline: While the follicular cells, are trapping I^-, they are also synthesising thyroglobulin (TGB). TGB is a large glycoprotein that is produced in the rough endoplasmic reticulum, modified in the Golgi complex, and packaged into secretory vesicles. The vesicles then undergo exocytosis, which releases TGB in the lumen of the follicle. The material that accumulates in the lumen is termed colloid.

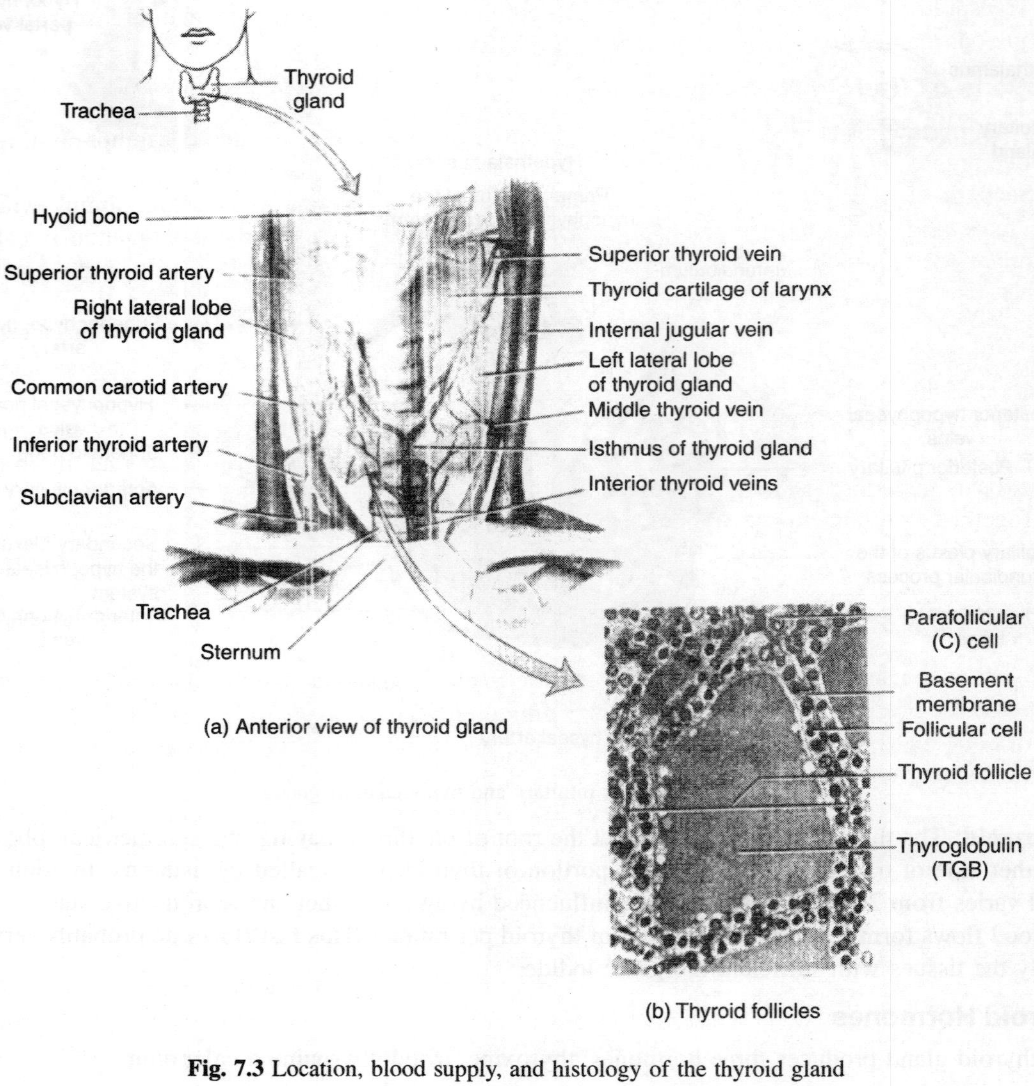

Fig. 7.3 Location, blood supply, and histology of the thyroid gland

Iodination of tyrosine: As iodine molecules (I_2) form, they react with tyrosines that are part of thyroglobulin molecules in the colloid. Binding of one iodine, atom yields monoiodyrosine (T_1), and a second iodination produces diodotyrosine (T_2).

Coupling of T_1 and T_2: During the last step in the synthesis of thyroid hormone, two T_2 molecules join to form T_4 or one T_1 and one T_2 joins to form T_3.

Pinocytosis and digestion of colloids: Droplet of colloid reenters follicular cells by pinocytosis and merge with lysosomes. Digestive enzymes breakdown TGB, clearing off molecules of T_3 and T_4.

Secretion of thyroid hormones: Because T_3 and T_4, are lipid soluble, they diffuse through the plasma membrane into interstitial fluid and then into the blood. T_4 normally is secreted into greater quantity than T_3, but T_3 is several times more potent. Moreover, after T_4 enters a body cell, most of it is converted to T_3 by removal of one iodine.

Transport in the blood. More than 99% of both the T_3 and the T_4 combine with transport proteins in the blood, mainly thyrosine-binding globulin (TGB).

Actions of Thyroid Hormones

Thyroid hormones increase basal metabolic rate (BMR), the rate of oxygen consumption at rest after an overnight fast, by stimulating the use of cellular oxygen to produce ATP.

Secondly thyroid hormones stimulate synthesis of additional sodium-potassium pumps, which then use more ATP, more heat is given off and body temperature rises. This phenomenon is called the calorigenic effect of the thyroid hormones. In this way, they play an important role in the maintenance of normal body temperature.

In the regulation of metabolism the thyroid hormones stimulate protein synthesis and increase the use of glucose and fatty acids for ATP production. They also increase lipolysis and enhance cholesterol excretion, thus reducing blood cholesterol level.

The thyroid hormones enhance some actions of the catecholamines because they up-regulate β-receptors. For this reason, symptoms of hyperthyroidism include increased heart rate, more forceful heart beats including increased blood pressure.

Together with human growth hormone and insulin, thyroid hormones accelerate body growth, particularly the growth of nervous tissue. Deficiency of thyroid hormones during fetal development, infancy causes stunted growth and severe mental retardation.

Calcitonin

The hormone produced by the parafollicular cells of the thyroid gland is calcitonin (CT). CT can decrease the level of calcium in the blood by inhibiting the action of osteoclasts, the cells that break-down bone matrix. Calcitonin's importance in normal physiology is unclear because it can be present in excess or completely absent without causing clinical symptoms.

When its blood level is high, calcitonin lowers the amount of blood calcium and phosphates by inhibiting bone resorption *i.e.*, breakdown of bone matrix by osteoclasts and by accelerating uptake of calcium and phosphates into bone matrix.

PARATHYROID GLANDS

Partially embedded in the posterior surface of the lateral lobes of the thyroid gland are several small round masses of tissue called the parathyroid glands. Usually one superior and one inferior parathyroid gland are attached to each thyroid lobe (Fig. 7.4).

Parathyroid hormone is the major regulator of the levels of calcium (Ca^{2+}), magnesium (Mg^{2+}), and phosphate (HPO_4^{2-}) ions in the blood. The blood calcium level directly controls the secretion of both calcitonin and parathyroid hormone via negative feed back loops that do not involve the pituitary gland.

A higher-than-normal level of calcium ions (Ca^{2+}) in the blood stimulates parafollicular cells of the thyroid gland to release more calcitonin.

Calcitonin inhibits the activity of osteoclasts, thereby decreasing blood Ca^{2+} level.

A lower than normal level of Ca^{2+} in the blood stimulates chief cells of the parathyroid gland to release more PTH.

PTH promotes resorption of bone matrix, which releases Ca^{2+} in the urine; raising the blood level of Ca^{2+}.

PTH also stimulates the kidneys to synthesize calcitriol, the active form of vitamin D.

Calcitriol stimulates increased absorption of Ca^{2+} from foods in the gastrointestinal tract, which helps to increase the blood level of Ca^{2+}.

PTH increases the number and activity of osteoclasts. The result is elevated bone resorption, which releases ionic calcium (Ca^{2+}) and phosphates (HPO_4^{2-}) into the blood. PTH also acts on the kidneys. First, it shows the rate at which Ca^{2+} and Mg^{2+} are lost from blood into the urine. Second, it increases loss of HPO_4^{2-} from blood into the urine. Because more HPO_4^{2-} is lost in the urine than is gained from the bones, PTH decreases blood HPO_4^{2-} level and increases blood Ca^{2+} and Mg^{2+} levels. A third effect of PTH on the kidneys is to promote formation of the hormone calcitriol, the active form of vitamin D. Calcitriol also known as 1, 25 – dihydroxy vitamin D_3, increases the rate of Ca^{2+}, HPO_4^{2-}, and Mg^{2+} absorption from the gastrointestinal tract into the blood.

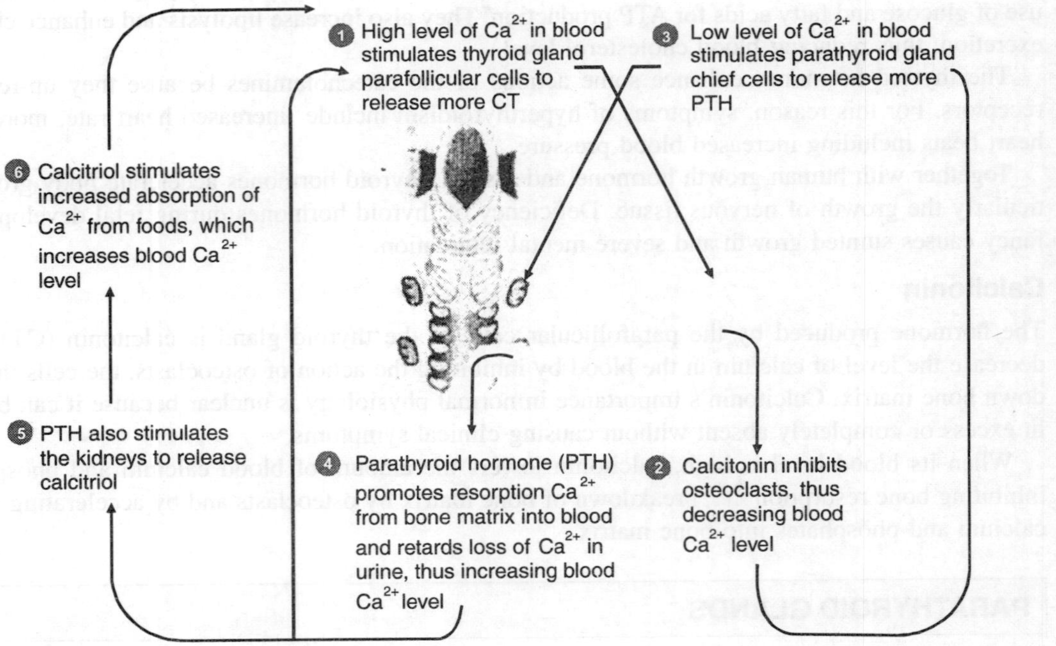

Fig. 7.4 The roles of calcitonin, parathyroid hormone and calcitriol in calcium homeostasis

PANCREAS

The pancreas (Pan = all; creas = flesh) is both an endocrine gland and an exocrine gland. Here only endocrine functions are discussed. The pancreas is a flattened organ that measure about 12.5–15 cm in length. It is located in the curve of the duodenum. Roughly 99% of the pancreatic cells are arranged in clusters called acini. The acini produce digestive enzymes which flow into the gastrointestinal tract through a network of ducts. Scattered among the exocrine acini are 1–2 million tiny clusters of endocrine tissue called pancreatic islets of Langerhans (Fig. 7.3).

Pancreatic Islets-cell Types

(*i*) Alpha or A cells constitute about 15% of pancreatic islet cells and secrete glucagon.
(*ii*) Beta or B cells constitute about 80% of pancreatic islet cells and secrete insulin.
(*iii*) Delta or D cells constitute about 5% of pancreatic islet cells and secrete somatostatin.
(*iv*) F cells constitute the remainder of pancreatic islet cells and secrete pancreatic polypeptide.

The interaction of the four pancreatic hormones are complex and not completely understood. Whereas glucagon raises blood glucose level, insulin lowers it. Somatostotin acts in a paracrine manager to inhibit both insulin and glucagon release from neighbouring beta and alpha cells. It may also act as a circulating hormone to slow absorption of nutrients from the gastrointestinal tract. Pancreatic polypeptide inhibits somatostatin secretion, gall bladder contraction, and secretion of digestive enzyme by the pancreas.

Regulation of Glucagon and Insulin Secretion

The principal action of glucagon is to increase blood glucose level when it falls below normal. Insulin on the other hand helps lower blood glucose level when it is too high. The level of blood glucose controls secretion of glucagon and insulin via negative feed back.

1. Low blood glucose level (hypoglycemia) stimulates the secretion of glucagon from alpha cells of the pancreatic islets.
2. Glucagon acts on hepatocytes (linear cells to accelerate the conversion of glycogen into glucose (glycogenolysis) and to promote formation of glucose from lactic acid and certain amino acids; (gluconeogenesis).
3. As a result hepatocytes release glucose into the blood more rapidly, and blood glucose level rises.
4. If blood glucose continues to rise, high blood glucose level (hyperglycemia) inhibits release of glucagon (negative feed back).
5. At the same time, high blood glucose (hyperglycemia) stimulates secretion of insulin by beta cells of the pancreatic islets.
6. Insulin acts on various cells in the body to accelerate facilitated diffusion of glucose into cells, especially skeletal muscle fibres; to speed conversion of glucose into glycogen (glycogenesis); to increase uptake of amino acids by cells and to increase protein synthesis; to slow glycogenolysis; and to slow glyconeogenesis.
7. As a result blood glucose level reduces.
8. If blood glucose level reduces below normal, low blood glucose inhibits release of insulin (–ve feeback).

Although high blood glucose level is the most important stimulator of insulin release, several hormones and neurotransmitters also regulate the release of insulin and glucagon. Whereas glucagon stimulates insulin release, insulin suppresses glucagon secretion. As blood glucose level declines and less insulin is secreted, the alpha cells are released from the inhibitory effect of insulin. Then, they secrete more glucagon. Indirectly human growth hormone (hGH) and adrenocorticotropic hormone (ACTH) stimulate secretion of insulin because they act to elevate blood glucose.

Insulin secretion is stimulated by:

(*i*) Acetylcholine, the neurotransmitter liberated from axon terminals of parasympathetic vagus nerve axons that innervate the pancreatic islets,

(*ii*) The amino acids Arginine and leucine, which would be present in the blood at higher levels after a protein-containing meal and

(*iii*) Glucose-dependent insulinotropic peptide (GIP), a hormone released by enteroendocrine cells of the small intestine in response to the presence of glucose in the gastrointestinal tract.

Thus, digestion and absorption of food containing both carbohydrates and proteins provide strong stimulation for insulin release.

Glucagon secretion is stimulated by:

(*a*) Increased activity of the sympathetic division of the ANS, as occurs during exercise and

(*b*) A rise in blood amino acids if blood glucose level is low, which could occur after a meal that contained mainly protein.

ADRENAL OR SUPRARENAL GLANDS

The paired adrenal (suprarenal) glands, one of which lies superior to each kidney (Fig. 7.4) have a flattened pyramidal shape. In an adults each, adrenal gland is 3–5 cm in height, 2–3 cm in width. A large, peripherally located adrenal cortex develops from mesoderm; a small centrally located adrenal medulla develops from ectoderm.

The adrenal cortex produces steroid hormones that are essential for life. Complete loss of adrenocortical hormones leads to death due to dehydration and electrolyte imbalances in a few days to a week, unless hormone replacement therapy begins promptly. The adrenal medulla produces three cate-cholamine hormones – norepinephrine, epinephrine, and a small amount of dopamine.

Mineralocorticoids

Aldosterone is the major mineralocorticoid. It regulates homeostasis of two mineral ions, namely sodium ions (Na^+) and potassium ions (K^+) and helps adjust blood pressure and blood volume. Aldosterone also promotes excretion of H^+ in the urine; this removal of acids from the body can help prevent acidosis (blood pH < 7.35).

The renin-angiotensin – aldosterone or RAA pathway controls secretion of aldosterone (Fig. 7.5).

1. Stimuli that initiate the renin-angiotensin – aldosteron pathway include dehydration, Na^+ deficiency or hemorrhage.
2. These conditions cause a decrease in blood volume.
3. Decreased blood volume leads to decreased blood pressure.
4. Lowered blood pressure stimulates certain cells of the kidneys, called juxtaglomerular cells to secrete the enzyme renin.
5. The level of renin in the blood increases.
6. Renin converts angiotensinogen, should remain plasma protein produced by the liver, into angiotensin I.

7. Blood containing increased levels of angiotensin I circulates in the body.
8. As blood flows through capillaries, particularly those of the lungs, the enzyme angiotensin-converting enzyme (ACE) converts angiotensin I into the hormone angiotensin II.
9. Blood level of angiotensin II increases.
10. Angiotensin II stimulates the adrenal cortex to secrete aldosterone.
11. Blood containing increased levels of aldosterone circulates to the kidneys.
12. In the kidneys, aldosterone increases reabsorption of Na^+ and water so that less is lost in the urine. Aldosterone also stimulates the kidneys to increase secretion of K^+ and H^+ into the urine.
13. With increased water reabsorption by the kidneys, blood volume increases.
14. As blood volume increases, blood pressure increases to normal.
15. Angiotensin II also stimulates contraction of smooth muscle in the walls of arterioles. The resulting vasoconstriction of the arterioles increases blood pressure and thus helps raise blood pressure to normal.
16. Besides angiotensin II, a second stimulator of aldosterone secretion is an increase in the K^+ concentration of blood. A decrease in the blood K^+ level has the opposite effect.

GLUCOCORTICOIDS

The glucocorticoids, which regulates metabolism and resistance to stress, include cortisole (hydrocortisone), corticosterone and cortisone. Among the three, cortisol is the most abundant, accounting for about 95% of glucocorticoid activity.

Low blood levels of glucocorticoids, mainly cortisol, stimulate neurosecretory cells in the hypothalamus to secrete 'corticotropin-releasing hormone (CRH)'. CRH promotes the release of ACTH from the anterior pituitary. ACTH flows in the blood to the adrenal cortex, where it stimulates glucocorticoid secretion. ACTH also stimulates secretion of aldosterone.

Glucocorticoids have the following effects:

(*a*) **Protein breakdown:** Glucocorticoids increase the rate of protein breakdown, mainly in muscle fibres, and thus increase the liberation of amino acids into the blood stream. The amino acids may be used by body cells for synthesis of new proteins or for ATP production.

(*b*) **Glucose formation:** Upon stimulation by glucocorticoids, liver cells may convert certain amino acids or lactic acid to glucose, which neurons and other cells can use for ATP production. Such conversion of a substance other than glycogen or another monosaccharide into glucose is called gluconeogenesis.

(*c*) **Lipolysis:** Glucocorticoids stimulate lipolysis, the breakdown of triglycerides and release of fatty acids from adipose tissue into the blood.

(*d*) **Resistance to stress:** Glucocorticoids work in many ways to provide resistance to stress. The additional glucose supplied by the liver cells provides tissues with a ready source of ATP to combat a range of stresses including exercise, fasting, fright, temperature extremes, high altitudes, bleeding, infection, surgery, trauma and disease. Because glucocorticoids make blood vessels more sensitive to other hormones that cause vasoconstriction; they raise blood pressure. This effect would be an advantage in cases of severe blood loss, which causes blood pressure to drop.

(*e*) **Anti-inflammatory effects:** Glucocorticoids inhibit white blood cells that participate in inflammatory responses. Unfortunately they slow wound healing. High doses can causes severe mental disturbances; glucocorticoids are very useful in the treatment of chronic inflammatory disorders such as rheumatoid arthritis.

(*f*) **Depression of immune responses:** High doses of glucocorticoids depress immune responses. For this reason, glucocorticoids are prescribed for organ transplant recipients to retard tissue rejection by the immune system.

ANDROGENS

In both male and female human beings, the adrenal cortex secretes small quantity of weak androgens. The major androgen secreted by the adrenal gland is dehydroepiandrosterone (DHEA). After puberty in males, androgens are also released in much greater quantity by the testes. Thus, the amount of androgens secreted by the adrenal gland is usually so low that their effects are insignificant. In females, however, adrenal androgens play important roles. They promote libido (sex drive) and are converted into estrogens (feminizing sex steriods) by other body tissues. After menopause, when ovarian secretion of estrogens ceases, all female estrogens come from conversion of adrenal androgens. Adrenal androgens also stimulate growth of axillary and pubic hair in boys and girls and contribute to the prepubertal growth spurt. Although control of adrenal androgen secretion is not fully understood, the main hormone that stimulates its secretion is ACTH.

Adrenal Medulla

The inner region of the adrenal gland, the adrenal medulla, is a modified sympathetic ganglion of the autonomic nervous system (ANS). It develops from the same embryonic tissue as all other sympathetic ganglia, but its cells lack axons and farm clusters around large blood vessels. Instead of releasing a neurotransmitter, the cells of the adrenal medulla secrete hormones.

The two major hormones synthesized by the adrenal medulla are epinephrine and norepinephrine (NE), also called adrenaline and noradrenaline; respectively.

These two hormones greatly augment the fight-or flight response. By increasing heart rate and force of contraction, adrenaline and noradrenaline increase the pumping output of the heart, which enhances blood pressure. They also increase blood flow to the heart, liver, skeletal muscles and adipose tissue; dilate airways to the lungs; and increase blood levels of glucose and fatty acids.

Sex Glands

Gonads are the organs that produce gametes – sperm in males and oocytes in females. The female gonads, the ovaries produce several steroid hormones including two estrogens (estradiol and estrone) and progesterone. These female sex hormones, along with FSH and LH from the anterior pituitary, regulate the menstrual cycle, maintain pregnancy and prepare the mammary glands for lactation. They also help in establishing and maintaining feminine secondary sex characteristics, such as big breasts and wide hips.

The ovaries also produce inhibin, a protein hormone that inhibits secretion of follicle-stimulating hormone (FSH). During pregnancy, the ovaries and placenta produce a peptide hormone called relaxin, which increases the flexibility of the pubic symphysis during pregnancy and helps dilate the uterine cervix during labour and delivery. These actions help ease the baby's passage by enlarging the birth canal.

The male gonads the testes, are oval glands that lie in the scrotum. The main hormone produced and secreted by the testes is testosterone, an androgen or male sex hormone. Testosterone regulates production of sperm and stimulates the development and maintenance of masculine secondary sexual characteristics, such as beard growth and deepening of the voice. The testes also produce inhibin, which inhibits secretion of FSH.

GASTROINTESTINAL MUCOSA

Once food reaches the stomach sensory receptors in the stomach initiate both neural and hormonal mechanisms to ensure that gastric secretion and motility continue. This is the gastric phase of gastric digestion. Food of any kind distends (stretches) the stomach and stimulates stretch receptors in its walls. Chemoreceptors in the stomach monitor the pH of the stomach chyme. When the stomach walls are distended or pH increases because proteins have entered the stomach and buffered acid in stomach.

The peristaltic waves mix the food with gastric juice, and a little quantity of chyme (10–15 ml) crosses the pyloric sphincter to reach duodenum. Chemoreceptors and stretch receptors monitoring these changes stimulate parasympathetic neurons to release acetylcholine, in turn acetylcholine stimulates secretion of the hormone 'gastrin' by G cells, enteroendocrine cells in the mucosa of the pyloric antrum. Gastrin enters the blood stream and finally reaches its target cells, the gastric glands.

When chyme containing fatty acids and glucose leaves the stomach and enters the small intestine, it triggers enteroendocrine cells in the small intestinal mucosa to release into the blood two hormones-secretin and cholecystokinin (CCK); the latter inhibits stomach emptying and former mainly decreases gastric secretions.

Thymus Gland

The thymus is located behind the sternum between the lungs. It is a bilobed organ located in the mediastinum between the sternum and the aorta (Fig. 7.5).

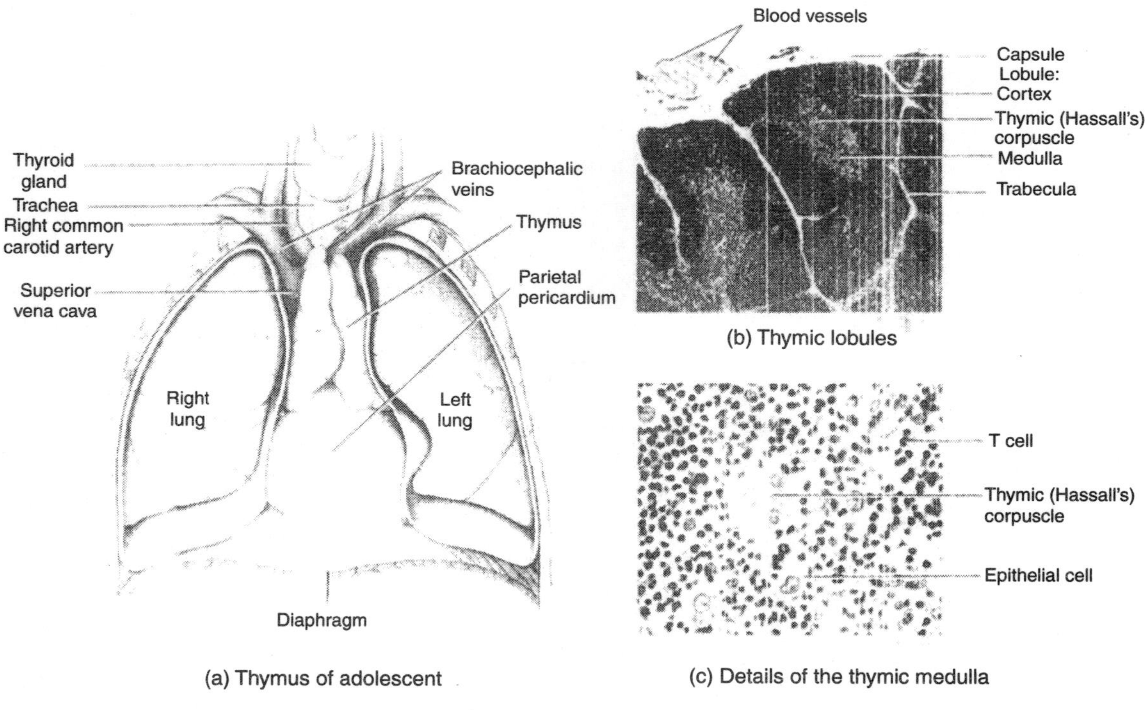

Fig. 7.5 Thymus

Each thymic lobule consists of a deeply staining outer cortex and a lighter-staining central medulla. The cortex is composed of large numbers of T cells and scattered dendritic cells, epithelial cells and macrophages. Immune T cells migrate from red bone narrow to the cortex of the thymus, where they proliferate and mature. Dendritic cells, so named because they have long, branched projections that resemble the dendrites of a neuron. The medulla consists of widely scattered, more mature T cells. T cells that leave the thymus via the blood are carried to lymph nodes, the spleen and other lymphatic tissues where they colonize parts of these organs and tissues.

The hormones produced by the thymus – thymosin, thymic humoral factor (THF), thymic factor (TF) and thymopoietin – promote the maturation of T cells, a type of white blood cell that destroys microbes and foreign substances and may retard the aging process.

Pineal Gland

The pineal gland is a small endocrine gland attached to the roof of the third ventricle of the brain at the mid line (Fig. 7.6). If is part of the epithalamus, positioned between the two colliculi and consists of masses of neuroglia and secretory cells called pinealocytes. One hormone secreted by the pineal gland is melatonin, an amine hormon derived from serotonin. More melatonin is released in darkness, and less melatonin is liberated in strong sunlight.

Melatonin contributes to setting the body's biological clock. During sleep, plasma levels of melatonin increase tenfold and then decline to a low level again before awakening. Small doses of melatonin given orally can induce sleep and reset daily rhythms, which might benefit workers whose shifts alternate between daylight and night time hours. Melatonin also is potent antioxidant.

UNIT 8

Reproductive System

The process of sexual reproduction in human beings can be approached with acknowledge of the male and female reproductive systems.

Female reproductive system includes ovary, Fallopian tubes, uterus and vagina ovary is paired and involves in production of Graafian follicle or ova. Fallopian tube is a connecting source helps in passage of egg to uterus and fertilization. Both the left and right fallopian tubes converge into a single uterus which is responsible for implantation and development of fetus. Vagina is the entrace to female reproductive system for facilitating the entry of sperm and emergence of child (Fig. 8.1).

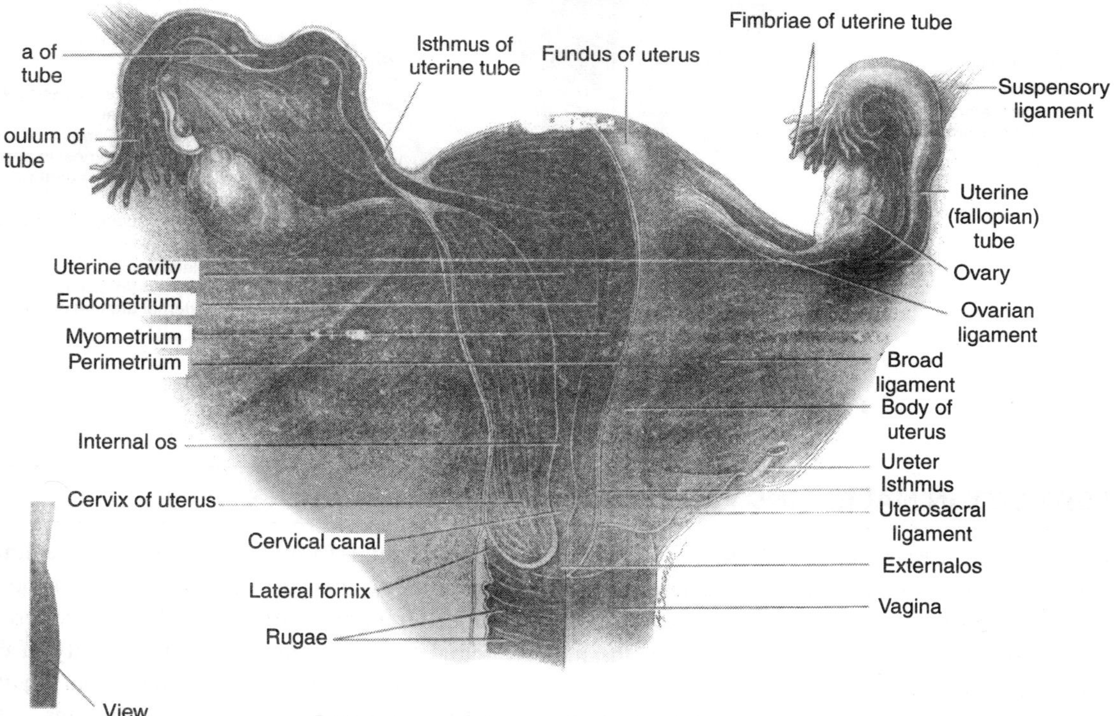

Posterior view of uterus and associated structures

Fig. 8.1 Relationship of the uterine (fallopian) tubes to the ovaries, uterus

Male reproductive system includes a pair of testis responsible for the production of sperms. Epididymus is a storage of sperms. From epididymus a sperm duct leads into penis, this sperm duct is termed vas deferens, aids in passage of sperm transfer. Seminal vesicle and prostate are glandular structures associated with the male reproductive system; as the latter secretes mucus and alkaline fluid – neutralizes acidity of vagina, activating the sperm, the former secretes mucus and alkaline fluid that contains fructose which forms energy source for sperm. Cowper's gland is a glandulor structure which secretes mucus and alkaline fluid into the urethra, neutralizing the acidity of any remaining urine. Further, penis is a highly vascularised and muscularised male intromittent, organ, which helps the process of introduction of a sperm into the female reproductive tract. And this process of copulation is called intercourse (Fig. 8.2).

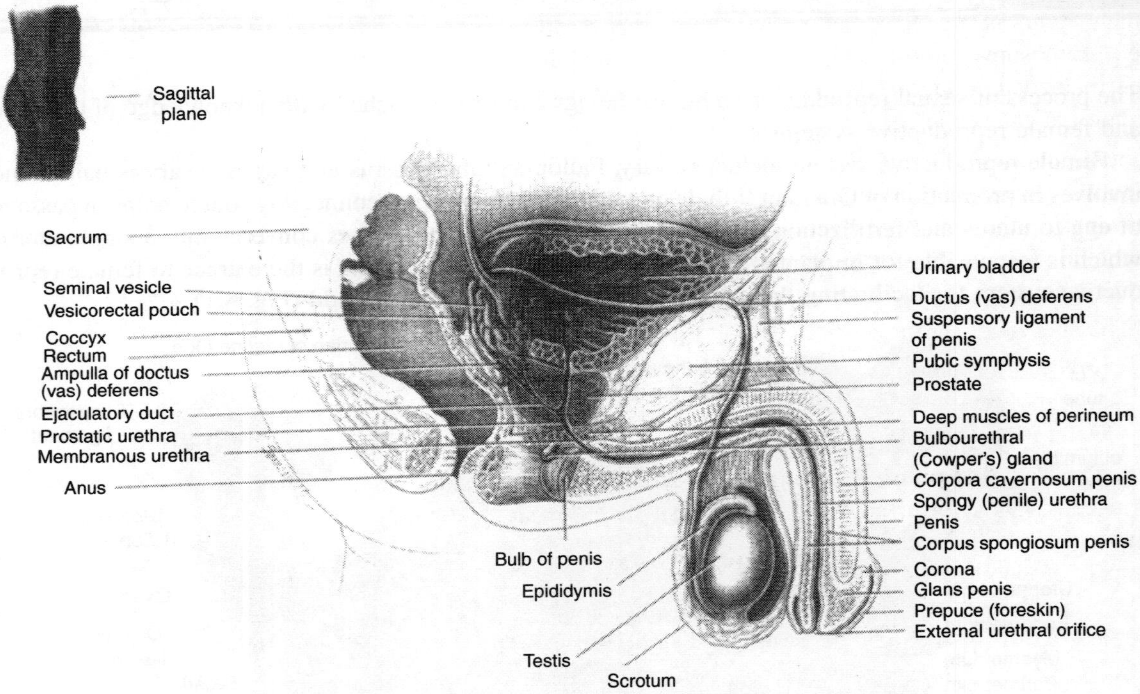

Fig. 8.2 Male organs of reproduction and surrounding structures

FERTILITY IN MALES AND FEMALES

In human beings both males and females are fundamentally bisexual. In males the masculine features predominate, the feminine features remain rudimentary. In females, it is just the reverse. The urine of both sexes contains both male and female hormones but in different proportions. It is more than probable that the physical and mental get-up of the male or female does not depend on the activity of the male or female does not depend on the activity of the male or female hormone only—but on the combined action of both the hormones. It seems that an optimum balance of these two hormones is the more important factor than anyone of them.

Among sexually reproducing multicellular organisms, two kinds of cells are recognised. These are somatic cells which are concerned with metabolic activities and germ cells which perform reproductive function. All higher organisms produce two kinds of germ cells or gametes, *viz.*, the male gametes or sperms and the female gametes or ova.

Specialised regions are demarked in the body for the manufacture of sexual cells. Such regions where sex cells are manufactured are called gonads. Two types of gonads are generally present male gonad–testis and female gonad-ovary. In primitive animals both gonads develop in the same animal and this condition is called hermaphrodite. Among higher animals, however, the testis, and ovary develop in different animals. Testis bearing animal is the male and ovary possessing animal is the female.

Gametogenesis: The development of gametes or sex cells from primordial germ cells is called gametogenesis. The maturation of germ cells is called gametogenesis. The maturation of germ cells, which occurs in the testis and ovary, is the direct result of meiosis (reduction division) which reduces the chromosome number from diploid (2n) to haploid (n). The development of the male germ cell (spermatozoan or sperm) is called the spermatogenesis. The formation of mature egg or ovum (female gamete) is called the oogenesis.

Spermatogenesis: Spermatogenesis means the formation of spermatozoa. It occurs in the testis. Testis contains seminiferous tubules. Inside the tubules spermatogenic cells are present which form the germinal epithelium, whose cells develop into spermatozoa during spermatogenesis. The germinal epithelial cells are separated by special cells called sertoli cells, which are believed to nourish the developing spermatozoa.

The spermatogenesis process occurs in three phases *viz.*,

- proliferation phase,
- growth phase, and
- maturation phase.

1. **Proliferation phase:** The cells in the seminiferous tubules are diploid. They are called primordial germ cells. They form the stem cells from which the spermatozoa emerge. These stem cells divide repeatedly by mitosis and increase in number. The newly formed cells are known as sperm mother cells or spermatogonia.
2. **Growth phase:** The spermatogonia during this phase grows in size. Subsequently their volume is doubled. At this stage they are designated as primary spermatocytes. The are diploid.
3. **Maturation phase:** The totally grown, diploid primary spermatocyte undergoes meiotic division. This division is a reduction division followed by an equational division. In reduction division, each diploid primary spermatocyte divides into two haploid secondary spermatocytes. In equational division each haploid secondary spermatocyte divides into two haploid cells known as spermatids.

Thus at the end of the maturation stage, a single diploid spermatogonium divides into four equal spermatids.

Spermiogenesis or Spermatogenesis

Spermatids are spherical cells. This stationary spermatid undergoes a series of drastic changes and finally emerges as a slim and trim spermatozoan, ideally suited to carry the haploid set of chromosomes of the male parent and to safely deposit it into the ovum. This process is known as spermiogenesis or spermatogenesis.

During this process, the nucleus shrinks and becomes compact. It becomes long and marrow to form the head. The Golgi complex forms a protective cap around the head. This is the acrosome. The centriole forms the axial filament of the sperm which becomes the tail. Cytoplasm forms a thin sheath around the proximal part of the axial filament. The two centrioles arrange one below the other along the longitudinal axis of the sperm. The nucleus at its base undergoes an unvagination to accommodate one of the centrioles. This is the proximal centriole; the other centriole is placed below the proximal centriole and is known as distal centriole. This distal centriole gives rise to the axial filament which continues into the tail.

ULTRA STRUCTURE OF HUMAN SPERM

Human spermatozoan is a long, microscopic, motile cell. The electron microscope distinguishes five definite regions of it; head, neck, middle piece, tail and end piece (Fig. 8.3).

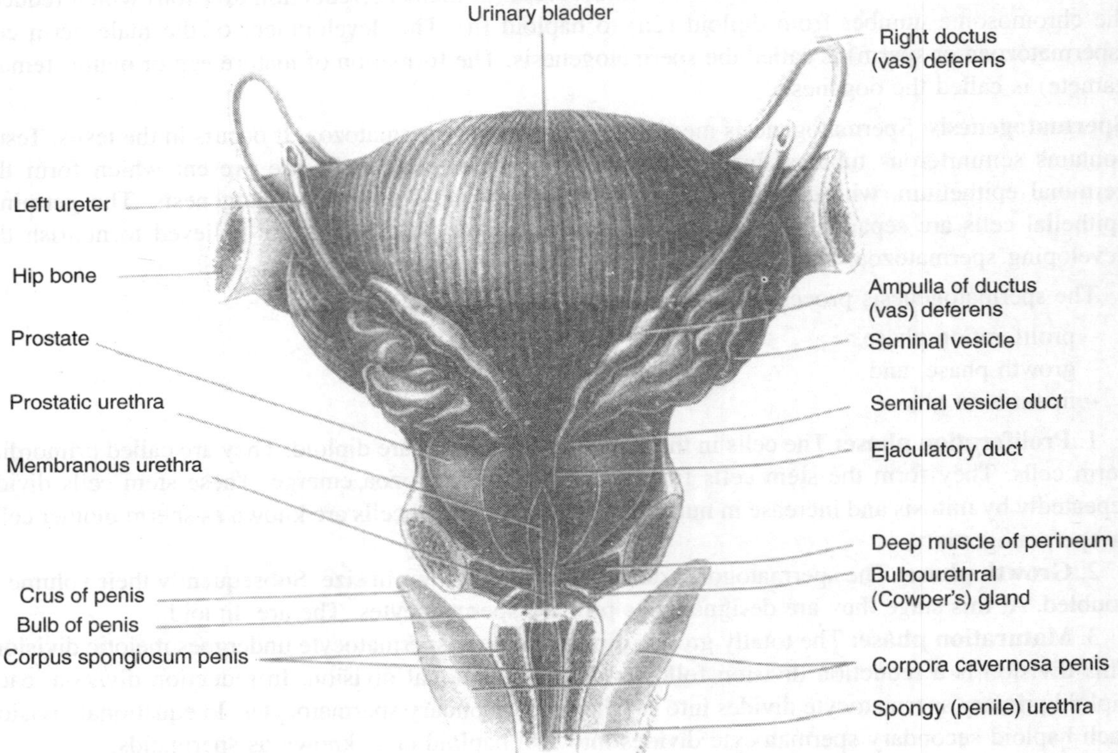

Fig. 8.3 Locations of several accessory reproductive organs in males

Oogenesis

Oogenesis occurs in ovary. Oogenesis occurs in three phases *viz.*,

- Proliferation phase,
- Growth phase and,
- Maturation phase.

1. **Proliferation phase:** The primordial germ cells are the stem cells. They multiply in number through repeated divisions and modify into diploid oogonia.

2. **Growth phase:** The oogonia enter a phase of growth as they increase in size. First of all there happens an increase in the volume of the nucleus and cytoplasm. Following this packaging of reserve food in the form of proteins, lipids and glycogen. All these together form the yolk or vitelline. Such fully grown oogonium is termed primary oocyte, which is diploid.

3. **Maturation phase:** The primary oocyte meiotic division. The first division is a reduction division and by which the chromosomal number is reduced to a haploid condition.

The first of the two divisions which are reduction divisions produces two unequal haploid cells. The larger cell is termed secondary oocyte and the smaller one is termed polar body or polocyte. The second equatorial division divides the secondary oocyte again into two unequal cells – a larger ootid and a smaller polocyte. Thus at the end of the maturation division, a single large ootid and three polocytes are formed. As the ootid develops into large ovum, the polocytes degenerate.

Ultra Structure of Ovum

A mature ovum is spherical and several limes larger than the sperm. It is bound by a plasma membrane which is termed oolemma. It is filled with cytoplams which is termed as ooplasm. A spherical haploid nucleus is present in the centre. Ooplasm contains yolk granules which form the deutoplasm. The oolemma is covered by a non living membrane which is secreted by the ovary. It is the primary egg membrane. This is called zona pellucida.

Fertilization

Fertilization can be defined as a process in which a haploid sperm fuses with a haploid ovum resulting a diploid zygote.

In human beings fertilization ocurs inside the body of the female. Hence it is known as internal fertilization.

In vitro Fertilization – Embryo Transfer (IVF – ET)

Test tube baby: In vitro fertilization is particularly useful for women with damaged Fallopian tubes which cannot be repaired. This method can also be employed in cases where there is low sperm count or abnormal sperm in males. If the female is unable to produce eggs, donor eggs can be used for this purpose.

This method requires highly training medical experts and paramedical staff. It needs hospital facility and a sophisticated laboratory facility which is expensive.

IVF method involves the stimulation of the ovary with fertility drugs to produce several eggs, collecting the mature eggs, fertilizing them in the laboratory (in vitro means in glass, *i.e.*, in glassware such as test tubes). The fertilized eggs are now known as pre-embryos. They are cultured and finally transferred to the uterus.

Treatment involves the following steps:

(*i*) The hormonal profile of the woman is assessed.

(*ii*) An exogenous hormone preparation or a fertility drug containing follicular stimulating hormone (FSH) is administered at the appropriate time. This triggers super ovulation *i.e.*, several follicles grow and the ova undergo maturation at the same time.

(*iii*) A dose of gonadotrophin is administered for ovulation.

(*iv*) A laproscope (a short telescope as long as a pencil) is introduced into the abdominal cavity through a slit. The abdominal cavity is filled with carbon dioxide gas to distend it and separte the body wall from the gut. The laproscope is illuminated thus enabling the surgeon to view the interior of the abdomen.

(*v*) With the help of an hollow needle, the mature follicles are punctured and the eggs are collected.

(*vi*) A special cutlure medium is taken in a petri dish in which the collected eggs are placed. They are kept at an incubator at body temperature.

(*vii*) Sperms from the husband are collected and washed in a culture fluid to remove the seminal fluid. A number of healthy sperms are added to each egg. The eggs are thus fertilized.

(*viii*) Then the well-developed pro-embryo is transferred along with the help a fine plastic tube. The remaining pre-embryos can be frozen for future use. The transplanted pre-embryos get implanted in the wall of the uterus and starts development. A pregnancy test is conducted about 14 days after implantation.

The success rate of this entire procedure is about 10%.

Part B
Biochemistry

UNIT

1

Basic Concepts

TYPES OF CHEMICAL REACTIONS

You come across various chemical reactions in your daily life. Iron rusts if exposed to humid atmosphere. A freshly cut apple exposed to atmosphere becomes brown after sometime. During summer, milk left at room temperature for a few hours becomes sour and is converted into a curdy thick mass. If dilute sulphuric acid (H_2SO_4), is added to barium chloride solution, a thick white precipitate of barium sulphate is obtained. Some chemical reactions you come across in the laboratory are slow and some are fast. Some chemical reactions are very fast. For example, when HCl is added to NaOH solution, the reaction occurs immediately.

Slow and Fast Chemical Reactions

If a clean magnesium wire is heated in a Bunsen flame, it catches fire and is burnt down with a flash in a few seconds. Here, magnesium is oxidised in air.

$$2\ Mg\ (s) + O_2\ (g) \rightarrow 2\ MgO\ (s)$$

Contrary to this, iron rusting, which is also a process of oxidation, occurs in several hours or even days. (Rust is chemically represented as $Fe_2O_3 \times H_2O$.) The reaction between an acid and a base [such as $HCl\ (aq) + NaOH\ (aq) \rightarrow NaCl\ (aq) + H_2O\ (l)$] occurs in micro seconds. The reactions that lead to the contraction of muscles and that transmit nervous impulses, are still more rapid contrary to this, weathering of rocks takes millions of years time.

We find in certain cases, rate of conversion of reactants into product or products is fast, and in certain cases, slow. Sometimes we increase or decrease the rate of a reaction to suit our requirements. For example, the pressure cooker is adopted to minimise the time required to cook food, and the food is preserved in a refrigerator to prevent spoilage.

The chemical reactions, rates are very important in industries. Industrialists and chemical engineers want to obtain the end product or byproduct rapidly, cheaply and easily, to the possible extent. Time and money are important in industry. It is always necessary to accelerate chemical reactions so that they are economically feasible.

It is also discussed in this chapter that temperature, pressure, concentration, and catalyst, affect the rates of reactions. At normal temperature and pressure, including in absence of a catalyst, ammonia cannot be obtained from nitrogen and hydrogen. It is observed that reasonable chemical reaction rate results when the process is carried out at 250 atmospheric pressure and 723 K in presence of a catalyst.

$$N_2(g) + 3H_2(g) \xrightarrow{\text{catalyst }(250\text{ atm. }723\text{ K})} 2NH_3(g)$$

Biological reactions in human body also depend on the presence of catalysts. Every chemical reaction in human body is controlled by one or more catalysts. Catalysts are involved in simple reactions like hydrolysis of starch to glucose. These biological catalysts are called enzymes. These enzymes are basically proteins.

Rate of Chemical Reaction

During a chemical reaction, reactants are converted into products. The rate of a chemical reaction expresses us how fast the reaction is occurring by indicating how much of a product is formed in a given time.

Therefore,

$$\text{Reaction rate} = \frac{\text{Change in concentration of a substance (reactant or product)}}{\text{Time consumed}}$$

Thus one can define reaction rate as the rate of change of concentration of a particular reactant or product.

If we express concentration in terms of molarity M, and time in seconds, we can express rate in unit of M/s or (mol L^{-1}) s or mol $L^{-1}s^{-1}$. Normally, concentration is expressed in moles per litre in aqueous reactions.

Example: When acidified hydrogen peroxide (H_2O_2) is added to a solution of potassium iodide (KI), iodine is liberated

$$H_2O_2\ (aq) + 2KI\ (aq) + H_2SO_4\ (aq) \rightarrow 2H_2O\ (l) + I_2\ (aq) + K_2SO_4\ (aq)$$

Here, concentration of iodine is zero initially. With the passage of time, it increases and the reaction solution becomes brownish. Concentration of iodine can be measured at different intervals of time by titration against sodium thiosulphate. If concentration of iodine rises from 0 to 10^{-5} mol L^{-1} in 10 seconds, we can write

$$\text{Reaction rate} = \frac{\Delta(I_2)}{\Delta t} = \frac{10^{-5}\,\text{mol L}^{-1}}{10^{s}} = 10^{-6}\,\text{mol}\,\text{L}^{-1}\text{s}^{-1}$$

Here symbol Δ (delta) represents a change and (I_2), molar concentration of iodine.

This change in concentration of the product (iodine) takes place in ten seconds, therefore, it can be said that the rate is the average reaction rate.

Rate of a reaction depends on the concentration of the reactants. With laps of time, concentration of reactants decreases and therefore, rate of the reactants also decreases.

Energy changes during a chemical reaction.

Most of the chemical reactions are accompanied by energy changes. In some reactions, energy is given in form of heat and in some cases, in form of light. The chemical reactions which release heat are exothermic reactions. All combustion reactions are exothermic reactions. Methane on combustion releases heat energy.

$$CH_4(g) + 2O_2(g) \rightarrow CO_2(g) + 2H_2O(l) + \text{Energy }(890.4\text{ kJ mol}^{-1})$$

One striking example of an exothermic reaction is 'thermite reaction'. Here Fe_2O_3 is reduced to iron. This reaction is being used to weld iron rails.

$$2Al(s) + Fe_2O_3(s) \rightarrow Al_2O_3(s) + 2Fe(l)$$

There is another category of reactions which absorbs heat. Such reactions are called 'endothermic reactions'. When barium hydroxide, Ba $(OH)_2$ and H_2O and ammonium thiocyanate, NH_4SCN, are mixed together in a beaker or boiling tube. This reaction mixture becomes so cold that moisture from the air forms a layer of frost on outer surface of the beaker. This reaction is expressed as

$$Ba(OH)_2 \cdot 8H_2O(s) + 2NH_4SCN(s) \rightarrow Ba(SCN)_2(aq) + 2N_3(aq) + 10H_2O(l)$$

There are certain reactions which occur in presence of light. Such reactions are photochemical reactions. Photosynthesis by plants is a photochemical reaction.

Photographic films which are normally coated with silver bromide (and very small amount of silver iodide) undergo chemical reaction when exposed to sunlight. This basic reaction is used in photography.

Reversible and Irreversible Reactions

Normally, we write most of the chemical reactions with the assumption that they get completed; are indicated them by using single arrow (→). For example, the reaction between N_2 and H_2 shall be written as

$$N_2(g) + 3H_2(g) \rightarrow 2NH_3(g)$$

pH. The pH of a solution is defined as the negative logarithm of the hydrogen ion concentration.

Physiological Importance of Maintenance of pH

A large number of living cells are extremely sensitive to changes in the pH of their extra cellular fluids. For the extra cellular fluid an useful requirement is that it would not be either too acidic or too alkaline (basic). The pH of human blood plasma is usually maintained between 7.36 and 7.45. If these limits change, *i.e.*, above pH 7.8 or below 6.8, fatal illness and perhaps death would ensue. Luckily, human blood plasma equalises this limit with substances called buffers which act to prevent any sudden shift in the pH.

Buffers

Buffers are usually a mixture of either weak acids, with their salts of strong bases or strong acids with their salts of weak bases. The buffer system of the body has the capability of resisting the change in pH by accepting either the (H^+) or (OH^-) ions. Buffering action is very important in the biological system in maintenance of normal acid-base balance of the body. It is most important in the action of enzymes, both in vitro and invitro.

Determination of pH of a Buffer Solution

Henderson-Hasselbalch equation can be used for determination of pH of a buffer solution or for achieving the normal pH.

The pH of the buffer solution prepared by mixing 35 ml of (N/10) acetic acid with 15 ml of (N/10) NaOH can be determined by the Henderson-Hasselbalch equation:

pH = pKa + log (salt/acid), where pKa is log (1/Ka). Ka is the dissociation constant of the acid.

$$HA \rightleftharpoons (H^+) + (A^-)$$

By law of Mass action $\frac{(H^+)(A^-)}{(HA)} = Ka$ or $(H^+) = \frac{Ka(HA)}{(A^-)}$

So $$\log(H^+) = \log Ka + \log(HA) - \log(A^-)$$

or $$-\log(H^+) = -\log Ka - \log(HA) + \log(A^-).$$

According to Sorensen –log (H^+) = pH, by similar reasoning

$$-\log Ka = pKa$$

Therefore, $$pH = pKa + \log\frac{(A^-)}{HA}.$$

When these two solutions are mixed up 15 ml of (N/10) NaOH will neutralise 15 ml of (N/10) acetic acid so as to form 15 ml of the salt, Na-acetate. So 20 ml of acetic acid will remain unneutralised.

So in the buffer solution, the salt and unneutralised acid ratio will be (15/20). The dissociation constant (Ka) of acetic acid is (1.86×10^{-5}) and so the pKa will be log (1/Ka), that is 4.73.

So in Henderson-Hasselbalch equation,

$$\begin{aligned} pH &= 4.73 + \log(15/20) \\ &= 4.73 + \log 0.75 \\ &= 4.73 - 1.8751 \\ &= 4.6 \end{aligned}$$

If the ratio of salt and unneutralised acid ratio becomes 1:1, then the pH value of the buffer mixture will be equal to pKa of the acid.

As for example, if 30 ml of (N/10) acetic acid and 15 ml of (N/10) NaOH are mixed up then 15 ml of salt —Na-acetate will be formed and 15 ml of acetic acid will remain as unneutralised acid. So the salt and acid ratio will be (15/15) = 1. So interpolating the value in the Henderson-Hasselbalch equation the result will be

$$\begin{aligned} pH &= pKa + \log(\text{salt/acid}) \\ &= 4.73 + \log(15/15) \\ &= 4.73 + \log 1 \\ &= 4.73 + 0 \\ &= 4.73. \end{aligned}$$

So the buffering action in this case is at the minimum and can react as acid or as base.

Concentration of Solution

The following are units of concentration of solutions:

1. *Percentage*: Weight in grams of solute per 100 grams of solution is per cent; by weight and the same amount dissolved in 100 ml is per cent; by volume.
2. *Molar solution*: Molecular weight in grams of a solute dissolved in 1000 ml is a molar solution. For example, molecular weight of H_2SO_4 is 98.016, so a molar solution is 98.016 grams per 1000 ml.

   Molal solution. Molecular weight in gram of a solute dissolved in 1000 grams of solvent.

A mol (M) or gram-molecular weight is the amount of substance in grams equal to the molecular weight of the substance.

One millimol (mM) is equal to one thousandth of a mol (M) and 1 micromol (μM), to one-thousandth of a millimol.

3. *Normal solution*: A normal solution is one which contains per litre, the amount equal to gram-molecular weight of the substance divided by equivalent weight or number of replaceable hydrogen atoms. For example, molecular weight of H_2SO_4 is 98.016 where $H^+ + H^+$, and SO_4^{--} ions are present. As number of replaceable hydrogen atoms is 2, *i.e.*, 49.008 grams in a litre.
4. Gram equivalent is the molecular weight in grams divided by the number of valence of ions present. One milliequivalent (mEq) is one-thousandth of such gram equivalent.

Ions, Electrolytes and Non-electrolytes

1. *Ions*: On an electric current being passed through solution of a compound in water, some elements migrate at the positive pole or anode and some at the negative pole or cathode. The charged particles are called ions. Negative ions (anions) move towards the anode and positive ions (cations) to cathode due to attraction of unlike charges.
2. *Electrolytes*: Electrolytes are compounds which can be dissociated into anions and cations on an electric current being passed; for example, acid, bases, salts, etc.
3. *Non-electrolytes*: Compounds which cannot be dissociated are called non-electrolytes.

STEREO-CHEMISTRY OF CARBON COMPOUNDS

Constitutional Isomers

Introduction

The term **isomers** refers to those compounds that have the same molecular form (*i.e.*, they have the same atoms); but differ in the arrangement of atoms. The three types of isomers are: *constitutional isomers, configurational isomers* and *conformational isomers.* **Constitutional isomers are** those isomers in which the atoms are linked together in a different skeletal framework and are different compounds. **Configurational isomers** have structures having the same atoms and bonds, but have different geometrical shapes that cannot be interconverted without breaking covalent bonds. Configurational isomers can be separated and are different compounds with different properties. **Conformational isomers** are different shapes of the same molecule and cannot be separated.

Definition

Constitutional isomers are those compounds that have the same molecular formula but have the atoms joined together in a different way. Thus, they have different carbon skeletons. Constitutional isomers differ in physical and chemical properties.

Alkanes

Alkanes having a particular molecular formula can exist as different constitutional isomers. For example, the alkane having the molecular formula C_4H_{10} can exist as two constitutional isomers—the straight chain alkane (butane) or the branched alkane (2-methylpropane; Fig. 1.1). These are different compounds with different physical and chemical properties.

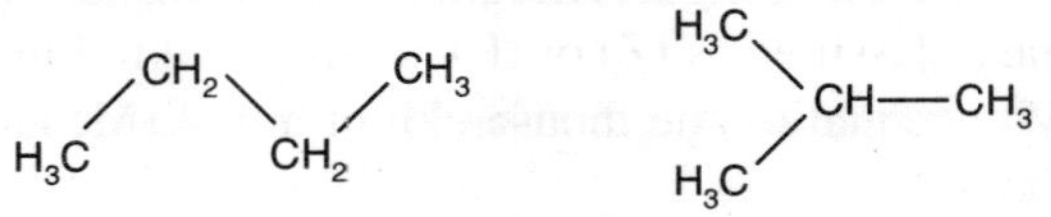

(a) (b)

Fig. 1.1 (a) Butane (C_4H_{10}); (b) 2-methylpropane (C_4H_{10})

CONFIGURATIONAL ISOMERS — ALKENES AND CYCLOALKANES

Definition

Configurational isomers are those isomers in which we find the same molecular formula and the same molecular structure. Thus, they have the same atoms and the same bonds. Such isomers are different because some of the atoms are arranged differently in space, and the isomers cannot be interconverted without breaking and remaking covalent bonds. Thus, configurational isomers are different compounds having different properties. Some examples of configurational isomers are *substituted alkenes and substituted cycloalkanes* in which the substitutents are arranged differently with respect to each other.

Alkenes—*cis* and *trans* Isomerism

Alkenes that have identical substituents at either end of the double bond can only exist as one molecule. However, alkenes with different substituents at each end of the double bond can exist as two possible isomers. For example, 1-butene (Fig. 1.2*a*) has two hydrogens at one end of the double bond and there is only one way of constructing it.

On the other hand, 2-butene has different substituents at both ends of the double bond (H and CH_3) and can be constructed in two ways. The methyl groups can be on the same side of the double bond (the *cis* isomer; Fig. 1.2*b*), or on, opposite sides (the *trans* isomer; Fig. 1.2*c*). The *cis* and *trans* isomers of an alkene are **configurational isomers** (also called **geometric** isomers) as they have different shapes and cannot easily interconvert. Since the double bond of an alkene cannot rotate. Therefore, the substituents are 'fixed' in space relative to each other. The structures are different compounds with different chemical and physical properties.

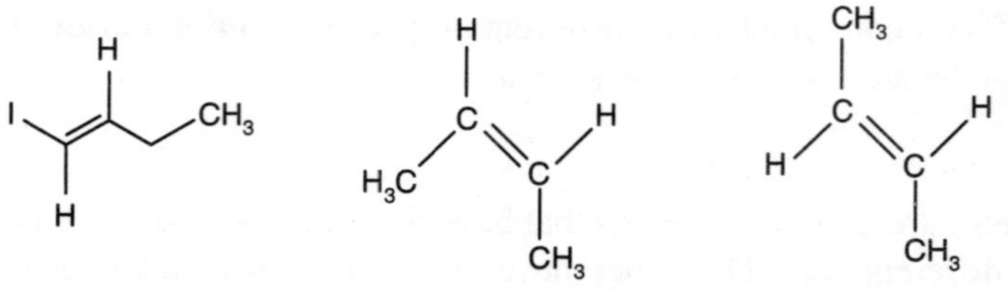

(a) (b) (c)

Fig. 1.2 (a) 1-Butene; (b) *cis*-2-butene; (c) *trans*-2-butene

Alkenes — (Z) and (E) Nomenclature

The *cis* and *trans* nomenclature for alkenes is an old method of classifying the configurational isomers of alkenes and is still in common use. However, it is only suitable for simple 1,2-disubstituted alkenes where we can compare the relative position of the two substituents with respect to each other. When it comes to trisubstituted and tetrasubstituted alkenes, a different nomenclature is needed.

The (Z)/(E) nomenclature permits a clear, unambiguous definition of the configuration of alkenes. The method whereby alkenes are classified as (Z) or (E) is illustrated in Fig. 1.3. First of all, the atoms directly attached to the double bond are identified and given their atomic number (Fig. 1.3*b*). Next we compare the two atoms at each end of the alkene. The one with the highest atomic number takes priority over the other (Fig. 1.3*c*). At the left hand side, oxygen has a higher atomic number than hydrogen and takes priority. At the right hand side, both atoms are the same (carbon) and we cannot choose between them.

Therefore, we now have to identify the atom of highest atomic number attached to each of these identical carbons. These correspond to hydrogen for the methyl substituent and a carbon for the ethyl substituent. These are now different and so a priority can be given (Fig. 1.4*a*). After identifying which groups have priority, we can now see whether the priority groups are on the same side of the double bond or on opposite sides. If the two priority groups are on the same side of the double bond, the alkene is designated as (Z) (from the German word '*zusammen*' meaning *together*). If the two priority groups are on opposite sides of the double bond, the alkene is called as (E) (from the German word 'entgegen' meaning *across*). In this example, the alkene is (E) (Fig. 1.4*b*).

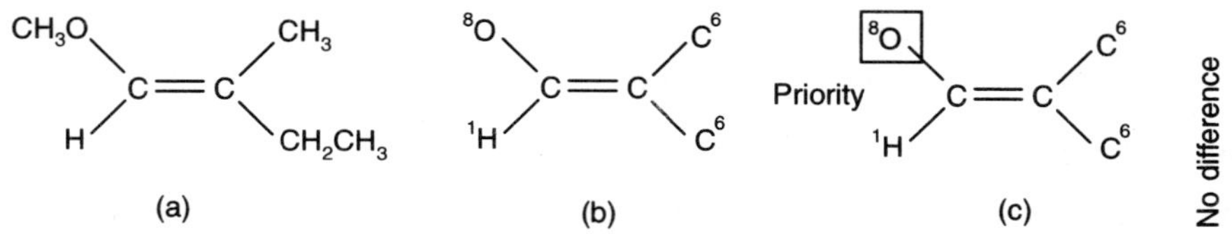

Fig. 1.3 (a) Alkene; (b) atomic numbers; (c) priority groups

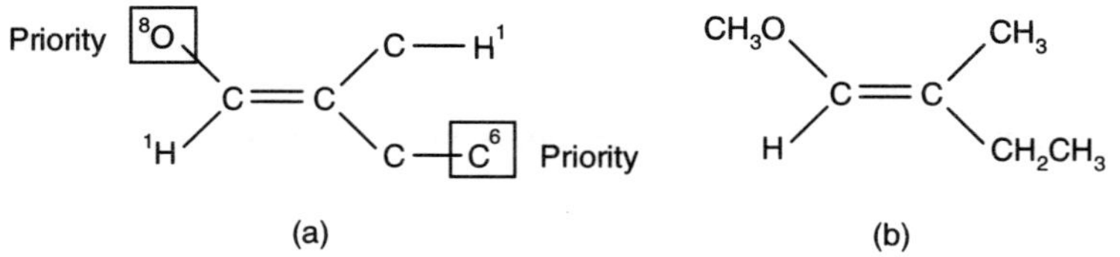

Fig. 1.4 (a) Choosing priority groups, (b) (E)-1-methoxy-2-methyl-1-butene

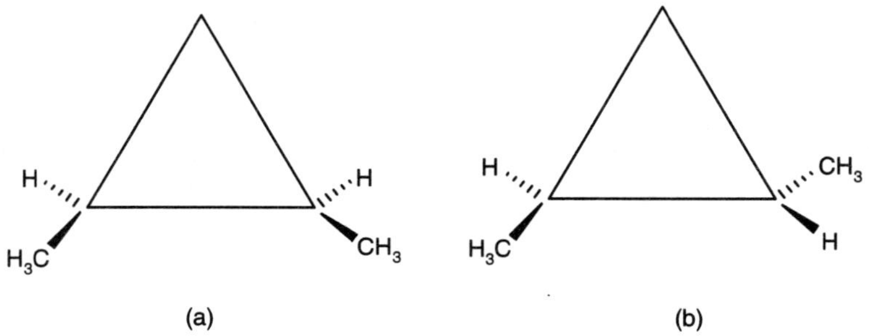

Fig. 1.5 (a) cis-1, 2-dimethylcyclopropane; (b) trans-1, 2-dimethylcyclopropane

Cycloalkanes

Substituted cycloalkanes are also capable of existing as configurational isomers. For example, there are two configurational isomers of 1, 2-dimethylcyclopropane depending on whether the methyl groups are on the same side of the ring or on opposite sides (Fig. 1.5). The relative positions of the methyl groups can be defined by bonds. A solid wedge indicates that the methyl group is coming out the page, whereas a hatched wedge indicates that the methyl group is pointing into the page. If the substituents are on the same side of the ring, the structure is classifed as *cis*. If they are on opposite sides, the structure is classified as *trans*.

CONFIGURATIONAL ISOMERS — OPTICAL ISOMERS

Definition

Optical isomerism is also a type of **configurational isomerism** and is so named because of the ability of optical isomers to rotate plane-polarized light clockwise or counterclockwise. The existence of optical isomers has very important consequences for life, because optical isomers generally have significant *differences in their biological activity.* Apart from their biological activity and their effects on plane-polarized light, optical isomers have identical chemical and physical properties.

Asymmetric Molecules

A molecule like chloroform ($CHCl_3$) is tetrahedral and there is only one way of fitting the atoms together. This is not the case for a molecule like lactic acid. For *lactic acid* there are two ways of constructing a model in such a way that the two structures obtained are nonsuperimposable and cannot be interconverted without breaking covalent bonds. Thus, they represent two different molecules that are configurational isomers. The difference between the two possible molecules is in the way the substituents are attached to the central carbon. This can be represented by the diagrams shown in Fig. 1.6 in which the bond to the hydroxyl group comes out of the page in one isomer but goes into the page in the other isomer.

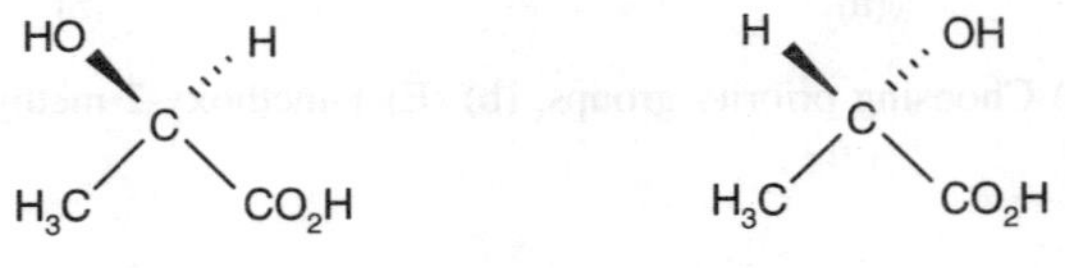

Fig. 1.6 Lactic acid

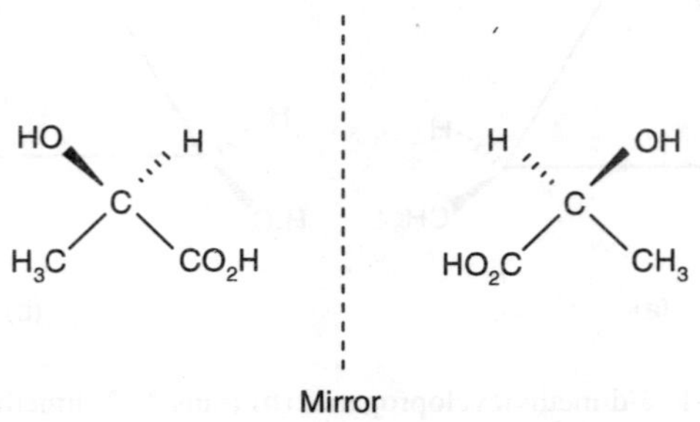

Fig. 1.7 Nonsuperimposable mirror images of lactic acid

The two isomers of lactic acid are *mirror images* (Fig. 1.7). A molecule that exists as two nonsuperimposable mirror images has optical activity if only one of the mirror images is present.

Lactic acid exists as two nonsuperimposable mirror images as it is **asymmetric** *i.e.* it has no element of symmetry. Asymmetric molecules can also be called a **chiral**, and the ability of molecules to exist as two optical isomers is called **chirality**. In fact, a molecule need not have to be totally asymmetric to be **chiral.** *Molecules containing a single axis of symmetry can also be chiral.*

Asymmetric Carbon Centers

A chiral molecule can be identified by what are known as **asymmetric carbon centers.** This is true for most chiral molecules, but it is not foolproof and that there are several cases where it will not work. For example, some chiral molecules have no asymmetric carbon centers, and some molecules having more than one asymmetric carbon center are not chiral.

Generally, a compound will have optical isomers if it has four different substituents attached to a central carbon (Fig. 1.8) in such cases, the mirror images are nonsuperimposable and the structure will exist as two configurational isomers known as **enantiomers**. The carbon center that contains these four different substituents is called a **stereogenic** or an asymmetric center. A solution of each enantiomer or optical isomer can rotate plane-polarized light. One enantiomer will rotate plane-polarized light clockwise while the other (mirror image) will rotate it counterclockwise by the same amount. A mixture of the two isomers (**a racemate**) will not rotate plane polarised light at all. In all other respects, the two isomers are identical in physical and chemical properties and are therefore indistinguishable. The asymmetric centers in the molecules shown (Fig. 1.9) have been identified with an asterisk. The structure lacking the asymmetric center is symmetric or **achiral** and does not have optical isomers. A structure can also have more than one asymmetric centre.

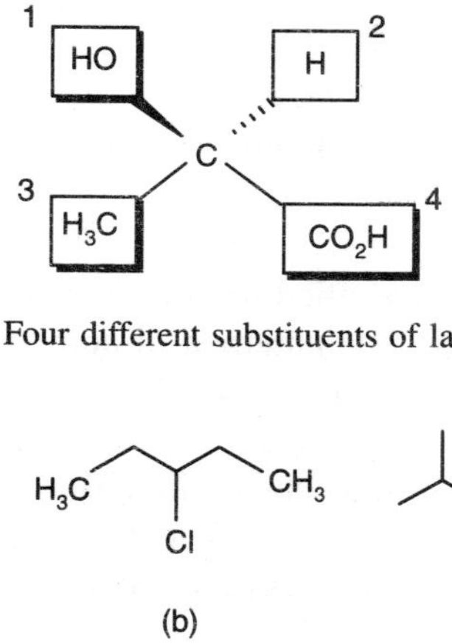

Fig. 1.8 Four different substituents of lactic acid

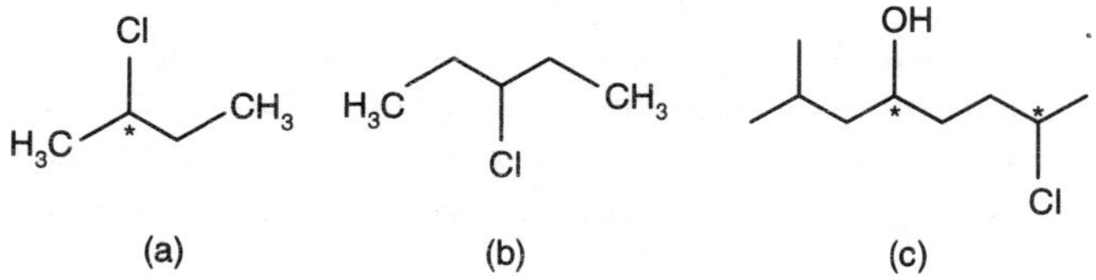

Fig. 1.9 Chiral and achiral structures: (a) chiral; (b) achiral; (c) chiral

Asymmetric centers are only possible on sp^3 carbons. An sp^2 center is planar and cannot be asymmetric. Similarly, an sp center cannot be asymmetric.

Fischer Diagrams

A chiral molecule can be represented by a *Fischer diagram* (Fig. 1.10). The molecule is drawn in such a way that the carbon chain is vertical with the functional group positioned at the top. The vertical

C—C bonds from the asymmetric center point into the page while the horizontal bonds from the asymmetric center come out of the page. This is generally drawn without specifying the wedged and hatched bonds.

The Fischer diagrams of **alanine** permit the structures to be defined as L- or D- from the position of the amino group. If the amino group is to the left, then it is the L-enantiomer. If it is to the right, it is the D-enantiomer. This is an old fashioned nomenclature and is only used for *amino acids* and *sugars.* The L- and D-nomenclature depends on the *absolute configuration* at the asymmetric center and not the direction in which the enantiomer rotates plane-polarized light. It is not possible to predict which way a molecule will rotate plane-polarized light and this can only be found out by experimentation.

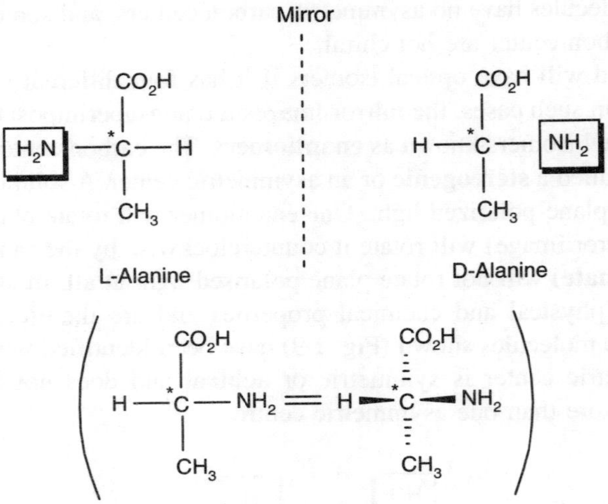

Fig. 1.10 Fischer diagrams or L-alanine and D-alanine

(R) and (S) Nomenclature

The structure of an enantiomer can be specified by the (R) and (S) nomenclature, determined by the **Cahn-Ingold-Prelog** rules. In this (Fig. 1.11) method of nomenclature, first of all, the atoms directly attached to the asymmetric center and their atomic numbers are identified. Then the attached atoms are assigned a priority based on their atomic numbers. In this example (Fig. 1.11) there are two carbon atoms with the same atomic numbers so they cannot be given a priority. In such a case the next atom of highest atomic number is considered (Fig. 1.12*a*). This means moving to an oxygen for one of the carbons and to a hydrogen for the other. The oxygen has the higher priority and so this substituents takes priority over the other.

Once the priorities have been decided, the structure is redrawn in such a way that the group of lowest priority is positioned 'behind the page'. In this example (Fig. 1.12*b*), the group of lowest priority (the hydrogen) is already positioned behind the page (note that hatched bond indicating the bond going-away from you). An arc is now drawn connecting the remaining groups, starting from the group of highest priority and finishing at the group of third priority. If the arc is drawn clockwise, the assignment is (*R*). (*rectus*). If the arc is drawn counterclockwise; the assignment is *(S) (sinister).* In this example, the arrow is drawn clockwise. Therefore, the molecule is (*R*)-lactic acid.

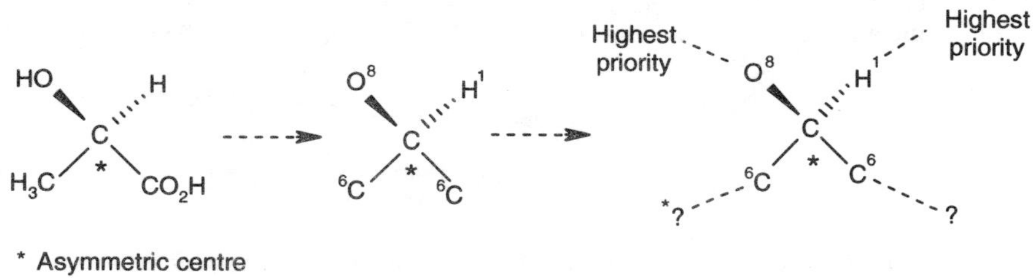

Fig. 1.11 Assigning priorities to substituents of an asymmetric center

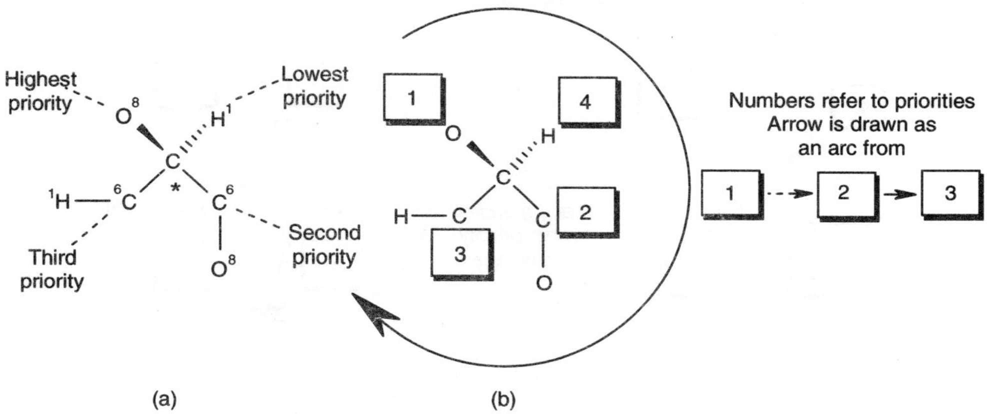

Fig. 1.12 (a) Assigning priorities to substitutes; (b) assigning an asymmetric center (R) or (S)

Another example (Fig. 1.13) that illustrates another rule involving substituents with double bonds. The asymmetric center is marked with an asterisk. The atoms directly attached to the asymmetric center are shown on the right with their atomic numbers. Now, it is possible to define the group of highest priority (the oxygen) and the group of lowest priority (the hydrogen). There are two identical carbons attached to the asymmetric center so move to the next stage and identify the atom with the highest atomic number joined to each of the identical carbons (Fig. 1.14). This still does not distinguish between the CHO and CH_2OH groups since both carbon atoms have an oxygen atom attached, next look at the second most important atom attached to the two carbon atoms. However, if there is a double bond present, you are allowed to 'visit' the same atom twice. The next most important atom in the CH_2OH group is the hydrogen. In the CHO group, the oxygen can be 'revisited' since there is a double bond.

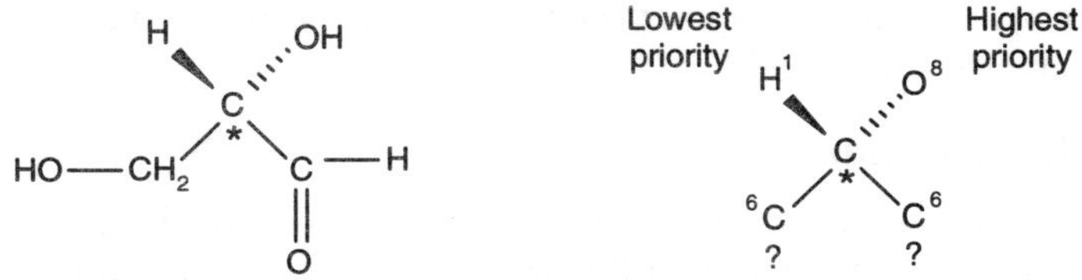

Fig. 1.13 Assigning priority to substituents of an asymmetric center

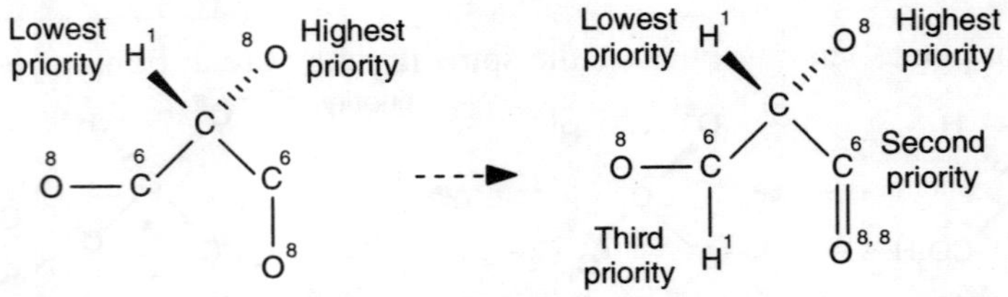

Fig. 1.14 Assigning priority to substituents of an asymmetric center

Therefore, this group takes priority over the CH_2OH group. The priorities have been determined, and thus the group of lowest priority is placed behind the page and three most important groups are connected to see if they are clockwise or counterclockwise (Fig. 1.15).

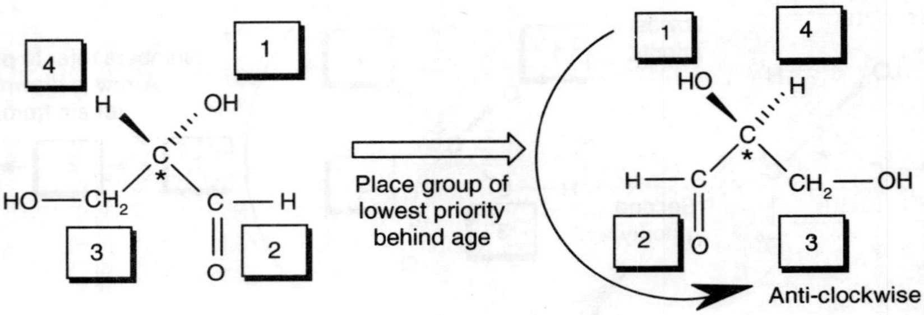

Fig. 1.15 Assigning an asymmetric center as (R) or (S)

(+) and (–)

The assignment of an asymmetric center as *(R)* or *(S)* has nothing to do with whichever direction the molecule rotates plane-polarized light. Optical rotation can only be determined experimentally. By convention, molecules which rotate plane-polarized light clockwise are written as (+) or *d.* Molecules which rotate plane-polarized light counterclockwise are written as (–) or *l.* The *(R)* enantiomer of lactic acid is found to rotate plane-polarized light counterclockwise and so this molecule is defined as *(R)*-(–)-lactic acid.

Optical Purity

The optical purity is a measure of enantiomeric purity of a compound and is given in terms of its **enantiomeric excess (ee)**. A pure enantiomer would have an optical purity and enantiomeric excess of 100%. A fully racemized compound would have an optical purity of 0%. If the enantiomeric excess is 90%, it signifies that 90% of the sample is pure enantiomer and the remaining 10% is a racemate containing equal amounts of each enantiomer (*i.e.* 5% + 5%). Therefore the ratio of enantiomers in a sample having 90% optical purity is 95.5.

Allenes and Spiro Compounds

All chiral molecules do not have asymmetric centers. For example, some substituted allenes and spiro structures have no symmetric center but are still chiral (Fig. 1.16). The substituents at either end of the allene are in different planes, and the rings in the spiro structure are at right angles to each other. The mirror images of the allene and the spiro structures are nonsuperimposable and are enantiomers.

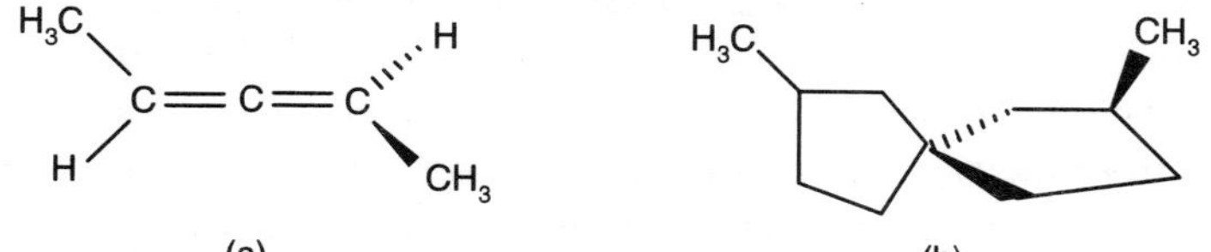

Fig. 1.16 (a) Alkene; (b) spiro structure

The rule for determining whether a molecule is chiral or not is to study the symmetry of the molecule. A molecule will be chiral if it is asymmetric (*i.e.* has no elements of symmetry) or if it has no more than one axis of symmetry.

Meso Structures

The molecule in (Fig. 1.17*a*) which there are two identical asymmetric centers but which is not chiral. The mirror images of such a structure are superimposable and so the compound cannot be chiral. This is because the molecule contains a plane of symmetry as shown in Fig. 1.17 in which the molecule has been rotated around the central C—C bond. A structure like this is known as a ***meso*** structure.

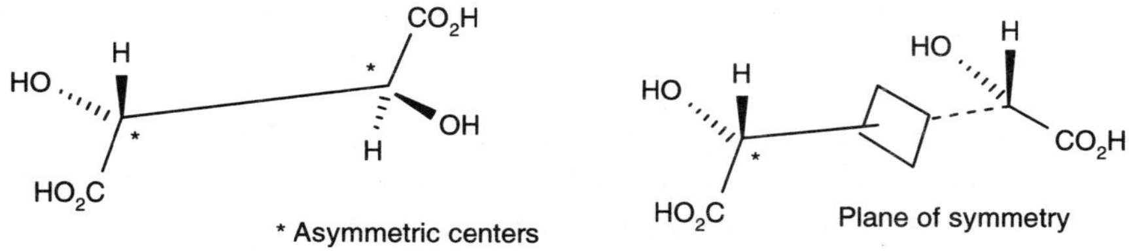

Fig. 1.17 (a) Meso structure showing asymmetric centers; (b) plane of symmetry in meso structure

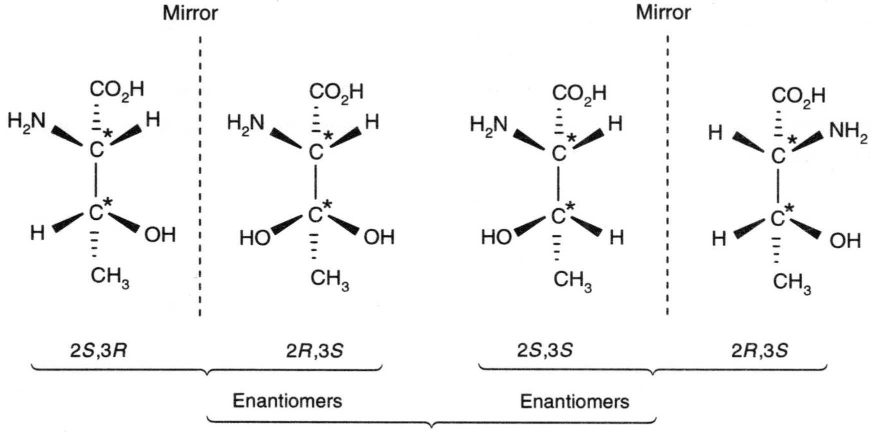

Fig. 1.18 Stereoisomers of threonine

Diastereomers

If a molecule has two or more asymmetric centers then there are various possible structures that are possible and we use terms like **stereoisomers, diastereomers** and **enantiomers** to discuss them. To illustrate the relative meaning of these terms, let us look at the possible structures of the amino acid *threonine* (Fig. 1.18). This molecule has two asymmetric centers. Thus, four different structures are possible which arises from the two different possible configurations at each center. These are shown with the asymmetric centers at positions 2 and 3 defined as (*R*) or (*S*). The four different structures are called as stereoisomers. The (2*S*, 3*R*) stereoisomer is a *nonsuperimposable mirror image* of the (2*R*, 3*S*) stereoisomer and so these structures are **enantiomers** having the same chemical and physical properties. The (2*S*, 3*R*) stereoisomer is the *nonsuperimposable mirror image* of the (2*R*, 3*S*) stereoisomer and so these structures are also **enantiomers** having the same **chemical** and physical properties. *Each set of enantiomers is called a diastereomer.* **Diastereomer** are not mirror image of each other and are completely different compounds with different physical and chemical properties. Thus, threonine has two asymmetric centers *i.e.* there are two possible diastereomers consisting of two enantiomers each, making a total of four stereoisomers. As the number of asymmetric centers increases, the number of possible stereoisomers and diastereomers increases. For a molecule having *n* asymmetric centers, *the number of possible stereoisomers is* 2^n and the number of diastereomers is 2^{n-1}.

CONFORMATIONAL ISOMERS

Definition

Conformational isomers are different shapes of the same molecule and it results from *rotation round* C–C *single bonds.* As rotation round a single bond can occur easily at room temperature. Conformational isomers are freely interconvertable and they cannot be separated.

Alkanes

Conformational isomers of alkanes arise from the rotation of C—C single bonds. Different shapes of alkanes can be adopted by a molecule like ethane by rotation around the C—C bond. The most distinctive ones (Fig. 1.19) are I and II are known as 'staggered' and 'eclipsed' respectively. In conformation I, the C—H bonds on carbon 1 are staggered with respect to the C—H bonds on carbon 2. In conformation II, they are eclipsed. **Newman projections** (Fig. 1.20) shows the view along the C_1—C_2 bond. In it carbon 1 is represented by the small black circle and carbon 2 is represented by the larger sphere. Viewed in this manner we can see that the C—H bonds on carbon 1 are eclipsed with the C—H bonds on carbon 2 in conformation II.

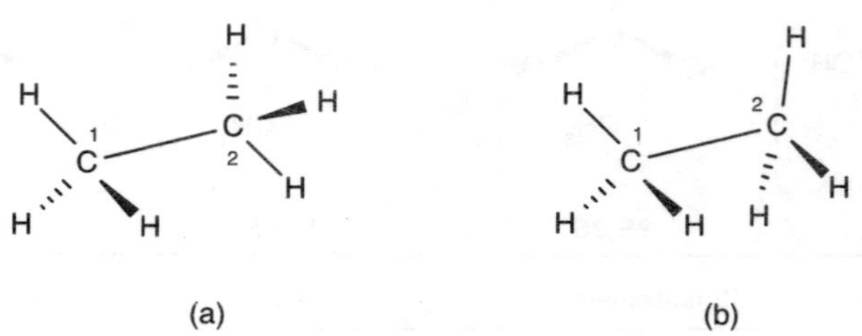

Fig. 1.19 (a) 'Staggered' conformation of ethane; (b) 'eclipsed' conformation of ethane

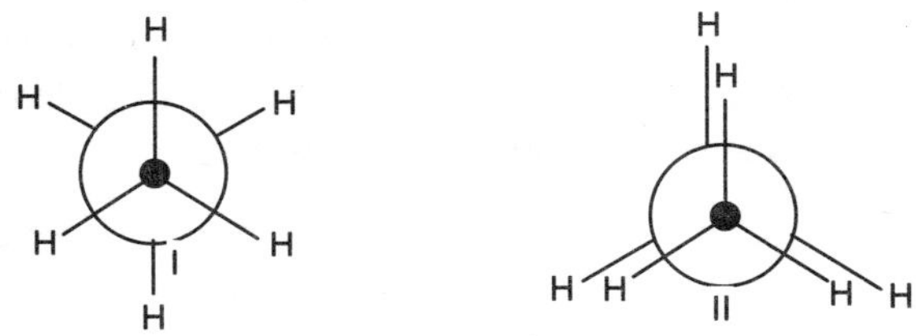

Fig. 1.20 Newman projections of the staggered (I) and eclipsed (II) conformations of ethane

Out of the two conformations, the *staggered conformation is the more stable* as the C—H bonds and hydrogen atoms are as far apart from each other as possible. In the eclipsed conformation, both the bonds and the atoms are closer together and it can cause strain due to electron repulsion between the eclipsed bonds and between the eclipsed atoms. Therefore, majority of the ethane molecules are in the staggered conformation at any point of time. Since the energy difference between the staggered and eclipsed conformations is small and so it allows each ethane molecule to pass through an eclipsed conformation (Fig. 1.21).

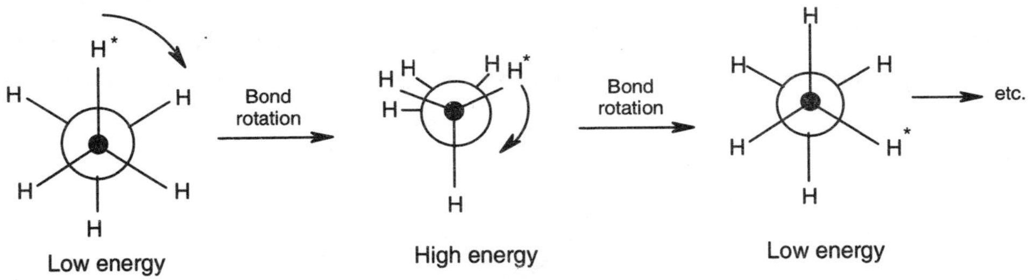

Fig. 1.21 Bond rotation of ethane

Ethane has only one type of staggered conformation, but different staggered conformations are possible with larger molecules like butane (Fig. 1.22). The first and the third C—C bonds in isomer are at an angle 60° with respect to each other when viewed along the middle C—C bond. In isomer II, these bonds are at an angle of 180°. This angle is called as the **torsional angle** or **dihedral angle**. Isomer II is more stable than isomer I because the methyl groups and the C—C bonds in this **conformer** are as far apart from each other as possible. The methyl groups are bulky and in conformation I they are close

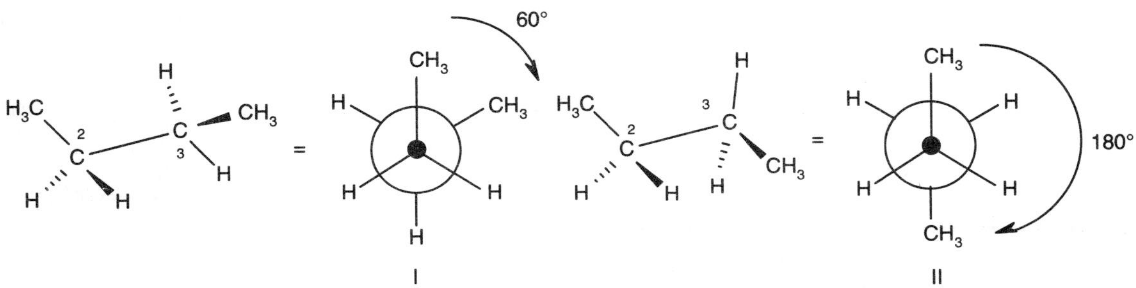

Fig. 1.22 Gauche conformation (I) and anti conformation (II) or butane

enough to interact with each other and lead to some strain. There is also an interaction between the C—C bonds in isomer I since a torsional angle of 60° is small enough for some electronic repulsion to exist between the C—C bonds. When C—C bonds have a torsional angle of 60°, the steric and electronic repulsions which arise are called as a **gauche interaction**.

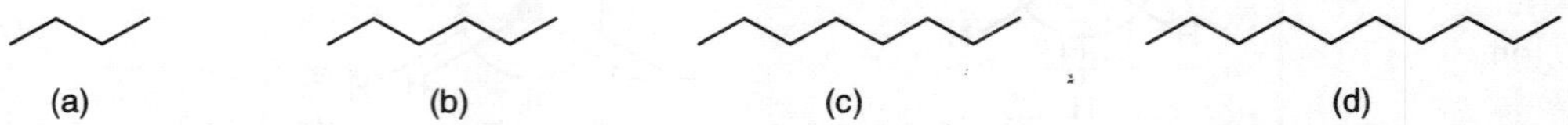

Fig. 1.23 Zigzag conformations of (a) butane; (b) hexane; (c) octane, (d) decane

Because of this, the most stable conformation for butane is that in which the C—C bonds are at torsional angles of 180° that results in 'zigzag' shape. In this conformation, the carbon atoms and C—C bonds are as far apart from each other as possible. The most stable conformations for longer chain hydrocarbons will also be zigzag (Fig. 1.23). However, since bond rotation is taking place all the time for all the C—C bonds, it is unlikely that many molecules will be in a perfect zigzag shape at any specific time.

Cycloalkanes

Cyclopropane (Fig. 1.24) is a flat molecule in respect of C-atoms, with the hydrogen atoms situated above and below the plane of the ring, so it has no conformational isomers. Cyclobutane (Fig. 1.25) can form three distinct shapes — *a planar shape* and *two 'butterfly'* shapes. Cyclopentane (Fig. 1.26) can

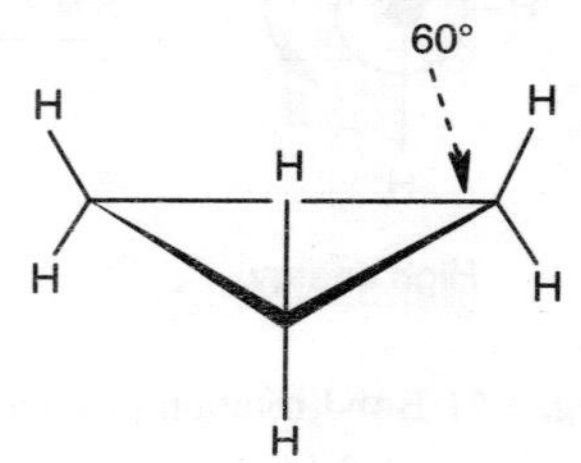

Fig. 1.24 Cyclopropane

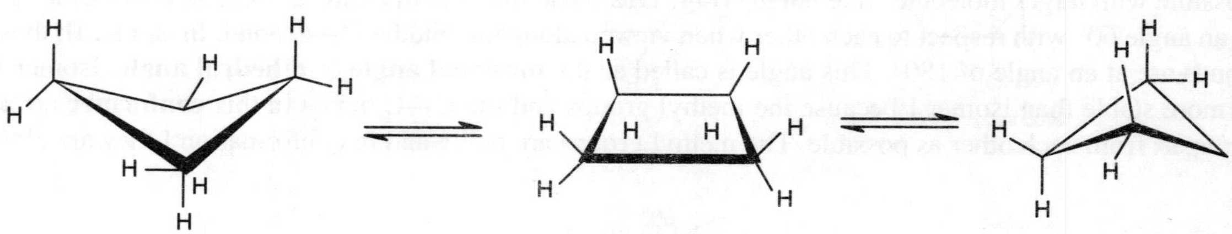

Fig. 1.25 Cyclobutane

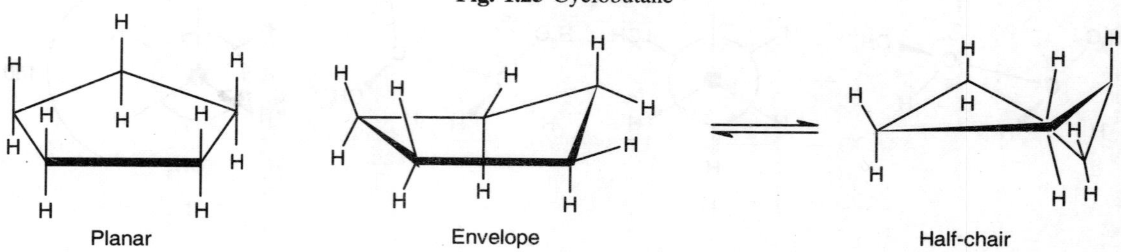

Fig. 1.26 Cyclopentane

also form a number of shapes for conformations. The planar structures for cyclobutane cyclopentane are too strained to exist in practice because of eclipse C—H bonds.

The two main conformational shapes for cyclohexane are can be the chair and the boat (Fig. 1.27). This chair form is more stable than the boat form since the latter has eclipsed C—C and C—H bonds can be observed in the Newman projections (Fig. 1.28) that has been drawn in such a way that we are looking along two bonds at same time–bonds 2-3 and 6-5. In the chair conformation, there eclipsed C—C bonds. But, in the boat conformation, bond 1-2 eclipse with bond 3-4, and bond 1-6 is eclipsed with bond 5-4. This indicates that the boat conformation is less stable than the chair conformation and the majority of cyclohexane molecules exist in chair conformation. However, the energy barrier is small and the cyclohexane molecules can pass through the boat conformation in a process known as **'ring flipping'** (Fig. 1.29). The ability of a cyclohexane molecule to ring-flip is important when substituents are present. Each carbon atom in the chair form has two C—H bonds, but these are not identical (Fig. 1.30). One of these bonds is called **equatorial** as it is roughly in the plane of the ring. The other C—H bond is vertical to the plane of the ring and is known as the **axial** bond.

When ring flipping occurs from one chair to another all the axial bonds become equatorial bonds and all the equatorial bonds become axial bonds. This does not matter for cyclohexane itself, but it becomes

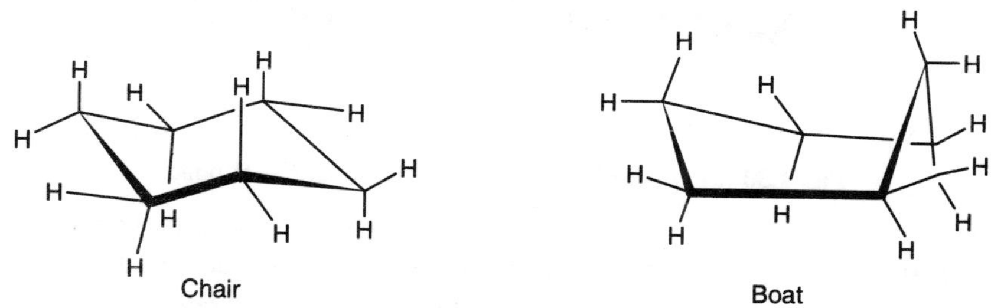

Fig. 1.27 Cyclohexane

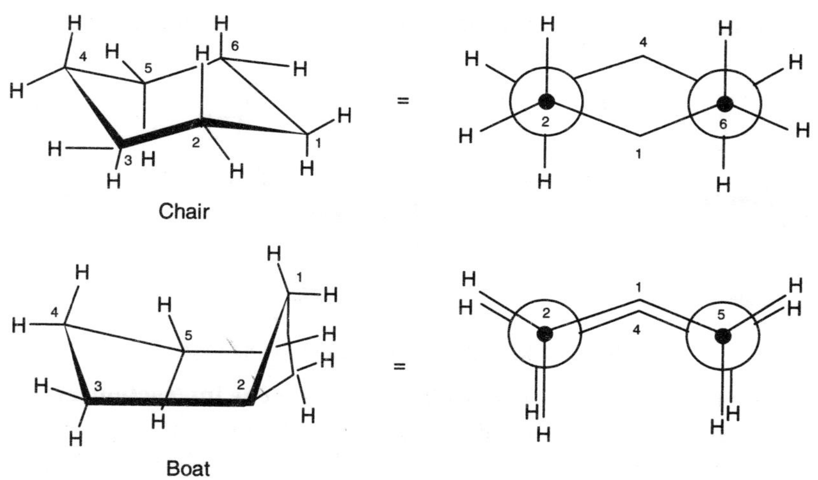

Fig. 1.28 Newman projections of the chair and boat conformations of cyclohexane

important when there is a substituent present in the ring. For example, methylcyclohexane can have two chair structures when the methyl group is either on an equatorial bond or an axial bond (Fig. 1.31).

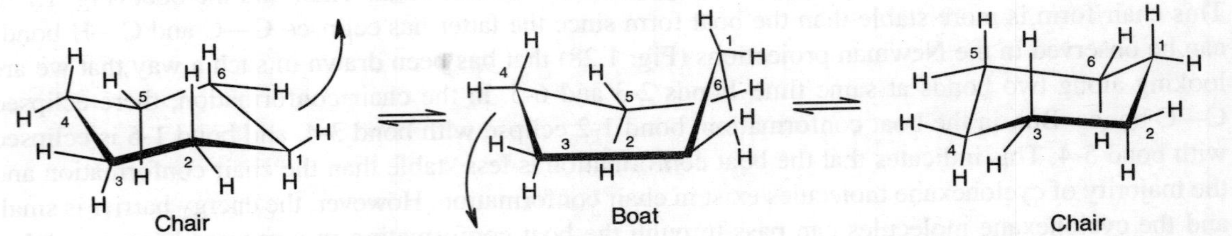

Fig. 1.29 Ring flipping of cyclohexane

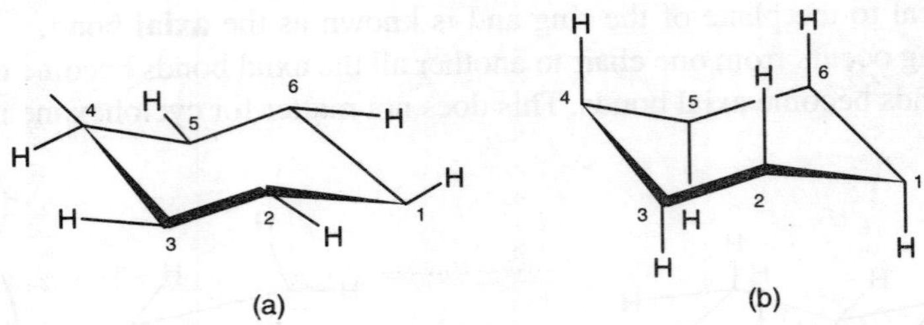

Fig. 1.30 (a) Equatorial C—H bonds; (b) axial C—H bonds

These are different orientation of the same molecule that are interconvertable because of rotation of C—C single bonds (the ring flipping process). The two chair structures are conformational isomers but they are not of equal stability. *The more stable conformation is the one where the methyl group is in the equatorial position.* In this position the C—C bond connecting the methyl group to the ring has a torsion angle of 180° with respect to bonds 5-6 and 3-2 in the ring. In the axial position, however, the C—C bond has a torsional angle of 60° with respect to these same two bonds. This can be shown by comparing Newman diagrams of the two methylcyclohexane conformations (Fig. 1.32).

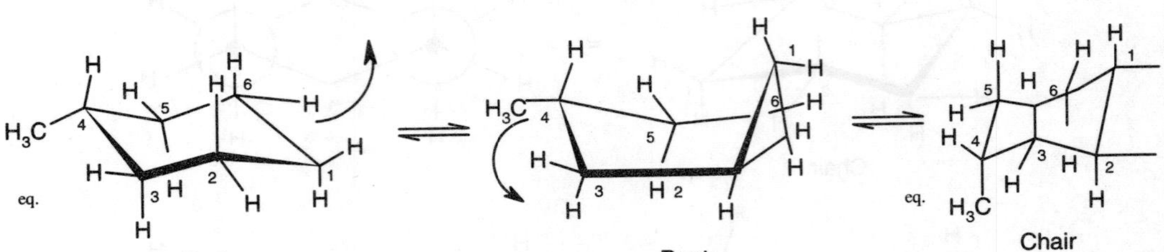

Fig. 1.31 Ring flipping of methylcyclohexane

A torsion angle of 60° between C–C bonds represents a **gauche interaction** and so an axial methyl substituents experiences two gauche interactions with the cyclohexane ring whereas the equatorial methyl substituents experiences none. Due to this, the latter chair conformation is preferred and about 95% of methylcyclohexane molecules are in this conformation at any point of time, compared to 5% in the other conformation.

Equatorial methyl

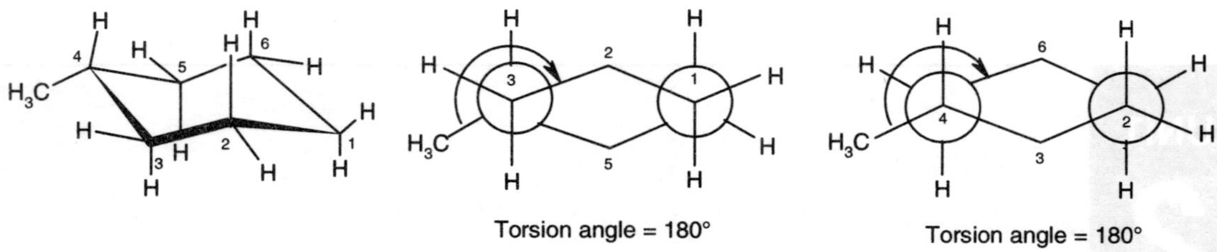

Axial methyl

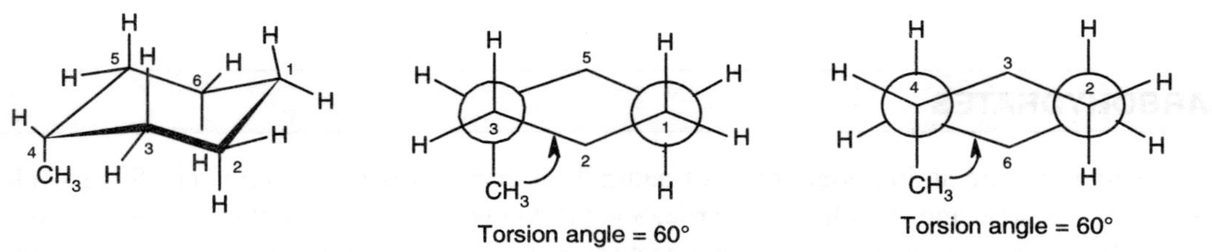

Fig. 1.32 Newman projections of the boat conformations of methylcyclohexane

UNIT

2

Introduction to Biomolecules

CARBOHYDRATES

The carbohydrates are an important class of naturally occurring organic compounds. Sugars (glucose, fructose, etc.) starch and cellulose are probably the known members of this group. Carbohydrate moieties also occur in a number of complex materials such as glycosides, mucoproteins, nucleic acids etc.

- Chemical analysis of sugar and other carbohydrates show that they are composed only the carbon, hydrogen and oxygen and they appear to be '***hydrates of carbon***'.

  $Cm\ (H_2O)n.$

  Glucose = $C_6H_{12}O_6 = C_6\ (H_2O)_6$

  Sucrose = $C_{12}H_{22}O_{11} = C_{12}(H_2O)_{11}$

- Later on it was found that certain compounds like formaldehyde (CH_2O), Acetic Acid ($C_2(H_2O)_2$) Lactic Acid ($C_3(H_2O)_3$) although contains general formula like carbohydrates which are not actually carbohydrates.

Now we know the structure of carbohydrates, they do not contain water molecules as such Carbohydrates are often referred to as ***Saccharides*** (Latin, Sacharum = Sugar) because of the sweet taste of the simpler members of the class, the ***sugars***.

- The carbohydrates are polyfunctional compounds, they contain the following functional groups.

  1. Alcoholic hydroxy groups, —OH
  2. Aldehyde group, $-\overset{\overset{\displaystyle O}{\|}}{C}-H$
  3. Ketone group, $-\overset{\overset{\displaystyle O}{\|}}{C}-$

A precise definition of the term ***'Carbohydrate'*** can be given as: ***'Polyhydroxyaldehydes (or) polyhydroxy ketones'***.

Example:

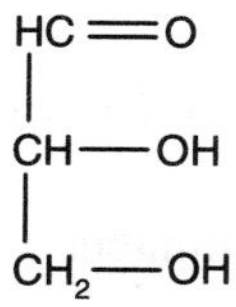

Glyceraldehyde
(Polyhydroxy aldehyde)

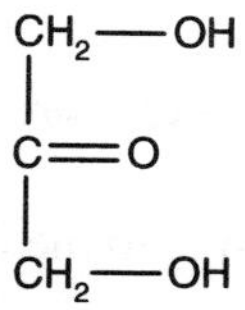

Dihydroxy acetone
(Polyhydroxy ketone)

The above definition of 'Carbohydrate' is not entirely satisfactory. It envisages the presence of free carbonyl group (—C═O) in the simple carbohydrate molecule which is not true.

A carbonyl compound (aldehyde (or) ketones) reacts with an alcohol to form a hemiacetal.

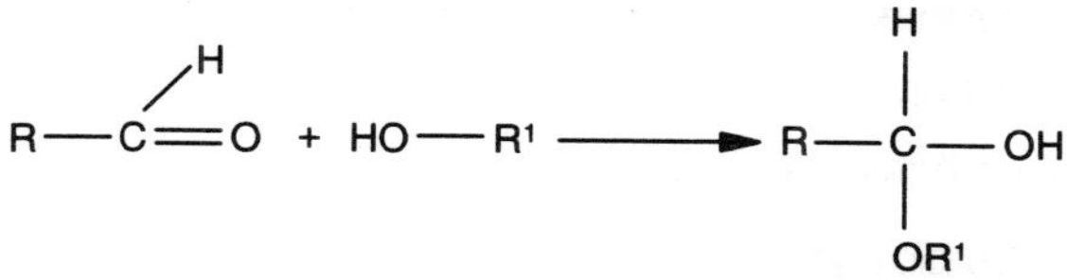

Hemiacetal

In carbohydrates, we have an aldehyde group which combines with an alcoholic OH of the same molecule to form an internal hemiacetal.

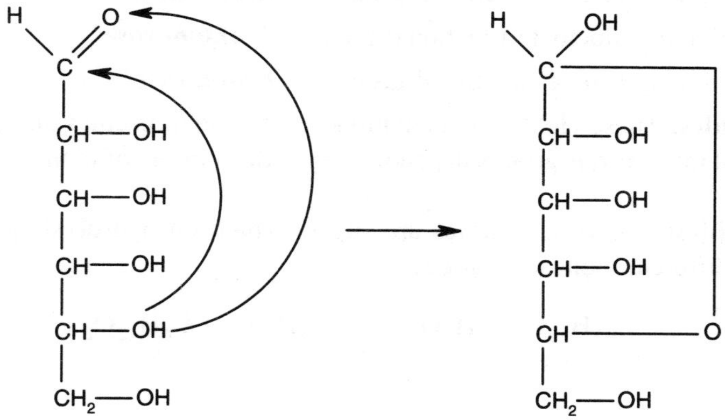

Glucose
(Polyhydroxy aldehyde)

An internal Hemiacetal
(Does not contain C═O)

Modern definition: A polyhydroxy compound that has an aldehyde (or) a ketone function present either free or as hemiacetal (or) acetal.

Classification: The carbohydrates are divided into two families:

1. Sugars 2. Non-Sugars

Sugars: The monosaccharides and oligosaccharides are soluble crystalline substances having a sweet taste. They are collectively known as sugars.

These are two types:

1. Monosaccharides
2. Oligosaccharides

Monosaccharides: The monosaccharides are polyhydroxy aldehydes (or) ketones which cannot be hydrolysed to simpler sugars.

These are again be classified according to the type of carboxy group and the number of carbon atoms in the molecule.

(A) Based on Number of Carbon Atoms

In this the carbon-chain build up by 3 to 7 carbon atoms

Trioses having - 3 - carbons

Tettroses having - 4 - carbons

Pentoses having - 5 - carbons

Hexoses having - 6 - carbons

Heptoses having - 7 - carbons, etc.

(B) Based on Type of Carbonyl

Aldoses — Having aldehydic functional group.

Ketoser — Having ketonic functional group.

5 carbon atoms having aldehyde functional group — ***'Aldopentose'***

5 carbon atoms having ketone functional group — ***'Ketopentose'***

6 carbon atoms having aldehyde functional group — ***'Aldohexose'***

6 carbon atoms having ketone functional group — ***'Ketohexose'***

2. **Oligosaccharides:** These yield two to ten monosaccharide molecules on hydrolysis and are thus again classified into various groups depending upon the number of monosaccharide units formed on hydrolysis.

   (*i*) **Disaccharides:** The disaccharides are sugars which on hydrolysis give two moles of the same (or) different monosaccharides.

   Example:
$$\underset{\text{(Sucrose)}}{C_{12}H_{22}O_{11}} + H_2O \longrightarrow \underset{\text{glucose}}{C_6H_{12}O_6} + \underset{\text{fructose}}{C_6H_{12}O_6}$$

$$\underset{\text{(Maltose)}}{C_{12}H_{22}O_{11}} + H_2O \longrightarrow 2C_6H_{12}O_6$$

   (*ii*) **Trisaccharides:** These give three moles of monosaccharides on hydrolysis.
   Example: Raffinose ($C_{18}H_{32}O_{16}$)

Non-Sugars (or) Polysaccharides: These are carbohydrates composes of several monosaccharide units and are polymers. On hydrolysis they yield a large number (>10) of monosaccharides. These are usually tasteless, amorphous solids and either insoluble in water (or) form colloidal suspensions.

Example: Starch, cellulose etc.

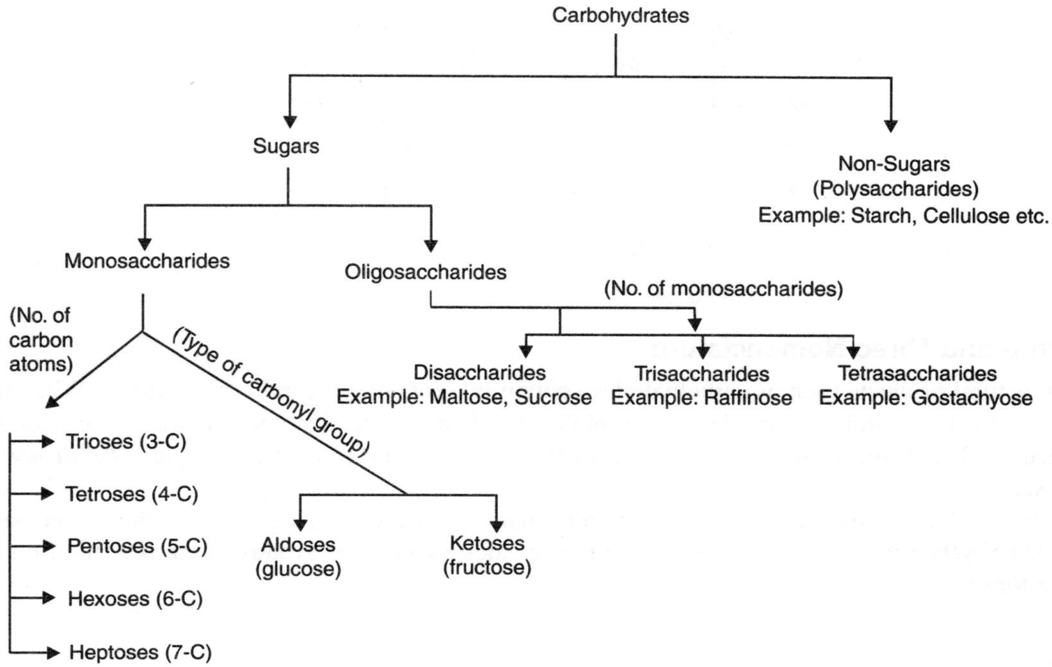

Configurational Nomenclature of Carbohydrates

D-L, Nomenclature:

Optical isomerism, the type of isomerism commonly found in carbohydrates, is usually encountered when a molecule contains one (or) more chiral carbon atoms with the existence of a large number of optical isomers in carbohydrates. The symbol (+) describes the sign of rotation (plane polarised light) and the letter D indicates absolute configuration shown for glyceraldehyde. This type (D) sugars belongs to the ***D-family***. Those enantiomers having the hydroxyl group, to the left belong to the ***L-family.***

The D-L system of nomenclature was proposed by ***Rosanoff in 1906***.

The sugars can be named on the bases of the absolute configuration of D- and L-glyceraldehyde.

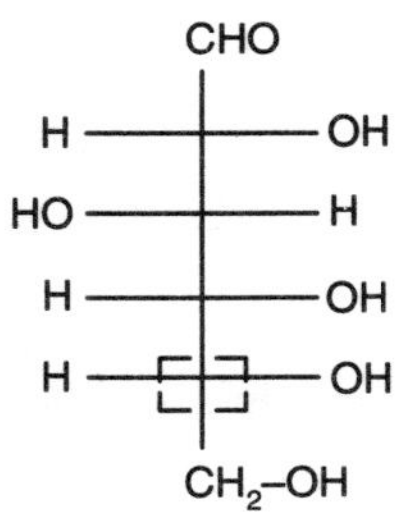

D-(+)-Glucose

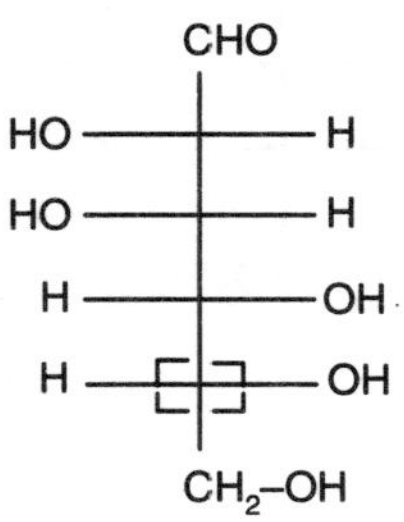

D-(+)-Mannose

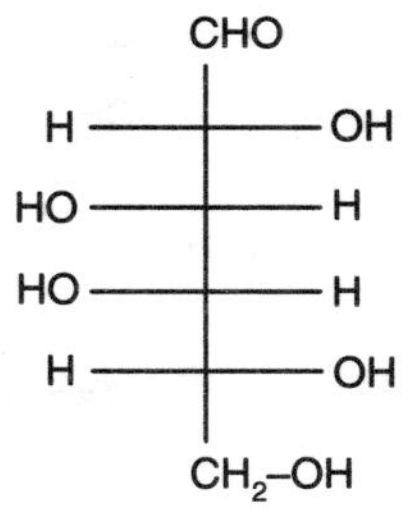

D-(+)-Galactose

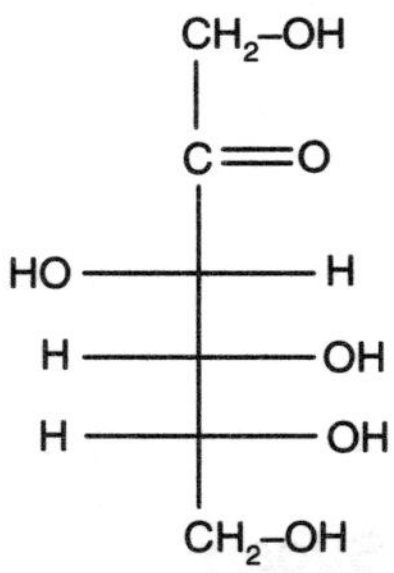

D-(–)-Fructose

All four sugars are D-sugars because they have the same configuration as D-glyceraldehyde on the penultimate carbon atom.

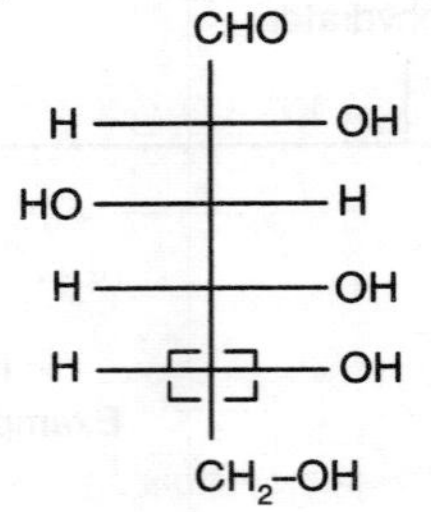

D-Glucose

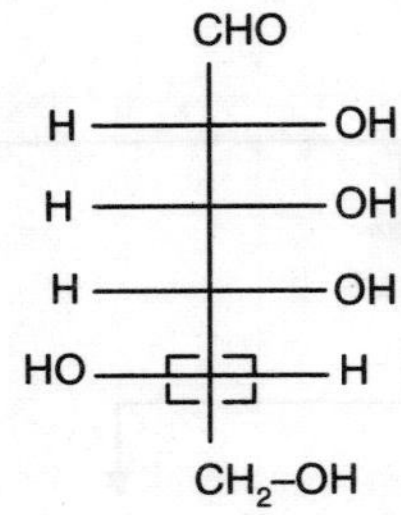

L-Glucose

Erythro and Threo Nomenclature

Many natural products such as carbohydrates contain two (or) more chiral carbon atoms. When these are two distinct chiral carbon atoms in a molecule, then a special nomenclature is used for such molecules. The nomenclature is derived from the names of the four-carbon sugars. ***Erythrose*** and ***threose.***

If the two like groups –OH in the projection formula are on the same side as in erythrose, the isomer is called ***Erythro form.*** If they are on the opposite sides as the –OH groups in ***threose***, the isomer is called ***threo form.***

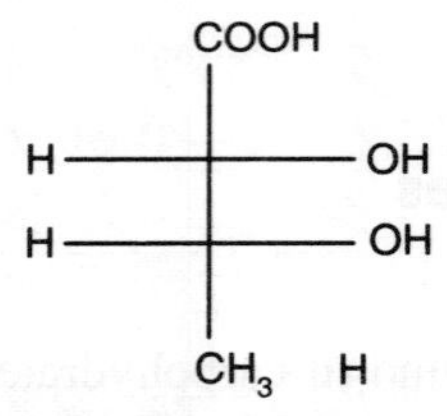

Erythro-2, 3-dihydroxy butanoic acid

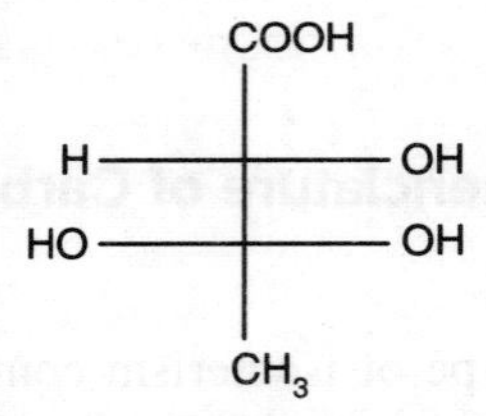

Threo-2, 3-dihydroxy butanoic acid

The term erythro and threo are generally employed only with molecules that do not have symmetric ends. In symmetric molecule such as tartaric acid the terms ***mero*** and ***d, l.***

Example:

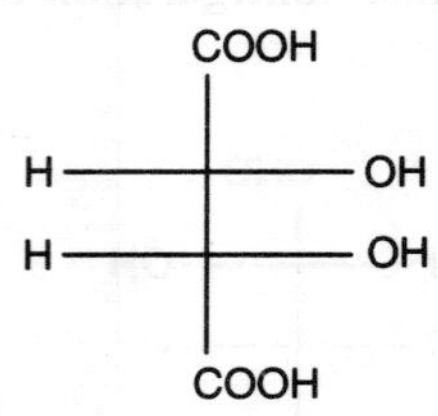

meso

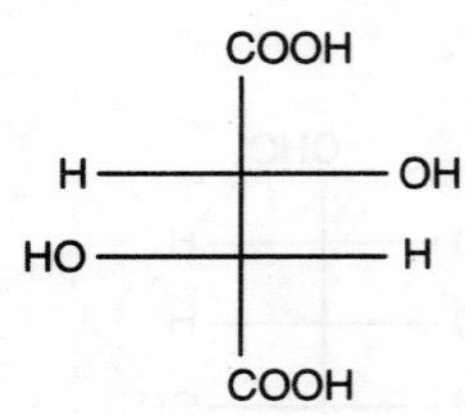

d,l.

Tartaric acid

Deoxy Sugars

Some naturally occurring sugars have a hydrogen atom in place of one (or) more of the –OH groups. In naming these compounds. The prefix, deoxy is used with the name of the oxygen containing analog to indicate the lack of oxygen and the class as a whole is known as deoxy sugars.

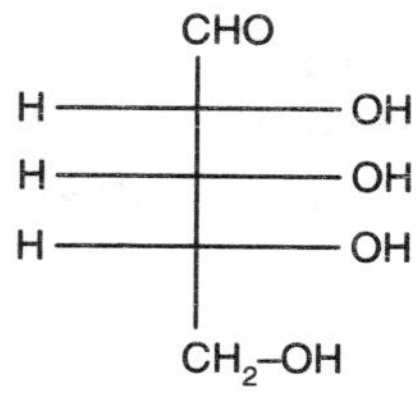

D-Ribose

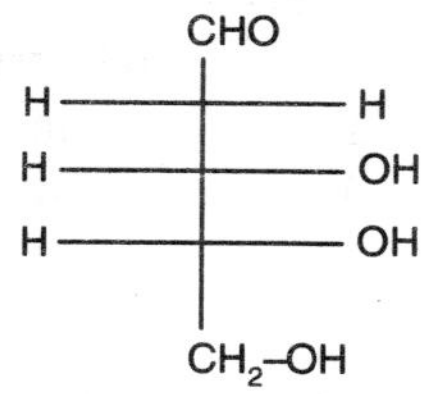

2-Deoxy-D-ribose

Occurrence of Carbohydrates

Carbohydrates are ultimately derived from the reduction of carbondioxide by living plants with the aid of the green pigment, chlorophyll. The energy is obtained from sunlight and this process is known as photosynthesis.

$$XCO_2 + YH_2O \xrightarrow{\text{Solar energy}} C_x(H_2O)_y + XO_2$$

Monosaccharides

The monosaccharides are those sugars which cannot be further degraded by hydrolysis. They include both aldoses and ketoses. The most important sugars of this class are ***glucose*** and ***fructose***.

Glucose

The name glucose is derived from a greek word for sweet wine and reflects the fact that it was originally isolated from grapes (***grape sugar***). It rotates plane polarised light to the right *i.e.* it is dextro rotatory and is called ***'Dextrose'***. and it is also present in blood (***Blood sugar***).

Glucose (*or*) ***Dextrose*** (*or*) ***Grape sugar*** (*or*) ***Corn sugar*** (*or*) ***Blood sugar.***

Preparation

(*i*) **From Sucrose:**

$$\underset{\text{Canesugar}}{C_{12}H_{22}O_{11}} + H_2O \xrightarrow[\text{50°C, 2 Hrs.}]{\text{Dil. HCl}} \underset{\text{Glucose}}{C_6H_{12}O_6} + \underset{\text{Fructose}}{C_6H_{12}O_6}$$

Glucose being almost insoluble in alcohol crystallizers on cooling leaving behind the more soluble fructose in solution. Thus the two are separated by filtration.

(*ii*) **From Starch:** Commercially, glucose is obtained by the hydrolysis of starch with hot, dilute mineral acids.

$$\underset{\text{Starch}}{(C_6H_{10}O_5)_n} + nH_2O \longrightarrow \underset{\text{Glucose}}{nC_6H_{12}O_6}$$

General Reactions of Monosaccharides

These monosaccharides contain only two kinds of functional groups, ($-\overset{O}{\overset{\|}{C}}-$) carbonyl and hydroxyls (–OH). They give most but not all of the characteristic reactions of these $-\overset{O}{\overset{\|}{C}}$, —OH groups.

- In some cases these monosaccharides behave as simple alcohols. For example hydroxyl groups of carbohydrate can be converted into esters and ethers.
- Monosaccharides are usually soluble in water because of their poly hydroxyl groups.
- Here we shall discuss only the reactions which having value in structural determination.

Oxidation

1. **Bromine water:** It is a mild oxidising agent it can oxidise the aldose to the corresponding aldonic acid which on evaporation gives a stable γ-lactone.

$$\text{Glucose} \xrightarrow{Br_2/H_2O} \text{Gluconic acid} \longrightarrow \gamma\text{-lactone}$$

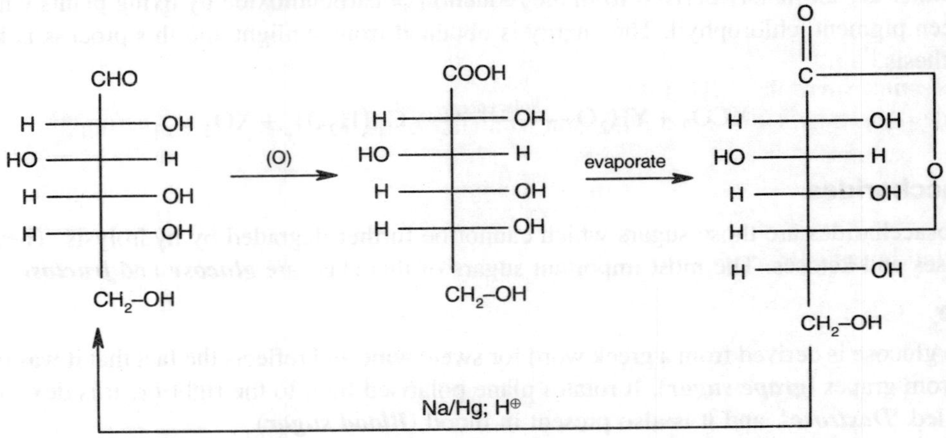

2. **Nitric acid:** Strong oxidising agent

It oxidise $C=O$ group as well as alcoholic group to give dicarboxylic acid.

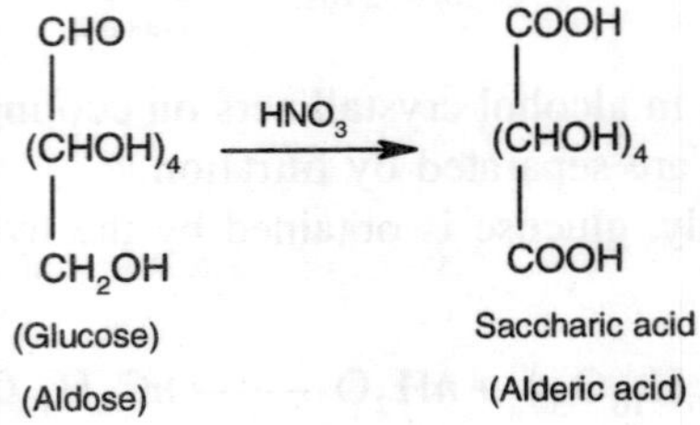

Lead Tetraacetate and Periodic Acid

With lead tetraacetate and periodic acid monosaccharides give dialdehydes which formed due to cleavage of C—C bond.

Reduction

Treatment of an aldose (or) a ketone with $NaBH_4$ reduces it to a poly alcohol called alditol.

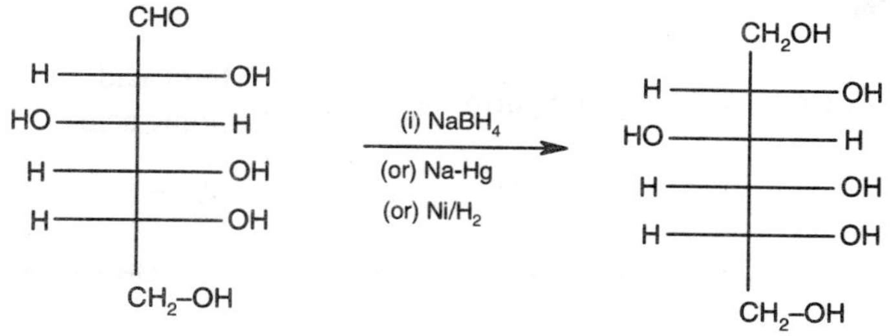

(Glucose) (Sorbitol)

Note: D-Sorbitol used as an artificial sweetener and sugar substituent in foods.

Structural Determination of Glucose

1. **Molecular formula:** $C_6H_{12}O_6$
2. **HI/P reduction:** Glucose on reduction with HI/P reduction gives *n*-hexane. It indicates glucose has 6-carbons.

$$\text{Glucose} \xrightarrow{\text{HI/P}} n\text{-hexane}$$

3. **Reaction with Hydroxyl amine** $(\ddot{N}H_2 - OH)$**:** On reaction with hydroxyl amine glucose forms an oxine it supports the presence of carbonyl group presence in glucose.

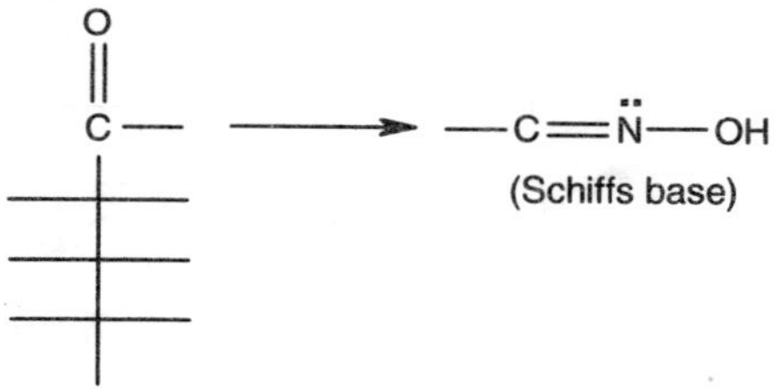

(Schiffs base)

4. $\textbf{Glucose} + \textbf{Br}_2/\textbf{H}_2\textbf{O} \longrightarrow \textbf{Gluconic acid}$

   It explains glucose has $\overset{O}{\overset{\|}{C}}$ group.

5. $\textbf{Glucose} + \textbf{Ac}_2\textbf{O} \longrightarrow$ **Penta acetyl derivative:** It suggests that glucose has 5-OH groups in its structure.
6. **Reaction with $NaBH_4$:** Glucose on reaction with $NaBH_4$ give glucito, it suggests US every carbon has one OH group.

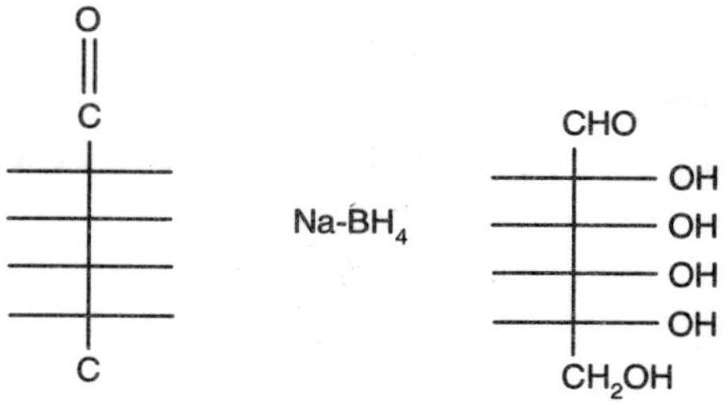

Glucose Sorbitol

Reaction with HCN

$$\text{Glucose} \longrightarrow \text{HCN} \longrightarrow \text{Cyanohydrin} \xrightarrow[\text{HI/Red P}]{H_3O^{\oplus}} \begin{matrix}\text{Mono acid}\\ \text{heptonic acid}\end{matrix}$$

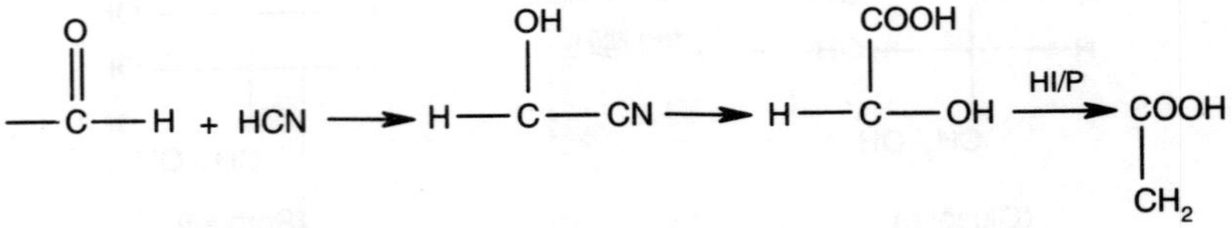

It gives an idea that glucose has carboxyl group at one end.

KILIANI-FISCHER SYNTHESIS

Most of sterio chemical studies in monosaccharides chemistry suggests is the actual conformations of monosaccharides.

One of the most important methods used was the Kiliani-Fischer synthesis which results in the lengthening of an aldose chain by one carbon atom. In this synthesis the C_1 aldehyde group of the sorting sugar becomes C_2 of the chain lengthened sugar.

Aldo pentose is converted to aldo hexose.

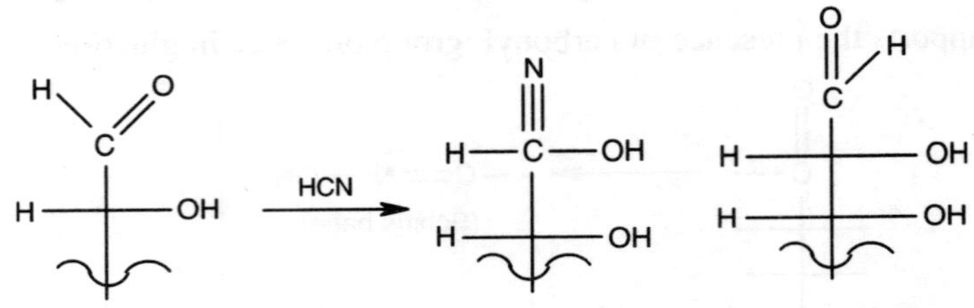

An aldose Cyanohydrine Chain lengthened aldose

Conversion of Arabinose [a pentose] to Glucose (a Hexose)

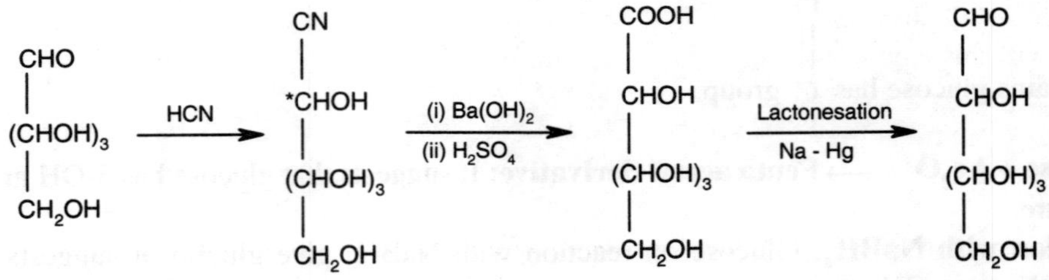

Glucose + Mannose

Descending in Aldose Series

The processes involves the removal of C_1 and the conversion of C_2 to an aldehyde group.

1. Ruff's method:

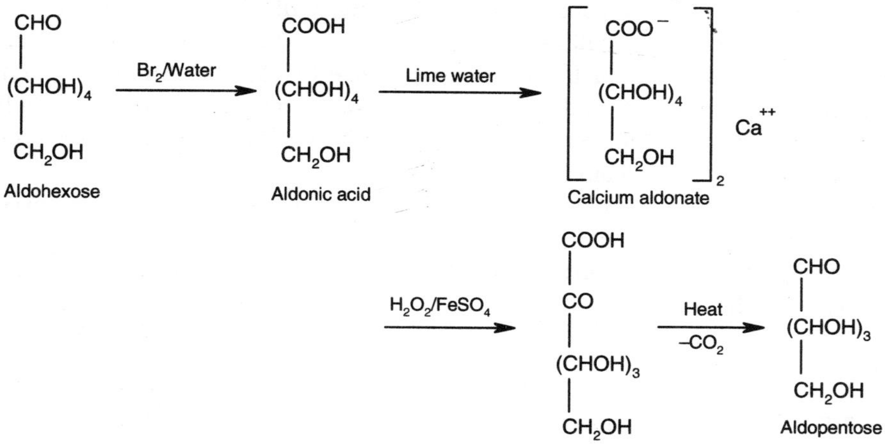

2. Wohls method:

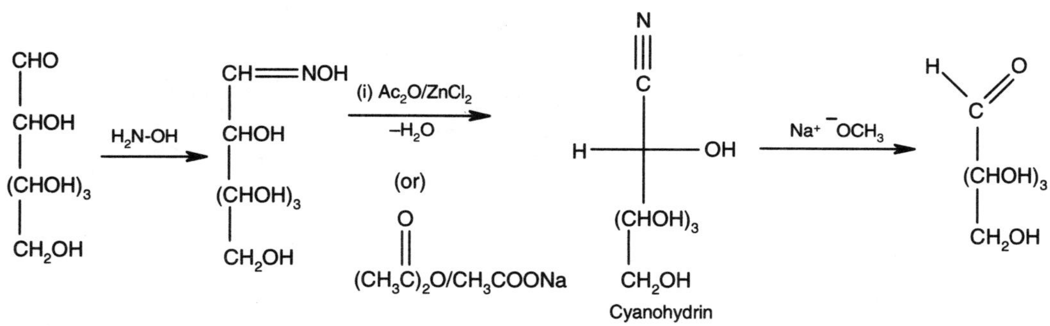

With the above all chemical reactions Bayer proposed open chain structure for Glucose.
Emil Fischer with his asymmetric carbon studies proposed Glucose's open chain structure as follows:

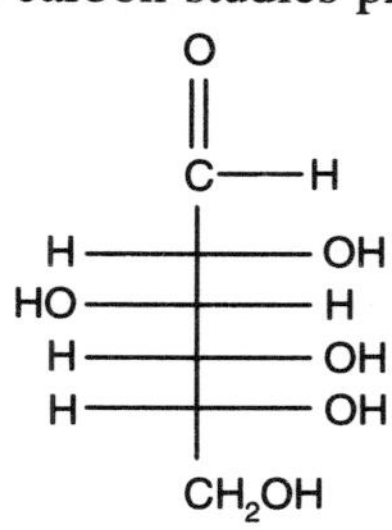

But

1. Glucose not reacting with $NaHSO_3$ even it contained $-\overset{\overset{O}{\|}}{C}-H$ group.
2. No reaction with schiffs reagent.
3. It cannot explain muta rotation.

MUTA ROTATION

When glucose is dissolved in water its specific rotation gradually changes until it reaches a constant value. The phenomenon, known as muta rotation is showed by almost all the reducing sugars.

Example: Glucose when prepared by crystallization from water below 50°C is dissolved in water its initial specific rotation of +113°C falls gradually to a constant value of +52.5°C. Similarly when the glucose crystallized from water above 95°C is dissolved in water its initial specific rotation of +19°C gradually rises to +52.8°C. The existence of the two isomeric glucoses and the range in specific rotation (muta rotation) is not explained by open chain conformation.

Muta rotation occurs by reversible ring opening of each anomer to the open aldehyde followed by reclo.

The above studies revealed that glucose open chain structure is not stable, but it is intermediate between 2 cyclic structures of glucose.

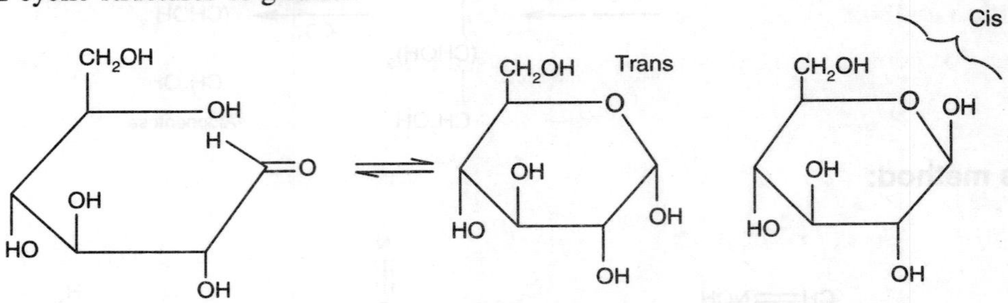

D-Glucose (0.1%)

α-D glucopyranose (36%) (α-anomeric OH and CH_2OH are Trans)

β-D glucopyranose (64%) (anomeric OH sand CH_2OH are Cis)

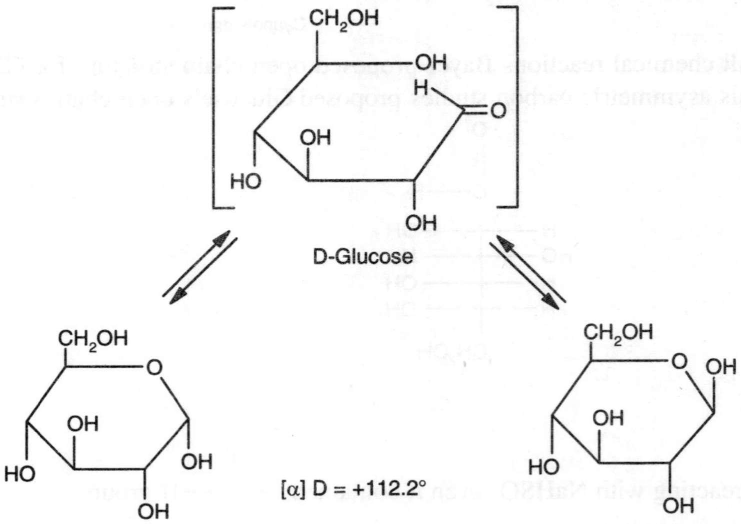

D-Glucose

[α] D = +112.2°

α-D glucopyranose (36%)

β-D glucopyranose (36%)
$[\alpha]_D = +18.7°$

Cyclic Structure [Haworth Structure]

The five and six membered oxide rings of monosaccharides bear a normal relationship to the heterocyclic compounds furan and pyran.

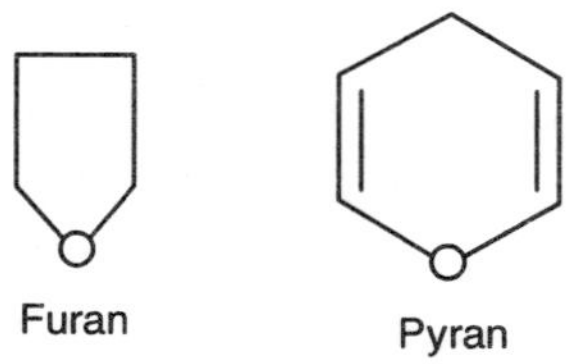

So the structure in sugars denoted by the terms furanose (5 membered) pyranose (6 membered).

Determination of Ring Size

Methylation method

The sugar first completely methylated then hydrolysed with dil. HCl, when only the glycosidic methyl group is hydrolysed and finally the product is oxidised. Identification of the product leads to the conformation of ring size in Glucose.

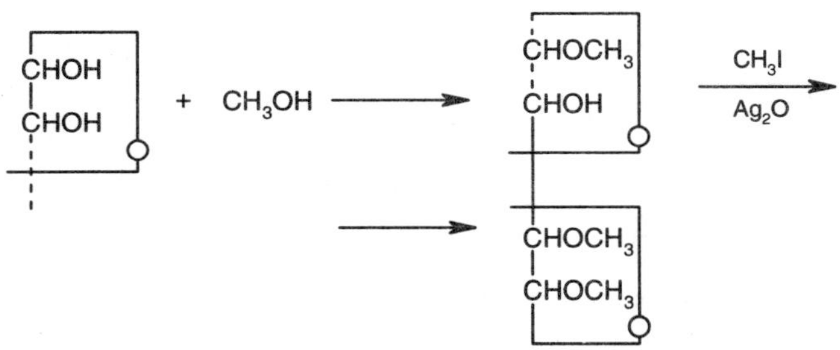

Nucleophillic Addition Reaction

Inter conversion of Fischer & Haworth projections of D-Glucose.

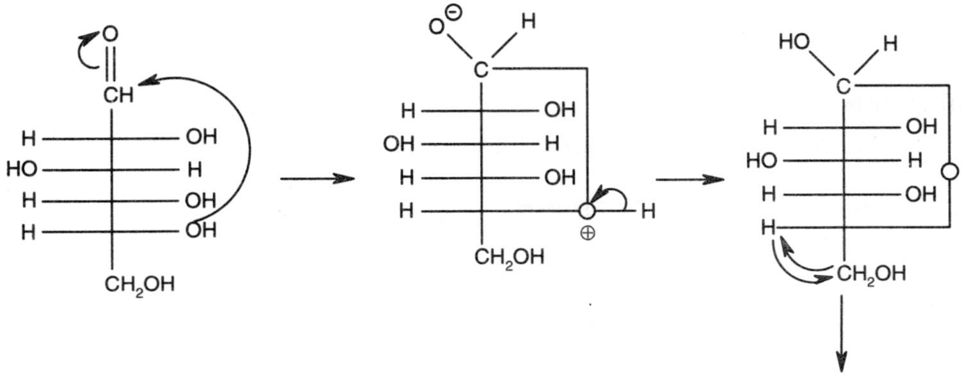

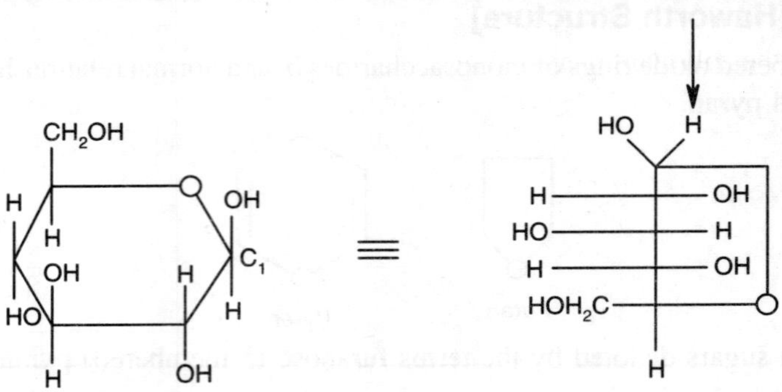

Haworth structure β-Gluco pyranose

***Note*:** To write cyclic structure to Haworth structure all left hand side groups are written up and right hand side groups are written down in the 6 membered pyranose ring structure.

Depending upon C_1 anomeric carbon we can name it has either α or β for a anomer C_1 OH present down or below.

α-anomer

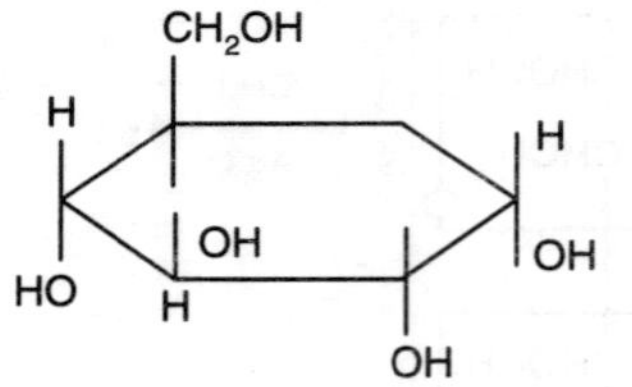

'Left - up'
'Right-down'

β another having all trans groups so it has highly stability.

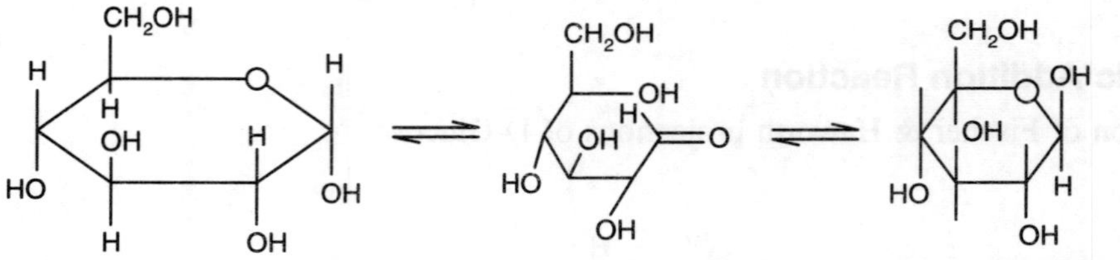

β-D glucopyranose 60%

Pyranose Structure

Sometimes it referred as δ-oxide, amylene or six membered ring.

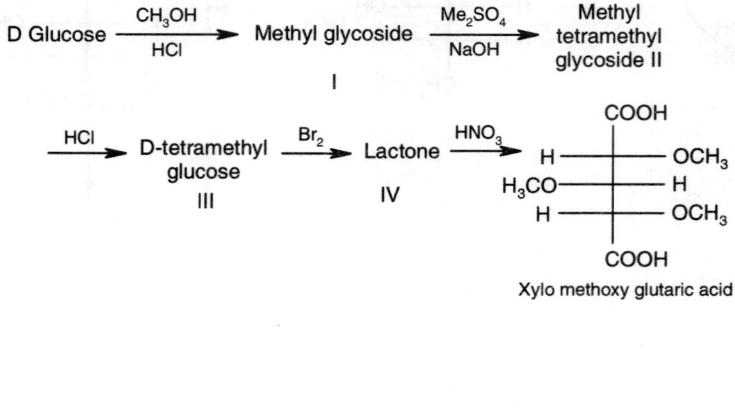

Xylo methoxy glutaric acid V

The structure of this dicarboxyllic acid (V) proved the key in determining the structure of the parent compound glucose is the following way.

One of the —COOH groups in structure V must be derived from Lactone (IV) only three hydroxyls are present in the ring of glucose because of structure V has only 3-OCH_3 groups.

Reactions for above proof

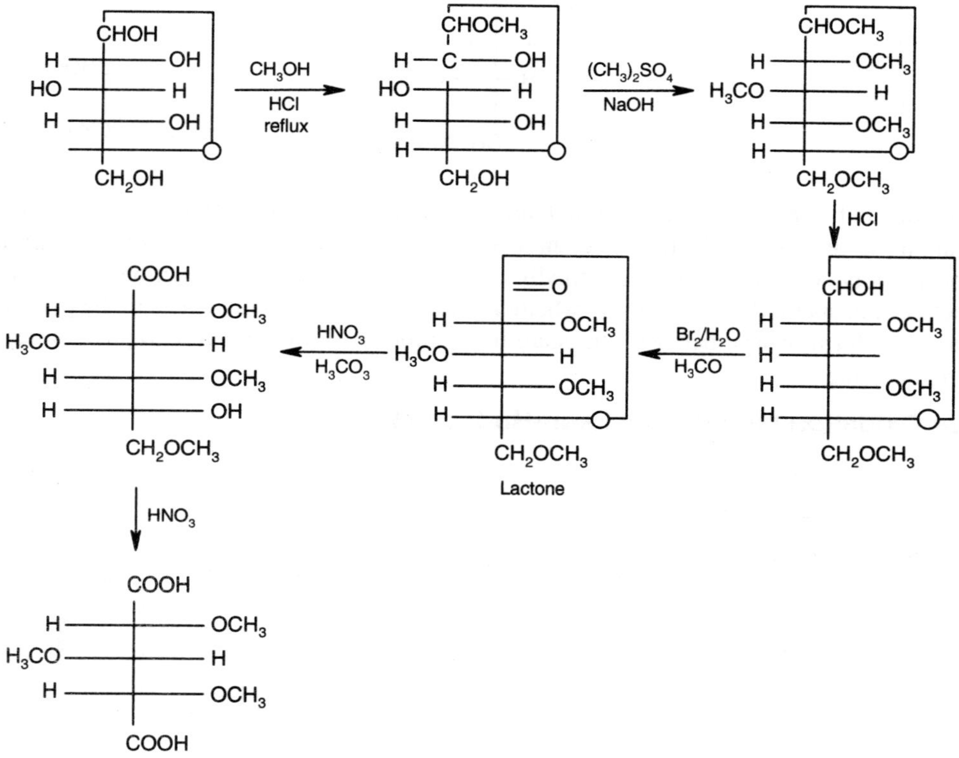

Furanose Structure

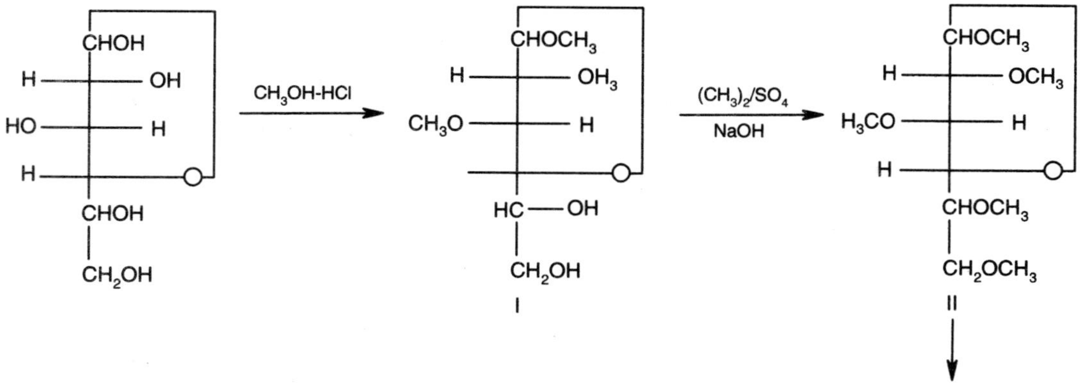

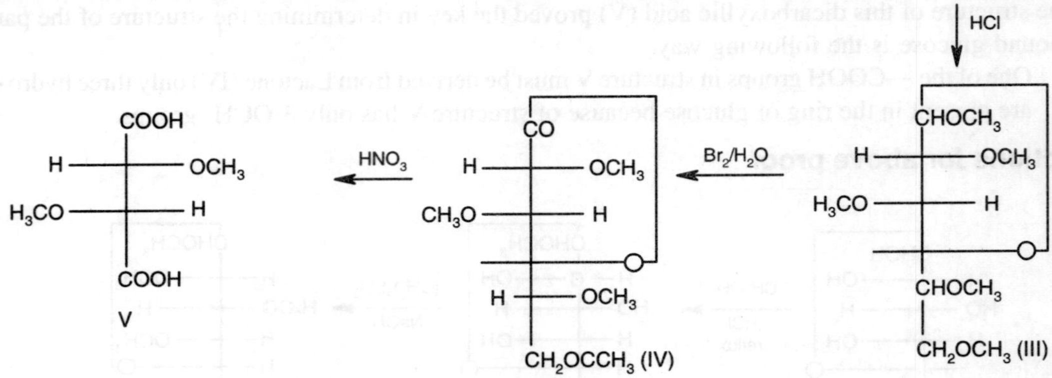

The structure and formation of the dimethyl tartaric acid from D-glucose indicate that within the ring of the sugar and its derivatives II, III and IV, there are only two hydroxyl (or) methoxyl groups and thus the ring must involve C_1 and C_4 which are finally oxidising during the set or reaction to the acid groups with the help of this method we cannot say whether glucose itself originally exists in the pyranose (1:5) or furanose (1:4) form or whether two forms are in equilibrium X-ray crystallography studies suggest the 6-membered ring.

Pyranose Oxidation Method: Jackson-Hudson Method

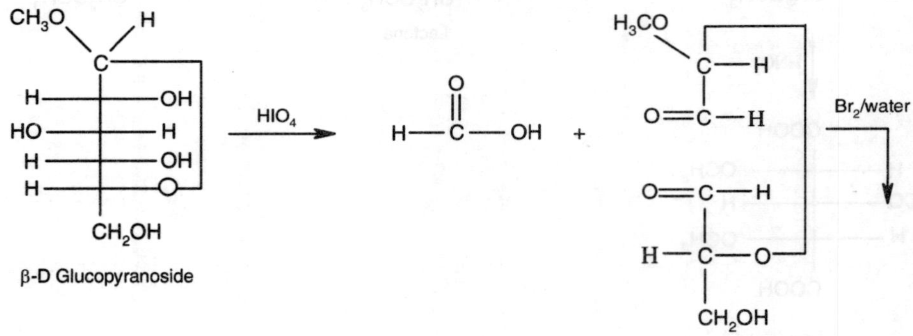

β-D Glucopyranoside

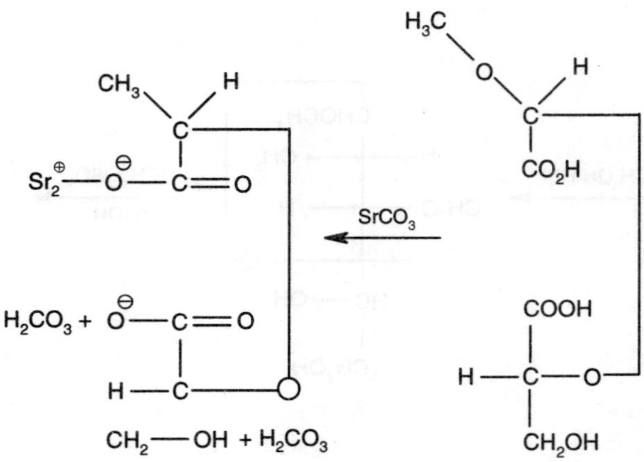

Strantium salt

Dionic acid

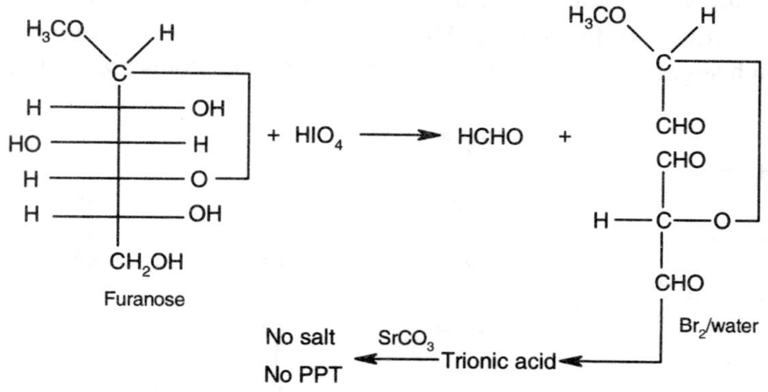

So the above reactions clearly indicating that glucose having pyranose structure. If it has a furanose structure it has to form a PPT with $SrCO_3$ but it can't actually Sr in +2 form if glucose has furanose structure trionic acid forms so it cannot react with Sr^{+2}. So there is no possibility of formation of PPT.

On the other hand, the pyranose structure reacting with HIO_4 giving a dial and after oxidation it is giving dioic acid which on reaction with $SrCO_3$ forms a PPT so this suggest the pyranose ring in the glucose.

Fructose

- The D (leavo) — fructose, also known as ***levulose*** or ***fruit sugar*** is found in any fruits and honey.
- The most important source of fructose is the polysaccharide insulin.
- M.P. 104°C.
- It exhibits muta rotation.
- Molecular formula $C_6H_{12}O_6$.
- Reaction with sodium acetate and acetic anhydride it gives a penta acetyl derivative suggesting that it has 5 OH groups in its structure.
- Forms cyanohydron with HCN and oxine with hydroxyl amine indicating the presence of carbonyl group.
- Reaction with HNO_3.

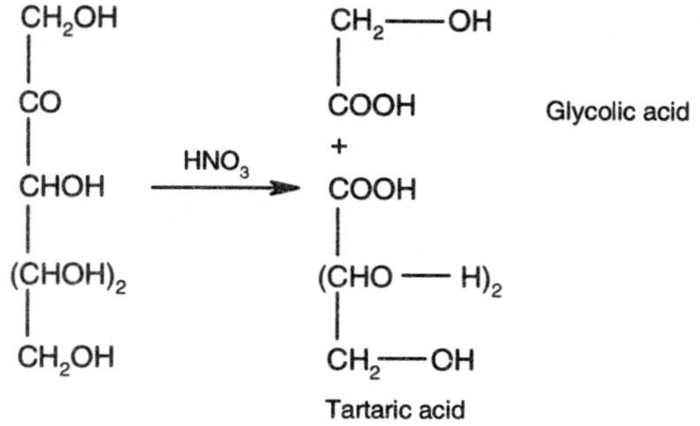

- Formation of glycolic acid and tartaric acids on reaction with HNO_3 suggests us fructose has ketonic group in its structure. The products which are formed on reaction with HNO_3 has one carbon less *i.e.* 5 carbons whereas fructose has 6 carbon atoms.

- On reducing with Na-Hg/water fructose give hexahydric alcohols, sorbitol and mannitol. Which on further reduction with hydroiodic (HI) acid and red phosphorous gives 2-Iodo hexane and *n*-hexane which suggest that in fructose molecule 6 carbon atoms are in straight chain.

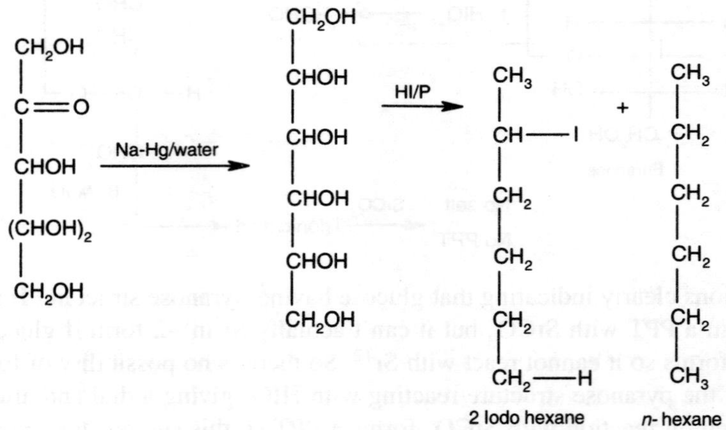

- Fructose on treatment with hydrogen cyanide gives cyanohydrin which on hydrolysis followed by reduction with HI/P gives *n*-butyl methyl acetic acid. Formation of *n*-butyl methyl acetic acid again indicates that ketonic group in fructose that is at C_2 position.

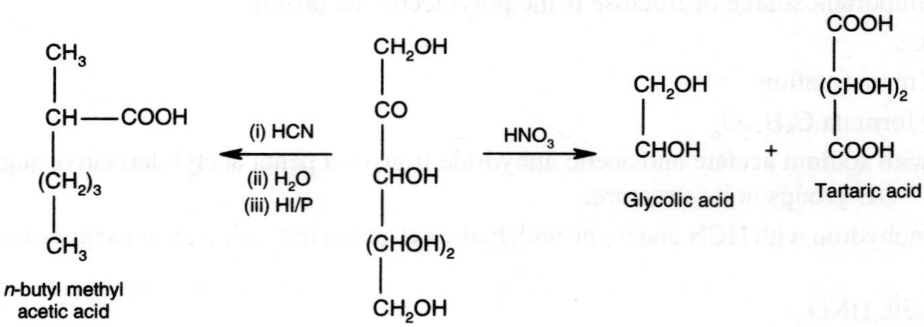

On the basis of above reactions and the reactions with phenyl hydrazine clearly give the structure of fructose.

Reaction with Phenyl Hydrazine

The exact configuration of fructose is established by the fact that it forms same osazone as glucose indicating there by that the configuration of C_3 to C_6 atoms of fructose is same as that of glucose.

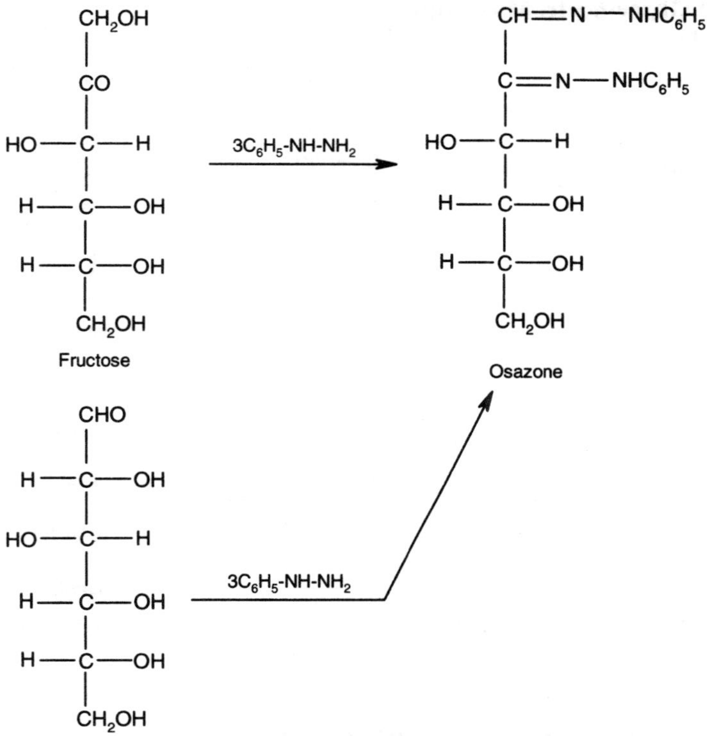

For this reaction 3 moles of phenol hydrazine is required.

Ring Structure of Fructose

Fructose also exhibits mutarotation and forms two isomeric methyl fructosides. So it is believed that fructose also exists in two isomeric α and β-forms.

- Fructose most probably the β-isomer with pyranose ring.
- In sucrose and insulin fructose is present in furanose ring which on hydrolysis converted into more stable pyranose form.

Significant Reactions

(1) Epimerization

Epimer: Epimers are the optical isomers which differ is the configuration at only one asymmetric carbon atom (in glucose C_2 is epimer).

Epimerization process

The process of converting one epimer into other *i.e.* inversion of configuration at one asymmetric centre (generally α) in a molecule containing several asymmetric centres is known as epimerization.

Epimerisation of Glucose to Mannose

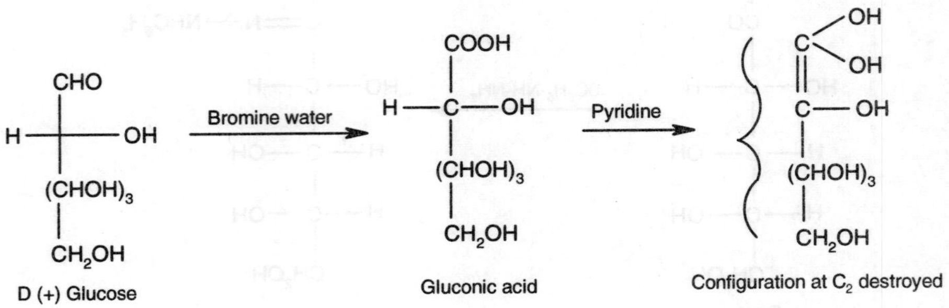

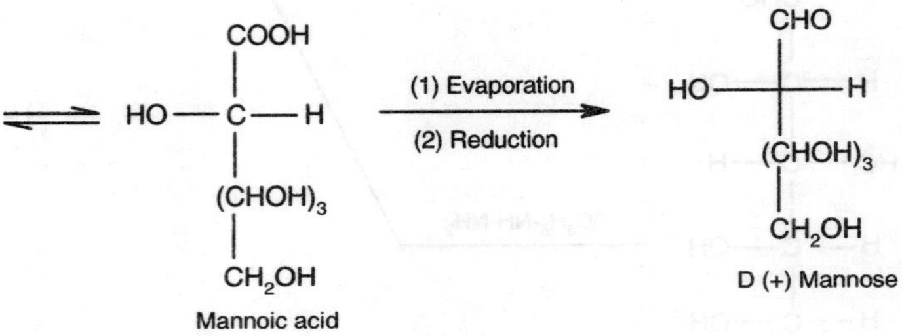

(2) Ene-diol rearrangement: [Brown-van Ekenstein rearrangement]

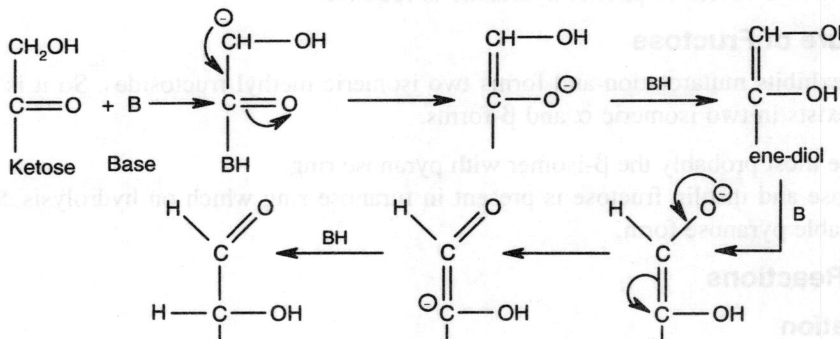

In the presence of base ketose through ene-diol rearrangement converting aldose (Unlike ketones in carbonyl compounds). They reaching with Tollens & Fehling's reagents and giving +ve tests. It is the significant property of carbohydrates.

- **Aldose ⟶ Ketose**

D-Glucose ⟶ D-Fructose

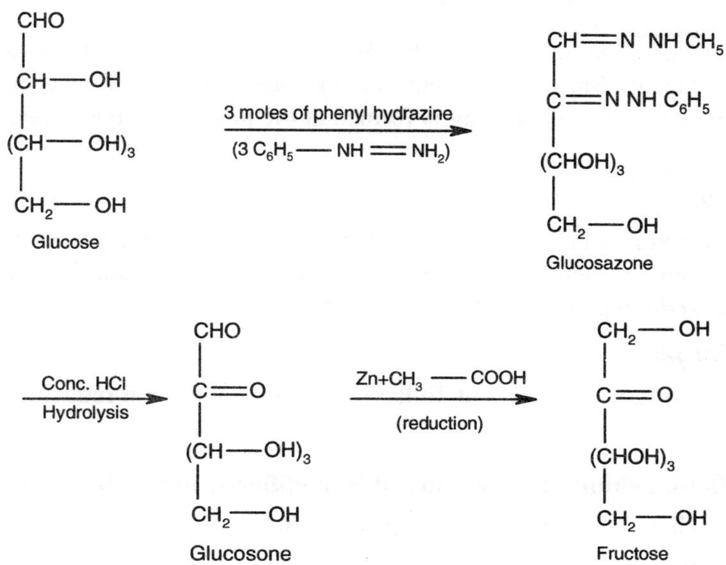

- **Ketose ⟶ Aldose**

Fructose ⟶ Glucose

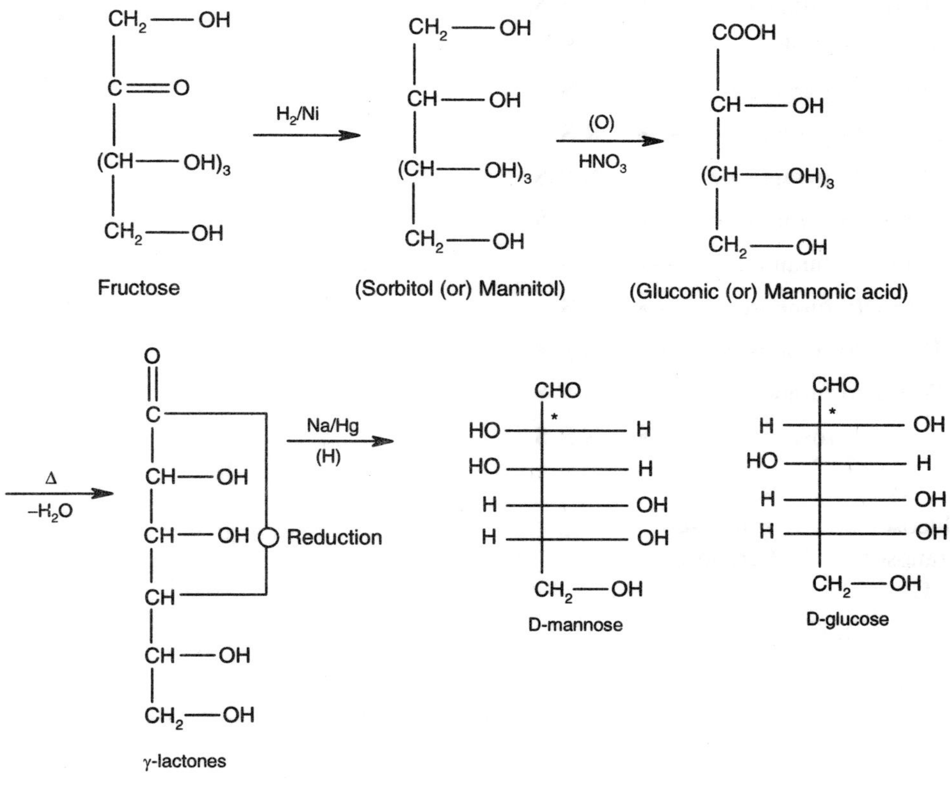

Molisch's Test

Take about 2 ml of glucose solution in a test-tube. Add to it a drop (or) two of 3% alcoholic ***α-naphthol solution*** (Molisch's reagent). Shake the contents and then add to it 1 ml of concentrated sulphuric acid along the sides of the test tube a violet ring is formed at the junction of two liquids. This test is given by ***all carbohydrates.***

Reducing Sugars

Glucose and Fructose reduces Fehling solution, on heating to form a red precipitate of cuprous oxide and there reduces Tollen's reagent, on heating to form a silver mirror. So, these sugars (Glucose and Fructose) are called ***'reducing sugars'.***

Non-Reducing Sugar

The sugars which do not give Fehling and Tollen's tests are called 'non-reducing sugars'.

Example: Sucrose.

- **Fructose give Tollens, Fehlings tests because it is modified to aldose by ene-diol rearrangement.**

1. Glucose open chain ⟶ Reducing sugar (R.S)
2. α-D-Gluco pyranose ⟶ R.S.
3. β-D-Gluco pyranose ⟶ R.S.
4. α-D-Gluco pyranoside ⟶ Non-reducing sugar (N.R.S)
5. β-D-Gluco pyranoside ⟶ N.R.S.
6. Fructose open chain ⟶ R.S.
7. α-D-Fructo pyranose ⟶ R.S.
8. β-D-Fructo pyranose ⟶ R.S.
9. α-D-Fructo pyranoside ⟶ N.R.S
10. β-D-Fructo pyranoside ⟶ N.R.S
11. α-D-Fructo furanose ⟶ R.S
12. β-D-Fructo furanose ⟶ R.S
13. α-D-Fructo furanoside ⟶ N.R.S
14. β-D-Fructo furanoside ⟶ N.R.S
15. Sucrose ⟶ N.R.S

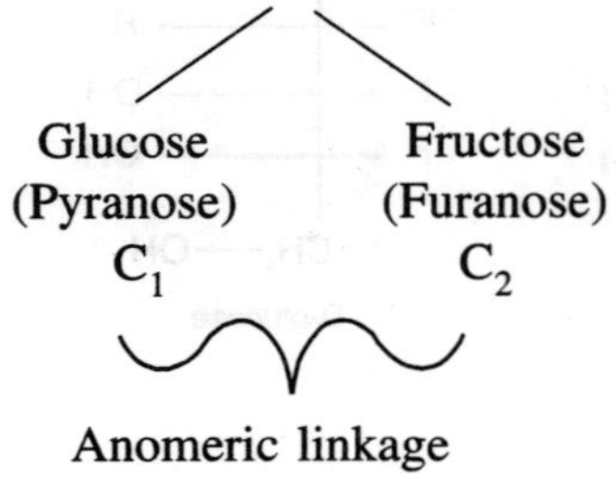

FATS AND LIPIDS

Due to the predominance of hydrocarbon chains in their structure, lipids have hydrophobic nature, which explains why they are insoluble or poorly soluble in water. However, they are readily soluble in non-polar solvents such as ether, chloroform and benzene.

Fats and oils are the principle stored forms of energy in many organisms, and phospholipid and sterols make up about half the mass of biological membranes. Other lipids, present in small quantities play crucial roles as enzyme co-factors, electron carriers, hight absorbing pigments, hormones, etc.

Fats and oils are derivatives of fatty acids (FA). FA's are hydrocarbon derivatives at about the same low oxidation state (*i.e.*, in highly reduced) as the hydrocarbons in fossil fuels.

Nomenclature: It specifies the chain length and number of double bonds, separated by a colon;

Example: 16:0 - Saturated 16-carbon palmitie acid

18:1 - Unsaturated 18-C oleic acid with one double bond

The position of double bonds are specified by superscript numbers following Δ (delta); a 20 C, FA with one double bond b/w C9 and C10 and another b/w C12 and C13 is designated as 20:2 ($\Delta^{9,12}$).

Example: 20 : 4 ($\Delta^{5,8,11,14}$) - Arachidonic acid, Double bonds between C5 and C6, C8 and C9, C11 and C12, C14 and C15 respectively.

Saturated fatty acids do not have any double bonds in the chain.

General formula: $—CH_3—(CH_2)_n—COOH$

n = Number of methylene groups $(—CH_2—)$

Unsaturated fatty acids have one or more double bonds. The double bonds of almost all naturally occurring unsaturated fatty acids are in *cis* configuration.

The physical properties of the FA's are largely determined by the length and degree of unsaturation of the hydro carbon chain. The longer the fatty acid chain and fewer the double bonds, the lower the solubility in water.

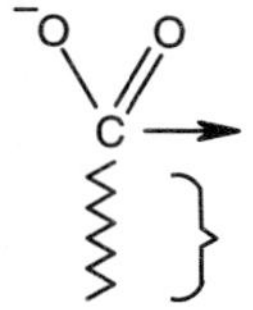

Carbonxyl group accounts for poor solubility in water.

Hydrocarbon chain accounts for insolubility in water.

A Fatty Acid

The saturated FA has waxy nature and unsaturated FA is oily in nature at 25°C. This is because, due to lack of double bonds in saturated FA's there is free rotation of C—C bonds and they are packed tightly in nearly crystalline arrays. In unsaturated FA's a *cis* double bond forces kinks and as a result, they can't pack tightly resulting in liquid/oily nature.

On vertebrate animals, free fatty acids (having a free carboxylate group) circulate in the blood bound to a protein carrier, serum albumin. However, FA's are present mostly as carboxylic acid derivatives such as esters or amides.

An Outlook of Typical Lipids

Lipid	*Primary Functions*
Fatty acids	Energy source, biosynthetic precursorses
Glycerides	Storage, transport, metabolic intermediate
Phospho glycerides	Membrane components
Ketone bodies	Energy sources
Sphingolipids	Membrane components
Eicasonoids	Modulators of physiologic activity
Cholesterol	Membrane component
Steroid hormones	Modulators of physiological activity

Fatty acids are detergent like due to their amphipathic nature; that is, they have non-polar (CH_3) and Polar (—COOH) ends. The polar end associates with water and non-polar end associates with the hydrophobic phase, in the biphasic systems.

Structural Lipids in Membranes

Membranes have double layers of lipids, these lipids are amphipathic; the orientation of their hydrophobic and hydrophilic region directs their packing into membrane bilayers.

Three general types of membrane lipids will be described:

Glycerophospho lipids: 2 FA's joined to Glycerol sphingo lipids: 1 FA's joined to sphingosine.

Sterols: Compounds having rigid system of 4 fused hydrocarbon rings.

Classification of Storage Lipids and Membrane Lipids

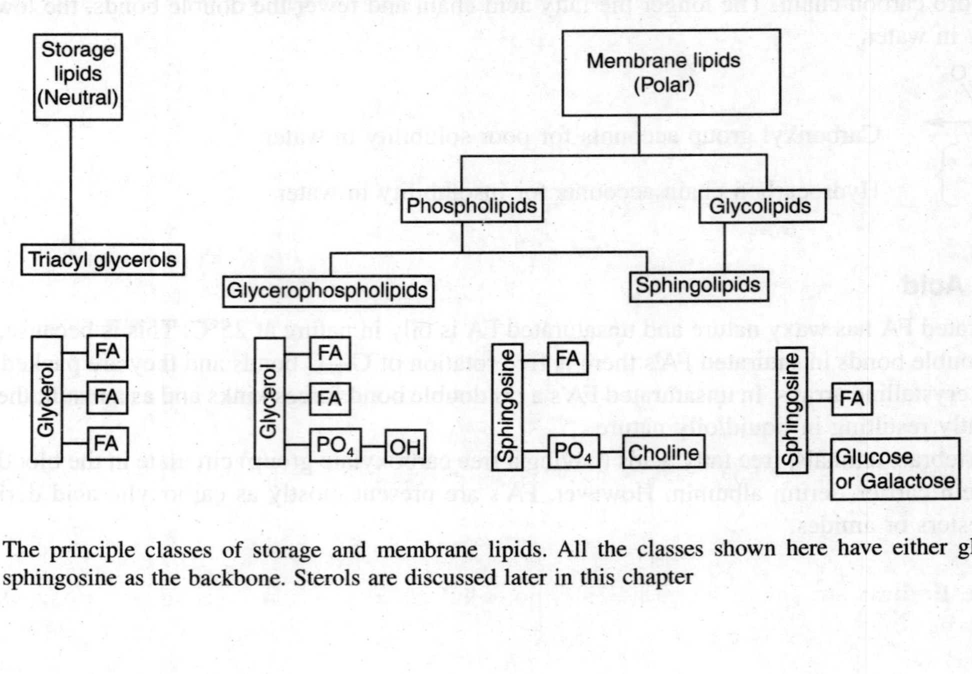

Fig. 2.1 The principle classes of storage and membrane lipids. All the classes shown here have either glycerol or sphingosine as the backbone. Sterols are discussed later in this chapter

Triacyl Glycerols are Fatty Acid Esters of Glycerol

Structure:

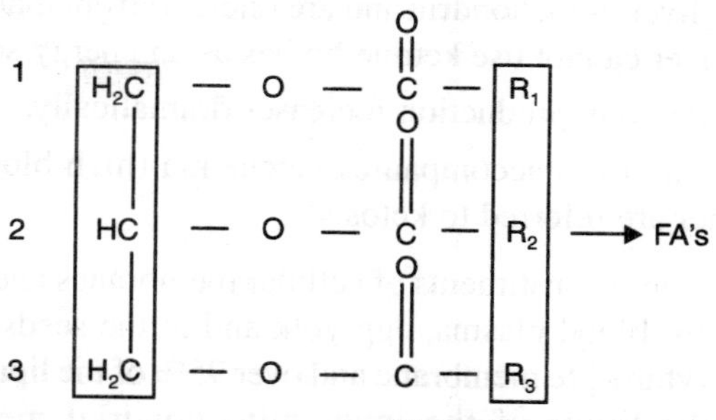

Glycerol backbone

They are also known as Triglycerides, fats or neutral fats.

Fats containing same kind of FA's in all three positions are called simple triacyl glycerols.

Example: Tristearin – 16 : 0
Tripalmitin – 18 : 0
Triolein – 18 : 1 respectively.

These provide stored energy and insulation. As stored fuels, fats have 2 advantages over glycogen and starch.

(1) The C-atoms of FA's are more reduced than those of sugars, and oxidation of fats yield more than twice as much energy compared to carbohydrates.

(2) As the fats are hydrophic and unhydrated, the organism that carry fats does not have to carry extra weight of water of hydration that is associated with stored polysaccharides.

Hydrolysis of Fats Produces Soaps

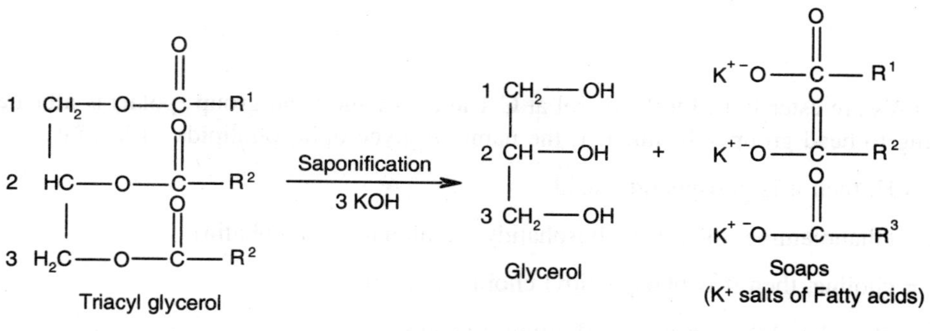

Lipolysis: The appearance of fatty acids in the blood during fasting is due to the mobilization of fat stores by the process of lipolysis.

- **Utilization of free FA's:** Used by most tissues except brain as metabolic fuel or as a source of carbon for biosynthesis of lipids.
- **Utilization of Glycerol:** Used by liver as a substrate for gluconeogenesis.

Ketone Bodies: are small, water soluble, potential units of acetate, acetone, acetoacetate and β-hydroxy butyrate, all of which are formed from FA's and carbohydrates.

Synthesis of Ketone Bodies

- They are synthesised in liver mitochondria and are released from that organ for used as metabolic fuel by other tissues. Liver cannot use ketone bodies as an energy source.
- During starvation, ketone body production increases dramatically.
- Ketonuria (high output in urine) accompanies ketonemia (high blood levels of ketone bodies); these conditions together are referred to ketosis.

Phospholipids are the major lipid constituents of cellular membranes and occur in high concentration in the lipids of glandular organs, blood plasma, egg yolk and in the seeds of legumes. They comprise about 40% of the lipids in the erythrocyte membrane and over 95% of the lipids in the inner mitochondrial membrane. About 20% of the lipids of the inner mitochondrial membrane are cardiolipin, a phosphoglyceride. Phospholipids are lipids that contain phosphorous and have backbone of glycerol (*e.g.* phosphoglycerides) or sphingosine (*e.g.* sphingomyclin).

Glycerophospholipids (or) Phosphoglycerides: are derivatives of phosphatidic acid.

General Structure

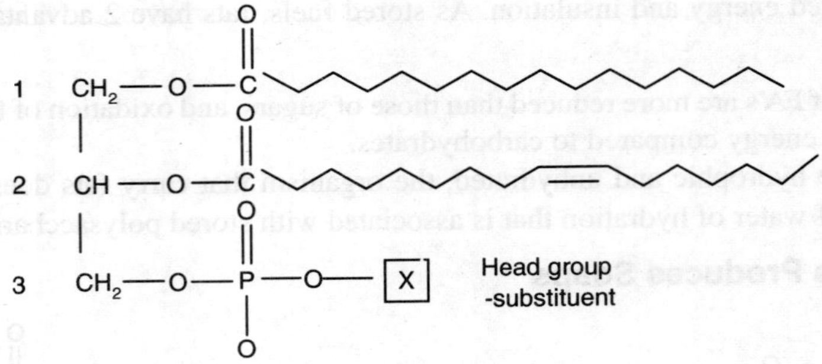

Here, 2 FA's are ester linked to Glycerol at C-1 and C-2 and head group (polar) is attached to C-3. According to head group substituent *x*, the name of glycerophospholipids varies, *i.e.*,

if X = H, then it is phosphatidic acid

X = ethaholamine, then it is phosphatidyl ethaholamine (cephalin)

X = choline, then it is phosphatidyl choline (lecithin)

X = phosphatidyl, then it is cardiolipin glycerol

In cardiolipin, 2 phosphatidic acids share a single glycerol.

In general, glycerophospholipids contain a saturated FA at C-1 and an unsaturated FA at C-2.

Phospholipids with Ether linked FA's

Example: Plasmalogens and platelet activating factor

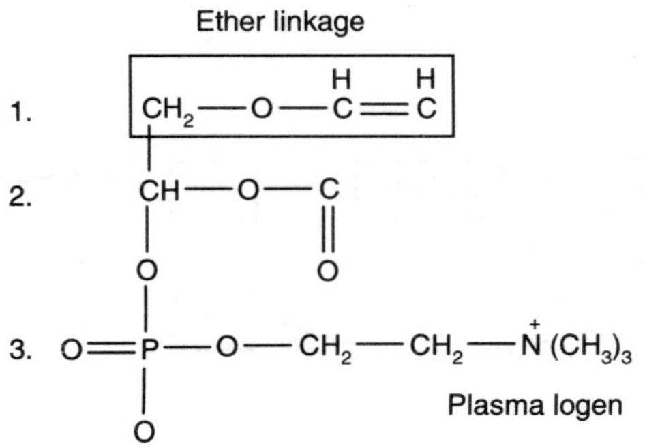

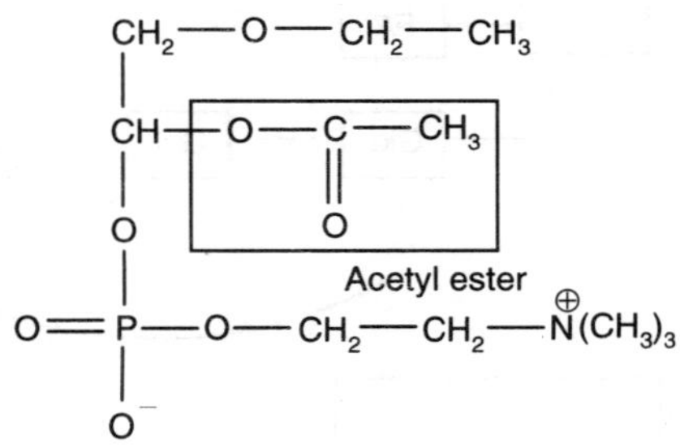

SPHINGOLIPIDS

They are derivatives of sphingosine, they are second large class of membrane lipids. They have a polar head and two non-polar tails. They contain no glycerol backbone.

General Structure

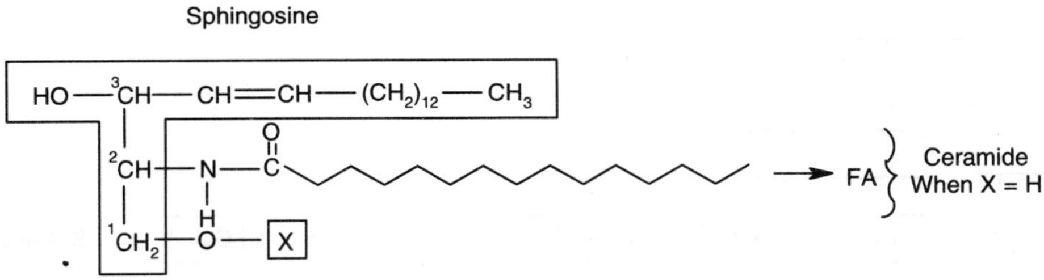

Name of X	*Name of sphingolipid*
H	Ceramide
Phosphocholine	Sphingomyclin
Glucose	Glucosylcerebrosides – A Glycolipid
Complex Oligosaccharide	Ganglioside GM_2

Sphingomyclins contain phosphocholine or phosphoethanolamine as their polar head group and are therefore classified as phospholipids, together with Glycerophospholipids.

Cerebrosides have single sugar linked to ceramide. Gangliosides, the complex sphingolipids, contain very large polar heads made up of several sugar units. One or more of terminal sugar units of gangliosides is N-acetyl neuraminic acid, also called sialic acid.

Sphingolipids are sites of biological recognition: Sphingolipids are now known to be involved in various recognition events at the cell surface.

Example: Glycosphingolipids are the determinants of human blood groups, A, B and O.

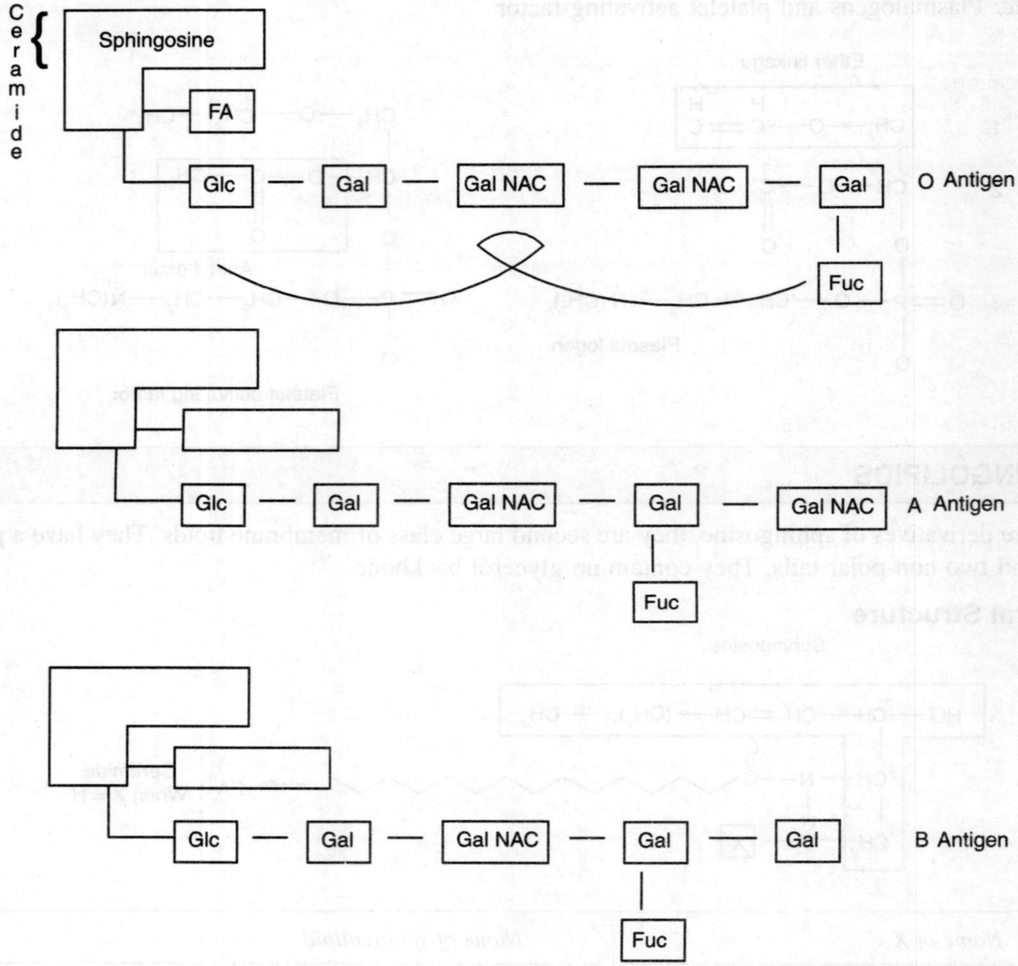

Glc — Glucose

Gal — Galactose

Gal NAC — N-Acetyl Galactosamine

Fuc — Fucose

Sphingolipidoses

There are a number of inherited genetic disorders referred to as lipid storage diseases, in which there is deficiency of an enzyme that is involved in the normal catabolism of a particular sphingolipid. Common sphingolipids are summarized in the Table 2.1 as shown.

Table 2.1: Sphingolipidoses

Disease	*Lipid accumulated*	*Enzyme deficiency*	*1° organ affected*
Niemann-pack	Sphingomyelin	Sphingonyelinase	Brain, Liver, Spleen
Gaucher's	Glucocerebroside	β-Glucosidase	"
Krabbe's	Galacto Cerebroside		Brain
Tay Sachs	Ganglioside, GM2	Hexosaminidase-A	Brain

EICAS ANOIDS are products of the metabolism of the 20-C poly unsaturated FA-Arachidonic Acid.

Prostaglandins (PG)

They are widely distributed in tissues, but their role is not yet fully understood. They have biological activities, include

- Smooth muscle contraction and relaxation
- Gastric secretion
- Platelet aggregation
- Inflammatory response
- Response to trophic hormones
- Sodium and water retention by kidney tubules.

Structure

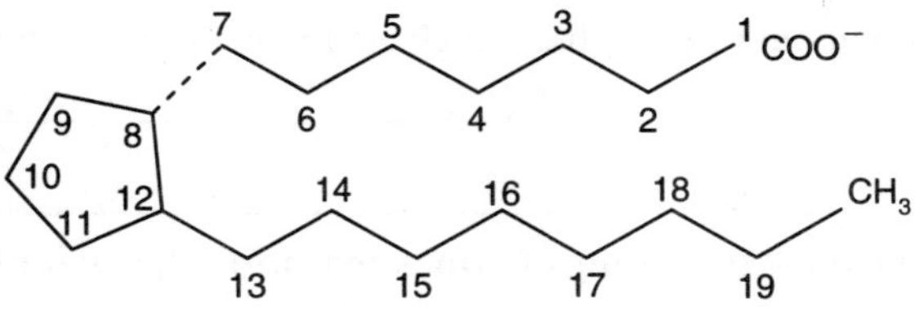

7-Carbon chain attached to C-8
& 12-carbon chain attached to C-12

Thrombaxanes

- Thrombaxane A2 (T × A_2) is produced by platelets; it causes contraction of arteries and triggers platelet aggregation.
- These are abbreviated as TX.

Structure

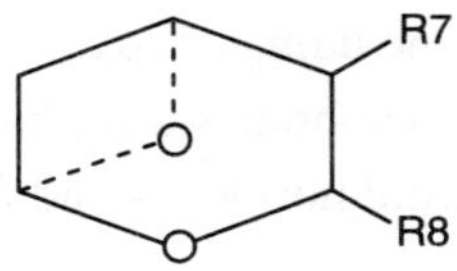

Thrombaxane (TXA)

They are analog of PG, but they possess a six-membered, oxygen containing ring.

Hydro peroxyeicasotetraenoic acids (HPETEs)

HPETEs are hydroxy FA derivatives of Arachidomic acid that do not contain a ring structure.

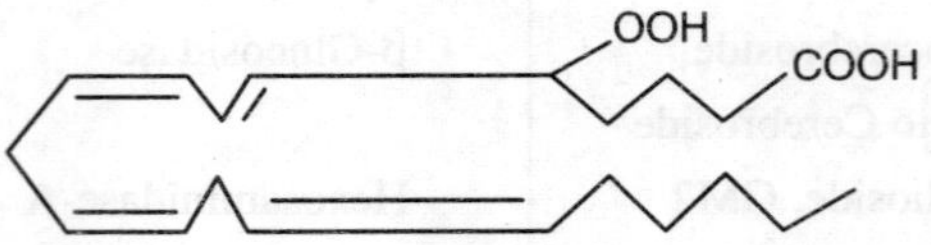

5-HPETE

It is not clear whether or not HPETE's are biologically active; however, they can be converted to active compounds like leukotrienes.

Leukotrienes

They are formed from HPETEs by Lipoxygenase and have common feature of 3 conjugated double bonds.

They are abbreviated as LT, are derived from Arachidonic acid.

They are involved in chemotaxis, inflammation and allergic reactions.

- Leukotriene D_4 (LTD_4) has been identified as the slow reacting substance of anaphylaxis (SRS-A) which causes smooth muscle contraction and is about 1000 times more potent than histamine in constricting pulmonary airways.
- LTB_4 attracts neutrophils and eosinophils which are found in large numbers at sites of inflammation.

STEROIDS

These are lipids which contain 4 fused rings C-rings that form the steroid nucleus, Cyclopentanoper hydrophenanthrene

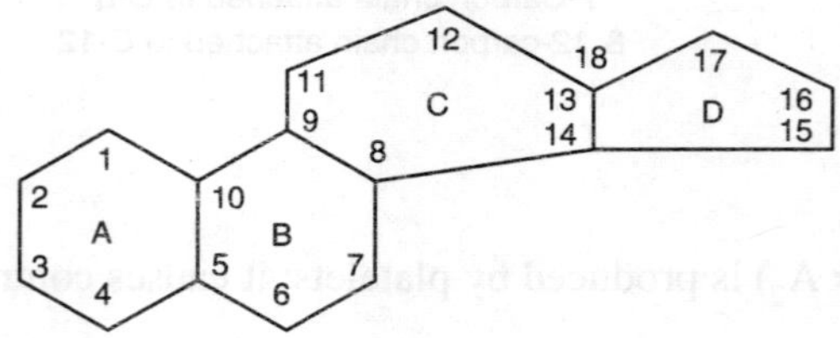

Sterols are a class of steroids characterized by

- A hydroxyl groups a C-3.
- An aliphatic chain of atleast eight carbons at C-17.
- Cholesterol is the major sterol in the human body.
- It is structural component of cell membranes and plasma lipo proteins.
- It is the precursor from which steroid hormones and file acids are synthesized.

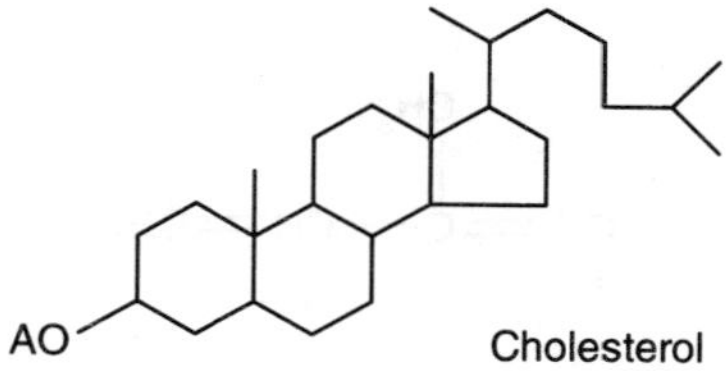

Cholesterol

Bile Acids: The predominant bile acids in humans are cholic, chenodeoxycholic and lithocholic acids.

Structure

The bile acids are C-24 steroids and are derived from cholesterol (a C-27 steroid) by scission of the side chain, which leaves

- A C-24 carboxyl group with the loss of three carbons.
- Saturation of the Δ^5 double bond of cholesterol.
- Hydroxylation of the steroid nucleus.
- Cholic and chenodeoxycholic acids are formed in the liver from cholesterol and are known as the primary bile acids. Deoxycholic acids and lithocholic acid are known as secondary bile acids because they are formed from the primary bile acids; this occurs in the intestine through the action of intestinal bacterial enzymes.
- The bile acids are conjugated to glycine or taurine in the liver to form glyco or taurocholate.

Continuous conversion of cholesterol to bile acids prevents the excessive accumulation of cholesterol in tissues. Bile acids are excreted in the faeces.

1° Bile Acids

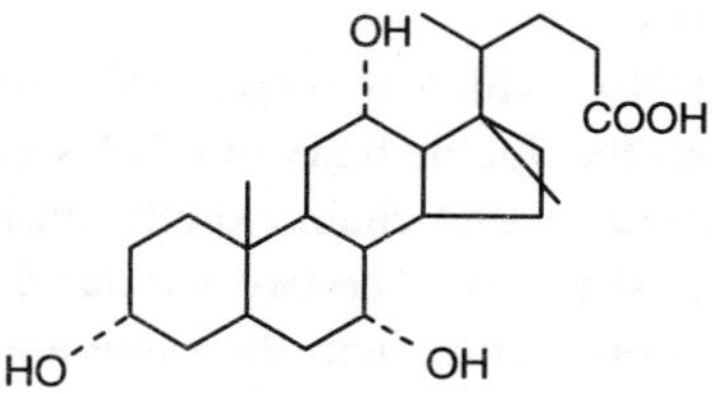

Cholic acid

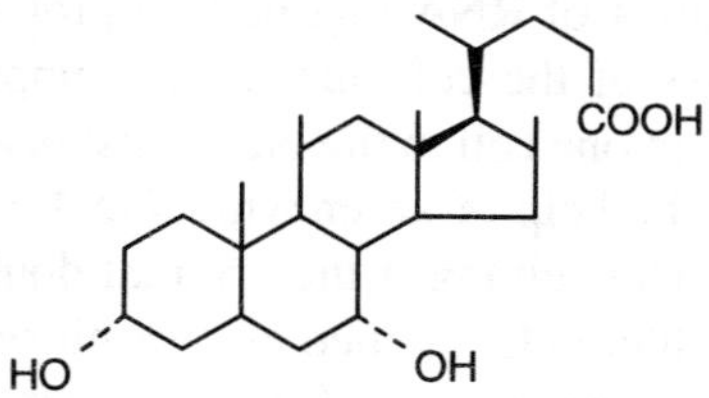

Chenodeoxycholic acid

2° Bile Acids

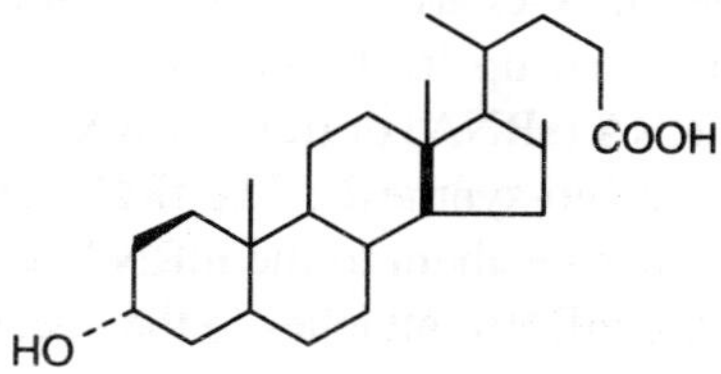

Lithocholic acid

All the steroid compounds that share the 4 ring steroid nucleus are synthesized from 5°C precursors related to isoprene.

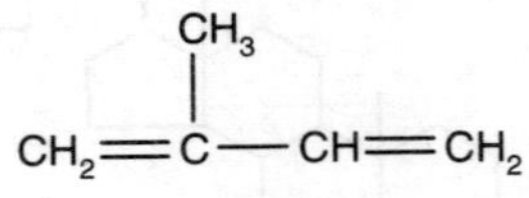

Purines, Pyrimidines

In the synthesis of purine, glycine provides carbon atoms at position 4 and 5 and, nitrogen atom at position 7 of purine nucleus.

In the synthesis of purine and pyrimidine β-carbon is the source of the methyl groups of thymine and of the carbons at positions 2 and 8 of purine nucleus.

In the synthesis of purine Glutamine donates nitrogen at positions 3 and 9 of purine nucleus. In the synthesis of purine esperatic acid takes part in donating the nitrogen at position 1 of purine nucleus.

In the synthesis of pyrimidine initially aspartic acid reacts with carbamyl phosphate and ureidosuccinic acid, uridine-5-phosphate is formed ultimately.

Metabolism of purine may take place under two conditions — first, when the purines remain combined as nucleoside and secondly, when they remain free as adenine and guanine. The end product of purine metabolism is chiefly uric acid and slightly hypoxanthine and xanthine. All purine bases undergoing metabolic changes in the body, are not excreted in these forms. A good deal of it is lost in some unknown way.

The intermediate metabolism of pyrimidine bases *e.g.*, cytosine, uracil and thymine is not clear. The catabolism of pyrimidines occurs mainly in the liver. Little is known about the fate of pentose. Probably they are oxidised.

Nucleic Acids: DNA and RNA

It is a well established fact that genetic information from one cell to the daughter one is carried by DNA and different forms of RNA that help in protein synthesis.

Chromosomes of the cell nucleus is composed of DNA, which is responsible for the passage of genetic code from one cell to the other. Just before mitosis the double helix of DNA separate from each other and with the help of an enzyme, DNA polymerase, a second chain exactly similar to the parent one is formed. Thus giving it the original double helical structure. The two similar double helixes go to the two daughter cells formed as a result of cell division giving them the same amount of DNA as the parent cell. In the process of meiosis such duplication of DNA in the parent cell does not occur so the daughter cell contains half the number of original chromosomes.

Two types of RNA are also formed in the nucleus. One of them is single stranded messenger RNA (mRNA) which has a smaller molecule than DNA. Its synthesis is catalyzed by the enzyme RNA polymerase. On being synthesized the mRNA comes out of the nucleus and becomes incorporated into the ribosomes of the cytoplasm. The mRNA carries the genetic information to determine the sequence in which the amino acids are to be lined up in the polypeptide chain. In this process of synthesis, another type of RNA called soluble RNA (sRNA) or transfer RNA (tRNA) is also formed in the nucleus to participate in the process of protein bio-synthesis. The tRNA attaches itself to the activated amino acids forming sRNA amino acid and carries them to the mRNA at the ribosomes. The sRNA finds out the definite code word or codon on the mRNA, attaches to the specific amino acids and is set itself free to be used again. The long mRNA thread has a number of ribosomal structures known as polysomes (polyribosomes, ergosomes) which are thus responsible for the synthesis of many amino acids forming a polypeptide. Finally the mRNA is set free from the polypeptide chain of the completed protein and renews its help in protein synthesis.

AMINO ACIDS

The most aboundant macromolecules in living cells are the Proteins. **The name protein is taken from the greek Proteios which means 'First' or 'foremost'.** The proteins are found in all cells in great variety, having many a diverse biological functions.

Chemically, proteins are polymers. They are polyamides and the monomers from which the diverse proteins are formed are the α-amino carboxylic acids, which are 20 in number.

All the 20 amino acids in proteins have a carboxyl group, an amino group bonded to the same carbon and they differ from each other in the 'R' group, which vary in structure, size, electric charge and solubility in water. **These 20 amino acids are referred to as the standard primary (or) normal amino acids to distinguish them from other kinds of amino acids occurring in living organisms but not in proteins.**

The standard amino acids have been given 3-letter abbreviation and one letter symbols as shown in Table 2.2.

Table 2.2

Amino acid	*Three letter abbreviation*	*One letter symbol*
Alanine	Ala	A
Arginine	Arg	R
Asparagine	Asn	N
Aspartic acid	Asp	D
Cysteine	Cys	C
Glutamine	Gln	Q
Glutamic acid	Glu	E
Glycine	Gly	G
Histidine	His	H
Isoleucine	Ile	I
Leucine	Leu	L
Lysine	Lys	K
Methionine	Met	M
Phenylalanine	Phe	F
Proline	Pro	P
Serine	Ser	S
Threonine	Thr	T
Tryptophan	Trp	W
Tyrosine	Tyr	Y
Valine	Val	V

All the standard amino acids except glycine have a chiral carbon/chiral center and thus show optical activity, *i.e.* they are capable of rotating the plane of polarized light, thus showing optical isomerism.

Laevorotatory (–) and Dextrorotatory (+) isomers are the two forms of enantiomers (optical isomers which are mirror images and not superimposable).

Threonine and isoleucine are the only amino acids which have 2 chiral center and hence have $2^n = 2^2 = 4$ stereoisomers.

Optical activity is expressed quantitatively by its specific rotation, which is related as follows.

$$[\alpha]_D^{25°C} = \frac{\text{observed rotation, deg}}{\text{length of tube, dmx concentration g/ml}}$$

dm = 0.1 m; D line is usually of sodium 589 nm. Naming of optical isomers is done by either of the two methods (a) D L system or Absolute configuration and (b) R S system.

All optically active amino acids in biological system exist in L-stereoisomer and the unique ability of cells to synthesise only L-forms is due to the stereospecific enzymes.

CLASSIFICATION OF AMINO ACIDS

Amino acids can be grouped into 4 classes based on the properties of their 'R' groups namely their *polarity*.

I. Amino acids with non-polar (or) hydroprobic 'R' groups

Eight amino acids have non-polar 'R' groups which includes five amino acids with aliphatic 'R' groups (alanine, valine, leucine, isoleucine and proline), two amino acids have aromatic rings (phenylalanine

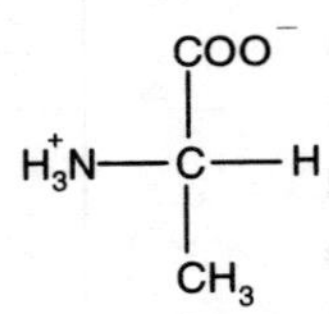

Ala

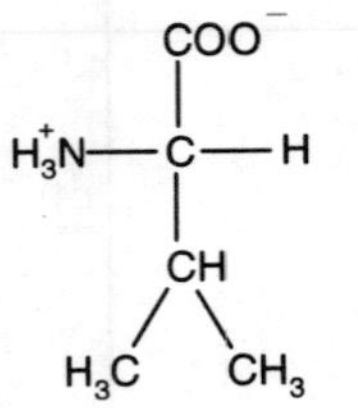

Val

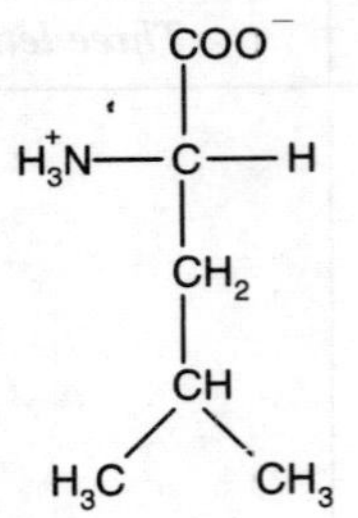

Leu

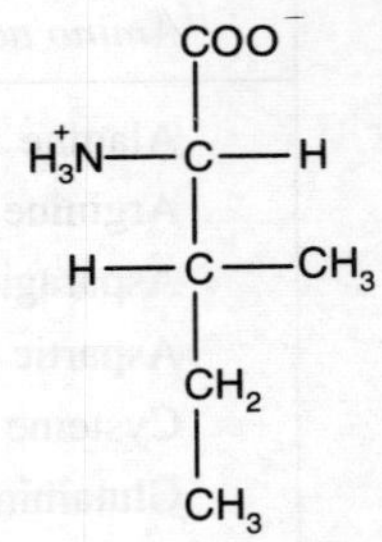

Ile

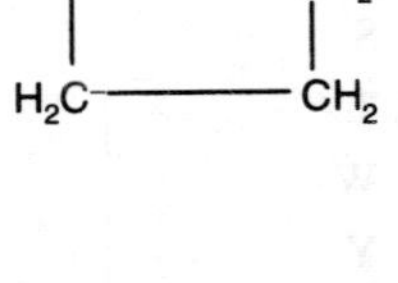

Pro

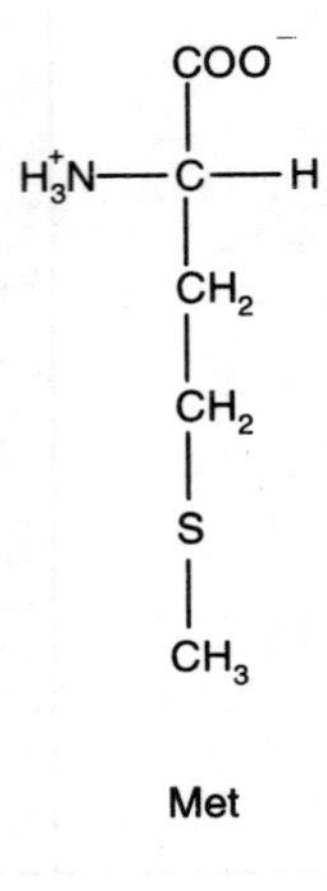

Met

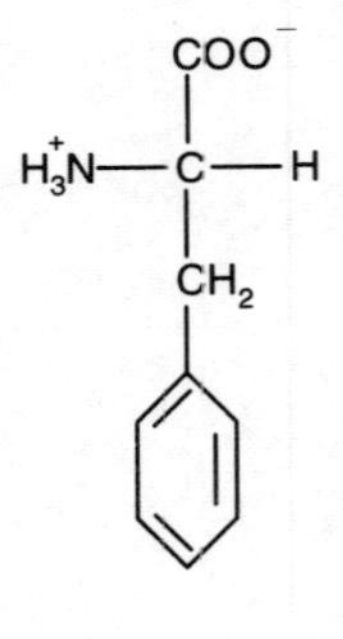

Phe

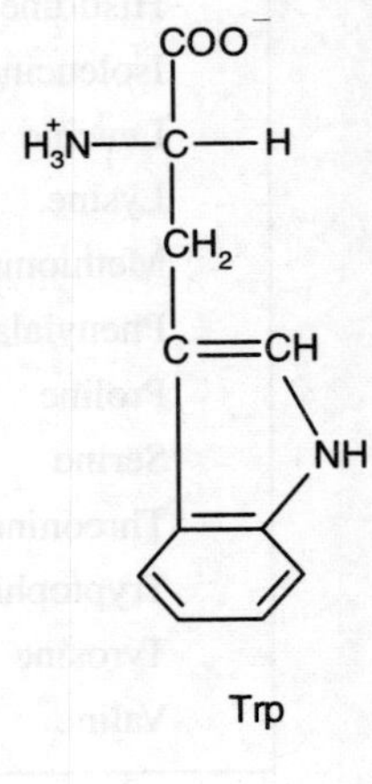

Trp

and tryptophan) and one has sulphur (methionine) proline requires a special mention in that its α-amino group is not free but substituted to yield a cyclic structure.

II. Amino acids with unchanged polar 'R' groups

Seven amino acids are more soluble in water than the non-polar amino acids, they include Glycine, which is too small to influence the polarity of amino and carboxyl groups. Serine, Threonine and Tyrosine are polar due to their hydroxyl group. While the amide groups of Asparagine and Glutamine contribute to their polarity, and cysteine has a thiol group.

Cysteine occurs in 2 forms in proteins either as cysteine itself or as **Cystine** in which 2 cysteine molecules are covalently joined by a disulphide bridge formed by oxidation of thiol group. Such cystine cross linkages are not usually found in intracellular proteins but are common in extracellular proteins.

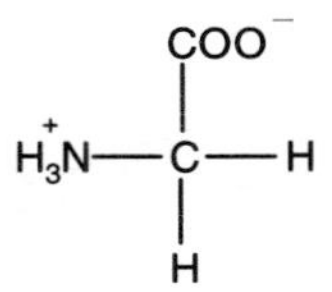
Gly

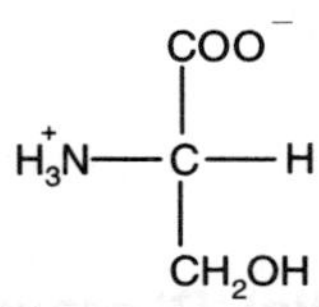
Ser

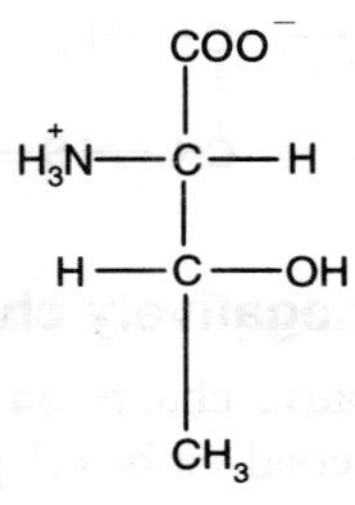
Thr

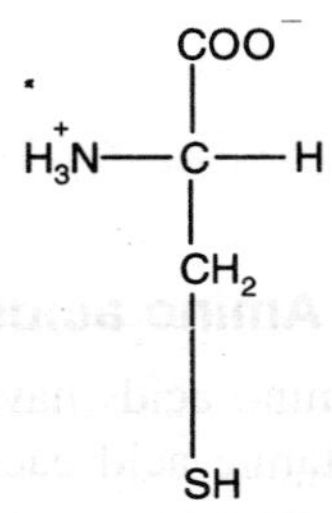
Cys

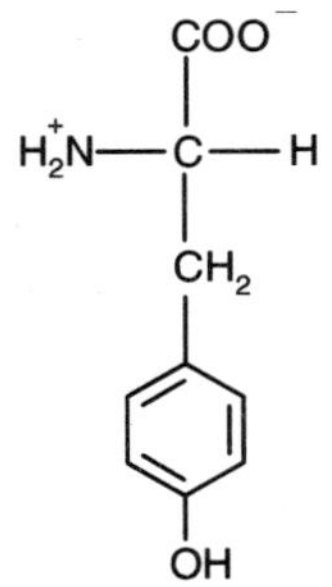
Tyr

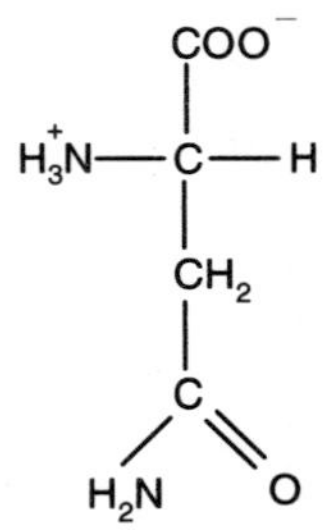
Asn

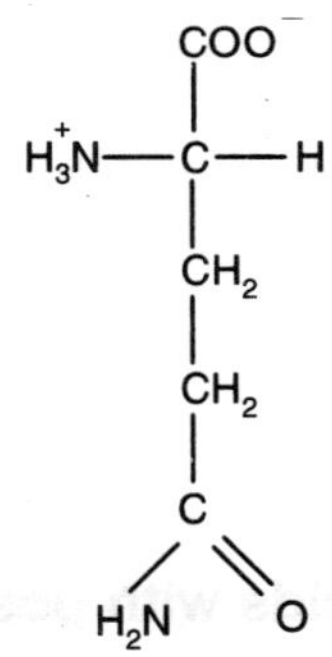
Gln

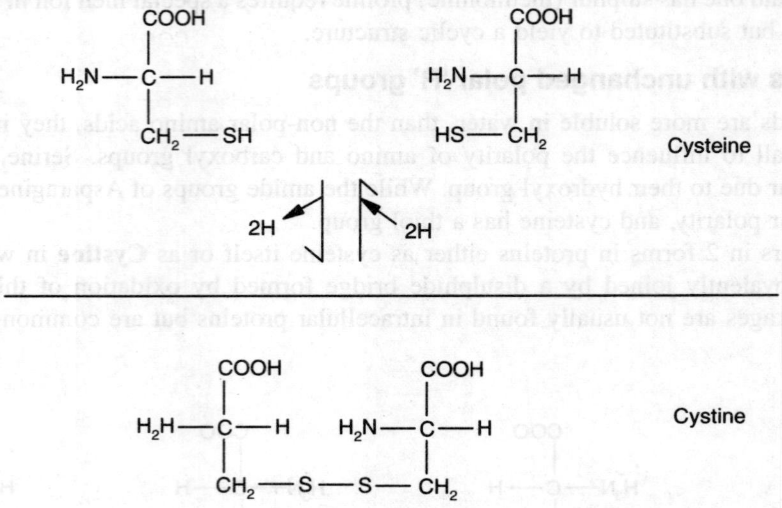

III. Amino acids having negatively charged (acidic) 'R' groups

2 amino acids have a net negative charge on their 'R' groups at pH 7.0. They are Aspartic acid and Glutamic acid each with a second carboxyl group. These are parent compounds of Asparagine and Glutamine respectively.

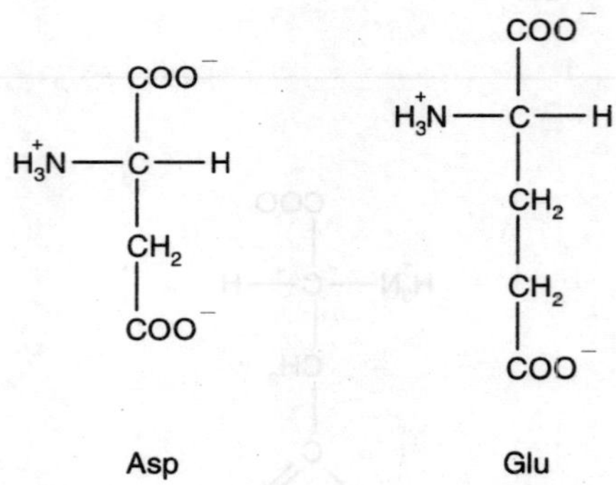

IV. Amino acids with positively charged (basic), 'R' groups

3 amino acids have a net positive charge on their 'R' groups at pH 7.0. Lysine has a second amino group. Arginine has a positively charged guanidino group and Histidine contains the imidazole group which is ionized.

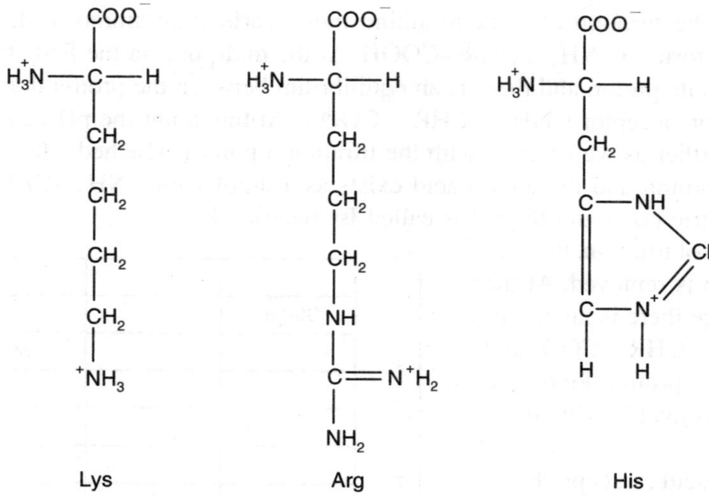

Characteristic Properties of α-Amino Acids

α-Amino acids having a single amino group and a single carboxyl group crystallize from neutral aqueous solutions as fully ionized species called Dipolar ion or zwitter ion. The dipolar nature was suggested by the fact that crystalline amino acids have melting points higher than organic molecules of similar molecular weight and size.

A fully protonated amino acid is diprotic as indicated below:

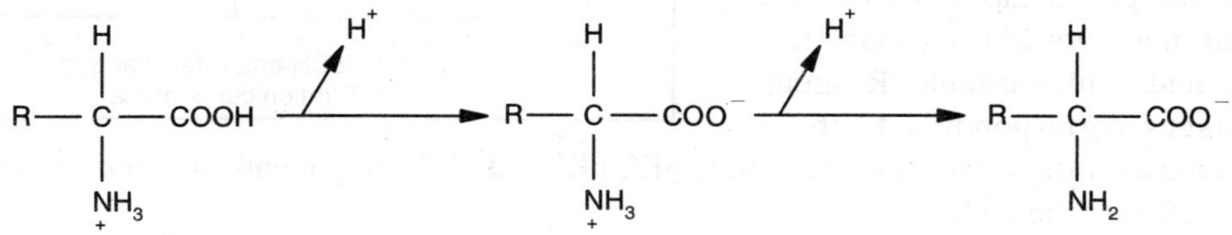

Amino acids are amphoteric and are called ampholytes as they can act, either as acids or as a base.

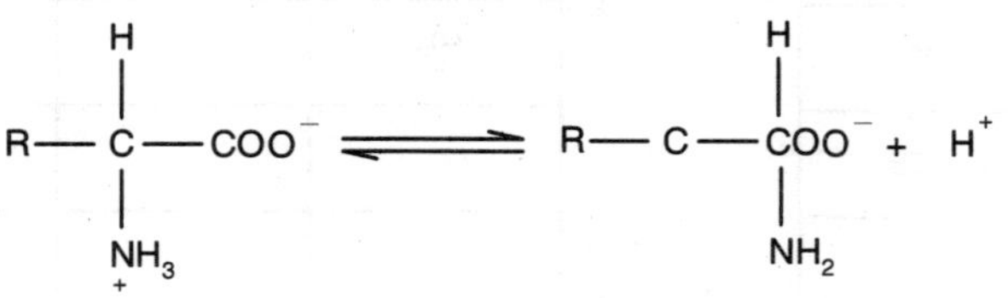

Amino acid as an acid

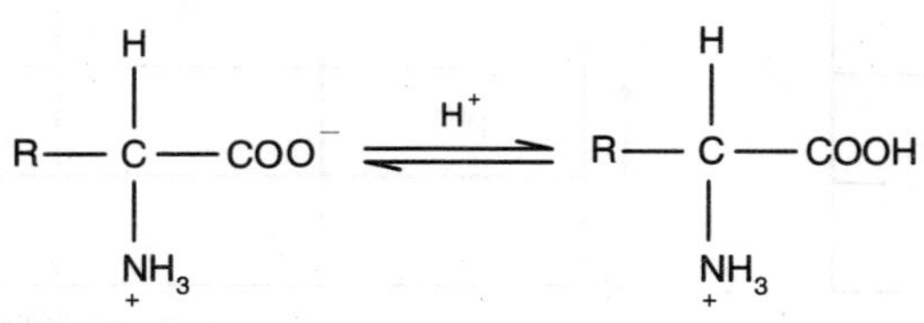

Amino acid as a base

Titration Curves of Amino Acids

At the beginning of the titration of a mono amino mono carboxylic amino acid, the amino acid is predominantly in the form of $^{+}NH_3$ – CHR –COOH. At the mid-point in the first stage of titration the carboxyl group losses its proton and there is an equilibrium between the proton donor $–NH_3$ – CHR – COOH) and the proton acceptor ($^{+}NH_3$ – CHR – COO^{-}). At this point the pH equals to its pk of the protonated group. Further as we proceed with the titration a point is reached where there is complete removal of the first proton and the amino acid exists as a dipolar ion $^{+}NH_3$ –CHR – COO^{-}. This is referred to as Isoelectric point and the pH is called Isoelectric pH.

In the second stage of titration, the proton from $^{+}NH_3$ group is removed. At the mid-point of this stage there is an equilibrium between $^{+}NH_3$ – CHR – COO^{-} and NH_2 – CHR – COO^{-} (proton donor and proton acceptor respectively). This point is referred to as pK_2.

Further the amino acid exists predominantly as NH_2 – CHR – COO^{-}.

The isoelectric points is the arithmetic mean of the 2 pK values.

$$pHI = 1/2\left(pK_1' + pK_2'\right)$$

At any pH above the isoelectric point the amino acid has a net negative charge and thus moves toward the anode. At any pH below the pH_1 it has a net positive charge and moves towards the cathode.

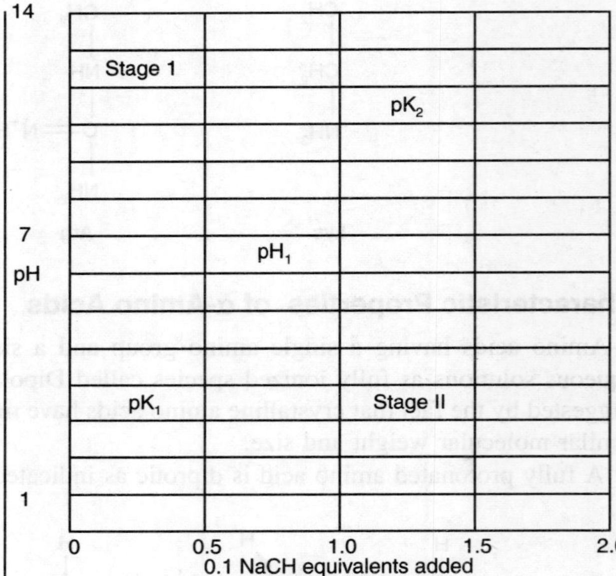

Titration curve of Ala

Amino acids with ionizable 'R' group have 3 stages corresponding to the 3 possible ionization steps, thus they have three pK′, pK_1' and pK_2' being identical to the above case. The 3rd stage, refers to the pK_R'.

Titration curves of Glutamic acid and Histidine are as shown below:

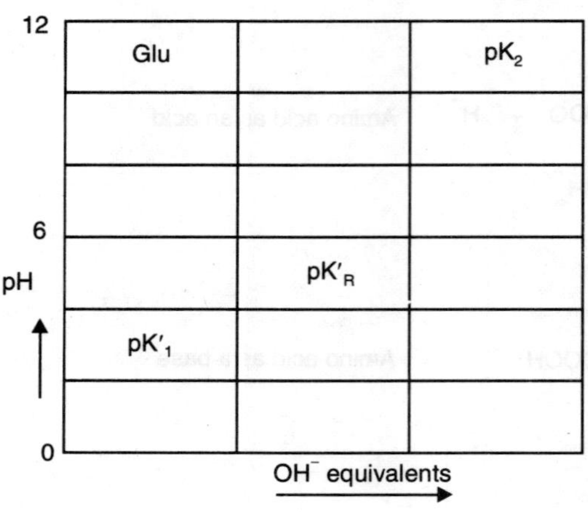

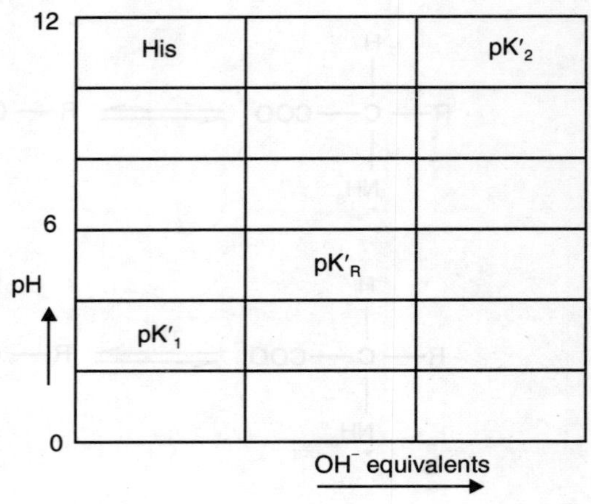

The isoelectric points of amino acids in these class reflect the type of 'R' groups. For example: Glutamic acid has an isoelectric point equal to the arithmetic mean of its pK_1' pK_R'

$$pH_1 = 1/2(pK_1' + pK_R')$$

which is considerably lower than that of Alanine.

Similarly isoelectric point of this is the arithmetic mean of pK_R' and pK_2'

$$pH_1 = 1/2(pK_R' + pK_2')$$

which is much higher than that of Alanine.

Though at their pK_1' or pK_2' amino acids should be having significant buffering action, they are not good buffering agents at the physiological pH of 7.0, since their pK_1' and pK_2' are far away from pH 7.0 to be effective buffers. Histidine is an exception, in which its pK_R' is 6.0 and thus provides for a good buffering action at pH 7.0, they occur in large amounts in hemoglobin and thus act as good buffer near pH 7.0, which is important for the role of RBCS.

The differences in the acid base properties of different amino acids, *i.e.* the magnitude and nature of sign at a given pH, forms the basis for the separation of amino acids from a mixture with the help of electrophoresis and ion-exchange chromatography.

pK′ values for ionizing groups of some amino acids at 25°C

Amino acid	pK_1' (–COOH)	pK_2' ($-NH_3^+$)	pK_R' (R group)
Gly	2.34	9.6	
Ala	2.34	9.69	
Leu	2.36	9.60	
Ser	2.21	9.15	
Thr	2.63	10.43	
Gln	2.17	9.13	
Asp	2.09	9.82	3.86
Glu	2.19	9.67	4.25
His	1.82	9.17	6.0
Cys	1.71	10.78	8.33
Tyr	2.20	9.11	10.07
Lys	2.18	8.95	10.53
Arg	2.17	9.04	12.48

Chemical reactions of amino acids.

1. Ninhydrin reaction

It is used to detect and quantitatively estimate amino acids. All amino acids except proline which have a free α-amino group when heated with Ninhydrin give a purple product while proline gives a yellow product.

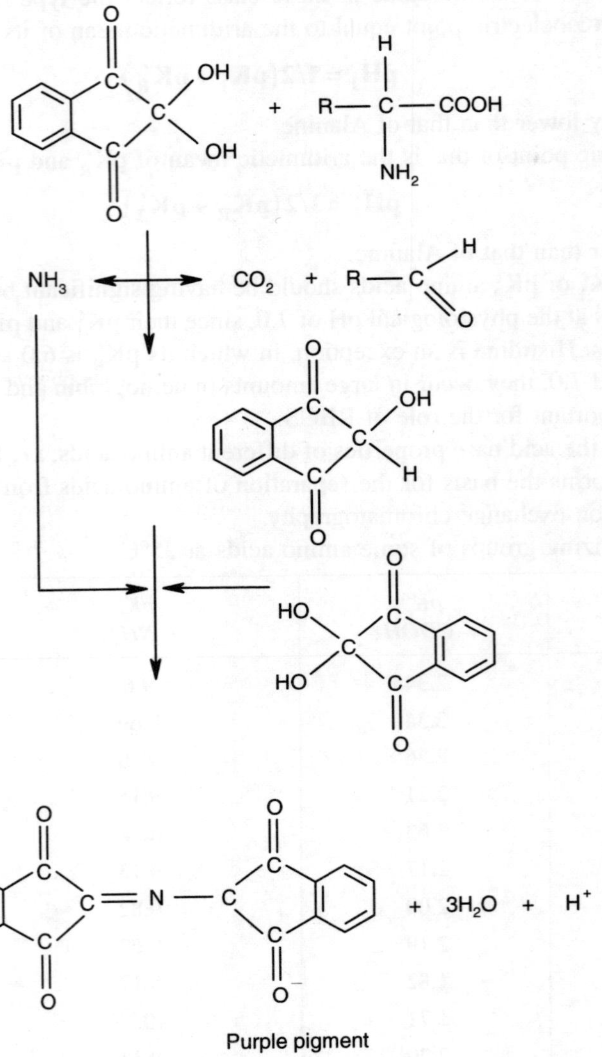

2. 1-Fluoro-2, 4-Dinitro benzene (FDNB) reaction

In mildly alkaline solution FDNB reacts with α-amino acids to yield 2,4-Dinitro phenyl derivatives which are useful in identifying individual amino acids.

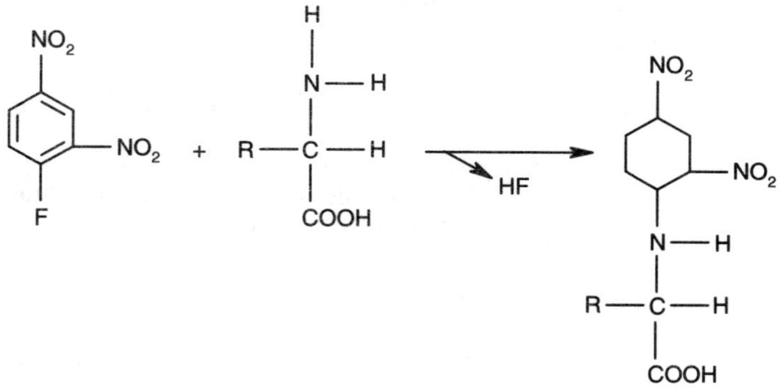

2,4-dinitrophenylamino acid

3. Peptide bond formation

2 amino acid molecules covalently join through a substituted amide linkage called a peptide bond to yield a dipeptide similarly 3 amino acids are joined by 2 peptide bonds to form a tripeptide.

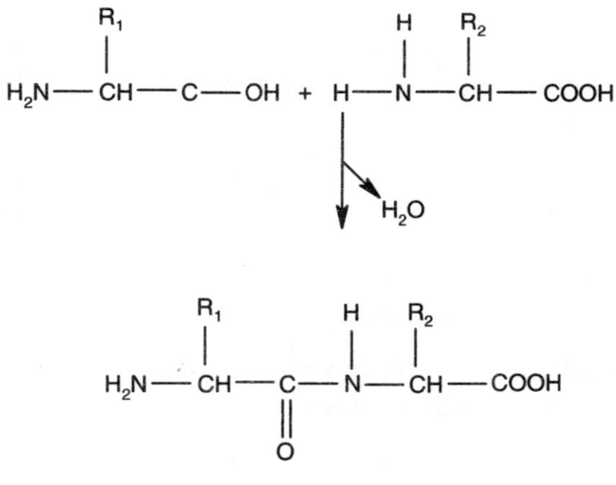

Similarly a polypeptide is formed which has one amino acid with a free α-amino group this is called amino-terminal residue or N-terminal residue and the residue at the opposite end has a free carboxyl group this is referred to as the carboxyl terminal @ C-terminal residue.

Peptides are named in a sequence beginning from N terminal residue towards the 'C' terminal residue.

PROTEINS

Proteins consist of extremely long chains of many amino acid units, called polypeptides. 2 amino acid molecules condense to form a substituted amide linkage called peptide bond. **The amino acid units in a peptide are called residues, and a polypeptide has a N terminal residue (with a free α-amino group) and a C-terminal residue (with a free carboxyl group).**

Many peptides occur in free form in living matter not associated with protein structure. Many such peptides have intense biological activities, as shown in table.

Polypeptide hormone	*Gland secreted in*	*Number of amino acid residues*	*Function*
Insulin	B cells in pancreas	2 polypeptide chains one 30 amino acid another 21 amino acid	Stimulates capacity of cells to use Glucose as a fuel
Oxytocin	Posterior pituitary	Nine amino acid	Stimulates uterine contraction
Bradykinin		Nine	Inhibits inflammation
Thyrotropin releasing factor	Hypothalamus	3	stimulates release of thyrotropin.

Proteins consist of polypeptides and are the instruments by which genetic information is expressed. Proteins have many different biological functions, based on which proteins may be classified as follows.

Classification of proteins according to biological function

Class	*Example*	*Class*	*Example*
Enzymes	Ribonuclease, trypsin	Transport protein	Hemoglobin, myoglobin
Nutrient & storage protein	Ovalbumin, caesin	Contractile (or) motile protein	Actin, myosin
Structural protein	Keratin, fibroin, collagen	Defense proteins	Immunoglobins, thrombin botulinus toxin snake venom
Regulatory proteins	Insulin corticotrophin		

Proteins are also classified into 2 broad classes:

- **Fibrous proteins (which are insoluble in water)**
- **Globular proteins (which are soluble in water)**

Molecules of fibrous proteins are long and thread like and lie side by side to form fibers, while molecules of globular proteins are folded into compact units that often approach spherical shapes.

Some proteins contain only amino acid and no other chemical groups. Example: ribonuclease, chymotrypsinogen, etc. which are called as simple proteins while others contain some other chemical group in addition to amino acids, they are called Prosthetic group and the proteins is referred to as conjugated proteins.

Example: Lipoproteins have lipids. Haemo proteins have Heme (Iron porphyrin). Metallo proteins have metals like zinc, iron, as their prosthetic group.

The multitude of proteins differ in their biological activity, but all are made from the same 20 standard amino acids, differing in the sequence of these amino acids.

Primary Structure of Proteins

The sequential arrangement of amino acids in a given protein is referred to as its primary structure. All the molecules of any given type of protein are identical in amino acid composition, sequence and length of the polypeptide chain.

Proteins which perform the same function in different species are called Homologous proteins. Such homologous proteins have polypeptide chains that are identical or nearly identical in length. **Moreover many positions of amino acid sequences of such proteins are occupied by same amino acid in all species and thus are called invariant residues and in other positions there may be considerable variation in amino acid from one species to another, these are called variable residues.**

Such a similarity in amino acid sequence of homologous proteins is called sequence homology, which suggests common evolutionary origin of homologous proteins.

Determination of Primary Structure if Polypeptide

The amino acid sequences of a polypeptide chain is determined by principles first developed by Sanger, which involves the following steps:

1. Determining the amino acid composition

Hydrolysis of the peptide bond of the purified polypeptide is undertaken and the amino acid mixture so formed is subjected to ion-exchange chromatography and analysed to determine the amino acids and their relative numbers.

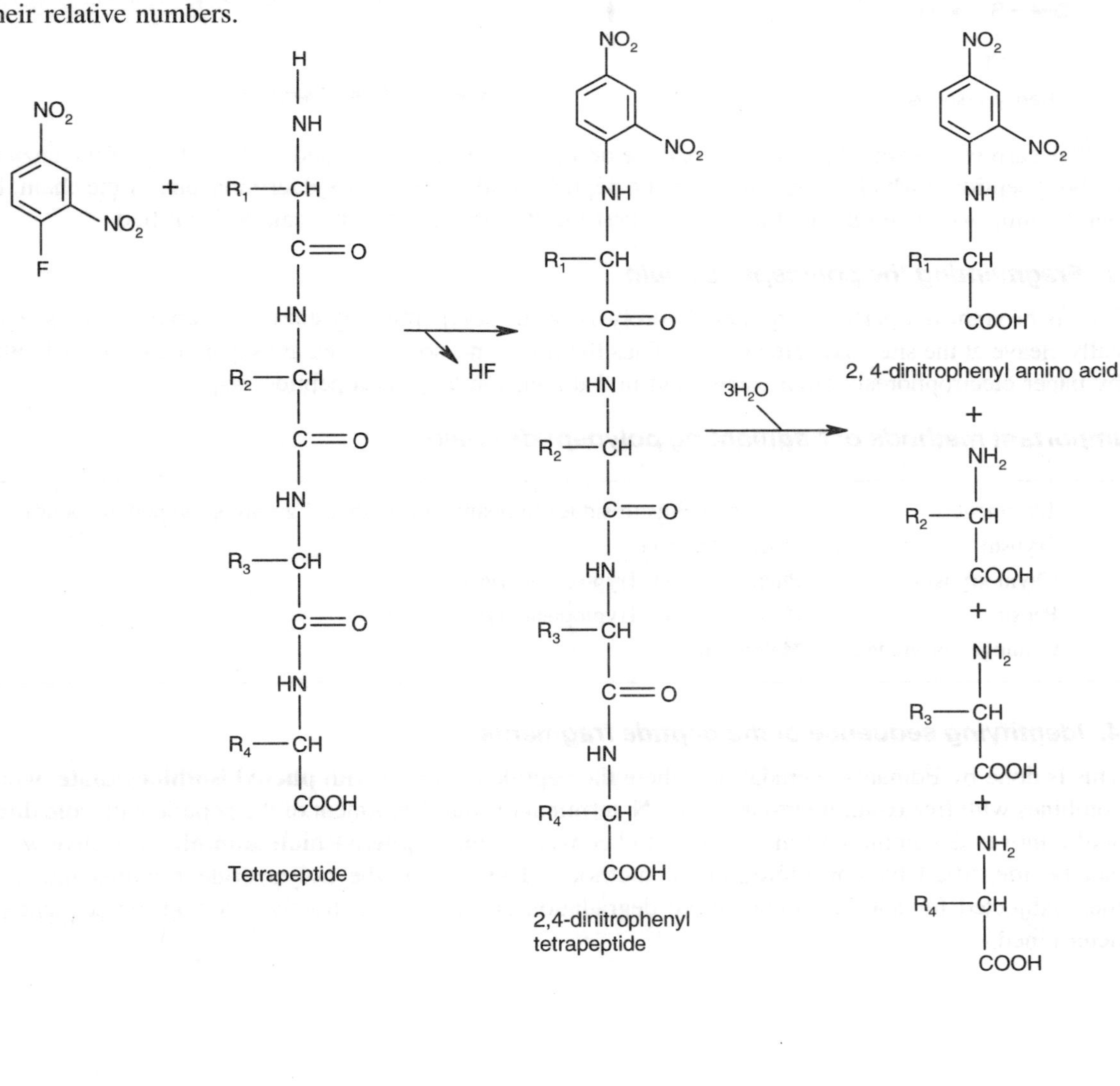

2. Identifying 'N' and 'C' terminal amino acids

The amino-terminal end is identified by with 1, fluoro- 2, 4-dinitro benzene, which labels the amino-terminal residue of the chain as the yellow 2, 4-dinitrophenyl derivative, which when subjected to acid hydrolysis all the peptide bonds are hydrolysed while the covalent bond between the 2, 4-dinitrophenyl group and the α-amino group of the amino-terminal residue is resistant, thus it occurs in the hydrolyzate as its 2, 4-dinitrophenyl derivative which can be separated and identified by chromatographic comparison with authentic dinitrophenyl derivatives of the different amino acids.

Another reagent used to label 'N' terminal residue is Dansyl chloride which reacts with the free α-amino group to yield dansyl derivative, which is fluroscent and can be detected easily after hydrolysis.

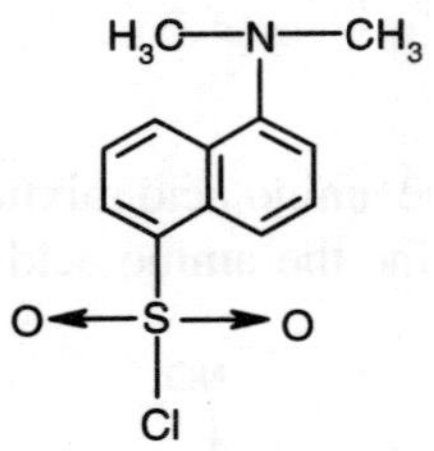

Dansyl chloride

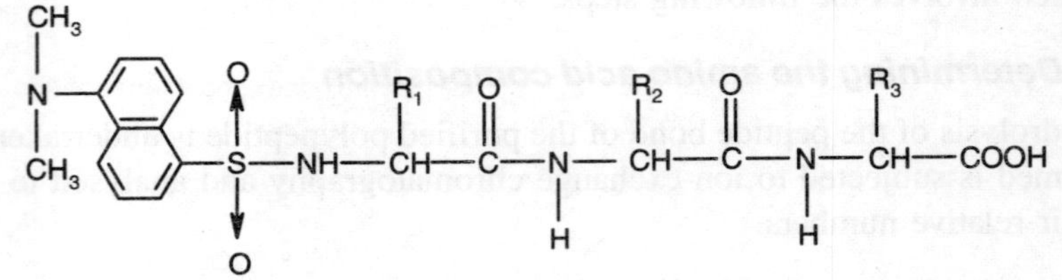

A dansyl labelled tripeptide

The carboxyl-terminal amino acid residue of a polypeptide is identified with the help of the enzyme carboxypeptidase which hydrolyses only the peptide bond at the carboxyl terminal end of the chain. By determining which amino acid is released first the 'C'-terminal residue can be identified.

3. Fragmenting the polypeptide chain

This is brought out partial enzymatic hydrolysis of the polypeptide by enzymes (table) which specifically cleave at the sites indicated (table). Thus the fragments so produced are separated from each other by paper electrophoresis often carried out in 2 dimension to yield a peptide map.

Important methods of fragmenting polypeptide chain

Treatment	cleavage points (residues furnishing the carbonyl group of the peptide bond)
Trypsin	Lysine, Arginine.
Chymotrypsin	Phenyl alanine, Tryptophan, Tyrosine.
Pepsin	Phenyl alanine, Tryptophan, Tyrosine etc.
Cyanogen bromide	Methionine.

4. Identifying sequence of the peptide fragments

This is done by Edman's degradation, where the peptide is treated with **phenyl isothiocyanate**, which combines with free α-amino group of the 'N' terminal residue. Treatment of the peptide with cold dilute acid removes the amino terminal residue and converts it into a **phenylthiohydantoin** derivative which can be identified by chromotographic methods. The rest of the polypeptide remains intact, is then subjected for another round of the degradation and by the subtractive method the sequence is determined.

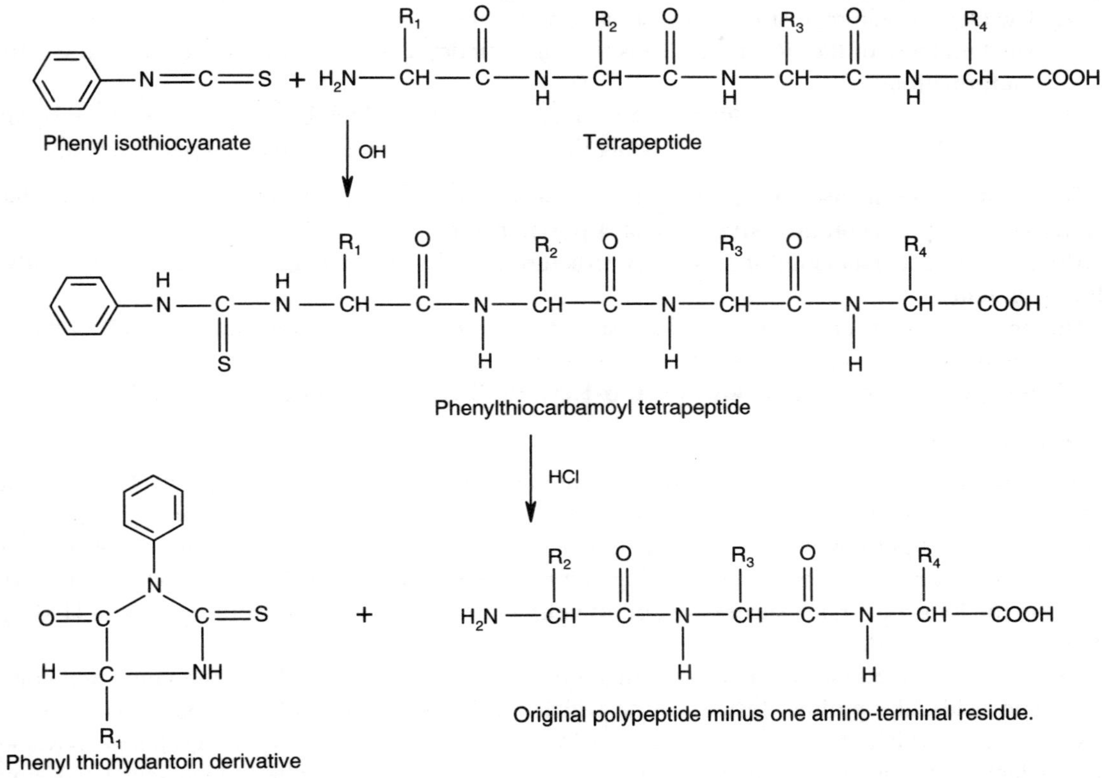

5. Cleavage of the original polypeptide chain by a second procedure

This is done to establish the order of the peptide fragments formed by one enzyme with now another enzyme or method and subjecting the fragments to edman's degradation.

6. Ordering peptide fragments by establishing overlaps

This is done by examining the sequences from the above fragments of the two procedures to establish continuity or overlaps. The 2 set of fragments check each other out for possible errors in determining the amino acid sequence of each fragment.

Secondary Structure

Polypeptides have single bonds right throughout their skeleton and thus we would expect the polypeptide to assume an infinite number of conformations in space. Further due to thermal motion and random rotation of segments of the chain around each of the single bonds in the back bone the conformations would constantly change.

However proteins in their native form have only one (or) very few conformations, which is accounted for by the following constraints on the conformation in space of a polypeptide:

1. **Rigidity of the peptide bond:** The C—N bond joining the 2 amino acids in a polypeptide is shorter than most other C—N bonds and it has a double bond character and thus cannot rotate freely as a result the Carboxyl Oxygen and the Hydrogen of the NH group are in *trans* position.

2. The electrostatic repulsion or attraction between amino acids with charged 'R' groups.
3. The bulkiness of the 'R' groups leads to stearic hindrances and hence restriction on the various conformations.
4. The occurrence of proline residues which has a rigid ring which lacks hydrogen on the N_2 atom as a result there is absence of 'H' bond, thus as a consequence it causes a kink or a bend.

The arrangement in space or conformation of successive amino acid residues in polypeptide chains is referred to as the **secondary structure of a polypeptide**.

Much of the understanding of secondary structure of polypeptide is being achieved by a study of fibrous proteins.

The fibrous proteins are arranged in long strands or sheets and these constitute one-half of the total body proteins in larger vertebrates. They provide external protection.

Fibrous proteins are of four kinds viz. α-Keratin, β-Keratin, collagen and elastin.

α-Keratin

They are the major fibrous proteins and constitute almost the dry weight of hair, wool, nails, claws, feathers, horns, tortoise shell and much of the outer layer of skin. It has an *α-helix* where the polypeptide backbone is tightly wound around the long axis of the molecule and the 'R' groups of the amino acid residues protrude outward. The repeating unit is a single turn of the helix which extends about 0.54 nm along the long axis. Each successive coil of helix is held to the adjacent coils by several Hydrogen bonds giving it considerable stability.

The α-Keratins are rich in amino acids that are compatible with the α-helix formation and contain few amino acids like proline, lysine, serine which are incompatible with it. α-Keratins are rich in cystine residues which provide disulphide cross-links between adjacent polypeptide chains. α-Keratins contain large amount of hydrophobic amino acids like val, met, ile, phe, which make α-Keratins insoluble in water. The polypeptide chains in α-Keratin are arranged in parallel fashion.

β-Keratin

The silk protein fibroin is stringy, insoluble, flexible and supple proteins which differ from α-Keratin in having a periodicity of 0.70 nm. In the β-conformation the backbone of the polypeptide chain is extended into a zigzag structure. The zigzag polypeptide chains are arranged side by side to form a series of pleats hence called a pleated sheet with no intrachain hydrogen bonds but has many interchain hydrogen chains. **The 'R' groups of amino acids protrude out from the zigzag structure and the adjacent polypeptide chains are antiparallel in nature**.

COLLAGEN AND ELASTIN

These are the two fibrous proteins which occur together in most connective tissues. Connective tissues consists of extra cellular structural and supportive elements of the body and occur as tendons, ligaments, cartilage, matrix of bones, etc. Tendors are largely composed of Collagen, while ligaments consist of Elastin. **The fibrous elements of connective tisue are embedded in the ground substance called Proteoglycans.**

The Collagen, Elastin and Proteoglycans are made by the fibroblasts and chondrocytes.

Collagen

It is the most abundant of proteins in the body. The collagenous connective tissue consists of fibers, which in inturn are made up of collagen *fibrils.* Which have a cross striated appearance, which are organized in different ways, depending on the biological function of the connective tissue. In the hide of the cow it forms an irregular, interlacing, very tough network, while cornea of the eye has sheets of collagen fibrils laid down in a criss cross manner.

Collagen contains about 35% *glycine* about 11% *alanine.* Apart from this it contains high content of proline and 4-hydroxy proline, together comprising 21%. The unusual composition of Collagen makes it, relatively poor in nutritional quality in the form of gelatin (soluble mixture formed by boiling Collagen in water).

Collagen fibrils consist of recurring polypeptide subunits called *tropocollagen*, arranged head to tail in parallel bundles. Tropocollagen sub-units consist of 3 polypeptide chains tightly called into a 3 strand rope. The 3 helically intertwined polypeptides are of equal length. In some, all three chains are identical in amino acid sequence but in others 2 chains are identical and the third differs. The polypeptide chains are tightly twisted and cross-linked to each other by Hydrogen bonds and by unusual covalent cross-link.

As we become older more and more covalent cross-links form in between tropocollagen unit rendering the collagen rigid and brittle and hence leads to associated problems of ageing like brittleness of bones, less transparency of cornea of the eye.

Elastin

The ligaments have greater elastin than Collagen apart from the connective tissue layer of large arteries. **The basic sub-unit of elastin is *tropoelastin*.** It is rich in Glycine and alanine and has a greater amount of lysine than proline. It forms a special type of helix where lengths of helix rich in Glycine are separated by short regions containing Lysine and Alanine. The regions containing lysine engage in the formation of covalent cross linkages. 4 lysine R groups form *desmosine* and *isodesmosine* which are capable of joining tropoelastin chains into arrays that can be stretched reversibly in all directions.

Tertiary Structure

In contrast to the fibrous proteins, the globular proteins are soluble in water and are folded into a compact Globular Shape. Two important pieces of evidences indicate the folded nature of the globular proteins:

(*a*) Native globular proteins undergo denaturation when heated, or exposed to pH extremes or treated with area. During which the covalent back bone remains intact but the polypeptide chain is unfolded into random, irregular and changing conformations in space, as a result it becomes insoluble in water and loses its biological activity.

(*b*) A comparison of the length of their polypeptide chains with their actual molecular dimensions revealed by physicochemical measurements indicate the highly folded nature of the globular proteins.

Tertiary structure thus designates the folded nature of the polypeptide chains of globular proteins into compact, spherical or globular shapes. Each globular protein has a distinctive tertiary structure, specialized for its particular biological function.

Tertiary structure is conferred by longer range amino acid sequence in contrast to the secondary structure which is determined by short-range amino acid sequence. Four kinds of interactions cooperate to hold the looped polypeptide chains of globular proteins in position under biological conditions of temperature, pH and ionic concentration.

1. **Hydrogen bonding between R groups of residue in adjacent, *loops* of the chain.**
2. **Ionic attractions between oppositely charged 'R' groups.**
3. ***Hydrophobic interactions* where 'R' groups of some amino acids shun an aqueous environment and thus tend to associate in the interior of the globular structure shielded from water.**

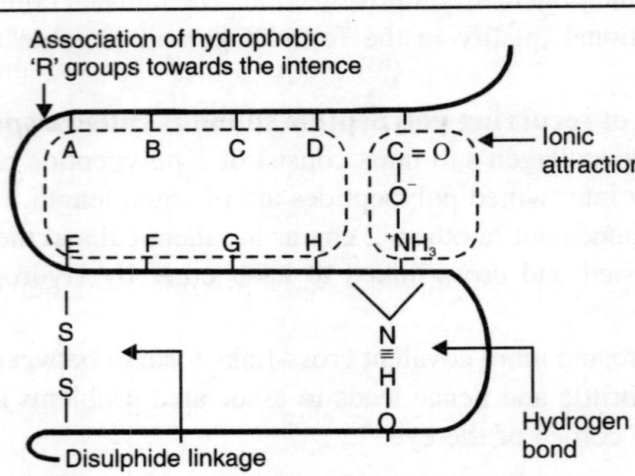

Factors maintaining the tertiary structure of Globular proteins

4. ***Covalent Cross linkages:* Many a polypeptide chains shows the formation of intra chain disulphide linkage, stabilizing the protein.**

Although the tertiary conformation of globular proteins are the stablest forms, it is not absolutely fixed and rigid and these proteins undergo changes in their conformation in the course of their biological function, for example, many enzyme molecules change their conformation when they bind their substrates as a part of their catalytic action. **The globular proteins have a certain amount of flexibility in their back bones and undergo short-range internal fluctuations, hence are considered to '*Breathe*'.**

Quarternary Structure

Proteins which are made up of two or more separate polypeptide chains are referred to as *Oligomeric proteins*. The polypeptides of which may be identical or different, for example Hemoglobin has 4 chains per molecule, pyruvate dehydrogenase complex has 72 units.

The term quarternary structure refers to the arrangement of the sub-unit chains in relationship to each other in the oligomeric protein.

Structure of Hemoglobin

The first oligomeric protein to be subjected to X-ray analysis was hemoglobin.

Hemoglobin consists of 4 polypeptide chains consisting of two α-chains (141 residues each) and two β-chains (146 residues each). This protein portion is referred to as Globin. In addition to the globin it has four heme prosthetic group in which iron atoms are in the ferrous state. Each of the four chains has a characteristic tertiary structure. The α and β chains contain several segments of α helix separated by bends (as in myoglobin). The four polypeptide chains fit together in an approximately tetrahedral arrangement, to constitute the characteristic quarternary structure. One heme group is bound to each chain. The hemes are rather far apart about 2.5 nm from each other and each heme is partially buried in a pocket lined with hydrophobic 'R' groups and is bound to the chain through a coordination bond of the iron atom to the 'R' group of a histidine residue. The sixth coordination bond of the iron atom of each heme is available to bind a molecule of O_2.

Many contact points exit between the α and β chains of the dissimilar chain pairs $\alpha_1\beta_1$ and $\alpha_2\beta_2$. These contact points consist largely of hydrophobic 'R' groups of amino acid residues. The $\alpha_1\beta_1$ and $\alpha_2\beta_2$ pairs, being made up of irregularly shaped polypeptide chains, do not fit each other precisely and have a central open channel, or cavity running right through the hemoglobin molecule.

Relationship between Myoglobin, α and β Chains of Hemoglobin

Chemical analysis and X-ray diffraction studies have revealed that the α and β chains of hemoglobin have nearly identical tertiary structure both have similar lengths of α-helical segments and the bends or turns are about the same angles.

Further tertiary structures of hemoglobin of many different vertebrates is approximately the same and there is a close resemblance of their quarternary structures.

The 3° tertiary structures of α and β chains of hemoglobin is very similar to that of myoglobin. Further they being homologus share a number of corresponding invariant residues also indicating a common descendency of myoglobin and hemoglobin from an anscestral oxygen-binding heme protein.

Transport of Oxygen by Hemoglobin

5 to 6 litres of blood occurs in adult human beings and 1/3 to 1/2 of the blood volume is contributed by the erythrocytes. The blood must carry about 600 litres of oxygen from the lungs to the tissues every day but little is carried by the blood plasma since oxygen is only sparingly soluble in aqueous solution and all the oxygen carried by whole blood is bound and transported by the hemoglobin.

Hemoglobin of RBC in arterial blood passing from lungs to the peripheral tissues is about 96% saturated with oxygen while in the venous blood it is only 64% saturated. Thus each 100 ml of blood passing through a tissue releases about 1/5 of oxygen it carries equal to about 6.5 ml of oxygen at atmospheric pressure and body temperature.

Effectiveness of hemoglobin as an oxygen carrier is best understood by comparing the oxygen binding affinity of myoglobin and hemoglobin.

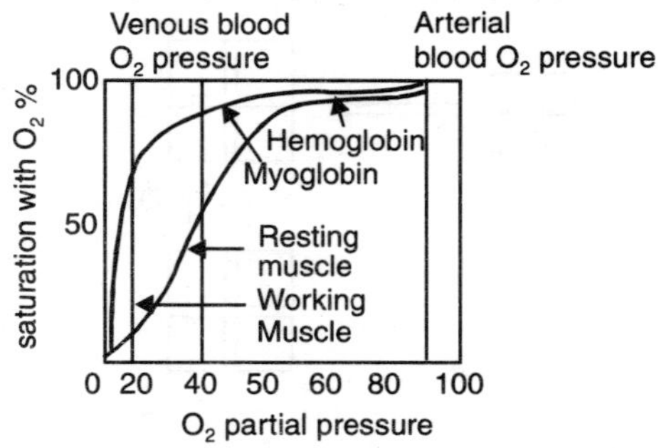

Oxygen saturation curves of Hemoglobin & myoglobin

From the oxygen-saturation curve shown in the figure, we note that myoglobin has a very high affinity for oxygen and is already 50% saturated with O_2 at 1 to 2 mm Hg pressure and at 20 mm Hg it is over 95% saturated. Thus it shows a simple hyperbolic curve.

In contrast affinity of hemoglobin for oxygen is much lower and the curve is a sigmoid curve indicating that affinity of hemoglobin for binding the first molecule of O_2 is relatively low, but subsequent O_2 molecules are bound with very much higher affinity. Thus the four heme-polypeptide sub-units are not identical and independent in their oxygen affinity.

The increasing affinity is understood to be due to cooperative interaction between the sub-units.

Binding of oxygen to hemoglobin is regulated by the concentration of H^+ (pH of solution), CO_2 and 2,3-diphosphoglycerate. An increase in concentration of any of the three has the same effect of decreasing affinity for oxygen, and vice versa.

The effect of the above factors was recognised by Bohr and is therefore called *Bohr effect*. The H^+ is bound to the R group of histidine residues while the CO_2 is bound by the α-amino group at the amino terminal end of each polypeptide chains of hemoglobin forming carbamino hemoglobin.

2,3-Diphospho glycerate occurs in relatively higher concentrations in the RBC whose equilibrium can be given as follows:

$$\text{HbDPG} + O_2 \rightleftharpoons \text{HbO}_2 + \text{DPG}.$$

The central cavity of Hemoglobin is lined with many positively charged R groups and it forms the binding site for DPG. Hemoglobin binds only one DPG but 4 O_2, 4 CO_2 and 4 H^+. In some birds inositol hexaphosphate is present instead of DPG which is more potent than DPG in reducing oxygen affinity of hemoglobin.

Oxygenation of Hemoglobin causes a change in its 3-d structure as Hemoglobin is considered to be a molecular computer which senses the changes in concentration of any of its four ligands transmitting the information through conformational changes of the molecule. Various theories have been proposed to account for the changes of which the following are noteworthy.

Symmetry Model and Sequential Model

In symmetry (all or none) model it is assumed that hemoglobin occurs in only 2 forms a low affinity form and a high affinity form with no intermediates.

In sequential model several intermediate conformations are said to occurring before the high affinity form is reached from the low affinity.

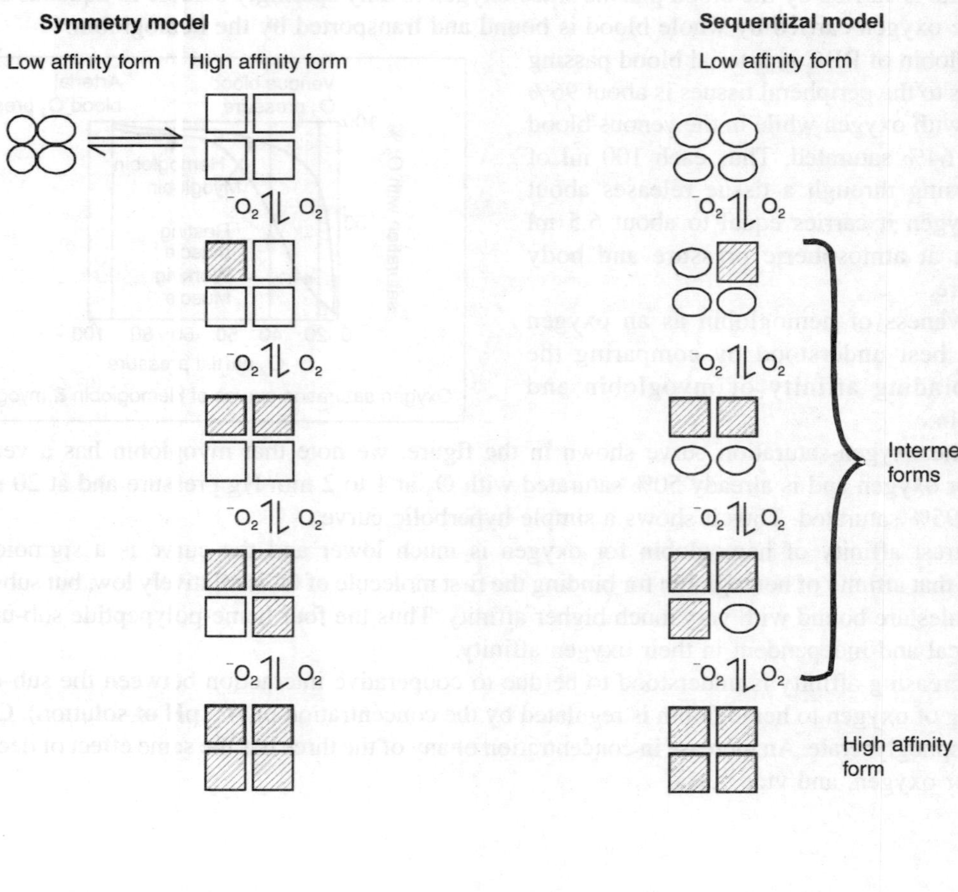

UNIT 3

Principles of Bioenergetics

Survival on the cellular level requires a constant exchange of matter and energy between the cell and its environment. Fuel molecules are constantly consumed by catabolic reactions, which harvest energy and biochemical building blocks. At the same time, the cell must continuously regenerate itself. Living cells harness the energy released by catabolism to fuel anabolic reaction sequences that synthesize molecules needed to maintain the fitness of the cell and contribute to growth. Cells also use captured energy for the performance of cellular work, such as transport across cell membranes and cell movement.

The thousands of enzyme catalyzed reactions that occur in living cells can be grouped into functional units called *metabolic pathways*. An important generality distinguishes catabolic from anabolic pathways. Catabolic routes usually funnel from a wide range of substances to a few end products, ultimately leading to the complete oxidation of biomolecules to CO_2 and H_2O. Anabolic pathways branch from a few starting point towards a large variety of end products.

Thermodynamic Principles Underlie the Study of Metabolism

Each of the reactions of metabolism involves the transfer of both matter and energy. A chemical reaction, a cell, or an entire organism can be thought as a *thermodynamic system*. The first law of thermodynamics states that the energy of the universe — that is, system plus surroundings — is constant. Consider a biochemical process in which energy from outside the system is used to convert two substrate (A + B) into two products (C + D)

$$\text{Energy} + A + B <=> C + D \quad ...(3.1)$$

The change in the internal energy of the system, ΔE, before and after the process can be expressed as

$$\Delta E = Q - W \quad ...(3.2)$$

where Q is defined as the heat absorbed by the system on its surroundings. It should be clear that a change in the internal energy of the system, ΔE, is simply an accounting of the heat or work transactions that take place across the boundary between the system and its surroundings.

Work can be defined mathematically as

$$W = P\Delta Y + W' \quad ...(3.3)$$

where P equals pressure, ΔV is the change in volume that occurs during the process, and W' represents work other than pressure-volume work. For most isolated biochemical reactions, $W' = 0$. Given this restriction, substituting equation $W = P\Delta V + W'$ (3.3) in equation $\Delta E = Q - W$ (3.2) and rearranging gives

$$Q_P = \Delta E + P\Delta V \quad \text{...(3.4)}$$

Where Q_p is the heat absorbed at constant pressure. Since any combination of the state functions must also be a state function, Q_p is therefore a state function, called the *enthalpy change* (ΔH)

$$\Delta H = \Delta E + P\Delta V \quad \text{...(3.5)}$$

The above equation relates the heat absorbed during a reaction at constant pressure to the change in the energy of the system, the pressure, and the change in the volume of the system. In biological systems, volume remains essentially constant. Thus, ΔH and ΔE are effectively equivalent.

The criterion for spontaneity is found in the 2nd law of thermodynamics. The 2nd law states that in every spontaneous process, the entropy (S) of the universe (System plus surroundigs) increases. *Entropy* can be considered a measure of disorder or randomness. In metabolic systems, the forces of entropy are countered by order-creating cellular processes. Metabolic pathways exists in a **steady state** that is far from the equilibrium state.

The Gibbs Free-Energy Change Combines Enthalpy and Entropy Changes

The 2nd law of thermodynamics establishes a direction for all process and thus also provides a criterion for the spontaneity of processes. A useful criterion for spontaneity that combines the enthalpy change and entropy change of a system was developed by J. Willard Gibbs. The new state function, termed the (Gibbs) *free-energy change* (ΔG), is defined in terms of H and S for processes that occur at constant pressure.

$$\Delta G = \Delta H - T\Delta S \quad \text{...(3.6)}$$

where it is the absolute temperature. The free-energy change, ΔG, is a measure of the energy available within a system to do work. The ability of a system to do work decreases as equilibrium is approached; at equilibrium, no energy is available to do work. The free-energy changes is a thermodynamic state function that defines the pressure. Since enthalpy reflects the first law of thermodynamics and entropy is a property of the 2nd law, the concept of free energy unites the 1st two laws of thermodynamics. ΔG is a practical function, since it can be determined by measurements of the system alone.

If the sign of ΔG for a species is –ve, the process is spontaneous; that is, it can be proceed in the absence of energy provided from outside the system. Such processes are said to be exergonic. If the sign of ΔG is positive, the process cannot proceed spontaneously. For such a process to proceed, enough energy must be supplied from outside the system to make the free energy change –ve. Processes with a +ve free-energy change are said to be endergonic, which is a way of saying the reverse process in thermodynamically favored when the process is at equilibrium, the free-energy change is zero.

The rotation $\Delta G°$ indicates the change in free energy under standard conditions. The chemical standard state has been modified slightly for biological chemistry. Since most biochemical reactions occur at a pH near 7, the standard concentration of hydrogen ions in the biological standard state is 10^{-+}M (pH = 7.0) rather than 10^{+} M (pH = 0.0). Free-energy change under the biological standard state is indicated by $\Delta G°'$.

An important relationship exists between the actual free energy change, or free energy change under nonstandard conditions. (ΔG), the standard free-energy change ($\Delta G°'$), and the equilibrium constant (K_{eq}). The free energy of a solution of substance A is related to its standard free energy by

$$GA = GA^{0'} + 2.303 \text{ RT} \log (A) \quad \text{...(3.7)}$$

where R is the universal gas constant (1.987 cal K^{-1} mol^{-1}) and 2.203 is a conversion factor (2.303 log = In). Free energy is expressed in units of kcal mol^{-1}.

For the reaction
$$A + B \rightleftharpoons C + D \quad ...(3.8)$$

The actual free-energy change is the sum of the free energies of the products minus the free energies of the reactants.

$$\Delta G_{\text{reaction}} = (G_C + G_D) - (G_A + G_B) \quad ...(3.9)$$

By substituting the relationship of actual free energy to standard free energy (Equation 3.7), we can obtain

$$\Delta G_{\text{reaction}} = \left(G_C^{0'} + G_D^{0'} - G_A^{0'} - G_B^{0'}\right) + 2.303RT\log\frac{[C][D]}{[A][B]} \quad ...(3.10)$$

$$\Delta G_{\text{reaction}} = \Delta G^{0'}_{\text{reaction}} + 2.303RT\log\frac{[C][D]}{[A][B]} \quad ...(3.11)$$

If the reaction has reached equilibrium, then the ratio of concentrations in the last term of equation 3.11 is by definition, the equilibrium constant, K_{eq}. When the concentrations of the reactants and products are at equilibrium, the rates of the forward and reverse reactions are the same and $\Delta G_{\text{reaction}} = 0$. Thus,

$$\Delta G^{0'}_{\text{reaction}} = -2.303RT\log K_{eq} \quad ...(3.12)$$

If we know the value of $\Delta G^{\circ\prime}$ for a reaction, equation 3.12 allows us to calculate K_{eq} and vice versa.

Actual Free-energy Change, and not Standard Free-energy Change, is the Criterion for Spontaneity in Cellular Reactions

For any enzymatic reaction within a living organism, the actual free-energy change (the free-energy change under cellular conditions) must be less than zero for the reaction to proceed. However, many metabolic reactions have standard free-energy changes that are +ve. The difference between ΔG and $\Delta G^{\circ\prime}$ depends upon cellular conditions. The most important condition affecting free-energy change in cells is the concentrations of substrates and products of a reaction. For equation 3.8 at equilibrium, the ratio of substrates and products is by definition the equilibrium const, K_{eq}, and the free-energy change is zero.

$$K_{eq} = \frac{[C][D]}{[A][B]} \qquad \Delta G = 0 \quad ...(3.13)$$

However, for a reaction not at equilibrium, a different ratio of products to substrates is observed, and the free-energy change is derived using equation 3.11.

$$Q = \frac{[C]'[D]'}{[A]'[B]'} \qquad \Delta G = \Delta G^{0'} + 2.303RT\log Q \quad ...(3.14)$$

where Q is the *mass action ratio*. It is this ratio relative to the ratio of products to substrates at equilibrium that determines the free-energy change for a reaction. In other words, the free-energy change is a measure of how far from equilibrium the reacting system is poised. Consequently, it is ΔG, and not $\Delta G^{0\prime}$, that is the criterion for assessing. The spontaneity of a reaction and therefore its direction within a particular system.

ATP is Principle Carrier of Biological Energy

Energy flow in metabolism depends on many coupled reactions in which ATP is common intermediate. ATP, a nucleoside triphosphate, contains one phosphate ester formed by linkage of the α-phosphate to the 5' oxygen of ribose and two phosphoanhydrides, formed by the α, β and β, γ linkages between phosphate groups. The free-energy associated with ATP is transferred when the phosphoanhydrides are cleaved to produce either adenosine diphosphate (ADP) and inorganic phosphate (Pi) or adenosine monophosphate (AMP) and inorganic pyrophosphate (PPi).

In vivo, the ratio of ATP to its hydrolysis products is maintained extremely far from equilibrium, and the free-energy-change associated with ATP hydrolysis in living cells is about –14 KCal mol^{-1}. Because of the free-energy change associated with cleavage of their phosphoanhydrides, ATP and the other nucleoside triphosphates—Uridine triphosphate (UTP), guanosine triphosphate (GTP), and Cytidine triphosphate (CTP)—are often referred to as *energy-rich metabolites*. Hydrolysis of ATP is highly exergonic under both standard and cellular conditions.

Solvation effects also make a significant contribution to the –ve energy of ATP hydrolysis. The products of hydrolysis. ADP and inorganic phosphate or AMP and pryophosphate—are better solved than in ATP itself. As ions are solvated (or in water, hydrated), they are separated from each other and electrically shielded. The decrease in the attraction of ions for each other is a thermodynamically favored result. Solvation of the products of ATP hydrolysis helps shift the equilibrium of the reaction towards hydrolysis.

Enzymes known as *Kinases* (also called phosphotransferases) catalyze transfer of the γ-Phosphoryl group from ATP (or less frequently, from another nucleoside triphosphate) to another substrate. Kinase reactions are typically unidirectional.

Several energy-rich metabolites release more free-energy upon hydrolysis than ATP under the same conditions. A variety of specific kinases catalyze the transfer of phosphoryl groups from such molecules to ADP, generating ATP.

Phosphoenolpyruvate, an intermediate of the glycolysis pathway has the most energy-rich phosphate bond known. The standard free-energy of phosphoenolpyruvate hydrolysis is –14.8 KCal mol^{-1}. Phosphoenolpyruvate is neither a phosphoanhydride nor a phosphoanide, but an *enol ester*. The hydrolysis of phosphoenolpyruvate can be divided into three steps. First, cleavage of the P–O bond of the phosphate ester produces inorganic phosphate and a resonance-stabilized enolate anion. The unstable enolate anion is then protonated to yield enol pyruvate. Finally, tautomerization of enol pyruvate yields pyruvate. The free energy associated with phosphoenol pyruvate can be understood by considering the molecules as an enol that is locked into position by attachment of the phosphonyl group. When the phosphonyl group is removed, the molecule can assume the much more stable keto form.

Transfer of a phosphoryl group from phosphoenol pyruvate to ADP is catalyzed by the enzyme pyruvate kinase. $\Delta G^{0\prime}$ for the reaction is –7.5 KCal $mo1^{-1}$. The reactions of catabolism result in net oxidation of food molecules. The oxidation of one molecule is necessary coupled with the reduction of another molecule.

The electrons released in biological oxidation reaction are often transferred enzymatically to one of two oxidizing agents, either nicotinamide adenine dinucleotide (NAD^+) or nicotinamide adenine dinucleotide phosphate ($NADP^+$). The reduced coenzymes NADH and NADPH are similar structurally, yet they have sharply different functions. NADH-produced during catabolic oxidation reactions — is converted to NAD^+ during respiration with concomitant production of ATP. This process is called *Oxidative Phosphorylation*. NADPH — generated by oxidation reactions in specialized pathways such as the pentose phosphate pathway — provides hydride anions for reductive biosynthetic processes such as fatty acid, amino acid, and nucleotide synthesis. The role of both of these enzymes can be described as supplying reducing power, which is measured quantitatively as *reduction potential.*

The standard reduction potential ($E^{0\prime}$) for the transfer of electrons from one molecular species to another is related to the standard free-energy change by the equation

$$\Delta G^{0\prime} = -nF\Delta E^{0\prime}$$

where n is the number of electrons transferred, F is Faraday's constant (23.06 KCal $V^{-1}mol^{-1}$), and $\Delta E^{0\prime}$ is the difference between the standard reduction potentials of the oxidised and reduced species.

$$\Delta G^{0\prime} = -2.303RT \log K_{eq}.$$

therefore,

$$\Delta E^{0\prime} = -2.303\frac{RT}{nF}\log K_{eq}.$$

THERMODYNAMICS

This means the flow of heat energy leading to mechanical work. The study of the flow of heat or any other form of energy into or out of a system as it undergoes a physical or chemical transformation is called thermodynamics

1. Thermodynamics is a science dealing with energy and its transformation.
2. It deals with equilibrium and feasibility of processes.
3. It deals with relations between heat and work and the properties of a system.

Laws of Thermodynamics

The principles of thermodynamics are summarized in four laws. They are

1. Zeroth law → deals with thermal equilibrium and gives the concept of temperature.
2. The Ist law of thermodynamics → deals about the conservation of energy and introduces the concept of internal energy.
3. The IInd law of thermodynamics → dictates the limits of converting the internal energy into work and introduces the concept of entropy. It also tells whether a particular process is feasible or not.
4. The IIIrd law of thermodynamics → provides a datum for the measurement of entropy with relation to temperature.

The laws of thermodynamics cannot be directly proved. They were deduced from experimental results through logical reasoning.

Classical Thermodynamics

This concept adopts macroscopic approach. In this approach

(*a*) Structure of matter is not considered
(*b*) Only a few variables are used to describe the state of matter
(*c*) The values of these variables can be measured.

Statistical Thermodynamics

This adopts microscopic approach — in this approach

1. a knowledge of structure of matter is essential

2. a large number of variables are needed to describe the state of matter
3. the values of these variables cannot be measured.

Systems and Surroundings

Systems

Any specified portion of matter under Thermodynamic study is known as a system

Or

A system is the part of universe which is under Thermodynamics study
Example: Water of polluted lake, a thermal power plant.

Surroundings

The remaining portion of universe apart from the system is called surroundings

1. A system is separated from the surroundings by a real or imaginary boundary
2. A system can exchange energy in the form of work and heat with its surroundings.

Homogeneous System

When a system is uniform throughout it is called a homogeneous system.
Example: Substance consists of only one phase.

Heterogeneous System

A system which consists of two or more phases is Heterogeneous system. It is non-uniform throughout.
Example: Ice, water in a lake.

Types of Systems

1. *Isolated system:* A system which is enclosed by a rigid and insulated (adiabatic) boundary cannot exchange energy (either heat or work) and also it cannot exchange matter with its surroundings. Such a system is called an isolated system.
   Example: Hot coffee in thermoflask (closed).
2. *Closed system:* The boundary may be closed but not insulated and thus only energy can be exchanged and not matter from the system to the surroundings and vice versa.
   Example: Hot tea present in a closed kettle.
3. *Open system:* The system which can exchange both matter as well as energy with its surroundings called an open system.
   Example: Hot coffee in a cup.
4. *Adiabatic system:* A system which is enclosed by insulated boundary cannot exchange energy as heat with its surroundings is called adiabatic system.
5. *Macroscopic system:* A system consisting of a large number of particles (like molecules, atoms, ions, etc.) is known as macroscopic system.

Property

1. The state or condition in which a system exists is specified by specifying its properties.
2. The properties can be classified into two groups as
   (*a*) *Relevant properties:* Those associated with energy and its transformation.
   Example: Pressure, temperature.

(*b*) *Irrelevant properties:* Those which are not associated with energy and its transformation *Example:* Colour, odour, taste.

In Thermodynamics one deals with relevant properties only.

3. A property is any characteristic feature which can be quantitatively measured and can be used to describe the state of a system.
4. The property is a state function and not a path function. Its differential must be exact.
5. Pressure (P), Volume (V), Temperature (T) etc. are examples of properties.

Extensive properties are those which depend upon the mass of the system. Volume is an extensive property. Intensive properties are those which do not depend upon the mass of the system. Pressure, temperature are intensive properties. The ratio of an extensive property to the mass of the system is called a specific property.

Example: Specific volume of a system $= \frac{V}{m} = \frac{\text{Volume of the system}}{\text{mass of the system}}$.

The ratio of an extensive property to mole number of the system is called a molar property.

Example: Molar volume $= \frac{V}{n} = \frac{\text{Volume of the system}}{\text{no. of moles of the system}}$.

THERMODYNAMIC EQUILIBRIUM

Equilibrium in mechanics means a body in a state of rest although a number of forces are acting on it. In thermodynamics equilibrium has a broader sense of meaning. A system in which the state variables have constant values throughout the system is said to be in a state of thermodynamic equilibrium. A system in which the state variables have different values in different parts of the system is said to be in a non-equilibrium state.

Criteria

1. *Thermal equilibrium:* Temp. of system must be uniform and must be same as the temp. of surroundings.

   Then it is said to be in thermal equilibrium. T = Constant.
2. *Mechanical/Dynamic equilibrium:* If pressure of system is uniform throughout the system and same as surrounding then it is said to be in mechanical equilibrium. P = Constant
3. *Chemical equilibrium:* The chemical composition of the system must be uniform with no net chemical change then it is said to be in chemical equilibrium.
4. If a system is satisfying all the three conditions then it is said to be in thermodynamic equilibrium.

   The equilibrium states of a system can be classified as

   (*a*) stable, (*b*) unstable, (*c*) meta stable.

   (*a*) *Stable equilibrium:* System always returns to its original state after it is disturbed, however large the disturbance may be.

   (*b*) *Unstable equilibrium:* System does not return to the original state after it is disturbed, eventhough the magnitude of disturbance is small.

   (*c*) *Meta stable equilibrium:* System returns to, if original state it subjected to small disturbances, but when the disturbance exceeds a certain magnitude, it settles in a new state.

Process

Whenever a system changes from one equilibrium state to another equilibrium state, it is said to have one undergone the process.

(*a*) *Quasi static process:* When a system changes its state very slowly under the influence of very very small driving forces (pressure differences, temperature differences), the system is said to undergo a quasi static process.

(*b*) *Reversible process:* When a system changes its state under such conditions the driving force is zero. ($dP = 0$, $dT = 0$) then it is said to undergo a reversible process. A reversible process can proceed in either directions.

A reversible process is always quasi static process, but the converse need not be true.

Criteria

A process is said to be reversible if both the system and its surroundings can be restored to their initial states by reversing the direction of process. The process is irreversible if it does not fulfill this criteria.

DIFFERENT THERMODYNAMIC PROCESSES

Isothermal: This is a thermodynamic process in which temp. remain constant

$$dT = 0 \Rightarrow dU = 0 \text{ where } U = \text{Internal energy}$$

Adiabatic: This is a thermodynamic process in which no exchange of heat from the system to surroundings and vice versa.

$$dQ = 0 \Rightarrow dH = 0 \text{ where } H = \text{enthalpy}$$

Isobaric: This is a thermodynamic process in which pressure remains constant.

$$\text{P} = \text{Constant} \Rightarrow dP = 0.$$

Isochoric: This is a thermodynamic process in which volume remains constant.

$$\text{V} = \text{Constant} \Rightarrow dV = 0$$

Cyclic Process

When a system in a given state goes through a number of different processes, and finally returns to its initial state, is called cyclic process, $dU = 0$.

Note: A polytropic process can be represented by the relation $PV^n = \text{Constant}$.

Depending on the value of 'n', the polytropic process reduces to other processes as

$n = 0$: Constant pressure process (P = Constant)

$n = 1$: Constant temperature process (PV = Constant)

$n = \gamma$: Reversible adiabatic process (PV = Constant)

$n = \alpha$: Constant volume process (V = Constant)

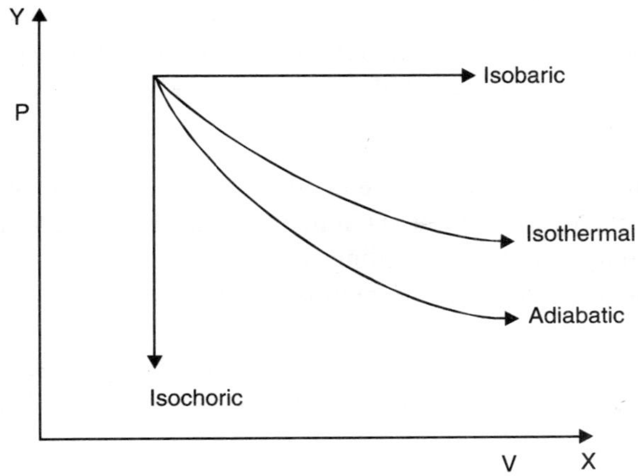

Fig. 3.1

Energy

1. A moving body can set into motion some other body and it has the ability to do work. This ability to do work is called energy. Thus we say that a moving object possesses some energy.
2. There are two different modes of energy namely macroscopic and microscopic.
3. The macroscopic modes of energy are kinetic energy (K.E.) and potential energy (P.E.).

   K.E. = 1/2 mv^2, P.E. = mgh
4. At a microscopic level matter consists of several particles—atoms, molecules etc. They are in continuous motion. A molecule possesses energy due to translational, rotational, vibrational modes of motion. The energy possessed by a molecule in the above modes cannot be evaluated through macroscopic measurements and hence they are named as microscopic modes of energy.
5. The total energy possessed by all the microscopic modes of motion is called internal energy (U).

Internal Energy

1. The inherent energy possessed by a body due to different modes of motion is called internal energy. Internal energy is a highly disordered form of energy. Mechanical energy is an ordered form of energy. It is an extensive property but not an intensive property.
2. The transfer of heat and performance of work constitute two methods for adding the energy or substracting energy from a system. Once the transfer of energy has occurred, system is said to have undergone a change in internal energy.
3. If dQ is the energy that has been added to a system by transfer of heat and dW is equal to the energy that has been extracted from the system by performance of work the difference $[dQ - dW]$ must represent internal energy change of the system. It follows that the internal energy change dU of a system is independent of path

$$dU = dQ - dW$$

4. When applying this equation it must be remembered that

   (*a*) all the three U, Q, W must be expressed in same units
   (*b*) dQ = Positive, when heat enters the system
   (*c*) dQ = Negative, when heat leaves the system

(*d*) dW = Positive, when system expands and does work on surroundings

(*e*) dW = Negative, when system contracts and work is done by surroundings.

For an ideal gas, internal energy of a system depends on temperature ($U \propto T$)

$$dU = c_v dT$$

5. When a quantity of heat dQ enters a system during a process, some of it remains in the system as increased internal energy (dU) while the remaining leaves the system again in the form of work done by the system against its surroundings (dW).
6. We cannot calculate absolute value of internal energy of a system, we can only calculate change in internal energy.
7. Internal energy is a state function and depends only on the initial state and final state of process but never depends on the path in which the process is carried out. Hence dU is always an exact differential

$$\int_1^2 dU = U_2 - U_1$$

Work

1. A system interacts with its surroundings or other systems in two ways. They are work and heat.
2. In mechanics work is defined as product of Force 'F' and distance 'dx' moved in the direction of force

$$dW = Fdx$$

3. Suppose the pressure of the gas inside the cylinder shown in figure is 'P' which is acting on a piston of cross-sectional area 'A'. Then the force acting on the piston is PA

$$F = P \times A.$$

(where P = Pressure, A = Cross-sectional area of the face of the piston)

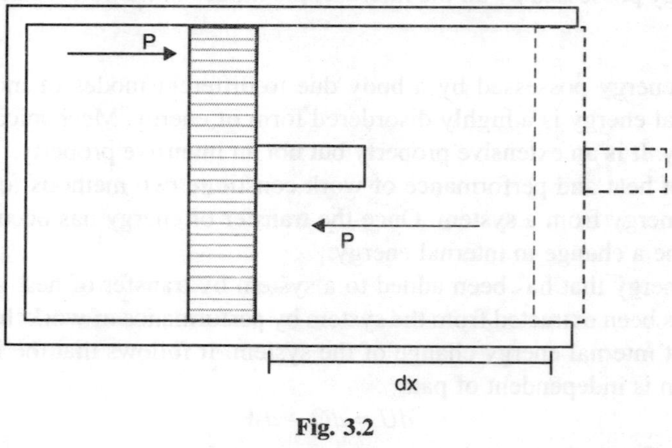

Fig. 3.2

4. If piston is allowed to move a distance dx under such conditions that the opposing force is exactly balanced by gas pressure, infinitesimal work done by the gas is given by

$$dW = Fdx = PAdx = P\,(Adx) = Pdv$$

where (Adx = change in volume = dv)

therefore, $dw = Pdv$

5. Total work done when the gas changes from state 1 to state 2 is given by

$$W = \int_1^2 dW = \int_1^2 Pdv$$

6. The work done, $W = \int_1^2 Pdv$ is the area under the curve on a pressure versus volume diagram.

Figure shows that the value of $\int_1^2 Pdv$ is different for different paths.

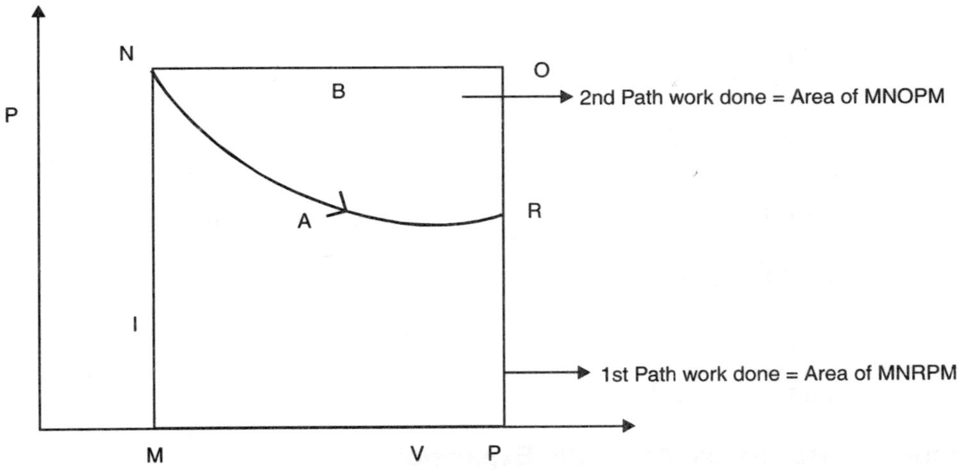

Fig. 3.3

7. Work done is a path function. It is not a state function. It dose not depend on the state of process. Work is not a property of system and its differential is not exact

$$\int_1^2 dW \neq W_2 \times W_1$$

8. Consider a vessel which is divided into two chambers. One chamber contains a gas at a known pressure while the other chamber is evacuated. If the partition is removed, the gas expands and occupies the entire container. In this case, the expansion of the gas is not restrained by an opposing force. (Since the other side is vacuum and the work done by the gas is equal to zero. Such an expansion which is not restrained is called a free expansion and it is an irreversible process. (Any process taking place in nature is generally an irreversible process.)

9. $W = \int PdV$ for a reversible process only, $W \neq \int PdV$ for an irreversible process or for free expansion.

10. $W \neq 0$, even if $dV = 0$

    Example: Work done by a paddle wheel on a quantity of liquid.

Work done in a Reversible Isothermal Expansion

Here we are carrying out an 'isothermal expansion' where volume is increasing from V_1 to V_2

Total work done is calculated as

$$W = \int_{V_1}^{V_2} dW = \int_{V_1}^{V_2} PdV$$

$$= \int_{V_1}^{V_2} \frac{nRT}{V} dV \qquad \boxed{\begin{array}{c} PV = nRT \\ P = nRT/V \end{array}}$$

$$= nRT \int_{V_1}^{V_2} \frac{dV}{V}$$

$$= nRT \left|\log V\right|_{V_1}^{V_2}$$

$$= nRT\left[\log V_2 - \log V_1\right] \qquad \because \log a/b = \log a - \log b$$

$$= nRT\left[\log V_2/V_1\right] \qquad \because V_2/V_1 = P_1/P_2$$

$$W = 2.303\,nRT\log\left(V_2/V_1\right)$$

$$W = nRT\log\left[p_1/p_2\right]$$

Work done in a Reversible Adiabatic Expansion

Here we are carrying out an adiabatic process where volume is increasing from V_1 to V_2

$$W = \int_{V_1}^{V_2} PdV = \int_{V_1}^{V_2} KV^{-n}dV \qquad \left(\because PV^{\gamma} = K\right)$$

$$= \frac{K\left(V_2^{1-\gamma} - V_1^{1-\gamma}\right)}{(1-\gamma)} = \frac{P_2V_2 - P_1V_1}{(1-\gamma)} = \frac{P_1V_1 - P_2V_2}{(\gamma - 1)}$$

$$= \frac{R\left(T_1 - T_2\right)}{(\gamma - 1)} = \frac{\left(C_P - C_V\right)\left(T_1 - T_2\right)}{\left(C_P/C_V - 1\right)}$$

$$= C_V\left(T_1 - T_2\right).$$

Indicator Diagram

The PV diagram of isothermal process is isotherm. The PV diagram of adiabatic process is adiabat. In isothermal expansion T is constant, the decrease of pressure occurs only due to increase in volume. In adiabatic expansion T is not constant, therefore the pressure falls not only due to the increase in volume but also due to the decrease in temperature. That's why Adiabat is always steeper than the Isotherm.

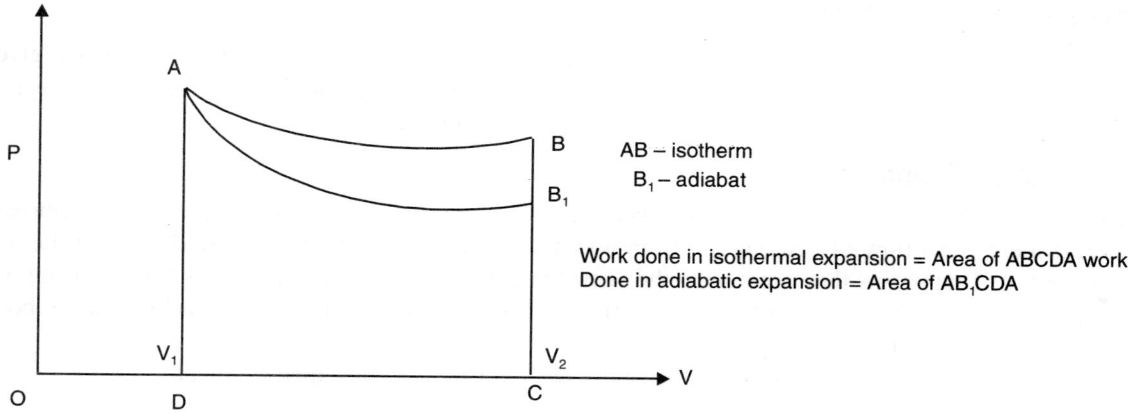

Fig. 3.4

From figure, work done in isothermal expansion is greater than work done in adiabatic expansion.

Heat

1. The mode of energy transfer which cannot be accounted as work from a macroscopic point of view is called heat interaction. Energy transfer as heat occurs by virtue of temperature difference across the boundary of the system.
2. Heat is energy in transit. It is also a path function. It is not a property of a system and its differential is not exact.

$$\int_1^2 dQ \neq (Q_2 - Q_1)$$

3. Work and heat interactions depend upon the choice of a system.

Zeroth Law of Thermodynamics

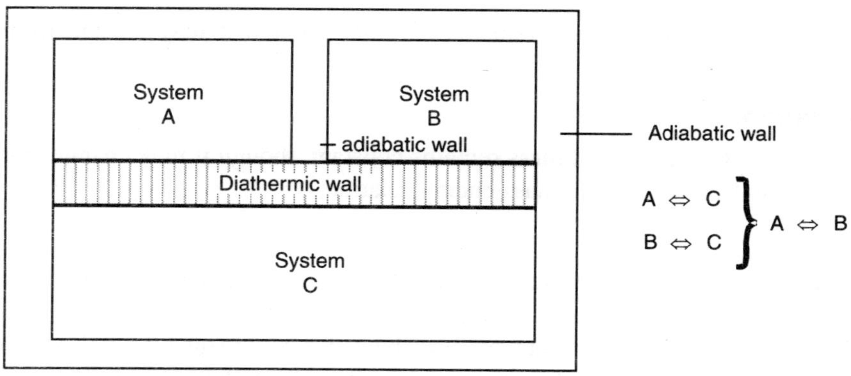

Fig. 3.5

If system A is in equilibrium with system C (A < = > C) and system B is in equilibrium with system C (B < = > C), then system A will be in equilibrium with B (A < = > B).

Statement

If each of the two given systems are in thermal equilibrium with a third system, then the two given systems are said to be in thermal equilibrium with each other. The zeroth law gives the concept of temperature and forms the basis for the measurement of temperature.

Concept of Temperature

The physical quantity that determines whether two bodies are in thermal equilibrium or not is temperature. If there is no flow of heat energy from one body to another body they are said to be in thermal equilibrium. That means they are having the same temperatures. If the heat energy is flowing from one body to another body depending on the direction of flow of heat we can say that which is a hot body and which is a cold body.

The zeroth law of thermodynamics forms the basis for temperature measurement.

Temperature Scale

A temperature scale is established by assigning numerical values to certain easily reproducible states. It is customary to use the two fixed points—the ice point and steam point.

Ice point

It is the equilibrium temperature of ice with air saturated water at a pressure of 101.325 kPa which is assigned a value of 0°C.

Steam point

It is the equilibrium temperature of pure water with its own vapor at 101.325 kPa which is assigned a value of 100°C.

The temperature scale with the above fixed points is called celsius scale.

First Law of Thermodynamics

The first law relates mechanical and heat energies and is nothing but law of conservation of energy principle.

Statement 1

The total energy of an isolated system remains constant, though it may change from one form to another.

Modified statement 2

The algebraic sum of total energy and mass remains constant, though it may change from one form to another. (This follows from Einstein's mass energy relation.)

Statement 3

The algebraic sum of total energy and mass of the universe remains constant, though it may change from one form to another.

- In Joule's paddle wheel experiment work was done on a certain quantity of liquid contained in an insulated vessel and then the energy transferred as heat from the liquid to the surroundings, such that the liquid is restored to its initial conditions. The quantity of energy transferred as work and heat were carefully measured and was found that

$$\mathbf{W \propto H}$$

∴ **W = JH** (J = Mechanical equivalent of heat)

- It was always observed that whenever a system undergoes a cyclic change the network done on the system was always equal to the amount of energy removed from the system as heat. This fact is known as the first law of thermodynamics.
- The first law of thermodynamics can also be expressed mathematically as given below:

$$\oint dQ = \oint dW \text{ (where } \oint \text{ denotes the integral over the cycle)}$$

- The first law is equally applicable to equilibrium and non-equilibrium processes.
- The first law is applicable to all the processes—reversible as well as irreversible processes.

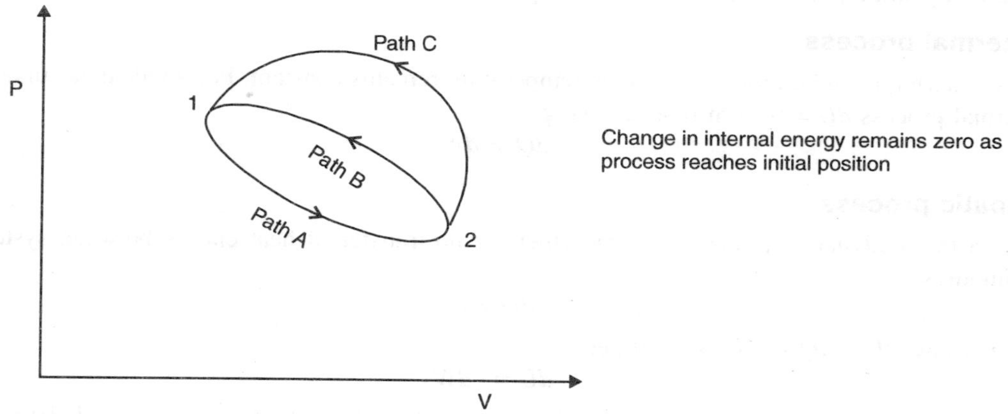

Fig. 3.6

- A process undergone by system, as a result of which the system is restored to its initial state is called a cycle.
- Suppose a system is taken a state one to state two by the path from A as shown in figure and it is restored to state one through another path B. Then the combination of the paths A and B *i.e.* the process 1A2B1 is a cycle and the system is said to have undergone a cyclic change. Similarly 1A2C1 is another cycle.

$$\int_{1A2} dQ + \int_{2B1} dQ = \int_{1A2} dW + \int_{2B1} dW \quad \text{...(1)}$$

$$\int_{1A2} dQ + \int_{2C1} dQ = \int_{1A2} dW + \int_{2C1} dW \quad \text{...(2)}$$

Subtracting the second equation from the first equation to obtain

$$\int_{2B1} dQ - \int_{2C1} dQ = \int_{2B1} dW - \int_{2C1} dW$$

$$\int_{2C1} dQ - \int_{2C1} dQ = \int_{2B1} dW - \int_{2B1} dW \quad \text{...(3)}$$

$$\int_{2C1} (dQ - dW) = \int_{2B1} (dQ - dW)$$

($\because$ $dQ - dW = dU$ whenever a system undergoes a process, the change in the energy of the system is given by $dU = dQ - dW$)

$$\int_{2C1} dU = \int_{2B1} dU$$

That is $(dQ - dW)$ is independent of path.
$dU = 0$ (In cyclic process)
Substituting in $dQ = dU + dW$, we get
$dQ = dW$ (Heat is totally converted into work).
Thermodynamic process should be a cyclic process for maximum work to be done.

Isothermal process

This is a thermodynamic process, in which temperature remains constant. For an ideal gas undergoing isothermal process $dU = 0$ from first law we get,

$$dQ = dW$$

Adiabatic process

This is a thermodynamic process, in which there is no transfer of heat energy between system and surroundings.

$$dQ = 0$$

from first law, $dU = dQ - dW$, we will get,

$$dU = -dW$$

when positive work is done internal energy decreases and temperature decreases. An adiabatic expansion is accompanied by cooling. When a negative work is done internal energy increases, temperature increases. Heating accompanies adiabatic compression.

Isochoric process

This is a thermodynamic process in which volume remains constant. Since this is a constant volume process $dV = 0 \Rightarrow dW = PdV = 0$, from first law we have $dU = dQ - dW$, we get
$dU = dQ$ is the equation of first law for isochoric process.

Isobaric process

This is a thermodynamic process in which pressure remains constant. Generally all the phase changes takes place at atmospheric pressure are examples of isobaric processes. Let us consider one such process, where water is converted to vapour changing its volume from V_1 to V_2. Let m kg of water is present and

$$dQ = mL \text{ Joules} = \text{latent heat of vaporization.}$$

$$dU = U_2 - U_1$$

$$dW = P(V_2 - V_1)$$

substituting the above values in first law equation, we get

$$mL = (U_2 - U_1) + P(V_2 - V_1)$$

Isolated system

An isolated system does not exchange energy either in the form of heat or work, that is $dQ = 0$ and $dw = 0$. From the equation of first law, $dQ = dU + dW$, we obtain $dU = 0$.

Hence the energy of an isolated system is conserved. A perpetual motion machine of the first kind (PMMFK) is a cyclically operating device which delivers work without requiring any interaction with the surroundings.

For a cyclically operating device $\oint dU = 0 = \oint dQ - \oint dW$. Since the device does not interact with the surroundings.

$\oint dQ = 0$ and hence $\oint dW = 0$. That is a PMMFK does not deliver any network. Therefore, the first law implies it is impossible to devise a PMMFK.

Limitations of 1st law

1. The 1st law does not indicate the direction in which processes of change in state occur. From the viewpoint of 1st law, for *e.g.* heat may pass either from a hot body to cold one or from a cold body to a hot body.
   It is only essential that energy given away by one body equal that received by the other one.
2. All natural processes for example, thermal conduction, free expansion, diffusion of gases etc., are irreversible processes. But the irreversibility of natural process does not follow, consequently, from the first law of thermodynamics.

Conclusion

The feasibility of a physical process or a chemical reaction is not discussed by 1st law *i.e.* it is unable to give the direction in which the process takes place.

Photosynthesis

Through the process of photosynthesis, carbohydrates are synthesized from atmospheric CO_2 and water. Photosynthesis is an energy requiring processs that is fueled by light energy. It completed the global carbon cycle-non-photosynthetic organisms give off CO_2 through the oxidation of carbohydrates, whereas photosynthetic organisms capture CO_2 and reduce it to the level of carbohydrates (CH_2O). Organisms capable of photosynthesis are called phototrophs, a diverse group that includes certain bacteria, cyanobacteria (blue-green algae), algae, non-vascular plants and vascular (higher) plants.

Photosynthesis Consists of Two Major Processes

The net reaction of photosynthesis is

$$CO_2 + H_2O \xrightarrow{\text{Light}} (CH_2O) + O_2$$

where (CH_2O) represents carbohydrate.

Photosynthesis encompasses two major processes that can be described by two partial reactions.

$$H_2O + ADP + Pi + NADP^+ \xrightarrow{\text{Light}} O_2 + ATP + NADPH + H^+$$

$$CO_2 + ATP + NADPH + H^+ \longrightarrow (CH_2O) + ADP + Pi + NADP^+$$

sum:

$$CO_2 + H_2O \xrightarrow{\text{Light}} (CH_2O) + O_2$$

In the first process, the so-called 'light reactions', protons derived from H_2O are used in the chemiosmatic synthesis of ATP from ADP + Pi, while a hydrogen atom from H_2O is used for the reduction of $NADP^+$ to NADPH. The reactions are characterized by the light dependent production of O_2 gas derived from the splitting of the water molecules.

The second process of photosynthesis involves the utilization of NADPH and ATP in a series of reactions leading to the reduction of gaseous CO_2 to carbohydrate. Because these reactions do not directly depend on light, but only on a supply of ATP and NADPH, they are referred to as the 'dark reactions'.

Photosynthesis in Algae and Green Plants Occurs in Chloroplasts

In algae and green plants, photosynthesis occurs in specialized organelles called chloroplasts. The main structural feature of the chloroplast is a membranous internal network called the thylakoid membrane, which is the site of the light—dependent reactions that lead to the formation of NADPH and ATP. The thylakoid membrane, also called the thylakoid lamellae, is a highly folded continuous membrane network suspended in the aqueous matrix of the chloroplast. This aqueous matrix called the stroma, is the site of the second portion of the photosynthetic process, the reduction of CO_2 to carbohydrate. The chloroplast is enclosed by a double membrane that is highly permeable to CO_2 and selectively permeable to other metabolites.

The aqueous space within the thylakoid membrane is called the lumen. The pumping of protons access the thylakoid membrane into the lumen creates the protons motive force that drives the synthesis of ATP. The thylakoid membrane is folded into a network of flattened vesicles that occur as pressed stacks called grana, or as simple, unstacked vesicles that traverse the stroma and connect grana. Regions of the thylakoid membrane located within grana and not in contact with the stroma are called granal lamellae, whereas regions in contact with the stroma are stromal lamellae. Membrane-embedded components involved in photosynthesis are differentially distributed between the granal and the stromal lamellae.

Chlorophyll and Other Pigments are Components of Photosystems

Among the components embedded in the thylakoid membrane are various pigments that capture light energy for photosynthesis. Chlorophyll is usually the most abundant pigment and has the most significant role in the harvesting of light for photosynthesis. Four different types of chlorophyll occur among phototrophic organisms. Chlorophyll *a* and chlorophyll *b* are the two types of chlorophyll found in higher plants. The *a* is more abundant than the *b*. The major pigments in photosynthetic bacteria are bacterio chlorophyll *a* and bacterio chlorophyll *b*. Two features common to the different chlorophylls are the hydrophobic phytol side chain, which anchors chlorophyll in the membrane, and the hydrophilic porphyrin ring. Together, these features make chlorophyll molecules amphipathic. The porphyrin ring is similar to the heme prosthetic groups of hemoglobin, myoglobin and the cytochromes. Unlike heme, which contains Fe^{2+}, chlorophyll contains Mg^{2+} chelated by the pyrrole nitrogen atoms of the ring. Chl *a* and Chl *b* both absorb light in the violet-to-blue region (absorption max 400–500 nm) and the orange to red region (absorption max 650–700 nm) of the electromagnetic spectrum. Because their structures are slightly different, however, the absorption maxima of chl *a* and chl *b* are not identical. Chlorophyll molecules are specifically oriented in the thylakoid membrane by non-covalent binding to proteins embedded in the membrane.

In addition to chlorophyll several accessory pigments may be present in the photosynthetic membranes. These are carotenoids, which are present in all phototrophs and the phycobilins, including phycoerythrin and phycocyanins which are found in same algae of cyanobacteria. Like the chlorophyll, these molecules contain a series of conjugated double bonds that give them the light-absorbing properties characteristic of pigments. The absorption maxima of the accessory pigments differ from these of the chlorophylls.

Two distinct protein complexes called photosystems are the functional units of the light dependent reactions of photosynthesis. Both photosystems named photosystem I and photosystem II are localised in the thylakoid membrane each contains a bed of pigments called an antenna complex that harvests light for the photosystem. Depending on the organism, each antenna complex may contain from 60 to 2000 pigment molecules. The actual site of photo chemistry is a multi subunit protein complex within each of the photosystems called a reaction center. The reaction center contains two chlorophyll molecules called the special pair, which are distinct from other chlorophylls in that light energy from the antenna complex is funnelled to them. The reaction center of a photosystem is named according to the absorption maximum of its special pair. The photosystem I (PSI) reaction center is called P 700 (P for pigment and 700 for the absorption maximum of its special pair); The PS II reaction center is called P 680. The two membrane-spanning photosystems work in series and although they are localized in different regions of the thylakoid membrane, they are linked by specific electron carriers. PS I is located in the stromal lamellae and therefore is exposed to the chloroplast stroma, whereas PS II is embedded in the granal lamellae away from the stroma. A large antenna complex called the chlorophyll *a/b* light harvesting complex (LHC) aids the photosystem in gathering light.

Non-cyclic Electron Transport Results in the Reduction of $NADP^+$ to NADPH

Photosynthetic electron transport follows one of two routes. Electrons can be transferred/linearly through a series of membrane-bound carriers, terminating with the transfer of electrons to $NADP^+$ or they can be transferred through many of the same carriers, but in a cyclic manner, so that there is no terminal electron transfer to $NADP^+$. During both non-cyclic and cyclic electron transport, protons are pumped across the membrane from the stroma to the lumen, generating the proton motive force used to drive the formation of ATP. However only non-cyclic electron flow yields NADPH. Reduction potential is the basis of electron transfer between membrane-bound electron carriers in the chloroplast. Electron transfer

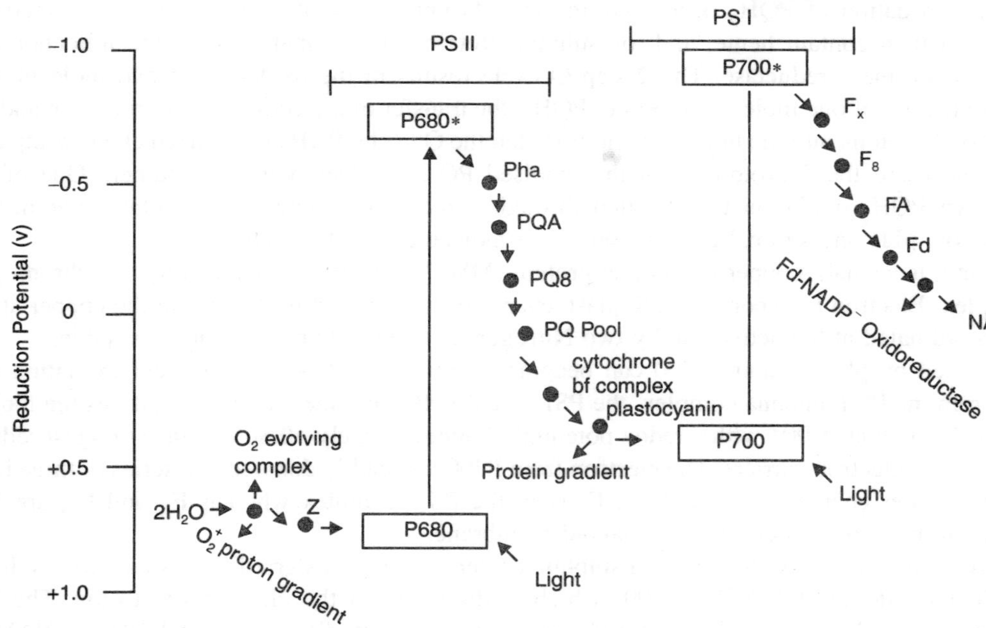

Fig. 3.7

proceeds from carriers of lower reduction potential (stronger reductants) to carriers of higher reduction potential (stronger oxidants).

When the reducing potentials of the series of photosynthetic electron transport components are plotted in sequence, the result is a zig-zag figure called the Z-scheme. The Z-scheme shows the rise in reducing power created by the excitation of the photosystem reaction centers.

The electrons needed for electron transport are obtained from water in a reaction catalysed by the oxygen-evolving complex. Sometimes called the water splitting enzyme, the oxygen-evolving complex is located on the lumen side of PS II and contains a cluster of four manganese ions. The manganese ions are specifically arrayed to accumulate and then sequentially transfer four electrons from two molecules of water to oxidized P680 ($P680^+$). Once the Mn cluster has given up its 4 electrons, the following events related to water splitting occur:

1. two molecules of H_2O are oxidized
2. four protons are released into the thylakoid lumen
3. one molecule of O_2 is evolved and
4. four electrons reduce the Mn cluster.

Electrons are transferred from the oxygen-evolving complex to $P680^+$ via a carrier called Z, which passes the reaction center for excitation to P680*. Once excited, P680* donates its electron to pheopytin a (pha), which is identical to chl *a* except that the mg ion of chl *a* is replaced by two protons. Electron transfer proceeds from reduced pheopytin *a* to PQ_A, a plastoquinone tighty bound to a PS II polypeptide.

The bound PQ_A transfers two at a time, one at a time, to a second plastoquinone molecule, PQ_B which is reversibly bound to a PS II polypeptide. PQ_B is thus reduced first to a semiquinone, PQH, and then to a Quinol, PQH_2. The fully reduced PQ_BH_2, which is soluble in the membrane due to its long hydrophobic tail, is released into the plastoquinone pool of PQH_2 and PQ.

The photosynthetic Q cycle, like the mitochondrial Q cycle, carries out the oxidation of reduced quinone. The oxidation of PQH_2 occurs via the cytochrome *bf* complex. This complex is made of cytochromes, which contain heme, and ion-sulphur proteins and is analogous to the mitochondrial ubiquinol-cytochrome *c* reductase. The 2-step Q cycle results in the oxidation of two molecules of PQH_2, the reduction of one molecule of PQ to PQH_2, the transfer of 2 electrons to plastocyanin and the net transfer of 4 protons into the lumen. In the first step the Q cycle, PQH_2 is oxidized at a site adjacent to cytochrome *f* and the 2 protons from the oxidised PQ are released into the lumen. One of the electrons from PQH_2 oxidation is funnelled through cytochrome *f* and reduces plastocyanin. The second electron, plus one stromal proton, converts a molecule of PQ to PQH.

Plastocyanin is a small, copper-containing protein (MW 11000) that is reduced by cytochrome *f* of the *bf* complex. Electron transport through plastocyanin occurs via redox changes in the copper atom, which is co-ordinated at the active site by two Nitrogen 2 sulphur atoms of 4 amino residues.

Once reduced by plastocyanin, p700 can become energised to P700* by pigment excitation that originates from the PS I antenna complex, the PSI bound LHC, or under certain conditions the mobile pool of LHC. Note that P700* with a redox potential of approximately –0.80V, is the strongest reductant in the chain of electron carriers. The electron from P700* is readily donated to a series of three iron-sulphur centers, designated F_x, F_B and F_A. F_x is in the PS I complex whereas F_A and F_B, are in a polypeptide on the stromal side of the thylakoid membrane.

The electron delivered to the last iron-sulphur center, F_A, is transferred to a small iron-sulphur protein called feredoxin (Fd, MW 10700) which is dissolved in the chloroplast stroma. The low reduction potential (E° = –0.43V) of feredoxin allows it to readily reduce $NADP^+$ to NADPH (E° = –0.32V). The reduction of $NADP^+$ by reduced feredoxin is catalysed feredoxin –$NADP^+$

oxidoreductase. The oxido reductase has the prosthetic group FAD, which is reduced to $FADH_2$, by reduced feredoxin in 2 one-electron transfers. $FADH_2$, is then reduces $NADP^+$ by donating two electrons and a proton in the form of a hydride ion. The pH difference across the thylakoid is increased by this step, since a proton from the stroma is consumed in the conversion of $NADP^+$ to NADPH. Formation of NADPH completes the non-cyclic electron transport sequence.

The membrane-bound ATP synthetase in the presence of a transmembrane proton gradient catalyzes the formation of ATP from ADP and Pi. Since this process is light-dependent in plants, photosynthetic ATP formation is termed photophosphorylation. The ATP synthase of chloroplasts is a multiprotein complex similar to its mitochondrial counterpart in that it consists of two major particles, CF_0, CF_1.

Cyclic Electron Flow Contributes to the Proton Gradient Across the Thylakoid Membrane

It is estimated that for every two electrons transferred to reduce one molecule of $NADP^+$ to NADPH, proton motive force develops across the thylakoid membrane sufficient to synthesize one molecule of ATP. This one to one ATP/NADPH stoichiometry of the light reactions would create an imbalance with the stoichiometry of the carbohydrate producing 'dark' reactions, of modified since, two molecules of NADPH and three molecules of ATP are consumed in the reduction of one molecule of CO_2 to (CH_2O). To compensate for this unbalanced coupling between the 'light' 'dark' reactions, a modified sequence of electron transport steps operates as a cycle to form ATP without the simultaneous formation of NADPH. This modified reaction sequence is called cyclic electron transport. As in non-cyclic electron transport, electrons are transferred via the Q cycle from plastoquinone to the cytochrome bf complex, and then on to PS I ferredoxin. In cyclic electron transport, however, the soluble ferredoxin donates its electron back to the PQ pool in a process mediated by the cytochrome bf complex via a cytochrome that is involved exclusively in cyclic electron flow. Through Q cycling, cyclic electron transfer supplements the proton motive force generated during non-cyclic electron flow.

Reactions of the Reductive Pentose Phosphate Cycle Assimilate CO_2 into Carbohydrates

The second major phase of photosynthesis is the reductive version of carbon dioxide into carbohydrates a process powered by the ATP and NADPH formed during the light reactions of photosynthesis. The formation of carbohydrates occurs in the chloroplast stroma and is accomplished by cycle of enzyme- catalyzed reactions that none three major consequences.

1. The fixation of atmospheric CO_2,
2. The reduction of CO_2 to carbohydrate, and
3. The regeneration of the molecule that accepts CO_2.

The metabolic pathway leading to carbon assimilation has several names, including the reductive pentose phosphate cycle, the C_3 pathway (indicating that the first intermediate of the pathway is a 3-carbon molecule), the photosynthetic carbon reduction cycle and the calvin, calvin-Benson or calvin-Bassham cycle. The pathway is referred to as the reductive pentose phosphate cycle (or) simply the RPP cycle.

The sequence of reactions following the reductive phase of the cycle ends with the regeneration of CO_2 acceptor molecule, ribulose 1, 5-biphosphate. Most of the reactions of the RPP cycle are part of the regeneration phase, which is stoichiometrically complex. Glyceraldehyde 3-phosphate is diverted into 3 different branches of the pathway and is interconverted between four-five-six and seven carbon phosphorylated sugars. This regenerative phase, the so-called 'sugar shuffle' is catalyzed by various isomerases, epimerases, aldolases, phosphotases and transketolase. The last step is the phosphorylation of ribulose-5-phosphate by ATP to form ribuiose 1, 5-biphosphate in a reaction catalyzed by phospho ribulokinase. All of the substrates, enzymes and reactions of the RPP cycle are summarized as follows:

Reaction	*Enzyme*
3 Ribulox 1,5-biphosphate + $3CO_2$ + $3H_2O$ $\rightarrow$ 6, 3 - phosphoglycerate + $6H^+$	RuBis Co
6, 3 - phosphoglycerate + 6 ATP 6, 1, 3 - bisphosphoglycerate + 6 ADP	Phosphoglycerate kinase
6, 1, 3-bisphosphoglycerate + 6 NADPH + $6H^+$ 6, Glyceraldehyde 3-phosphate + 6 $NADP^+$ + 6Pi	Glyceraldehyde 3-phosphate dehydrogenase
2 Glyceraldehyde 3-phosphate 2 Dehydroxyacetone phosphate	Triose phosphate isomerase
Dihydroxyacetone phosphate + Glyceraldehyde 3-phosphate Fructose 1,6-biophosphate	Aldolase
Fructose 1,6-biophosphate + H_2O $\rightarrow$ Fructose 6-phosphate + Pi	Fructose 1,6-bisphosphates
Fructose - 6-phosphate + Glyceraldehyde 3-phosphate Erythrose 4-phosphate + xylulose 5-phosphate	Transketolase
Erythrose 4-phosphate + Dihydroxyacetone phosphate Sedoheptulose 1,7- biophosphate	Aldolase
Sedoheptulose 1,7 biophosphate + H_2O $\rightarrow$ Sedoheptulose 7-phosphate + Pi	Sedoheptulose 1,7-biophosphatase
Sedoheptulose 7-phosphate + Glyceraldehyde 3-phosphate Xylulose 5-phosphate + Ribulose 5-phosphate	Transketolase
2 xylulose 5-phosphate 2 ribulose 5-phosphate	Ribulose 5-phosphate 3-epimerase
Ribulose 5-phosphate Ribulose 5-phosphate	Ribulose 5- phosphate isomerase
3 Ribulose 5-phosphate + 3 ATP $\rightarrow$ 3 Ribulose 1, 5-biophosphate + 3 ADP	Phosphoribulokinase
Net Reaction	

3 CO_2 + 9 ATP + 6 NADPH + 5 H_2O_4 $\rightarrow$
9 ADP + 8 Pi + 6 $NADP^+$ + triosephosphate
(Glyceraldehyde 3-phosphate or dihydroxy acetone phosphate).

Rubisco Catalyzes the Oxygenation of Ribulose 1, 5-biophosphate

As its complete name indicates, ribulose 1, 5-bisphosphate carboxylase-oxygenase catalyzes not only carboxylation but also oxygenation of ribulose 1, 5-bisphosphate. The two reactions are competitive — CO_2 and O_2 compete for active sites on RubBisCo. Although the affinity of RuBisCo for CO_2 (k_m = 12 μM) is much greater than for oxygen (k_m = 250 μM) and CO_2 is more soluble in the stroma than O_2, the atmospheric concentration of oxygen (21%) is much higher than the level of carbon dioxide (0.03%). As a result, both reactions contribute significantly to the consumption of ribulose 1, 5- bisphosphate in vivo. Under normal conditions, there is a carboxylation to oxygenation ratio of 3:1 to 4:1.

The products of the oxygenation reaction catalyzed by RuBisCo are one molecule of 3-phosphoglycerate and one molecule of phophoglycolate. Two molecules of 3-phosphoglycerate enter the RPP cycle directly and two molecules of 2-phosphoglycolate are metabolized to one molecule of CO_2 and one molecule of 3-phosphoglycerate, which enters the RPP cycle. Photorespiration, the light dependent uptake of O_2 and release of CO_2 is the name given to the process of oxygen uptake by RuBisCo and the metabolism of phosphoglycolate.

The metabolism of phosphoglycolate occurs in 3 different cellulae compartments of mesophyll cells — the chloroplast, the peroxisome and the mitochondrion. Phosphoglycolate is dephosphorylated to glycolate in the chloroplast, glycolate then enters the peroxisome, where it reacts with O_2 to form glyoxylate and hydrogen peroxide. Hydrogen peroxide is readily converted to O_2 and H_2O by the enzyme catalase and glyoxylate is transaminated to glycine. Glycine exists in the peroxisome and enters the mitochondria, where in one pathway it undergoes oxidative decarboxylation in a cleavage reaction that yields 5,10-methylene tetrahydrofolate, CO_2 and NH_3. CO_2 is either photosynthetically refixed or released form the leaf and NH_3 is reassimilated to form glutamate. 5, 10-methylene tetrahydrofolate donates its methylene group to a second molecule of glycine (also from glyoxylate) to form serine. Serine exits the mitochondrian and enters the peroxisome, where it is deaminated to hydroxy pyruvate; hydroxy pyruvate is then reduced to glycerate. Glycerate enters the chloroplast, where it is phosphorylated by ATP to 3-phosphoglycerate. 3-phosphoglycerate can be incorporated into the RPP cyle. In summary, two oxygenation events form 2 molecules of 3 phosphoglycerate (C_3) and two molecules of phosphoglycolate (C_3). The two phosphoglycolate molecules are metabolized to one molecule of 3-phosphoglycerate and one molecule of CO_2.

The C_4 Pathway Minimizes the Oxygenase Activity of RuBisCo by Concentrating CO_2

In several plant species, a second pathway for carbon fixation occurs in conjunction with the RPP cycle. In this pathway, the initial product of carbon fixation is a four carbon acid, rather than a three carbon acid, and two distinct cell types are involved. Plants with this pathway have essentially not photo respiratory activity. Since the initial product of CO_2 fixation is a C_4 acid, the metabolic route is called the C_4 pathway, just as the RPP cycle is also called the C_3 pathway. C_4 plants include such economically important species as maize (corn), sorghum and sugarcane and many of the most trouble some weeds.

The C_4 pathway, which may be considered a prelude to the C_3 pathway has two phases, carboxylation and decarboxylation, which occur into different cell types within the leaf. In the initial phase, CO_2 is fixed to C_4 acids in the mesophylls cells (typically the site of the RPP cycle in C_3 plants). In the second phase C_4 acids are decarboxylated in the bundle sheath cells, where CO_2 is released, refixed by RuBisCo and incorporated into RPP cycle. Two major structural features related to carbon assimilation distinguish C_4 from C_3 leaf types. First, in the C_4 leaf, mesophyll cells completely surround the bundle sheath cells, which surround the vascular tissue. Second, the C_4 bundle sheath cells are large and contain chloroplasts. In contrast, C_3 bundle sheath cells are non-photosynthetic.

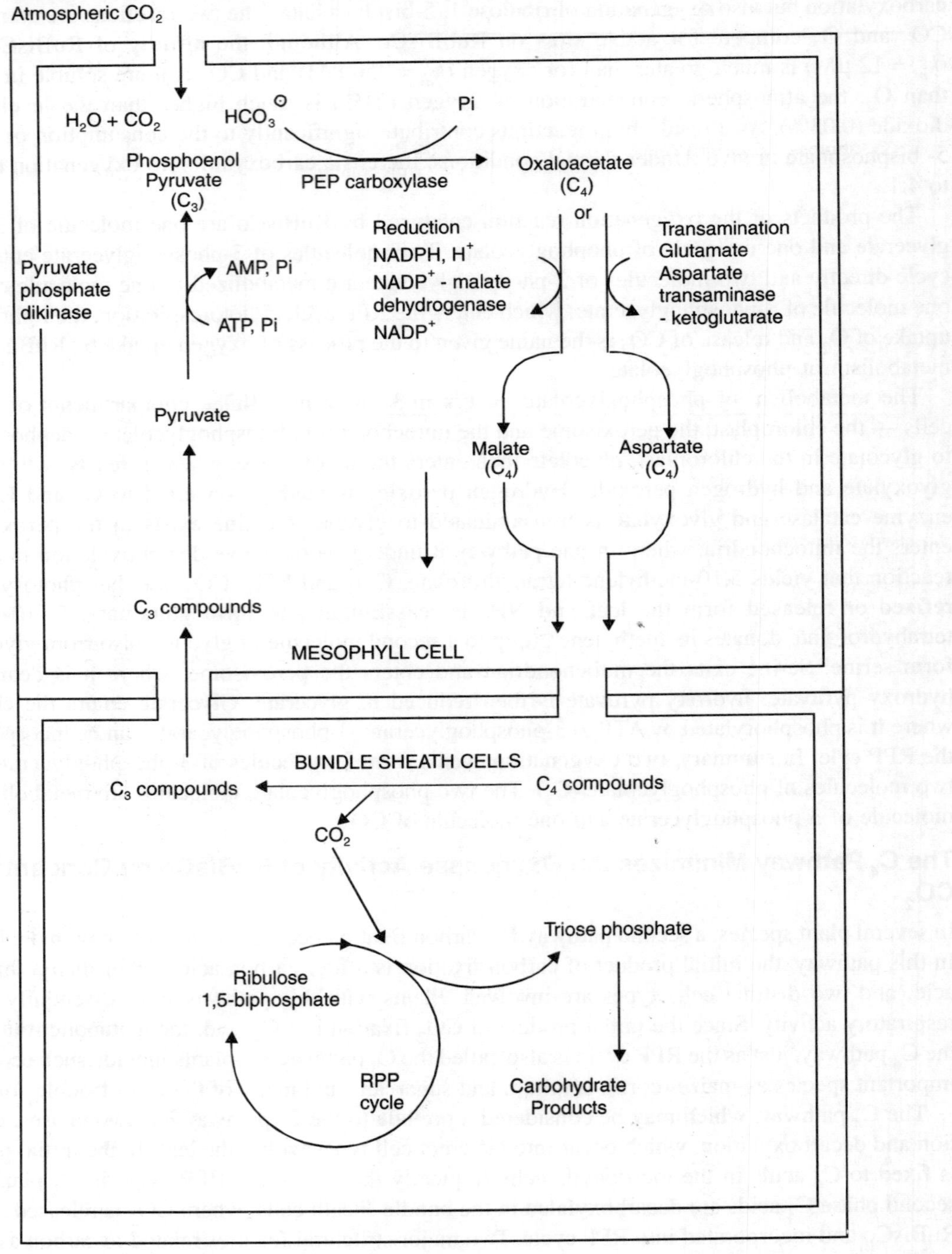
Atmospheric CO_2
$H_2O + CO_2$
$HCO_3^{\ominus}$
Pi
Phosphoenol
Pyruvate
(C_3)
PEP carboxylase
Oxaloacetate
(C_4)
or
Reduction
NADPH, H^+
$NADP^+$ malate
dehydrogenase
$NADP^+$
Transamination
Glutamate
Aspartate
transaminase
- ketoglutarate
Pyruvate
phosphate
dikinase
AMP, Pi
ATP, Pi
Pyruvate
Malate
(C_4)
Aspartate
(C_4)
C_3 compounds
MESOPHYLL CELL
BUNDLE SHEATH CELLS
C_3 compounds
C_4 compounds
CO_2
Triose phosphate
Ribulose
1,5-biphosphate
RPP
cycle
Carbohydrate
Products

Fig. 3.8

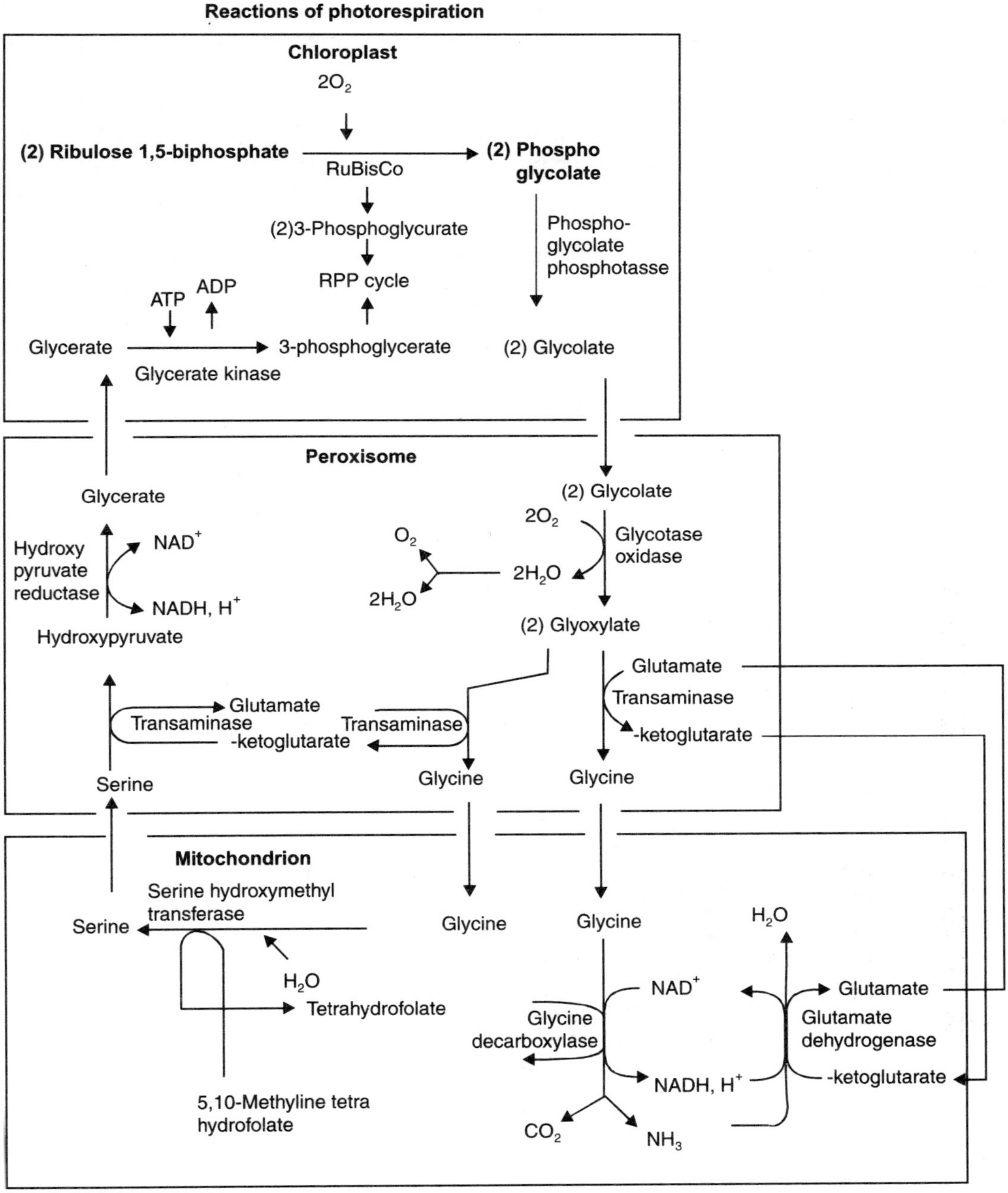

Fig. 3.9

CO_2 is hydrated to bicarbonate (HCO_3^-) in the mesophyll cytosol. Bicarbonate and phosphoenol pyruvate (PEP) are the substrates for carboxylation catalyzed by phosphoenol pyruvate carboxylase, a cytosolic enzyme that has no oxygenase activity. Mesophyll cells of C_4 plants contain PEP carboxylase but no RuBisCo, whereas bundle sheath cells contain RuBisCo but not PEP carboxylase.

Depending on the C_4 species, the oxaloacetate formed by the action of PEP carboxylase is either reduced to malate or transaminated to aspartate. C_4 acids are transported via plasmodesmata (inter cellular connections) to the adjacent bundle sheath cells.

Since the cell walls of bundle sheath cells are quite impermeable to gases. C_4 acid decarboxylation greatly increases the CO_2 concentration inside the cells and create a high ratio of CO_2/O_2. Because RuBisCo absent in the mesophyll cells and the ratio of CO_2/O_2 is extremely high in the bundle sheath cells, the oxygenase activity of RuBisCo is minimized. Thus, there is essentially no photorespiratory activity in the C_4 pathway.

Although there is an extra energy cost to form phosphoenol pyruvate for C_4 carbon assimilation, the absence of photorespiration gives C_4 plants a significant advantage over C_3 plants. C_4 plants have significantly higher quantum yields (molecules of CO_2 fixed per photon absorbed) than C_3 plants under their respective optimal photosynthetic conditions. For reasons related to the speed of intercellular metabolite movement, requirements form light activation and the temperature optima certain C_4 enzymes, C_4 plants are most photosynthetically efficient under conditions of high light intensity and high temperature.

UNIT 4

Transport Mechanism

BIOLOGICAL MEMBRANE STRUCTURE

The structure that separates the cell content from the external environment is the Plasmamembrane. It is a thin film (6 to 10 nm thick) constituting of continuous *lipid bilayer* with proteins intercalated in or adherent to both surfaces. The plasma membrane can be resolved only with the electron microscope, which reveals its numerous infoldings and differentiations as well as the different types of junctions that establish connections with neighboring cells. The main function of the plasma membrane is to control selectively the entrance and exit of the materials. The plasma membrane is covered and reinforced by the cell wall in plant cells and by the cell coat in animal cells. The cell wall is responsible for the rigidity of most plant tissues. This structure consists mainly of cellulose fibres into which other substances (such as lignin, the main component of timber) may be incorporated. There are funnels running through the cell wall *i.e.* plasmodesmata which allow communication with the other cells in a tissue.

In most animals cells the plasma membrane is covered by a cell coat made of glycoproteins, glycolipids and polysaccharides that may extend the thickness of the cell membrane and continue far beyond it. The cell coat has several functions in addition to that of protection. It is involved in molecular recognition between cells, contains enzymes and antigens, and is fundamental in the association of cell in a tissue.

The term *cell membrane* was coined by *Nageli* and *Cramer* (1865) and *Plowe* coined the term *plasmalemma* to plasma membrane.

The early theories on the molecular structure of the membrane were generally based on direct information. Since substance soluble in lipid solvents penetrate the plasma membrane easily, in 1902 *Overton* postulated that the plasma membrane is composed of a thin layer of lipid. In 1926 *Gorter and Grendel* found that the lipid content of hemolyzed erythrocytes was sufficient to form a double layer of lipid molecules over the entire cell surface. This theory was supported by electrical measurements that indicated a high impedance at the plasma membrane, which is due to the fact that it is difficult for ions to penetrate a lipid layer.

Other indirect information comes from the study of interfacial tension of different cells. Tension at a water-oil interface is about 10–15 dyn/cm, whereas surface tension of cells is almost nil. The low tension is due to the presence of protein among the lipid components. Infact, when a very small amount of protein is added to a model lipid-water system, the surface tension is lowered comparably.

To explain all these, many molecular models of cell membrane were proposed by many cell biologists. Some of them are:

(*i*) **Lamellar model:** *Danielle* and *Davson* proposed that the plasma membrane is sheath like and composed of two layers of lipid which are sandwiched between two continuous layers of proteins. This is known as Sandwich model or Trilamellar model (1935).

(*ii*) **Unit membrane model:** *Robertson* (1959) proposed his famous unit membrane model. According to him, the unit membrane consists of bimolecular lipid leaflet sandwiched between outer and inner layers of proteins organized into the pleated sheet configuration (protein-phospholipid-phospholipid-protein). Such an arrangement was presumed to be basically the same in all cell membranes. The observation of thin sections by electron microscopy has led to the concept of trilayered plasma membrane which is also called the *unit membrane model.* This is interpreted as indicating that the electron dense outer layers corresponds to the protein, and the less dense middle layer corresponds to the hydrocarbon chains of the lipids. The unit membrane concept, however, is certainly an over simplification.

(*iii*) **Micellar model:** According to this model a central core (micelle) of hydrophobic phospholipids is surrounded by the hydrophilic protein molecules.

(*iv*) **Protein crystal model:** According to this model the phospholipids are intercalated in the gaps or cavities of crystal like protein structure.

All these models and the unit membrane model, however are certainly an over simplification. Numerous fine details suggest that the molecular organization of the membrane is much more complex. Recent observations indicate that the unit membrane image is mainly artifactual and does not account for the many proteins traversing the membrane. However *Gorter* and *Grendel* in 1934, performed an experiment to study the arrangement of the components of the plasmamembrane. They selected RBC, which were ideal for the study of plasma membrane as they lacked any other membranous inclusions within the cell.

A known number of RBC were taken and lysed by treating the RBC with hypotonic solutions. This was done in beakers containing water. After the lysis of the cells, the air-water interphase shows a layer of phospholipid molecules spread over the surface of water. As they already knew the number of RBC they could approximately measure the area occupied by the membranes of the RBC on the air-water interface which now has a single layer of phospholipid molecules. However, they found that the area occupied by the single layer of phospholipid molecules on the air-water interface was double than the area occupied by the membranes in RBC as per calculation. This led to conclusion that the phospholipids were not present in a single layer but a double layer.

The Fluid Mosaic Model: At present, the most widely accepted model of cell membrane is the Fluid-mosaic model proposed by *Singer and Nicholson* (1972). This postulates that the lipid and proteins are dispersed in a kind of mosaic arrangement and that biological membranes are Quasi-fluid structure in which both lipids and the proteins are able to move with in the bilayer, *i.e.* the proteins are embedded in a lipid bilayer in such a way that *Proteins float in a lipid seal.* The lipid molecules mainly consist of phospholipids which are arranged in two layers. Each phospholipid molecule is made up of a *Glycerol head* connected to a tail of *long chain fatty acid* (Hydrocarbons), which always contain 1 or 2 double bonds and give the tail rather a kinked structure.

Many types of proteins namely *extrinsic, intrinsic and trans-membrane* proteins are present interspersed with the lipid molecules. Extrinsic proteins are the proteins which are superficially attached to the lipid molecules. The integral proteins of the membrane are intercalated to a greater extent into a rather continuous lipid bilayer. These integral proteins are amphipathic with polar regions (hydrophilic) protruding from the surface and nonpolar (hydrophobic) regions embedded in the hydrophobic interior of membrane. This arrangement may explain why different enzymes and glycoproteins may have their active sites exposed to the outer surface of the membrane. It is well recognized that a protein of appropriate size or a cluster of protein subunits may pass across the entire cell membrane (Trans-membrane proteins). Such traversing proteins may be in contact with the aqueous phase on both sides of the membrane. Proteins are involved in transport and act as receptors.

Proteins and phospholipids are not tightly associated with each other and have freedom to move along the surface of plasma membrane. Phospholipid molecules may also move along the cell membrane thus making the plasma membrane a rather viscous fluid rather than a rigid structure. The fluidity is not only due to noncovalent interactions. In fact, the cohesive forces between lipids and proteins are relatively weak interactions, *i.e.*, ionic, hydrogen bonds, and mainly hydrophobic in character. Hence the fluidity is maintained by hydrogen bond and hydrophobic interactions. Some of the lipids and proteins on the outer surface are frequently associated with sugars to form Glycolipids and Glycoproteins.

COMPOSITION

In general, the plasma membrane is composed of

1. lipids (Phospholipids, Cholesterol)
2. proteins
3. oligosaccharides bound to lipids (glycolipids) and to proteins (glycoproteins).

Lipids: The main lipid components of the plasma membrane are *phospholipid, cholesterol* and *glycolipids;* their proportions varies in different cell membranes.

Phospholipids: They are loosely arranged and form 20–79% of the cell membrane. The major proportion (50%) of membrane phospholipid is represented by

(i) Phosphatidylcholine ⎤
(ii) Phosphatidyl ethanolamine ⎥ *Neutral*
(iii) Sphingomyelin ⎦
(iv) Phosphatidyl serine → Acidic; –ve charge

Among these Phosphatidylcholine, Phosphatidyl ethanolamine and Sphingomyelin have no net charge at neutral pH (*i.e.* neutral phospholipids) and tend to pack tightly in the bilayer. (This property is also shared by cholesterol.) 5–20% of phospholipids are acidic, including phosphatidyl inositol and phosphatidylserine and cardiolipin and sulphalipin and phosphatidyl glycerol. They have a net negative charge Phosphatidyl choline and sphingomyelin which are neutral are distributed more on the outer layer of plasma membrane.

Phosphatidyl ethanolamine, Phosphatidyl serine and Phosphatidyl inositol are distributed more in the inner leaflet. Because of the presence of these negatively charged phospholipids, the inner layer plasma membrane gets a net –ve charge on it. These acidic phospholipids are negatively charged and in the membrane are associated principally with proteins by lipid-protein interactions. Because of the many different arrangements of the fatty acids in the phospholipid molecule, it is understandable that a mammalian cell membrane may contain 100 or more species of phospholipids. Though the phosphatidyl inositol is present in very less proportion, *i.e.* 1–2%, it is helpful in cell signalling. Certain glycolipids are also present mainly in the outer leaflet. They contain carbohydrate group which is exposed to the outside because of its hydrophilic nature and fatty acids are attached to this carbohydrate head. Using reagents that do not cross the membrane and various phospholipases (*i.e.* enzymes that hydrolyze different parts of phospholipid molecule). It has been demonstrated that the distribution of the phospholipids is partially asymmetrical; furthermore, the glycolipids are mainly in the outer half of the bilayer. It is assumed that this a symmetry is rather stable and there is little exchange of lipids across the bilayer.

Cholesterol: It is the sterol present in the plasma membrane and gives stability or acts as strengthening agent to preserve the structure of plasma membrane. Cholesterol is present equimolar to the phospholipids in cytoplasmic membrane. Cholesterol is having a rigid ringed structure which is rather inflexible unlike the phospholipid molecule and cannot form a membrane of its own. Hydrophilic part of cholesterol is always situated towards the outer surface or exterior just besides the hydrophilic heads of phospholipids. The function of cholesterol is temperature dependent. At higher temperature cholesterol interferes with movement of fatty acid chain and prevent the membrane from being broken and makes its impermeable to solutes. At low temperature cholesterol interferes with hydrophobic interactions avoiding freezing of the membrane.

Basic functional ability of plasma membrane depends upon the structure phospholipid molecules:

(*i*) Phospholipids have *hydrophilic heads* on either surfaces and *hydrophilic tails* are present in centre. Because of this arrangement the water soluble ions or molecules cannot enter though plasma membrane.

(*ii*) The hydrophobic tails are not straight but are kinked. These long chains of fatty acids can freely move due to which tight packing of these chains is not possible. This confers the fluidity to the membrane. Phospholipids are able to rotate on their own axis.

MEMBRANE PROTEINS—PERIPHERAL OR INTEGRAL

Proteins represent the main component of most biological membranes. They play an important role, not only in the structure of the membrane, but also as carriers or channels, serving for transport; they may also be involved in regulatory or ligand-recognition properties. Numerous enzymes, antigens, various kind of receptor molecules are present in plasma membrane. These proteins were identified by Singer and Nicholson. The membrane may consists of 30–50% of proteins. For every 50–100 molecules of lipid 1 protein molecule is present. Membrane proteins have been divided into *peripheral* (*extrinsic*) *and integral proteins* according to the degree of their association with the membrane and the methods by which they can be solubilized.

Peripheral proteins: They are present only on the surface of the plasma membrane, because of their hydrophilic nature, and are bound to the plasma membrane by ionic bonds and protein-protein interactions. They are easily dissociated by mild treatments like high pH, hypotonic solutions, high ionic strength, metal ion chelating agents etc. The dissociated proteins are soluble in water and are usually free of lipids, and molecularly dispersed in neutral aqueous buffers. Their dissociation will not cause actual disruption of the plasma membrane. For example, (*i*) Spectrin may be removed from red cell ghosts by chelating agents. (*ii*) Cytochrome-c found in mitochondria is easily removed in high salt solutions.

Integral proteins: They represent more than 70% of the membrane proteins. They pass through the membrane and are associated with hydrophobic regions of the lipid bilayer also. They are exposed to either side of the membrane and when present throughout the membrane, they are called as *trans-membrane proteins* and consists of both hydrophilic and hydrophobic regions *i.e.* they are *amphipathic.* They require drastic procedures for isolation. They are dissociated from the plasma membrane by surface acting agents like detergents, Hydrophobic bond-breaking agents, organic solvents, chatropic agents. For example, the dissociation of these proteins from the membrane by detergents is as follows: The detergent is also amphipathic. The hydrophobic regions of the detergents displaces the lipids in plasma membrane and associates or interacts with the hydrophobic region of protein. They are usually associated with lipids when solubilized. They are insoluble or aggregated in neutral aqueous buffers when dissociated from the membrane. Their dissociation is accompanied by the dissociation of the whole plasma membrane.

Example: Most membrane bound enzymes; Histocompatibility antigens, Drug and Hormone receptors.

Structure: The peripheral or hydrophilic ends of the integral proteins are globular and act as receptors. While the hydrophilic region of the trans-membrane proteins are made up of 20–25 amino acid and are α-helical in structure. The hydrophilic region on the outer surface is often associated with carbohydrates mainly oligosaccharides to form Glycoproteins. The clear exception is porin which has a β-pleated sheath structure *i.e.* they are present as a β-pleated barrel. These proteins are synthesized by Ribosomes attached to ER. Then these proteins are sent into ER where they are glycosylated and then are sent in vesicles into cytoplasm which then bind to golgi bodies and are further glycosylated if any Glycosylation is left. Then they are sent to the plasma membrane as secretory vesicles where they burst open and the Glycosylated proteins *i.e.* Glycoproteins are incorporated in the plasmamembrane. Mostly these glycoproteins form the peripheral proteins and are present mostly towards the outerside of the membrane. The carbohydrate groups are mainly oligosaccharides and account for 1–5% of the plasma membrane.

Other proteins: There are surface proteins acting as integral proteins but are not true integral proteins. In these type of proteins, the oligosaccharide group of glycoprotein formed from the ribosomes attached to ER integrates with phosphatidyl inositol present in the inner leaflet of membrane by GPI anchor (Glycosyl Phosphatidyl Inositol Link) to form a GPI molecule. These are known as GPI linked proteins. Other than these some more proteins are seen in plasma membrane formed or synthesized by free ribosomes. These include certain +vely charged proteins or amino acids linked to –vely charged phosphatidyl serine or phosphatidyl inositol on the inner leaflet of the membrane, and also certain amino acids like cystine with sulphur residues may link to the phospholipid of the plasmamembrane, by the formation of a strong bond between the lipids and sulphur residues. In erythrocyte or leucocyte membrane, the glycoproteins consists of Hexose, Hexosamine, Fucose, NAG and NANA $\rightarrow$ N-acetylnueraminic acid (sialic acid). The proteins which contain sialic acid in their glyco group are called sialo proteins and these proteins are recognized by endothelial receptors. (P-selectin of endothelial cells). Because of the amphipathic nature of the trans-membrane proteins, they cannot be easily crystallized and hence their 3 dimensional structure cannot be identified.

The structure of bacteriorhodopsin, which is present in the membrane of halophilic bacteria was first identified in Rhodopseudomonas, the trans-membrane protein has 3 regions; The middle portion is α-helical structure, above and below it, peripheral proteins are present which accept light and send to L, M, N portions of α-helical structure. It consists of a single polypeptide chain of 247 amino acids, which traverses 7 times the lipid bilayer. Each of the traversing segments is an α-helix with some 25 amino acid residues.

The oligosaccharides of the proteins and lipids on the outer side of the membrane form a layer around the membrane called as Glycocalyx in eukaryotes and its basic function is protection and also work as binding sites or receptors.

Movement of Proteins in the Plasma Membrane

The concept of membrane fluidity refers to the fact that both lipids and proteins may have considerable freedom of lateral movement within the bilayer. Vectorial movement across the membrane is severely constrained (meaning that a lipid or protein in the outer half of the bilayer cannot pass into the inner half). The proteins are able to move along the membrane but not across the plasma membrane.

This protein movement was proved by an experiment done by *Larry, Frye and Micheal Edidin.* In their experiment, a mouse cell and a human cell having different surface antigens labelled with fluorescent antibodies are fused by sendai virus. The heterokaryon has each half of its surface covered with different types of antigens. After 40 minutes, mixing of antigens occur because of their movement in the plane of membrane. However, epithelial cells are polarized *i.e.*, different parts of the cell has different functions. The proteins at apical regions are involved in the absorption of nutrients and the proteins of basal portion are involved in the transport of nutrients out of the cell *i.e.* into blood vessels. Thus in these cell the protein movement is restricted by the presence of tight junctions between the cells which do not allow the proteins to move along the cell membrane freely.

Functions of the Plasma Membrane

(*i*) Cell permeability.

(*ii*) Proteins present on cell membrane may act as carriers for transport of materials and may help in cell recognition or may act as receptors for informational molecules.

Cell permeability

Permeability is fundamental of the functioning of the living cell and to the maintenance of satisfactory intracellular physiological conditions. This function determines which substances can enter the cell, many of which may be necessary to maintain its vital processes and the synthesis of living substances. It also regulates the outflow of excretory substances and water from the cell.

The presence of a membrane establishes a net difference between the intracellular fluid and extra cellular fluid. This may be fresh or salt water in unicellular organisms growing in ponds or seas, but in multicellular organism, the internal fluid *i.e.* blood, lymph and especially the interstitial fluid is in contact with the outer surface of the cell membrane. One of the functions of the cell membrane is to maintain a balance between the osmotic pressure of the intracellular fluid and that of the interstitial fluid.

(*i*) **Passive transport:** Passive transport occurs along the lipid membrane depending upon the concentration gradient and does not involve any proteins or energy. The molecules move across the lipid membrane from higher to lower concentration. Molecules like water can pass through passive transport. Small polar molecules and hydrophobic molecules are permeable by passive transport whereas the charged molecules, larger molecules and certain hydrophilic molecules cannot pass through passive transport.

Small molecules like O_2, CO_2 polar molecules like water, Non-polar solvents like Benzene, Alcohol are permitted. Amino acids are not permitted as they are large and charged. Glucose is also not allowed due to its larger size. Small ions like H^+, Ca^+, Cl^-, Na^+, K^+ are also not allowed by passive transport due to their charge.

(*ii*) **Facilitated transport (diffusion):** Occurs by two different ways:

1. Occurs along the concentration gradient.
2. Occurs against the concentration gradient.

Proteins are involved in this transport. Two types of protein transporters are seen.

(*i*) Carrier proteins

(*ii*) Channel proteins.

Carrier proteins: The transport of many different molecules across the membrane shows a high degree of specificity, *i.e.*, the permeability of a molecule is related to its chemical structure. While one molecule may readily enter the cell, another of the same size but a slight different molecular structure may be completely avoided. A clear example of this selectivity is provided by the transport of two isomers (glucose and galactose) into a bacterial cell. Although the only difference between them is in the position of –OH group at carbon 4, these two sugars penetrate the membrane by different transport mechanisms. This type of selectivity is attributed to transport proteins or carrier proteins or permeases. It is thought that a permease functions in a manner somewhat similar to an enzyme or a receptor in having a binding site able to recognize the molecule to be transported. Permeases accelerate the transport process, provide for special selectivity, and are recycled, meaning that they remain unchanged after having assisted in the entry or exit of a molecule. Some permeases can transport only if there is a favorable concentration gradient while others can do so aganist it. *Example:* Glucose is transported along the concentration gradient. Glucose concentration is always less inside the cell as it is actively metabolized inside the cell, except in liver where active outward flow of glucose molecules is seen.

The carrier mechanism has also been postulated to mediate the entrance of the glucose into intestinal cells. This mechanism is activated by Na^+. The carrier ion in this model has polar amino acids or binding site for Glucose, and should exist in two conformational states, one with low affinity for glucose and with high affinity. A carrier model can be postulated in which the glucose site is initially with a high affinity state and binds the glucose molecules at the surface of the cell. In a second step, the carrier is translocated and the glucose site changes to a low affinity releasing the molecule into the cell. The two affinities are modulated by Na^+ conc. inside and outside the cell. This carrier mechanism follows fixed pole mechanism in which the carrier is represented by integral proteins that traverse the membrane and which, once bound to the molecule to be transported, undergo conformational changes.

Channel proteins: Special channels meant for the transport of water are present in the membrane. *Example:* Aquaporins, proteins are present in the outer membrane of Eukaryotes. The transport through these channel proteins is much faster than the passive diffusion.

Ion channels: Ion channels form the most important means of transport of various ions and water. They are more effective and efficient due to the following reasons:

1. Faster rate of movement of ions *i.e.* upto a million ions can pass through it per second.
2. They are specific to charge and size of ions.
3. They are not always open.

Depending upon the type of signals they require to close/open they are of two types:

(*i*) Ligand gated channels
(*ii*) Voltage gated channels.

(*i*) **Ligand gated channels:** They open and close in response to the binding of neurotransmitters or some other signalling molecules.
Example: The binding of acetylcholine to the acetyl choline transport molecule determines the opening or closing of the negatively charged hydrophobic gates. Hence acetyl choline acts as signalling molecule and when it binds to the molecule, the gates open due to conformational change and allow only +vely charged ions like Ca^+, K^+, etc. to pass through them. But –vely charged ions are not allowed to pass through them because of –vely charged hydrophobic gates.

(*ii*) **Voltage gated channel:** The opening and closing of these voltage gated channel depends upon the electron potential or charge of the cell. *Example:* Na^+ pump, K^+ pump.

Na^+ pump: Generally Na^+ are pumped outside the cell and the Na^+ pump aids in this. The Na^+ size is small and appropriate to the size of the Na^+ pump and so passes through it along with water. But as K^+ are larger than Na^+, K^+ cannot pass through the Na^+ pump.

K^+ pump: Generally K^+ are pumped inside the cell and the $^+$channel or K^+ pump has a binding site and so can pass through it along with water. But Na^+ size is very small to bind with the active site in K^+ pump and hence only K^+ ions can pass through them.

Groups translocation: Translocation of a molecule inside the cell which involves the chemical alteration of that molecule is called Group translocation. Group translocation always occur against the concentration gradient but without true expenditure of energy. *Example:* Glucose is transported by group translocation into cell as glucose-6-phosphate by addition of a phosphate molecule outside the cell. Here the ATP is used up in the phosphorylation of glucose *i.e.* for biochemical reaction but not actually for transport. This glucose-6-phosphate may follow glycolysis in the cell. Sugars are generally transported by this method.

Active transport: Active transport involves transport of molecules via the cell membrane with *expenditure of energy.* Energy procured from ATP hydrolysis of electron potential of the cell, without the chemical alteration of the molecule *against the concentration gradient. Example:* Na^+–K^+ Pump. It includes protein molecules with 3 binding sites for Na^+ and two binding sites for K^+ and can pump 3 Na^+ outside and $2K^+$ inside the cell by utilization of only one ATP molecule. In normal configuration, the Na^+ binding sites are exposed to the cytoplasm and the Na^+ which are always pumped outside the cell bind to these binding sites when the protein molecule is phosphorylated by the hydrolysis of ATP. As soon as the Na^+ binds to the active sites, conformational change takes place and the Na^+ binding sites are exposed to outside and the Na^+ are released outside the cell. During this release of Na^+, K^+ binding sites are exposed to the cytoplasm and so K^+ from outside the cell binds to these active sites which causes dephosphorylation of protein molecule of thus leading to conformational change. The K^+ are released inside the cell. So only one ATP molecule is utilized to pump out $3Na^+$ and to pump in $2K^+$ ions by active transport.

Ca^{2+} pump: Ca^{2+} pump is also similar in structure to the Na^+, K^+ pump and helps in active transport of Ca^{2+} ions outside the cell by ATP hydrolysis. The conc. of Ca^{2+} inside the cell = 0.1 mm. The conc. of Ca^{2+} outside the cell = 1 mm. But the Ca^+ are transported outside the cell. During this transport some other ions may be transported inside the cell simultaneously.

***Cystic fibrosis:** It is a disease caused by malfunctioning of ion channels namely CFTR *i.e.* Cystic Fibrosis Transport Regulator Channel.

Ion Gradient Active Transport

The concentration of ions inside or outside the cell is called ionic gradient. The cell always maintains the ionic gradient by pumping of ions like Cl^-, Ca^+, Na^+, H^+, K^+. If +ve charge is formed outside of the cell. OH^- ions are always present inside the cell having –ve charge. This is the stable gradient always present around the cell. Any cell maintains an ion gradient across its membrane by active transport. Cells contain a net +ve ion gradient on the outer surface, where large amount of H^+ and +ve charged ions are accumulated. The inner surface of the cell membrane maintain a –ve gradient by the accumulation of OH^- ions. The ion gradient keeps fluctuating across the membrane due to the movements of ions into or outside the cell against the concentration gradient by active transport. When Na^+ and Ca^+ is transported out of the cell, against the conc. gradient, there is a reduction in net ionic gradient; across the cell membrane. The movement of cation across the cell membrane is thus compensated by the inward flow of equal number of cations. This mechanism of maintaining ionic equilibrium in the cell membrane is called *Donnans Equilibrium.*

Transporters: Three types of transporters are seen in the cell membrane namely:

(*i*) Uniporter
(*ii*) Symporter
(*iii*) Antiporter.

(*i*) **Uniporter:** These are that type of transporters which transfer only one specific molecule either outside or inside the cell through it. *Example:* Na^+ Pump.

(*ii*) **Symporter:** These are that type of transporters, which transport one type of ions and some other ion or molecule can also pass along these ion in the same direction. *Example:* Glucose enters into the cell along with Na^+ ions through Na^+ channel along the conc. gradient.

(*iii*) **Antiporter:** These are that type of transporters which allow one type of ions inside and other type of ions outside the cell simultaneously.

Example: Na^+ –K^+ pump → Na^+ outside, K^+ inside

K^+ –Ca^+ pump → K^+ inside, Ca^+ outside the cell.

- The cells are very sensitive to the Ca^+ concentration. Ca^+ acts as a cell signalling molecule.
- H^+ pump more frequent in plant cells and bacteria.
- Na^+, K^+ pumps are frequent in animal cells.
- The entry and exit of the water molecules in the cell is governed by the *Turgor pressure* which develops, due to the presence of a rigid cell wall.

Other Functions

Endocytosis: Endocytosis is related to the activity of plasma membrane. It may not take place in all the cells. It includes *phagocytosis* and *pinocytosis.*

Phagocytosis: The ingestion of the food particles inside the cell by the formation of a phagosome or vesicle by specialized cells called phagocytes is called *phagocytosis.* Macrophages and neutrophils are called professional phagocytes which help in the immune system of the body.

Pinocytosis: The ingestion of liquid or semisolid particles by the formation of pinosomes or vesicles is called *pinocytosis.*

Process: Once the phagosome or pinosome are formed, they are associated with lysosome to form phagolysosome and pinolysosome into which the hydrolytic enzymes are released and active digestion takes place and simpler compounds or waste materials are formed in the phagolysosome. The useful material is secreted in the cell and finally the phagosome with debris or waste material is sent or vomitted out of the cell by a process known as *Exocytosis, Emiocytosis* (or) *Cell Vomiting.*

Vesicles: Vesicles are globular structures that bud off from the cell membrane during endocytosis. They may be of two types:

(*i*) Non-specific vesicles
(*ii*) Specific vesicles.

- Non-specific vesicles are more or less equal to phagosomes or pinosomes, which may later combine with lysosome, E.R. or golgi bodies depending upon the type of molecules or on the nature of the residue in vesicles.
- If it contains complex material → It associates with lysosome for digestion into simpler compounds.

- If it contains proteins → They are passed to E.R. for sorting, marking or for Glycosylation. If the proteins present are useful to the cell, then they are targeted to specific spaces otherwise, they are thrownout if they are not required.
- Chances of union of vesicles with Golgi bodies is rare.
- Sometime these vesicles may also combine with endosome (which is vesicle that is present in the cell for a long time, before forming a secretory vesicles.

Specific vesicles: Specific vesicles contain receptors which facilitate them to capture particular material. *Example:* Intake of cholesterol. There are certain pits in the mammalian cells called as *clatherin coated pits* coated by a substance called clatherin which contains LDL receptors. Cholesterol is ingested in the form of LDL-*Low density lipoprotein*. The LDL gets bound to the LDL receptors and slowly the pit will be filled with LDL, then it forms a clatherin coated vesicles which will be transferred specifically to the lumen of Endoplasmic reticulum to get degraded to form cholesterol. This cholesterol is then passed to a specific site of via E.R., Golgi apparatus. Sometimes LDL may be sent outside the cell or may be loaded on the innerwalls of blood vessels and thus reducing the width of the blood vessels causing high B.P.

Specialized Structures of the Cell Membrane

The differentiations of the cell membrane correspond to regions specially adapted to different functions such as absorption, secretion, fluid transport, mechanical attachment or interactions with neighboring cells.

(*i*) **Microvilli:** In the intestinal eqithelium microvilli are prominent and form a compact structure that appears under the light microscope as a striated border. They represent the cycloplasmic processes covered by plasma membrane or the projections of the plasma membrane. They are mainly involved in the absorption of nutrients and are supported by cytoskeletal elements like microtubules and microfilaments. The actin filaments are attached to the tip of the microvilli by α-actinin, their function is to produce contraction in the microvilli. The outer surface of microvilli is covered by a coat of filamentous material (fuzzy coat) composed of glycoprotein macromolecules.

(*ii*) **Interdigitation:** These are structures formed by folding of the plasma membrane into finger like or digit like structures which are predestined to form other specialized structures.

(*iii*) **Special invaginations:** Special invagination of plasma membrane are seen forming the chambers between them and mitochondria are present in these chambers helping in the production of energy for transport.

(*iv*) **Zonula occludens (tight junction):** Tight junctions are specially differentiated regions that seal the intercellular space, thus preventing the passage of fluid to and from the lumen. These junctions are present just below the apical border of the cell. In a polarized epithelium, one of the functions of the tight or occluding junctions is the maintenance of different intercellular environments in the apex and basolateral regions. The tight junction could act as a barrier to the diffusions of macromolecules or lipids in the bilayers, so as to differentiate the apical from the basolateral portion of the cell membrane, each having their own particular composition and physiological characteristics.

Mechanical adhesion between cells is mainly done by the *desmosomes,* of which there are two types, belt and spot desmosomes. The difference between them resides in their localization, extension and relationship with the microfilaments.

Belt desmosome (zonula adherence or terminal bars): Intermediary junctions are generally found at the interphase between columnar cells, just below the region of tight junctions. They form a band that girdles the inner surface of the cell membrane. This band contains a wed of (6 nm) actin filaments and another group of interwoven intermediate filaments of 10 nm. Actin microfilaments are contractile and intermediate filaments play a structural role. The plasma membranes of the adjacent cells come closer to each other and the cytoplasm gets thickened into a gel in this region.

Macula adherence (sport desmosome): They appear as darkly stained bodies on the cell surface under the light microscope. They represent localized circular areas of contact about 0.5 mm in diameter, in which the plasma membranes of the two adjacent cells are separated by a distance of 30 to 50 nm. The electron microscope reveals a midline structure or intercellular core and two parallel desosomal plasma membranes. Under each plasma membrane there is a disocidal intracellular plaque, towards which numerous filaments called *Tonofibrils* or *Tonofilaments* converge. These tonofibrils are not contractile and belong to common type of intermediate keratin filaments. They are rigid structures.

Hemidesmosomes: Along the basal surface of some epithelial cells septate desmosomes called hemidesmosomes may be observed. These are similar to desmosomes in fine structure, but represent only half of them, the outer surface frequently being substituted with collagen fibrils. Some tonofibrils may be present.

Gap junction: The gap junction or *nexus* or communication junction appears as a plaque-like contact in which the plasma membrane of adjacent cells are in close-apposition or tightly connected, separated by a space of only 2 to 4 nm. The cytoplasm is absent between plasma membrane in this region. It mainly binds the cells together and help in cell communication.

Inter membrane canaliculus or inter cellular canaliculus: They are hexagonal structures formed by the infolding of the plasma membrane. Cytoplasm is thickened at this region. It basically helps in maintaining communication between cells in the form of ions.

Cell wall: The cell wall constitutes a kind of exoskeleton that provides protection and mechanical support for the plant cell. This includes maintenance of a balance between the osmotic pressure of the intracellular fluid and the tendency of water to penetrate the cell. The cell wall of fungi is made up of chitin, a polymer of N-acetyl glucosamine. The exoskeleton of crabs, insects is also made up of chitin.

Cell wall of plants: The wall consists of a microfibrillar network lying in a gel like matrix of interlinked molecules. The microfibrils are mostly cellulose, consisting or straight polysaccharide. These are the glucose chains, which by intra and intermolecular hydrogen bonding produce the structural unit known as microfibril. Microfibrils in turn bind among themselves with non-cellulosic polysaccharides and proteins to form the cell wall. In addition to protein, the gel matrix contains some polysaccharides and lignin. The major polysaccharide fractions are *Pectin substances* (containing Galactose, Galacturonicacid, arabinose, sometimes Rhamnose may be present) which are soluble in water and *Hemicelluloses* (consisting of glucuronic acid, xylose, fucoses mannose, glucose, etc.) which are extracted with alkalis. *Lignin* is found only in mature cells and is made up of an insoluble aromatic polymer resulting from the polymerization of phenolic alcohols. *Suberin* consists of gums and resins along with some aromatic residues and suberic acid.

Development of primary and secondary walls and plant cell differentiation

Cell walls are complex and are highly differentiated in certain plant tissues. To a greater extent, cell wall determines the shape of the cells and serves in the classification of plant tissues. In general, primary and secondary walls develop in a special sequence and are differentiated by the deposition of the cellulose microfibrils and by the composition of the matrix. Development of the cell wall starts with the middle lamella formed by the cell plate after cell division. This layer mainly contains pectin substances. After cell division each daughter cell deposits outer layer that constitutes the primary cell wall which is composed of pectin, hemicellulose, and a loose network of the cellulose microfibrils, most of which are oriented transverse to the long axis of the cell.

When the cell has enlarged during maturation, usually with the development of large vacuoles, the secondary cell wall appears. This consist of material that is added to the inner surface of the primary wall, either as a homogeneous thickening, as in cells forming sieve tubes (phloem) or as localized thickenings, as in Xylem vessels. In both cases, this material consists mainly of cellulose and hemicellulose, with little pectin.

The characteristic of most plant cells is the presence of plasmodesmata. Plasmodesmata are bridges of cytoplasm that establish communications between cells across the cell wall, and they may play a role in cell differentiation.

UNIT 5

Carbohydrate Metabolism

CORI-CYCLE

Synthesis of Glucose from Lactate during the recovery period, takes place at a high rate. Much of the extra oxygen consumed during the recovery period, *oxygen debt*, is used to generate ATP necessary for resynthesis of blood glucose and muscle glycogen from the Lactate formed anaerobically. The Lactate is removed from the blood by the liver and converted into blood glucose which returns to the muscles where it is rebuilt into glycogen.

The process of Gluconeogenesis is especially active in Ruminant animals, where ingested plant flood is subjected to bacterial fermentation in the rumen where the different bacterial species cooperate to degrade plant components, esp. cellulose, and ferment all the glucose to form lactate and other products, Lactate formed in the rumen is absorbed into the blood and converted into glucose by the liver.

GLYCOGEN SYNTHESIS

In animals glycogen synthesis occurs in virtually all the tissues but is prominent in liver and skeletal muscles.

Glucose 6-phosphate is reversibly converted into Glucose 1-phosphate by *phosphoglucomutase.*

The Glucose l-phosphate reacts with UTP in the presence of Glucose 1-*phosphate uridyltransferase* forming *UDP-glucose*

$$\text{UTP + Glucose - 1 - phosphate} \rightarrow \text{UDP-Glucose + PPi}$$

Hydrolysis of pyrophosphate causes the reaction to proceed to the right.

UDP-Glucose is the immediate donor of Glucose residues in the formation of Glycogen by the action of *Glycogen synthase*, which transfers the glucosyl residue from UDP-glucose to a non-reducing end of the branched glycogen molecule, establishing $[1 \rightarrow 4]$ linkage. The enzyme requires as a primer an α $[1 \rightarrow 4]$ polyglucose chain having at least 4 glucose residues.

The making of α $[1 \rightarrow 6]$ bonds found in glycogen at the branch points is by the glycogen—branching enzyme *glycosyl* $[4 \rightarrow 6]$ *transferase*, catalyses transfer of terminal oligosaccharides fragment of six or seven glucosyl residues from the non-reducing end of a glycogen branch having at least eleven residues to the 6-hydroxyl group of a glucose residue creating a new branch.

Glycogen synthase occurs in two covalent forms active dephosphorylated form and an inactive phosphorylated form. The conversion of the less active Glycogen synthase *b* into the active form is promoted by *phosphoprotein phosphatase* while the inactive form is obtained by the action of *protein kinase*.

Thus Glycogen phosphorylase and Glycogen synthase are reciprocally regulated. Further since Glycogen synthase *b* depends upon Glucose 6-phosphate for activity it is also called *Glycogen synthase D*. While the active form is not stimulated hence also called *Glycogen synthase* I.

Disorders of Carbohydrate Metabolism

Some of the important problems/disorders of carbohydrate metabolism are—

1. Hyperglycaemia, 2. Hypoglycaemia and 3. Glycosuria.

These problems are physiologically satisfied by a regulating mechanism existed in the human body which is known as regulation of normal blood sugar level.

CARBOHYDRATE METABOLISM

Although the products of carbohydrate digestion absorbed into the blood, are mainly hexone monosaccharides, *i.e.*, glucose, fructose, galactose, as well as partly pentose sugars but animal cells utilize mostly glucose so monosaccharides are converted into glucose by the cells mainly for their oxidation and partly for their storage as glycogen. As fructose and galactose are utilized to a lesser extent in the body so they are reconverted from glucose as evident from the following:

(*a*) Large quantities of galactose are found in the brain tissue.

(*b*) The brain probably uses carbohydrates as galactose only.

(*c*) During the normal process of glucose utilization, fructose diphosphate is formed as an intermediary step.

(*d*) When liver function is deficient, fructose and galactose fail to be converted into glucose and are turned out into the blood stream as such. Due to their low renal threshold they are excreted through the urine.

Since carbohydrate is utilised by the cells of human being mainly as glucose so carbohydrate metabolism is meant essentially as metabolism of glucose and other substances, convertible to glucose or vice versa.

Xylose, arobinose and ribose like pentose sugars in the diet may be absorbed but their fate is not known whereas, α-ribose and α-2-deoxyribose after absorption are utilized in the body for the synthesis of nucleoproteins.

Glucose metabolism normally not only supplies major amount of energy for the body but also

(*i*) other forms of monosaccharides, disaccharides and polysaccharides,

(*ii*) reserve depot fats,

(*iii*) tissue glycolipids, mucopolysaccharides, and

(*iv*) amino acids are formed from glucose and reversely protein and lipids are metabolized through glucose pathway. Thus glucose metabolism holds a key position in carbohydrate metabolism which is closely associated with the metabolism of protein and fat.

The metabolism of carbohydrate discussed under the following points:

1. Storage (by the process of glycogenesis) when glucose is not immediately required by the tissues, it remains stored. The chief form of storage is glycogen. The main store houses are liver and muscles in which glycogen is almost equally distributed. Some glucose may remain stored as such temporarily in the skin and subcutaneous tissue.
2. Sources of energy (by the process of glycolysis and oxidation of pyruvic acid through TCA cycle)—carbohydrates are oxidized in the tissues and supply energy. The grain yields 4 C of energy. The major part of daily energy requirement under normal conditions is derived from carbohydrate oxidation. It is the most readily available source of energy and its combustion involves the least oxygen requirement.
3. Maintenance of blood sugar (homeostasis)—Glucose remains in blood and its level is always kept within a narrow range. Normal blood sugar varies from 80–120 mgm per 100 ml whenever it tends to fall, glucose is mobilized from glycogen and blood sugar is maintained.
4. Synthesis of hexosephosphate—This is an intermediary step formed during oxidation of glucose, absorption of glucose from intestine and reabsorption from kidney tubules. It is a very important form in which hexose exists in the body. It is present in large quantities in the muscles, intestine, liver etc.
5. Synthesis of lactose—in the lactating mother lactose is synthesized from blood glucose. The mammary glands convert glucose into galactose and then unite the latter with another molecule of glucose to form lactose.
6. Synthesis of complex fats containing sugar—For instance, cerebrosides are synthesized by the nerve cells from galactose and fats. Galactose is locally synthesized from blood glucose.
7. Synthesis of fat (Lipogenesis)—it is an established fact that the body can convert carbohydrates into fats.
8. Synthesis of proteins—Simple amino acids may be formed by uniting ammonia with pyruvic acid, etc., which may easily be derived from carbohydrates.
9. Synthesis of glycoproteins—in glycoproteins, the protein molecule remains combined with a carbohydrate radicle, containing nitrogen, viz., hexosamines: glucosamine or galactosamine. They are also known as aminosugars. Glycoprotein contains less than 4 per cent carbohydrate. The carbohydrate part is a tetrasaccharide, being made up of two molecules of acetyl hexosamine, which are united through two molecules of glucuronic acid.

Excretion of glucose: Glucose is not excreted from the body in normal health and Fehling's test is negative. It is a high threshold substance (170 mgm per 100 ml of blood). In metabolic disorders like diabetes mellitus, glucose appears in the urine and Fehling's test is positive. When blood sugar level gives higher value than renal threshold value, the glucose appears is the urine, this condition is known as glycosuria. In lactating mother, lactose may be found in the mothers urine. Pentose is found in a condition known as pentosuria.

Glycolysis

It is the central pathway of Glucose catabolism. The glycolytic sequence of reactions differs from one species to another only in the rate and in the subsequent metabolic fate of the pyruvic acid.

In aerobic organisms pyruvic acid is oxidized, with the loss of carboxyl group as CO_2, to form Acetyl group of Acetyl Coenzyme A, which is then oxidized completely to CO_2 and H_2O by the citric acid cycle.

In Lactic acid fermentation, as well as in some animal tissues like vigorously contracting skeletal muscle, the pyruvic acid is reduced to lactate. In skeletal muscle this process is called anaerobic glycolysis.

In some microorganisms like brewer's yeast, the pyruvic acid is converted anerobically into Ethylalcohal and CO_2, by alcoholic fermentation.

The breakdown of the 6-carbon glucose into two molecules of 3-carbon pyruvic acid is brought about by the sequential action of 10 enzymes. The first five steps of which constitute the preparatory phase. It serves to collect the carbon chains of all the metabolized hexoses in the form of one common product—glyceraldehyde 3-phosphate.

In the second phase energy is freed during the conversion of Glyceraldehyde 3-phosphate to pyruvic acid (which is conserved by formation of ATP from ADP).

Each of the intermediates between glucose and pyruvic acid is a phosphorylated compound where the phosphate group performs the following functions:

A. It is completely ionised at pH 7, giving each of the intermediates of glycolysis a net negative charge making them impermeable.
B. They are the essential components in the enzymatic conservation of metabolic energy since they are ultimately transferred to ADP to form ATP.
C. It serves as recognition or binding group required for the proper fit of the glycolytic intermediates to the active sites of their corresponding enzymes.

First Phase

In the first step the D-glucose molecule is primed by its phosphorylation at the 6 position to yield *Glucose-6-phosphate* at the expense of ATP. This is a irreversible reaction catalyzed by hexokinase requiring Mg^{2+}. *Hexokinase* also phosphorylates common hexoses such as D-fructose and D-mannose.

Hexokinase occurs in different isozyme forms in different organisms or tissues. These isozymes differ with respect to their kinetic properties.

Hexokinase is a regulatory enzyme in which it is allostearically inhibited by its product.

In the liver, there is another enzyme called Glucokinase which is specific only for D-glucose and is not inhibited by glucose-6-phosphate. Glucokinase has a higher k_m for glucose than hexokinase. It is responsible for the conversion of excess of glucose to glycogen after a meal rich in sugar. This enzyme is deficient in diabetes mellitus.

The glucose-6-phosphate is acted upon by *Phosphoglucoisomerase* which catalyses the reversible isomerization to yield fructose-6-phosphate.

The fructose-6-phosphate is acted upon by *Phosphofructokinase* which undertakes the second priming, requiring ATP and Mg^{2+} for the phosphorylation to yield fructose 1, 6-diphosphate. This reaction is an irreversible reaction and is the second control point in glycolysis. This enzyme is a regulatory enzyme, its activity is accelerated whenever the cells supply of ATP becomes depleted or there is an excess of ADP and AMP. Its activity is inhibited by ample supply of ATP, citrate or fatty acids.

Fructose 1, 6-diphosphate is cleaved reversibly to yield 2 different trioses, Glyceraldehyde 3-phosphate and Dihydroxy acetone phosphate by the action of *fructose diphosphate aldolase* (aldolase).

The DHAP so formed is immediately and reversibly converted into Glyceraldehyde 3 phosphate by *triose phosphate isomerase* whereby carbon atoms 1, 2 and 3 of the starting Glucose become indistinguishable from carbon atoms 6, 5 and 4 respectively.

Second Phase

The second phase of Glycolysis contains the energy conserving phosphorylation steps in which the free energy of Glucose molecule is conserved in the form of ATP. The conversion of 2 molecules of Glyceraldehyde 3 phosphate into 2 molecules of pyruvic acid is accompanied by the formation of four molecules of ATP from ADP, however since 2 ATPs were invested in the first phase of Glycolysis to phosphorylate the 2 ends of the hexose molecule, the net gain of ATP is 2 per molecule of Glucose degraded.

The first of the two energy conserving reactions is the step catalysed by *Glyceraldehyde phosphate dehydrogenase* which catalyses the dehydrogenation of Glyceraldehyde 3-phosphate to Acyl phosphate which is a super high energy phosphate called 3-phosphoglyceroyl phosphate. NAD^+ is the Hydrogen acceptor and a inorganic phosphate is required for this reaction.

Glyceraldehyde dehydrogenase contains 4 identical subunits and has a molecular weights of 140,000 and it is inhibited by iodoacetate which combines with the essential-SH group of the enzyme.

In the next step the high energy phosphate group of 3-phosphoglyceroyl-phosphate is transferred to ADP forming ATP and leaving 3-phosphoglycerate which is catalyzed by the enzyme *phosphoglycerate kinase.* The generation of ATP coupled to an enzyme transformation of a substrate is spoken as *substrate-level phosphorylation.*

The 3-phosphoglycerate is acted upon by the enzyme *phosphoglycerate mutase* and thus the inter-molecular shift of its phosphate group to the second position forming *2-phosphoglycerate.*

The enzyme *enolose* catalyses the reversible removal of a molecule of water from 2 *phosphoglycerate* to yield phosphoenol pyruvate.

In the last step of Glycolysis there is the transfer of high energy phosphate group from Phosphoenol pyruvate to ADP catalyzed by the enzyme *Pyruvate Kinase* to yield enolpyruvate which is rapidly and non-enzymatically rearranged to yield the keto form of pyruvate. This step of yielding ATP constitutes substrate level phosphorylation. The reaction is essentially irreversible and the enzyme pyruvate kinase is a regulatory enzyme.

Feeder Pathways

Apart from the free glucose entering into the Glycolytic cycle, many other carbohydrates ultimately enter the Glycolytic sequence. The most important are the storage polysaccharides Glycogen, starch. The disaccharides maltose, lactose and sucrose and the monosaccharides fructose, mannose and galactose.

Glycogen and starch gain entrance into the glycolytic pathway through the sequential action of *Glycogen Phosphorylase* (in plants starch phosphorylase) and *Phosphoglucomutase.*

Glycogen phosphorylase cleaves the Glycogen at the $\alpha[1 \rightarrow 4]$ linkages. The terminal $\alpha[1 \rightarrow 4]$ glycosidic linkages at the non-reducing end of the glycogen branch undergoes phosphorolysis, to yield α-D-Glucose-1-phosphate. The action is a repeated one until it reaches $\alpha[1 \rightarrow 6]$ branch point.

Further degradation occurs by the action of $\alpha[1 \rightarrow 6]$ *glucosidase*, which makes another length of glycogen available for glycogen phosphorylase. The glucose 3 phosphate so formed is converted to glucose-6-phosphate by *phosphoglucomutase* which is a reversible reaction.

D-fructose present in many fruits can be phosphorylated by hexokinase. This is a major pathway in muscles and the kidney. In the liver, Fructose gains entry into glycolysis by a different pathway fructokinase catalyses the phosphorylation of fructose at carbon 1 and not at carbon atom 6. The fructose 1-phosphate is then cleaved to form D-glyceraldehyde and DHAP by aldolase. DHAP is converted into glyceraldehyde 3-phosphate. D-glyceraldehyde is phosphorylated by ATP and triose kinase to D-glyceraldehyde 3-phosphate.

D-galactose derived from lactose by action of lactose, also enters the gycolytic process following phosphorylation at 1 position at the expense of ATP by the enzyme *galactokinase* to yield *D-galactose 1-phosphate*. This reacts with UDP-glucose in the liver to yield UDP-D-galactose and Glucose 1 phosphate. This reaction is catalyzed by the enzyme UDP-*glucose α-D-galactose 1-phosphate uridylyl-transferase.*

The galactose residue of UDP-D-galactose is then enzymatically epimerized at carbon 4 to yield UDP-D-glucose by UDP-*glucose 4-epimerase*. UDP-glucose is then cleaved by UDP-glucose pyrophosphate to D-glucose 1-phosphate which is converted to glucose 6-phosphate by phosphoglucomutase. This sequence is used in the reverse in the mammary gland for the synthesis of D-galactose.

In galactosemia VDP-glucose α D-galactose 1-phosphate uridylyltransferase is genetically defective as a result D-galactose and D-galactose 1-phosphate accumulates in the blood tissues, with the enlargement of the liver, formation of cataract leading to vision impairment and mental retardation.

D-mannose formed by digestion of various polysaccharides and glycoproteins is phosphorylated at the 6 position by *hexokinase* and the mannose 6-phosphate so formed is isomerized by *phosphomanno-isomerase* to yield *fructose 6-phosphate*.

Regulation of Glycolysis

Glucose enters the glycolytic pathway in the form of free glucose or from glycogen. Entry of free glucose is regulated by Hexokinase which phosphorylates glucose to glucose 6-phosphate. Hexokinase is inhibited by its product, thus showing allostearic regulation.

The entry of glucose from glycogen is brought by the key enzyme glycogen phosphorylase which is also a regulatory enzyme, showing both covalent regulation and allosteric regulation.

The enzyme occurs in 2-forms, a active phosphorylated form (phosphorylase *a*) and a muchless active dephosphorylated form (phosphorylase *b*). The interconversion is brought about by specific enzymes phosphorylase *a* is converted into the less active form by the enzyme *phosporylase a phosphatase*. The inactive phosphorylase *b* is converted to its active form by the enzyme *phosphorylase b kinase*. Further the inactive form phosphorylase *b* can be stimulated by its allosteric modulator AMP which increases in concentration during ATP breakdown, while ATP is a negative modulator. The secretion of adrenaline stimulates the formation of phosphorylase *a* from phosphorylase *b*, in the liver, thus resulting in an increase in the concentration of glucose in the blood preparing the organism for emergency.

In addition to regulation of the entry of glucose into the glycolytic sequence. Glycolysis rate is controlled within the sequence itself by the 2 major enzymes *Phosphofructokinase and pyruvatekinase.*

PFK is a complex allosteric enzyme. In skeletal muscle its rate is regulated by concentration of ATP and fructose 6 phosphate (substrates) and its product ADP and fructose.

1, 6-diphosphate (products) AMP, citrate, Mg^{2+}, phosphate are also important modulators. ATP and citrate are the most important inhibitory modulators, while AMP and fructose 1, 6-diphosphate are the most active stimulatory modulators.

The *pyruvate kinase* is the second enzyme which is also an allosteric enzyme. It is inhibited by high ATP concentration, Acetyl coA, long chain fatty acids, which are important fuels of TCA. Thus whenever the cells already has a high concentration of ATP or whenever ample fuels are already available for energy yielding respiration, Glycolysis is inhibited by the action of phosphofructokinase or pyruvate kinase.

Anaerobic Respiration

Organisms which do not require oxygen for the further degradation of pyruvate formed by Glycolysis are called Anaerobes which may be facultative or obligatory. In these organisms pyruvic acid is converted to lactic acid or to ethyl alcohol.

In active skeletal muscles or in lactic acid bacteria NADH generated by glycolysis cannot be reoxidised by oxygen, but must be reoxidised to NAD^+ by pyruvate in turn forming lactic acid. The reduction of pyruvate is catalysed by *Lactate dehydrogenase. Lactate dehydrogenase* occurs in five different isozyme form, which differ in their k_m values for pyruvate and their V_{max}.

$$\text{Pyruvate} + \text{NADH} + \text{H}^+ \rightleftharpoons \text{Lactate} + \text{NAD}^+$$

In yeast and other microorganisms that ferment glucose to ethanol and CO_2, the enzymatic pathway takes place by *pyruvate decarboxylase* and *alcohol dehydrogenase. Pyruvate decarboxylase* brings the decarboxylation of pyruvic acid to acetaldehyde. *Acetaldehyde* is reduced to ethanol with NADH derived from glyceraldehyde 3-phosphate dehydrogenation by the action of *alcohol dehydrogenase.* This is a reversible reaction.

$$\text{Glucose} + 2\text{pi} + 2\text{ADP} \rightarrow 2\text{ ethanol} + 2\text{CO}_2 + 2\text{ATP} + 2\text{H}_2\text{O}$$

CITRIC ACID CYCLE

Most animals and plant cells normally are aerobic and oxidize their organic fuels completely to CO_2 and water. Cellular respiration occurs in three major steps.

A. Mobilization of acetyl CoA from the various organic fuel molecules—carbohydrates, fatty acids and amino acids.
B. In the second stage the acetyl groups are fed into the citric acid cycle where enzymatic degradation yields energy-rich hydrogen atoms and free CO_2.
C. In the last stage the hydrogen atoms are separated into protons and energy rich electrons which are transferred along the electron transport chain to molecular oxygen which reduces to H_2O. During the electron transport a lot of energy is released and conserved in the form of ATP, by oxidative phosphorylation. Prior to the entry of organic molecules into the TCA, their carbon backbones must be degraded so that they yield acetyl group of *acetyl CoA.* Pyruvic acid formed by glycolysis is dehydrogenated to yield Acetyl CoA and CO_2 by the *pyruvate dehydrogenase complex* located in the mitochondria of Eukaryotic cells and in the cytoplasm of prokaryotes.

Pyruvic acid undergoes dehydrogenation and decarboxylation, involving a sequential action of three different enzymes, pyruvate dehydrogenase (E_1), *Dihydrolipoyl transacetylase* [E_2] and *dihydrolipoyl dehydrogenase* [E_3] and 5 different coenzymes namely thiamine pyrophosphate (TPP), flavin adenine dinucleotide (FAD), coenzyme A (CoA), nicotinamide adenine dinudeotide (NAD^+) and *Lipoic acid.*

In addition to these enzymes two other enzymes that regulate pyruvate dehydrogenase activity are also present in this pyruvate dehydrogenase complex.

The sequence of enzymatic actions are as discussed below:

1. 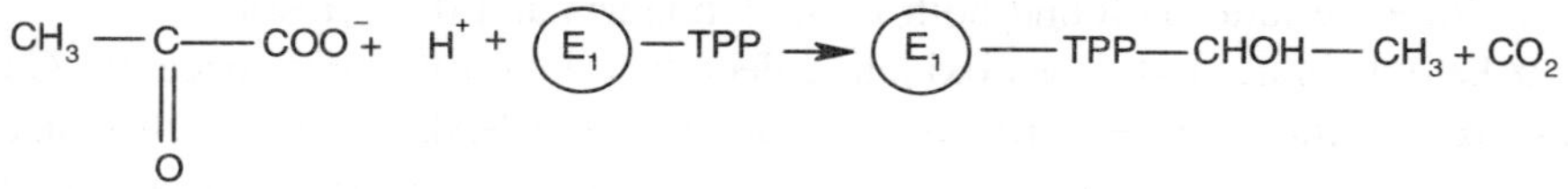

2. 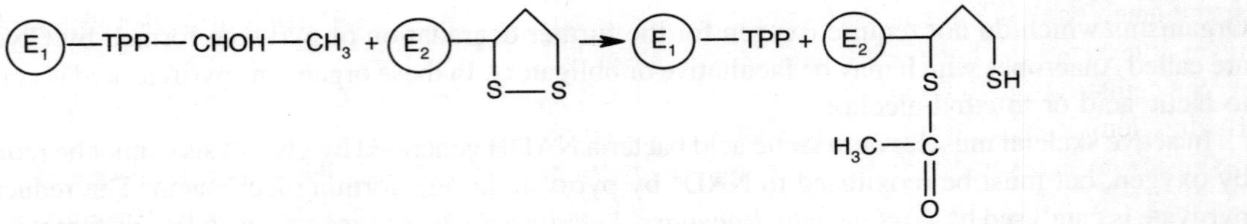

3. 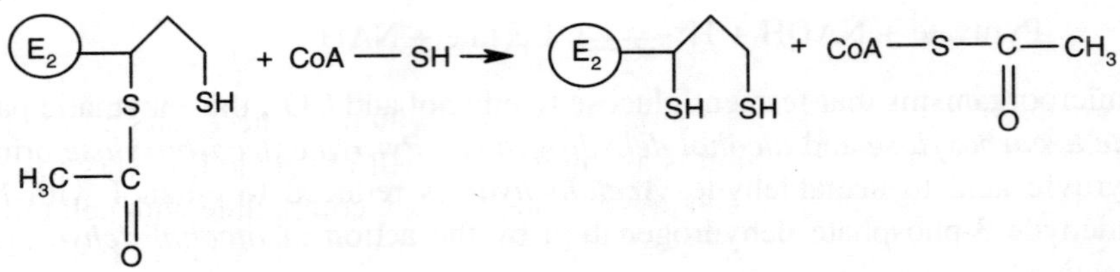

4. 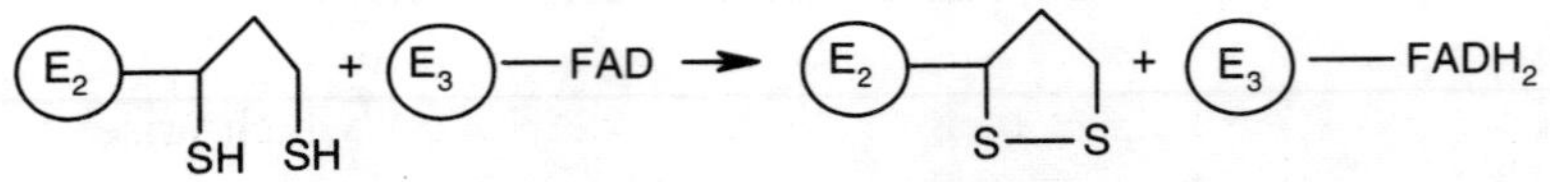

5. E_3 — $FADH_2$ + NAD^+ → E_3 — FAD + NADH + H^+

In the first step pyruvate loses its carboxyl group as it reacts with the bound thiamine pyrophosphate of E_1 to form the hydroxy ethyl derivative of the thiazole ring of TPP. In the second step the H atoms and the acetyl group of TPP are transferred to the oxidised form of Lipoyllysyl prosthetic group of E_2, to form the partially reduced form. In step three a molecule of CoA—SH reacts with the acetyl derivative of E_2 to yield Acetyl-5-CoA and the fully reduced or dithiol form of the Lipoyl groups. In the next step the reduced form of E_2 is acted upon by E_3 which promotes the transfer of Hydrogen atoms to the FAD prosthetic group of E_3. In the last step the reduced FAD transfers Hydrogen to NAD^+ forming NADH.

The Citric Acid Cycle

Following simple experiments and logical reasoning krebs postulated the citric acid cycle also called tricarboxylic acid cycle. The entire set of cyclic reactions take place within the mitochondria.

The first reaction of the cycle is the condensation of Acetyl CoA with oxaloacetate to form citric acid by citrate synthase which is a regulatory enzyme.

Citrate is enzymatically transformed into isocitrate through the intermediate *cis*-aconitate by the enzyme *Aconitase,* which contains iron and acid-labile sulphur atoms arranged in clusters called *Iron-sulphur center.*

Isocitrate is dehydrogenated to α-ketoglutarate and CO_2 by isocitrate dehydrogenase. There are two different *isocitrate dehydrogenase.* One requiring NAD^+ and the other $NADP^+$ as the electron acceptor. Mitochondria contain both NAD-linked and NADP-linked isocitrate dehydrogenase, the former is found only in mitochondria while the latter is found both in mitochondria and the cytosol.

In the next step α-ketoglutarate undergoes oxidative decarboxylation to form succeinyl CoA and CO_2, by the action of α-ketoglutarate dehydrogenase complex where NAD^+ is the electron acceptor.

The succeinyl CoA which is a high energy compound undergoes an energy conserving coupled reaction in which cleavage of the thioester bond is accompanied by the phosphorylation of Guanosine diphosphate (GDP) to Guanosine triphosphate. The enzyme succeinyl CoA synthetase yields free succeinate. The GTP formed may then donate its terminal phosphate group to ADP forming ATP by reversible action of nucleoside dephosphokinase.

Succeinate is dehydrogenated to fumarate by the flavoprotein succeinate dehydrogenase which contains flavin adenine dinucleotide as the hydrogen acceptor.

Fumarate is reversibly hydrated to malate by Fumarate hydratase also called fumarase.

Malate is dehydrogenated to form oxaloacetate by the NAD linked Malate dehydrogenase, which is a reversible reaction, but pulled in the forward direction by the removal of oxaloacetate by the action of citrate synthase reaction.

Logic of TCA

The krebs cycle involves successive 6, 5 and 4-'C' intermediates which are necessary to oxidize the simple 2 carbon acetyl group. Though acetic acid is a simple and small molecule it is very resistant to chemical oxidation. Of its methyl carbon atom and drastic conditions, incompatible with cell milieu are required to oxidize acetic acid to CO_2. Thus the round about pathway, with a lower free energy of activation.

Regulation of TCA

Rate of citric acid cycle is regulated by the rate of formation of Acetyl CoA which arises from the oxidation of pyruvic acid and from the oxidation of fatty acids.

Pyruvate dehydrogenase shows covalent regulation in existing in two forms, active form which is dephosphorylated and the inactive form which is phosphorylated. When ATP concentration in mitochondria is relatively high and there are ample amounts of Acetyl CoA and intermediates of krebs cycle ATP serves as a stimulatory modulator and activates pyruvate dehydrogenase kinase, which phosphorylates a serine residue of pyruvate dehydrogenase to yield its inactive form pyruvate dehydrogenase phosphate.

However if the demand for ATP increases the inactive form is reactivated by another enzyme, Pyruvate dehydrogenase phosphate phosphatase. It is stimulated by a rise in concentration of Ca^{2+}, which rises at times of ATP need.

The pyruvate dehydrogenase complex is also regulated through allostearic modification. It is strongly inhibited by ATP, Acetyl CoA and NADH.

The citric acid cycle is also regulated. The rate of citrate synthase reaction is controlled by the concentration of Acetyl CoA, and by concentration of oxaloacetate. Apart from these, whenever succinyl CoA increases it inhibits citrate synthase. In some cells citrate and NADH are inhibitors of citrate synthase.

Oxidation of isocitrate to α-ketoglutarate and CO_2 by isocitrate dehydrogenases is also regulated where ADP is a stimulatory while NADH and NADPH are inhibitory α-ketoglutarate dehydrogenase complex is inhibited by its product succeinyl CoA.

Citric acid cycle is an amphibolic pathway and functions not only in the oxidative catabolism of organic molecules but also as the first stage in many a biosynthetic pathways providing the precursors. Intermediates of the TCA namely α-ketoglutarate, succinate and oxaloacetate are removed from the cycle to serve as precursors of amino acids. And as expected rate of citric acid cycle declines due to the drain of its intermediates. These intermediates are replenished by another set of anzymes such that the concentration of the intermediates remains relatively constant.

This special enzymatic system replenishing the intermediates constitute the *Anaplerotic system.*

The most important of such system in animal tissues is the activity of *pyruvate carboxylase* which catalyses the reversible carboxylation of pyruvic acid in the presence of Mg^{2+} and ATP to yield oxaloacetate. Thus when the TCA is deficient in oxaloacetate or any of the other intermediates, the above reaction takes place. *Pyruvate carboxylase* is a very complex enzyme, having a molecular weight of 650,000 containing 4 prosthetic groups each with a molecule of the vitamin *biotin*. It is a regulatory enzyme and Acetyl CoA is its positive allostearic modulator, helping in the formation of oxaloacetate whenever Acetyl CoA is in excess.

While the pyruvate carboxylase reaction is important anaplerotic reaction in liver and kidney, in the heart and muscle *phosphoenol pyruvate carboxykinase* carboxylates PEP to yield oxaloacetate and requires GTP for the same.

In plants and some microorganisms the citric acid cycle operates in two modes.

(*a*) *the standard cyclic reactions.*

(*b*) *the specialized modification called the glyoxylate cycle.*

The *Glyoxylate cycle* may also be considered as an anaplerotic pathway and it consists of the following sequences:

Acetyl Co-A condenses with oxaloacetate to form citrate by *citrate synthase.* Citrate converts to isocitrate by Aconitase. The isocitrate is acted upon by *isocitrate lyase* which forms *succinate* and *Glyoxylate.* The glyoxylate condenses with Acetyl Co-A to yield malate by the action of malate synthase. The malate is dehydrogenated to oxaloacetate which condenses with another molecule of *Acetyl CoA* to start another turn of the cycle. The succinate may be converted via fumarate and malate into oxaloacetate which can then be converted into PEP by the reversal action of PEP carboxykinase. The PEP so formed is then used as a precursor of biosynthesis of Glucose.

Animals lack the enzymes isocitrate lyase and *malate synthase* and hence the absence of Glyoxylate cycle in them. This cycle however is very active in germinating plant seeds where acetyl residues derived from the fatty acids of their storage triacylglycerols are converted into glucose. This takes place in the *glyoxysomes.*

SECONDARY PATHWAY OF GLUCOSE CATABOLISM

The minor catabolic pathways for specialized purposes constitute the *secondary pathways.* They are not the mainling pathway. Two such pathways of *Glucose catabolism* are

(*a*) *Pentose phosphate pathway.*

(*b*) *Production of Glucuronic acid and Ascorbic acid.*

Pentose Phosphate Pathway

Production of NADPH and Ribose 5-phosphate for nucleotide biosynthesis are the functions of PPP. NADPH is a cellular reductant that is used in fatty acid and cholesterol synthesis. It also serves as the reductant in a mechanism that protects against oxidative damage arising from the non-enzymatic formation of O_2 and H_2O_2 from O_2. In the RBCs the concentration of oxygen is very high and the potential for oxidative damage is great. Consequently there is a large demand for NADPH and the PPP in red blood cells accounts for about 10% the total consumption of glucose. In other cells for little of the overall consumption of glucose.

The first reaction of the PPP is the enzymatic dehydrogenation of glucose 6-phosphate by *glucose 6-phosphate dehydrogenase* to form 6-*phosphogluconate* where $NADP^+$ is the electron acceptor. In the next step 6-*phosphogluconate* undergoes dehydrogenation and decarboxylation by 6-*phosphogluconate* dehydrogenase to form D-*ribulose 5-phosphate* which generates another NADPH molecule. *Phosphopentose isomerase* then converts D-ribulose 5-phosphate to *D-Ribose 5 phosphate* which can be used for the biosynthesis of Ribonucleotides and deoxyribonucleotides.

Defective or inactive enzymes of PPP, causes the RBCs to undergo haemolysis and it is worsened in the presence of antimalarial drugs like *primaquine.*

Glucose to Glucuronic Acid and Ascorbic Acid

Glucuronate is an important product of secondary metabolism of glucose in detoxification and excretion of foreign organic compounds.

The glucose 1-phosphate is first converted to UDP-glucose by reaction with UTP. The glucose part is enzymatically dehydrogenated to yield UDP-glucuronate by *UDP-glucose dehydrogenase.* The UDP-glucuronate can be then used to detoxify drugs or foreign compounds. It also is the precursor of D-glucuronate residue of acid polysaccharides as hyaluronic acid, heparin. It is also the precursor of *L-Ascorbic acid.* First it gets reduced by NADPH to the 6-'C' sugar acid. *L-Gulonate* which is converted into its lactone *L-Gulonolactone* which is then dehydrogenated by *Gulonolactone oxidase* to yield *L-Ascorbic acid.*

Due to lack of gulunolactone oxidase in humans, guinea pigs, monkeys etc., L-Ascorbic acid is an essential vitamin.

Oxidative Phosphorylation

The culminating event of cell respiration is the electron transport and oxidative phosphorylation, in which the electron flows from organic substrates to oxygen yielding energy stored in the form of ATP.

In each turn around the citric acid cycle. Four pairs of hydrogen atoms are removed, which donate their electrons to the electron transport chain and become H^+ ions which escape into the aqueous medium. The electrons are transported along a chain of electron carrying molecules until they reach cytochrome aa_3 which transfers the electron to oxygen, which simultaneously accepts two H^+ forming H_2O.

The respiratory chain consists of a series of proteins with prosthetic groups capable of accepting and donating electrons. These along with the enzymes synthesising ATP from ADP and Pi embedded in the inner membrane of mitochondria.

The electrons are transferred from one molecule to another in any of the following ways:

(*a*) *they are transferred directly as electrons*

$$Fe^{2+} + Cu^{2+} \rightarrow Fe^{3+} + Cu^{+}$$

(*b*) *in the format hydrogen atoms*

$$AH_2 \rightleftharpoons A + 2e^- + 2H^+$$

where AH_2 and A constitute conjugate redox pair, which can reduce an electron acceptor B. By transferring H_2 atoms

$$AH_2 + B \rightleftharpoons A + BH_2$$

(*c*) in the form of *hydride ion* (H^-) as is case of NAD-linked dehydrogenases.

(*d*) direct combination of an organic reductant with oxygen

$$R - CH_3 + 1/2\ O_2 \rightarrow R - CH_2 - OH.$$

The tendency of a given conjugate redox pair to lose an electron is given by the *standard oxidation reduction potential* E_0' which is defined as the emf in volts given by a responsive electrode placed in a solution containing both the electron donor and its conjugate acceptor at 1.0 m concentration at 25°C and pH – 7.0.

Systems having increasing tendency to lose electrons are assigned increasingly negative values while positive values are given to systems having a tendency to accept electrons.

The E_0' values of various redox couples allow us to predict the direction of flow of electrons from one redox couple to another. The direction of flow of electrons in the respiratory chain is as shown below:

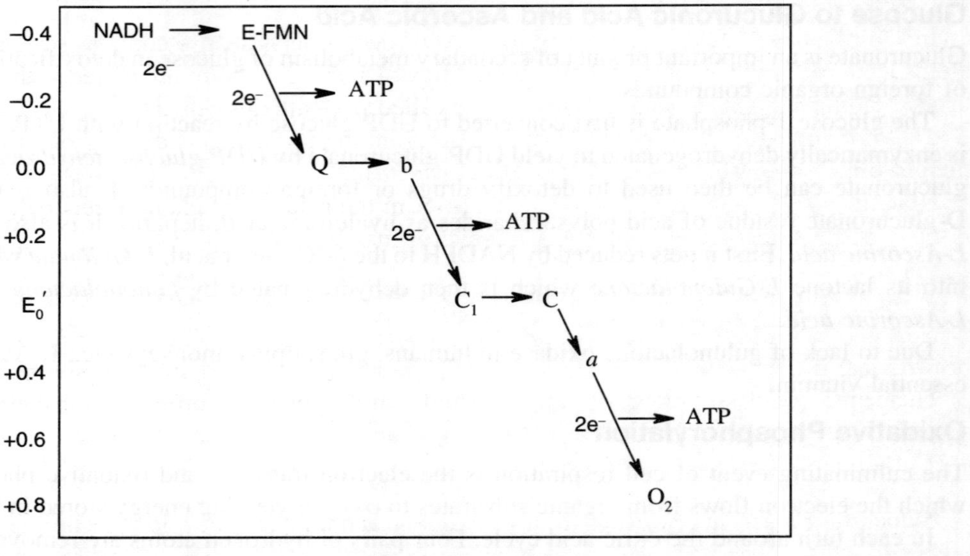

The tendency for electrons to flow from electronegative towards electropositive systems is due to the loss of free energy. The greater the difference in the standard potentials between two redox pairs the greater the free energy loss.

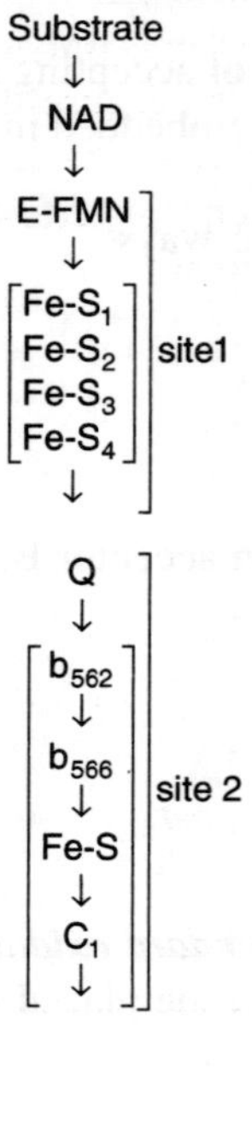

The standard free energy change is given by

$$G^{0\prime} = -nF \, \Delta \, E_0'$$

where $\Delta G^{0\prime}$ is the standard free energy change in calories, '*n*' is the number of electrons transferred.

F is a constant—Faraday (23, 0 62 cal/V mol). $\Delta E_0'$ is the difference between the standard potential of the electron donor system and the acceptor system.

The complete set of electron carriers of the respiratory chain.

Most of the electron pairs entering the respiratory chain arise from the action of dehydrogenases that use the coenzymes NAD^+ or $NADP^+$ as electron acceptors.

The majority of the pyridine linked dehydrogenases are specific for NAD^+ and few others require $NADP^+$. The cytosolic and mitochondrial pools of NAD and NADP are separated by the inner mitochondrial membrane which is impermeable to these coenzymes.

NAD^+ can collect reducing equivalents from substrates acted upon by NADP-linked dehydrogenases. This is made possible by *pyridine nucleotide transhydrogenase.*

C
↓
a
a_3
Cu^{2+}
Cu^{2+}
↓
O_2
site 3

$$NADPH + NAD^+ \rightleftharpoons NADP^+ + NADH.$$

In the next step in the electron transport, a pair of reducing equivalents is tranferred from NADH to NADH *dehydrogenase* where the tightly bound prosthetic group flavin mono nucleotide (FMN) becomes reduced

$$NADH + H^+ + EFMN \rightleftharpoons NAD^+ + E\text{-}FMNH_2.$$

In addition to its FMN prosthetic group, NADH dehydrogenase also contains several non-heme iron atoms which are grouped into several clusters, in which the iron atoms are paired with an equal number of acid-labile sulphur atoms such clusters are called iron-sulphur centers.

The iron atoms transfer reducing equivalents from the $FMNH_2$ prosthetic group to *ubiquinone* and therefore it is also called *NADH-ubiquinone reductase.*

Ubiquinone or coenzyme Q is the next carrier in the respiratory chain it gets reduced to ubiquinol or QH_2 with the regeneration of NADH dehydrogenase.

$$E - FMNH_2 + Q \rightleftharpoons E - FMN + QH_2$$

Next in the sequence are the *cytochromes*, which are iron containing electron transferring red or brown proteins. They are heme proteins, in which iron is present as an iron-porphyrin, prosthetic group.

There are three classes of cytochromes *a*, *b* and *c*, distinguished by differences in their light absorption spectra. The cytochromes in the respiratory chain are arranged in the sequence

$$b \rightarrow c_1 \rightarrow c \rightarrow aa_3$$

Cytochrome *b* accepts electrons from Ubiquinone and transfers them to cytochrome C_1 which in turn give it to cytochrome C. The last electron carrier is cytochrome aa_3 also called cytochrome oxidase which is capable of donating electrons to oxygen. Cytochrome aa_3 has two essential copper atoms which undergo cuprous-cupric redox changes. Thus by a complex step four electrons are passed simultaneously to O_2 to yield two H_2O with the uptake of four H^+ from the aqueous medium.

If oxygen is partially reduced by accepting two electrons the product is H_2O_2. If O_2 accepts only one electron the product is superoxide radical (: O_2^-). These are extremely toxic to the cells because they attack the unsaturated fatty acid components of membrane lipid thus damaging membrane structure. Aerobic cells protect from such effects by the action of superoxide dismutase which converts the superoxide radical into H_2O_2. The enzyme catalase converts H_2O_2 into molecular oxygen.

$$2O_2^- + 2H^+ \rightarrow H_2O_2 + O_2$$

$$2H_2O_2 \rightarrow 2H_2O + O_2$$

Inhibitors of E.T.C.

Inhibitor	*Source*	*Point of action*
1. Rotenone	Plant	NADH to ubiquinone
2. Antimycin A	antibiotic isolated from streptomyces	ubiquinone to cytochrome C
3. Cyanide		cytochrome aa_3 to O_2

Oxidative Phosphorylation

The free energy of the electrons is trapped in the three stages of the electron transport and conserved in the form of ATP. The ATP synthesizing enzyme system of the inner mitochondrial membrane is a complex known as ATP synthetase of F_0F_1 ATPase. The F_1 component resembles a door knob protruding from the inner membrane into the matrix. It is attached to the stalk F_0 which is embedded in the inner membrane.

Various theories have been proposed to explain the transfer of energy between electron transport and ATP synthesis of which the chemiosmotic theory is a novel one. This hypothesis is supported by the following characteristic properties of oxidative phosphorylation:

A. No high energy intermediates linking electron transport to ATP synthesis have been found.

B. Oxidative phosphorylation requires intact inner membrane structure.

C. Existence of a different ionic concentration (*a*) composition across the inner membrane of respiring mitochondria essential for ATP synthesis.

D. Prevention of oxidative phosphorylation by uncoupling agents like 2, 4-dinitrophenol.

The chemiosmotic hypothesis proposes that the function of electron transport, is to pump H^+ from the matrix to the outer medium, thus generating an acid outside gradient of H^+. Such a gradient of H^+ in which H^+ concentration outside of the mitochondrion is greater than interior contains potential energy. These H^+ ions flow back into the mitochondrial matrix through specific H^+ channel in the F_0 F_1 ATPase molecule. The free energy released as H^+ flows back through the ATPase causes the coupled synthesis of ATP from ADP and phosphate.

Two specific transport systems make possible the transport of H^+, OH^-, K^+, ADP^{3-}, $H_2PO_4^-$ and ATP^{4-}.

The adenine nucleotide translocase transports one molecule of ADP^{3-} inward in exchange for one ATP^{4-} coming out. This translocase is very specifically inhibited by atractyloside.

The phosphate carrier or translocase is the 2nd membrane transport system cotransporting $H_2PO_4^-$ ion from the outside with an H^+ ion into the matix.

BIOSYNTHESIS OF CARBOHYDRATES

Catabolism and anabolism proceed simultaneously in a dynamic state. The pathway taken in the biosynthesis process is not totally identical with the catabolic one, it is controlled by different regulatory enzymes and it is obligatorily coupled to the energy yielding breakdown of ATP.

In animals formation of D-glucose from noncarbohydrate precursors is called gluconeogenesis. The important precursors being lactate, pyruvate, glycerol, most amino acids, intermediates of citric acid cycle. Gluconeogenesis occurs in the liver and to a smaller extent in kidney cortex.

Seven steps in gluconeogenesis are the same as in glycolysis which are all freely reversible. Three steps in glycolysis are essentially irreversible and are therefore bypassed by alternate set of enzymes.

The first bypass is the conversion of pyruvic acid into phosphoenol pyruvate. The first step in this bypass is catalyzed by pyruvate carboxylase of mitochondria a biotin containing enzyme which catalyses formation of oxaloacetate from pyruvate.

$$\text{Pyruvate} + CO_2 + \text{ATP} \xrightarrow{\text{Acetyl CoA}} \text{oxalo acetate} + \text{ADP} + \text{Pi}$$

It is a regulatory enzyme which is almost inactive in the absence of acetyl CoA.

The oxaloacetate formed in the mitochondria is then reversibly reduced to malate by mitochondrial malate dehydrogenase at the expense of NADH.

$$NADH + H^+ + Oxaloacetate \Leftrightarrow NAD^+ + Malate$$

The malate leaves the mitochondria into the cytosol where it is reorganized by the cytosolic NAD linked malate dehydrogenase to yield extra mitochondrial oxaloacetate.

$$Malate + NAD^+ \longrightarrow Oxaloacetate + NADH + H^+$$

The oxaloacetate so formed is acted upon by phosphoenol pyruvate carboxykinase to yield phosphoenol pyruvate in which GTP serves as the phosphate donar.

$$Oxaloacetate + GTP \Leftrightarrow PEP + CO_2 + GDP.$$

The second bypass is achieved by the enzyme fructose diphosphatase which carries the hydrolysis of 1-phosphate group to yield fructose 6-phosphate.

$$Fructose\ 1,6\text{-diphosphate} + H_2O \rightarrow Fructose\ 6\text{-phosphate} + Pi$$

The enzyme is a regulatory one being inhibited by AMP but stimulated by ATP.

The third bypass is the dephosphorylaton of glucose 6-phosphate to yield glucose which is delivered from the liver into the blood. The enzyme catalyzing this reaction is glucose 6-phosphatase found in the endoplasmic reticulum of the liver and is absent in muscle and in brain.

The citric acid cycle intermediates namely—citrate, isocitrate, ketoglutarate, succinate, fumarate, malate are the chief precursors. All may undergo oxidation to form oxaloacetate which is converted into PEP. However only three carbon atoms of each of the cycle intermediate can be converted into glucose.

Most of the amino acids are glucogenic, where some or all of the carbon atoms of amino acids are converted into either pyruvate or to other intermediates of citric acid cycle and thus undergo a net conversion to glucose and glycogen.

In diabetes mellitus the net conversion of the glucogenic amino acids into glucose is a very active process.

Ala, Ser, Cys, Gly are converted to pyruvic acid while Asn, Asp to oxaloacetate, val, thr, met to succinyl CoA. Glu, Gln, Pro, Arg and His to –ketoglutarate.

Compensatory reactions of hypoglycaemia

Hypoglycaemia stimulates hypothalamus which in its return promotes the secretion of ACTH and other hormones which oppose the actions of insulin and restore the blood sugar level to normal. Hypoglycaemia also stimulates the secretion of epinephrine which stimulates glycogenolysis and raises the blood sugar level.

Hyperglycaemia

Hyperglycaemia is a condition in which blood sugar increases above the normal level, *i.e.*, above 120 mg per 100 ml. When the blood sugar level exceeds the renal threshold (180 mg per 100 ml), sugar appears in the urine. Persistent hyperglycaemia occurs when there is diminished utilization of glucose, and discharge of excess sugar from the liver. Hyperfunction of some of the endocrine glands, viz. anterior pituitary, adrenal cortex, etc., causes hyperglycaemia. Lack of diminished secretion of insulin is the main factor which produces the hyperglycaemia.

Glycosuria

Glycosuria is a condition when the blood glucose level exceeds 180 mg glucose per 100 ml blood above the normal blood glucose level (60–100 mg) glucose per 100 ml blood. At this time the renal tubule cells are not able to reabsorb all the glucose.

Hypoglycaemia

Hypoglycaemia is a condition, in which blood sugar level is present below the normal level, *i.e.*, below 80 mg per 100 ml. In diabetic patients whose tissues are accustomed to high blood sugar, hypoglycaemic symptoms depend on three factors:

(*a*) The actual blood sugar level,
(*b*) The rapidity of blood sugar reduction,
(*c*) The previous blood sugar level, *i.e.*, that level with which the tissues were accustomed.

Since nerve cells have very little stored food and since they use sugar mostly at the sole source of energy, hypoglycaemia will, therefore, affect the nerve cells first. For instance, a feeling of fatigue, weakness and hunger exists. Secondly extreme anxiety and irritability and abnormal behavior as in alcohol poisoning arises. Tremors develop and fine movements are not possible. Vasomoter disturbances, such as flushing or palor, perspiration and chilliness occur. Later on, there may be delirium, diabetic coma and convulsions and loss of deep reflexes.

Hypoglycaemic symptoms are relieved by the administration of glucose. Other monosaccharides are less effective reaches the urinary bladder and glycosuria results. In normal condition, the renal tubule cells are able to transfer all the glucose back into the blood by active transport, and no glucose passes to the urinary bladder. But the tubule cells are limited in the quantity of glucose that they can transport back to the blood in a given time.

UNIT 6

Lipid Metabolism

BIOSYNTHESIS OF FATTY ACIDS (FA'S)

- The major site of fatty acid synthesis is in the liver. The FA's may be synthesized from dietary glucose via pyruvate. The extent to which FA biosynthesis occurs depends on the amount of fat in the diet and the need to convert excess glucose to FA's for storage.
- The enzymes that synthesize FA's localized in the cytosol and are completely different from the mitochondrial enzymes that catalyse FA degradation.
- The major pathways for synthesis involves the polymerization of the 2-C units derived from acetyl coenzyme to form a 16 C-Saturated FA, palmitic acid.
- Reduced NADPH and ATP are required for FA's synthesis.

Sources of Acetyl COA and NADPH

1. Cytosol

- By oxidation of glucose via glycolysis by pyruvate dehydrogenase.
- NADPH is generated by pentose phosphate pathway.

2. Mitochondrial

- Acetyl COA is not able to cross the inner mitochondrial membrane and is transported into the cytosol via this shuttle. NADPH is also produced in the cytosol during the following reactions.
- Citrate synthetase catalyzes the reaction of acetyl COA with OAA to form citrate in mitochondria.
- Citrate is transported into the cytosol by a tricarboxylic acid transport system.
- Citrate reacts with COA in the cytosol to form acetyl COA and OAA.
- OAA is converted to malate.
- Malate is decarboxylated to form pyruvate. This reaction, which is catalyzed by the malic enzyme, also forms NADPH from $NADP^+$.
- Malate is decarboxylated to form pyruvate.
- Pyruvate is transported into mitochondrion by an active transport system.
- OAA can then be regenerated from pyruvate by the action of pyruvate carboxylase, which requires ATP.

The acetate COA shuttle system

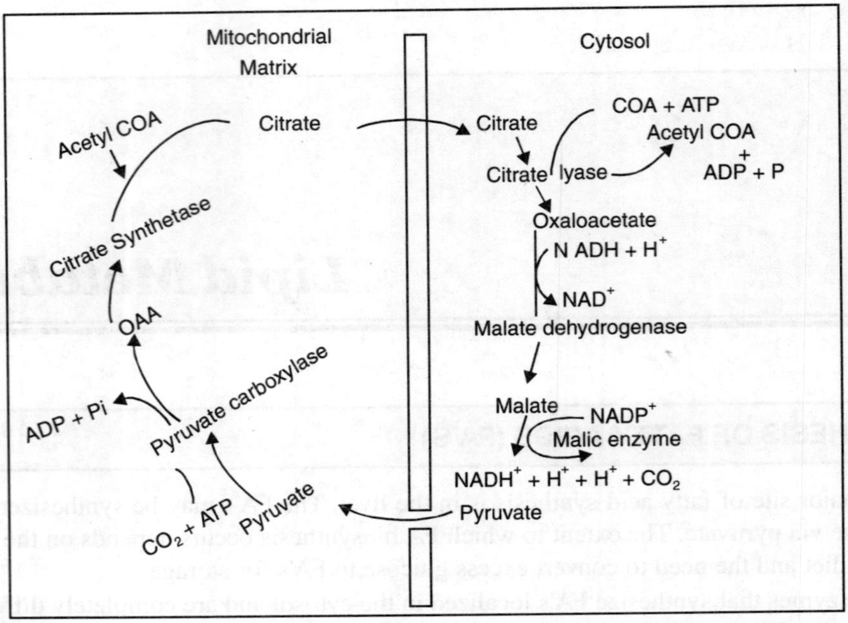

Fig. 6.1

Reaction sequences in FA biosynthesis:

(1) Acetyl COA + HCO_3 + ATP → malonyl COA + ADP + P

(*a*) *Enzyme:* The reaction is catalysed by acetyl COA carboxylase, which requires biotin as a cofactor to perform the carboxylation reaction. Actyl COA carboxylase is one of the enzymes that is inactive in the condition known as multiple carbon deficiency, which is done to an inability to utilize biotin.

(*b*) This is the key regulatory site for FA synthesis.

1. Activators	(*a*) Citrate	(*b*) Insulin
2. Inhibitors	(*a*) Palmitoyl COA	(*b*) Glucagon

(2) Acetyl COA + acyl carrier protein (ACP) → Acetyl–ACP + COA. This reaction as well as reactions 3-7 are catalyzed by FA synthase enzyme complex.

(*a*) This step is catalyzed by the acyl transferase component.

(*b*) The ACP possesses a phosphopantathiene prosthetic group, which is derived from the vitamin pantothenic acid and is identical to that portion of the COA molecule. The acyl groups of the substrate molecule are covalently bound to the ACP via a thioester linkage with the terminal sulphydryl group.

(3) Malonyl COA + ACP → malonyl-ACP + COA

This reaction is catalyzed by the malonyl transacylase component of the fatty acid synthase enzyme reaction.

(4) Malonyl-ACP + acetyl ACP → Acetoacetyl-ACP + ACP + CO_2

This reaction is catalyzed by the 3-keto acyl synthase component of the fatty acid synthase enzyme complex.

(5) Acetoacetyl-ACP → -3-OH butyrl-ACP.

(*a*) This reaction is catalyzed by the 3-Keto acyl reductase component of the fatty acid synthase enzyme complex.

(*b*) NADH is required for this reaction.

The Principal Reactions of FA Biosynthesis

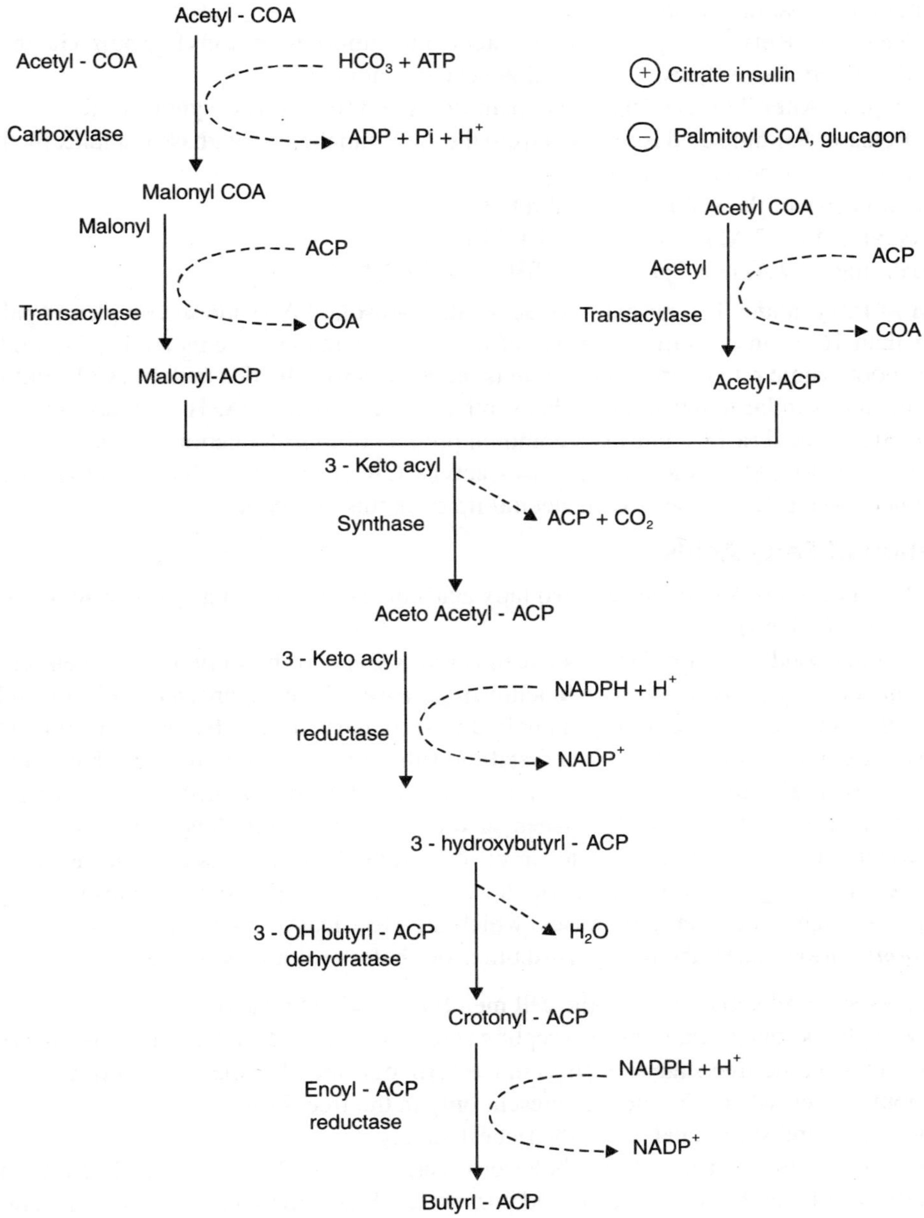

Fig. 6.2

(6) 3-OH butyrl-ACP → crotonyl-ACP: This reaction is catalyzed by the 3-hydroxy butyrl dehydratase component of the FA synthase enzyme complex.

(7) Crotonyl-ACP → butyrl-ACP

(*a*) This reaction is catalyzed by the enoyl-ACP reductase component of the FA synthase enzyme complex.

(*b*) NADPH is required for this reaction.

(8) Formation of palmitic acid

(*a*) Reaction-Butyrl-ACP (4c-unit) reacts with another malonyl group via the reaction described above (Secs. 1 to 3) and the cycle continues.

(*b*) Repeat: After 7 turns of the cycle, palmitic ACP (16-c unit) is synthesized.

(*c*) Release: Palmitic acid is released from the ACP and the FA synthase complex by the enzyme palmitoyl deacylase (thioesterase).

(9) Stoichiometry of the synthesis of palmitate

8-Acetyl COA + 7 ATP + 14 NADPH + 14 H^+ + H_2O
→ palmitate + 7 ADP + 7pi + 8 COASH + 14 $NADP^+$

Elongation of fatty acids: The normal product of the cytosolic FA synthase complex is palmitic acid FA's longer than 16-c can be formed via one of the two elongation systems adding 2-c units.

(1) The endoplasmic reticulum (ER) system is the most active. It adds malonyl COA onto palmitate in a manner similar to the action of FA synthase, except that COASH is involved rather than the ACP. Stearic acid (a 18-c unit) is a common product of this elongation system.

(2) A mitochondrial elongation system uses actyl COA units, rather than malonyl COA units to elongate FA for the synthesis of structural lipids in this organelle.

Desaturation of Fatty Acids

(1) The two most common monosaturated fatty acids in mammals are palmitoleic acid (16:1:Δ9) and oleic acid (18:1:Δ9).

(2) The double bonds are introduced between carbons 9 and 10 by fatty acid oxygenase in the ER. The enzyme, oxygenase is mixed in function oxygenase, which requires molecular O_2 and NADPH.

Cholesterol: Cholesterol is a complex monohydric secondary alcohol being a very important member of the sterol class. It is a stable white crystalline substance, insoluble in water but readily soluble in chloroform, ether, alcohol and other fat solvents. The crystals have a rhombic or rectangular shape, with one corner broken off. Cholesterol when mixed with fats or oils helps them to absorb a large volume of water due to formation of water in oil emulsion. This sterol is present in high amount in nervous tissues and being a poor conductor of electricity, serves the purpose of an insulating cover over the structures through which nerve impulses, which are electrical in nature, are conveyed.

Distribution: Various facts about the distribution of cholesterol are as follows:

1. It is present in all cells—both in the cell membrane and cytoplasm.
2. All body fluids contain cholesterol excepting cerebrospinal fluid, in which the amount is negligible.
3. It may exist in the free state as well as in the form of esters, but these two forms are not equally distributed everywhere. In bile it is present only in the free form.
4. Brain (17%) and suprarenals have the richest supply.
5. *Blood cholesterol*: Normal blood cholesterol varies from 0.15% to 0.2%. The total quantity is equally distributed between plasma and corpuscles. But in the corpuscles it is present chiefly in the free form, while in the plasma the major part remains as esters.

6. The free cholesterol content of a tissue is characteristic and normally remains very constant but the cholesterol esters may vary in quantity.
7. The distribution of cholesterol in different types of muscles is directly proportional to their degree of activity. The cardiac and smooth muscles have a richer cholesterol content than the voluntary muscles. Moreover, the amount of cholesterol in any tissue is roughly proportional to their degree of activity.
8. Cholesterol and phospholipids always remain together. Added to this they always remain in definite proportions.

Cholesterol is very closely related to several other sterols of immense physiological, importance, for instance

(*a*) the active principles of adrenal cortese,
(*b*) the male sex hormones,
(*c*) the female sex hormones,
(*d*) vitamin D and ergosterol,
(*e*) cholic acid etc.

Excepting in the case of cholic acid, it has not been proved that cholesterol acts as the mother substance for the other sterols.

Cholesterol: Average diet supplies about 0.5 gram cholesterol a day. Another 0.5 gram is synthesized in the body. The complex ring structure is shown to be synthesized from acetate. 15 of the 27 carbons coming from methyl group and the remaining 12 from carboxylic group of acetate. The liver and the intestine are the principal sources for blood cholesterol and synthesize about 1.5 gram each day. The skin, adrenal glands, tests and ovary also synthesize small amounts of the substance mainly for the formation of specialized substances like vitamin D_2 and sebum from skin, steroid hormones from others. They do not contribute to plasma cholesterol. Adipose tissue, aorta, muscle and brain also synthesize small amounts. The synthesis occurs mainly in the microsomal and soluble fractions of the cell.

Different steps in the biosynthesis of cholesterol are as follows:

1. *Formation of acetyl CoA:* A molecule of acetic acid is activated by combining with coenzyme A. The reaction requires energy from the breakdown of ATP and is catalyzed by the enzyme acetyl-CoA synthetase. Magnesium ions are also required.
2. *Formation of acetyl-CoA:* Two molecules of acetyl CoA condense to form an acetoacetyl CoA molecule.
3. *Formation of HMG-CoA:* The acetoacetyl-CoA now condenses with one more molecule of acetyl CoA to form beta hydroxy, beta methyl glutaryl-CoA (abbreviated as HMG-CoA). The enzyme which mediates this reaction is called HMG-CoA synthetase.
4. *Formation of mevalonic acid:* The HMG-CoA is reduced by an NADPH + H^+ depending reductase to form mevalonic acid.

   The HMG-CoA is reduced by an NADPH + H^+ depending reductase to form mevalonic acid.
   The HMG-CoA reductase enzyme exists in an active, dephosphorylated form and an inactive, phosphorylated form and is subject to hormonal influence through the cascade system of enzymes as in the case of glycogen synthase and phosphorylase. Epinephrine and glucogon favor phosphorylation and inactivation of the enzyme and thus inhibit cholesterol biosynthesis. Increase

in the levels of mevalonate and cholesterol also inhibit the enzyme. Insulin has an opposite effect and stimulates cholesterol biosynthesis.

5. *Formation of isopentenyl pyrophosphate:* In a series of 3 steps involving phosphorylation by 3 ATP and loss of one of the 6 carbons of mevalonic acid as CO_2 a five carbon compound, isopentenyl pyrophosphate, is formed. It is also known as the isoprenoid unit. The isoprene units can exist in more than one isomeric form.
6. Two isoprene units condenses with geranyl pyrophosphate (10 carbons).
7. One more isoprene unit condenses with geranyl pyrophosphate to form farnesyl pyrophosphate (15 carbons).
8. Two farnesyl units combine to form squalene (30 carbons)
   Other polyprenoid derivatives: condensation of isoprene units can also form several other biologically important polyprenoid derivatives:
   (*i*) Dolichol, a long chain (95 carbons) alcohol, takes part in glycoprotein synthesis.
   (*ii*) Ubiquinone plays an important role in respiratory chain.
   (*iii*) The side chains of vitamins A and E.
   (*iv*) Vitamin D and its hormone derivative calcitriol.
9. Closure of the ring occurs to form lanosterol.
10. Lanosterol is then converted through 14-desmethyl lanosterol, zymosterol, Δ7, 24-cholastadienol, 24-dehydrocholesterol (desmosterol) and finally to cholesterol.

The reductive reactions in the cholesterol synthesis require NADPH + H^+ and they are dependent on its production through HMP pathway of glucose metabolism.

Synthesis of Phospholipids

The steps are same up to formation of phosphotidic acid (α, β-diglyceride phosphate). This combines with cytidine diphosphate of CTP to form CDP-diglyceride. The CDP-diglyceride can now interact with inositol (or active choline or active ethanolamine) to form inositol phosphatide or lecithin or ethanolamine phosphatide and CDP is released. Activation of choline and ethanolamine CDP-choline or CDP-ethanolamine after preliminary activation by ATP.

Glycolipids: Glycolipids contain sphingol, a carbohydrate-galactose and fatty acid. They donot contain phosphoric acid. Hence they are not phospholipids, but are called galactolipids or glycolipids. They are present in large amounts in the white matter of the brain and in the myelin sheaths of nerves. In Gancher's disease they accumulate in large amounts in the liver and spleen.

LIPID METABOLISM

Oxidation of Fatty Acids

Fatty acid oxidation to acetyl COA is the central energy yielding pathways in animals, protists and some bacteria. The $\overline{e}$s removed from FA's oxidation pass through electron transport chain (ETC) resulting in ATP synthesis. The Acetyl COA produced may be completely oxidized to CO_2 via citric acid cycle, resulting in further energy conservation. Here, the oxidation is called β-oxidation (a 4 step process) by which FA's → Acetyl COA.

The C—C bonds in fatty acids are stable and this is overcomed by activation of the carboxyl group at C-1 by attachment to coenzyme A, which allows stepwise oxidation of fatty acyl group at C-3 (β-position) and so, the name is β-oxidation.

Apolipoproteins are lipid binding proteins in the blood responsible for transportation of triacyl glycerols, phospholipids, cholesterol between organs after ingestion of dietary fats.

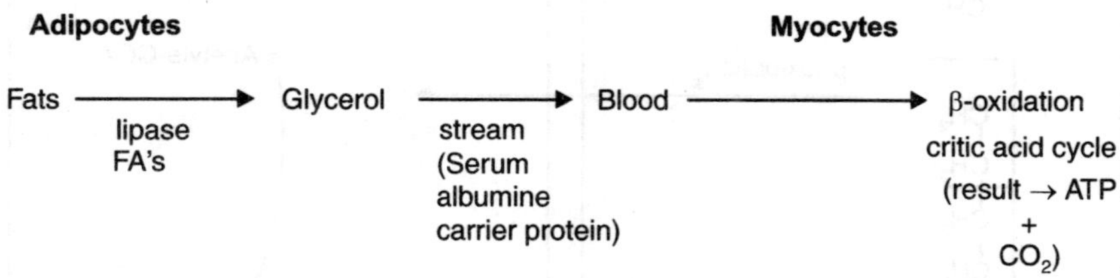

Glycerol derived from lipase action enters Glycolysis pathway in the form of D-glyceraldehyde 3-phosphate.

The Process and Enzymes

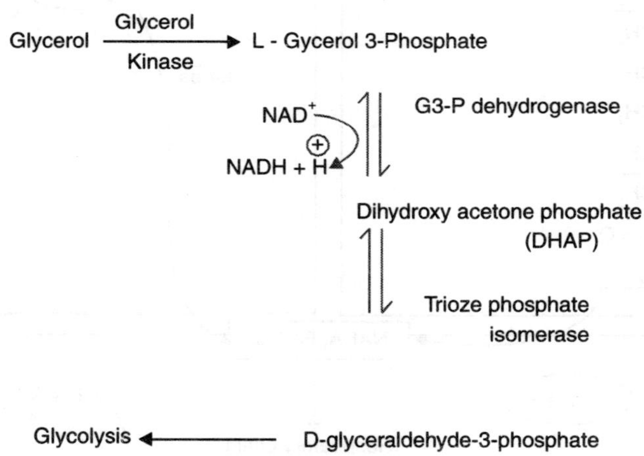

The enzymes of fatty acid oxidation in animal cells are located in the mitochondrial matrix. The free FA's that enter cytosol have to undergo a series of 3 enzymatic reactions.

(1) $\text{Fatty acid} + \text{COA} + \text{ATP} \xrightarrow[\text{Synthetases}]{\text{Acyl CoA}} \text{Fatty acyl-COA} + \text{AMP} + 2\text{P}; \ \Delta G^{0\prime} = -32.5 \ \text{kJ/mol}$

First reaction occurs in the other mitochondrial membrane. The fatty acyl-COA esters formed in the outer mitochondrial membrane do not cross the inner mitochondrial membrane intact. Instead, the fatty acyl group is transientty attached to the hydroxyl group of carnitine and fatty acyl carnitine is carried across the inner membrane by specific membrane transporter. This processes is assisted by carnitine acyl transferase I, FA – COA + carn itine → F Acyl carnitine + COA. The fatty acyl carnitine ester crosses the inner mitochondrial membrane into the matrix. In the 3rd step of entry process.

$$\begin{matrix}\text{Fatty acyl carnite} \\ + \text{COA}\end{matrix} \xrightarrow[\text{acyl transferase}]{\text{carnitine}} \begin{matrix}\text{Fatty acyl COA} \\ + \\ \text{Carnitine}\end{matrix}$$

Once inside mitochondrion, the fatty acyl-COA is ready for oxidation (βeta)

β-Oxidation

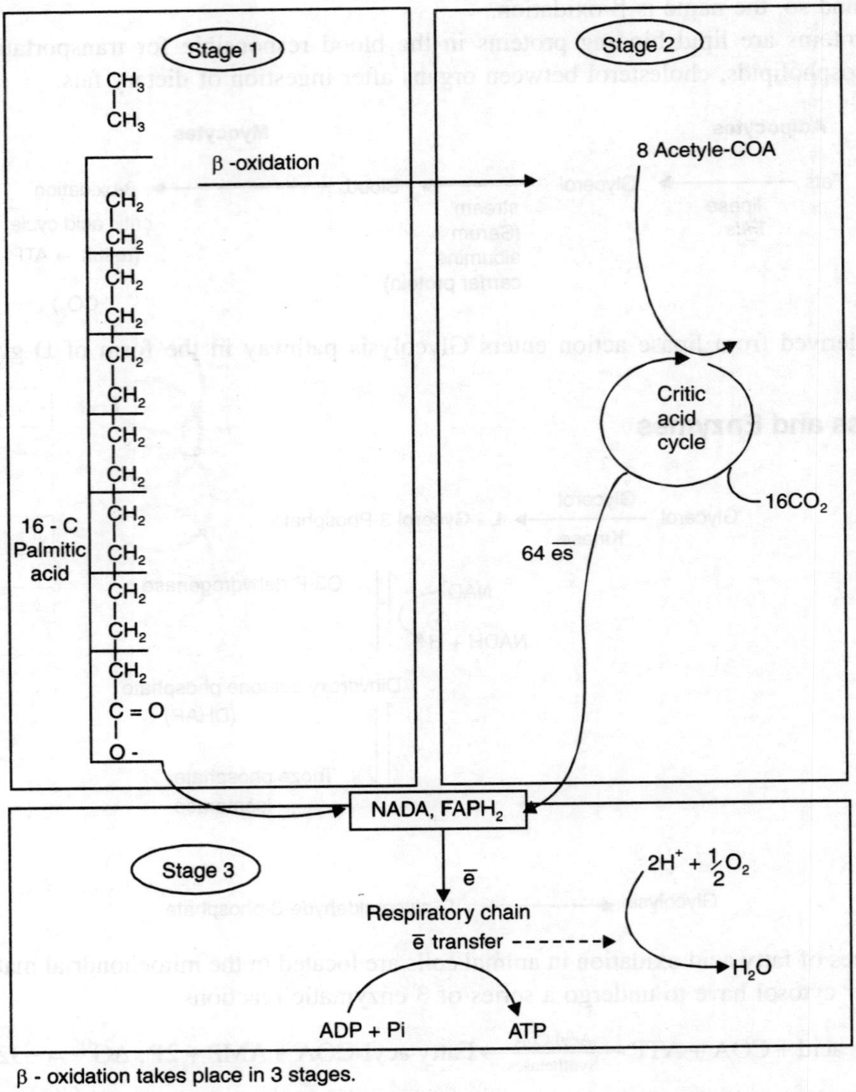

Fig. 6.3

Stage 1: The FA's undergo oxidation to remove two successive-carbon units in the form of acetyl COA, starting from the carboxyl end of the fatty acyl chain.

Result: 16C FA → 8 Acetyl COA

Formation of each molecule of acetyl COA requires removal of 4 H-atoms (2 pairs of $\overline{e}s$ and 4 H^+) from the fatty acyl moiety by the action of dehydrogenases.

Stage 2: In this stage of oxidation the acetyl residues of acetyl-COA are oxidized to CO_2 via citric acid cycle.

First 2 stages result-production of NADH & $FADH_2$ which is 3rd stage donate $\overline{e}s$ to ETC leading to formation of ATP. The energy released by FA oxidation is conserved as ATP.

R^h sequences

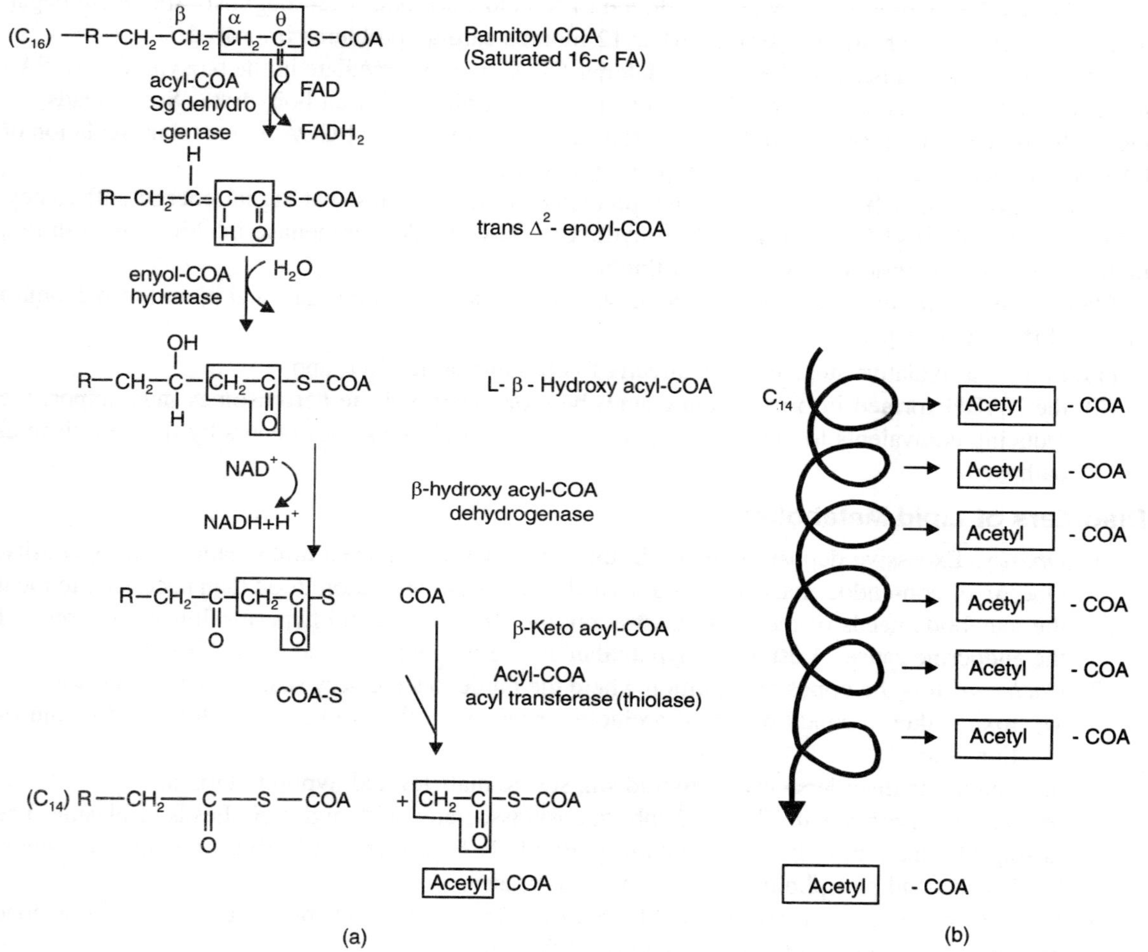

Fig. 6.4 a and b - β-oxidation pathway

(*a*) In each pass through this sequence, one acetyl residue (in box) is removed in the form of acetyl - COA from the carboxyl end of palmitate (C_{16}), which enters as palmitoyl-COA.

(*b*) 6 more passes through the pathway yield seven more molecules of acetyl-COA, the seventh arising from the last two carbons of C-16 chain. 8 molecules of acetyl-COA are formed in all.

Fats bears carry on β-oxidation in their hibernation (about 7 months), at this time, they do no eat, drink, urinate or defecate for months at time.

Bears store an enormous amount of body fat in preparation for their long hibernation periods.

Clinical Aspect

Carnitine deficiency: This disease is of 2 types systemic and myopathic in the former type, the carnitine levels are reduced in all tissues and in the later case only in muscle tissue, including the heart.

A systemic carnitine deficiency case was observed in a boy having an enlarged heart while being examined for cough. Cardiomegaly case increased developing symptoms of congestive heart failure for sometime in that body. After muscular fiopsy (skeletal), showed lipid deposition and low carnitine levels. Plasma levels of carnitine were also found to be below normal. After diagnosis, treatment began with 174 mg/kg of 1 creatinine per day. After 12 months cardiac problems improved.

Fatty acid oxidation is tightly regulated Malonyl-COA, first intermediate in the biosynthesis of FA's from acetyl COA, increases when the animal is well supplied with carbohydrate. So, it leads FA biosynthesis. The inhibition of carnitine acyl transferase and by malonyl-COA assures that oxidation of FA's is inhibited whenever the liver is supplied with excess glucose.

Two of enzymes of β-oxidation are also regulated by metabolites that signal energy sufficiency. When the [NADH]/[NA$^+$D] ratio is high, β-hydroxy acyl-COA dehydrogenase inhibited, in addition, high concentrations of acetyl COA inhibit thiolase.

Plant peroxysomes also carry out β-oxidation. The animal (mitochondrial) and plant (peroxisomal) system differs in 2 respects.

(1) In the 1st oxidative step, $\overline{e}$s pass directly to O_2, generating H_2O_2 and

(2) the NADH formed in β-oxidation cannot be reoxidized and the peronisomes must export the reducing equivalents to the cytosol. FA's oxidation in glyoxysomes occurs by the peroxisomal pathway.

Disorders of Lipid Metabolism

1. *Obesity:* Excessive deposition of fat in the depots with decreased mobilization causes obesity. Obesity of non-endocrinal causes is always due to ingestion of more food than necessary to meat the metabolic needs of the adult. Castration, hypothyroidism and hyperinsulinism are some of the endocrine causes. Lesions of hypothalamus increase appetite and cause obesity.
2. *Cachexia:* It is the opposite condition where the fat depots are very scanty. The causes are:
   (*i*) toxic—due to conditions like carcinoma, prolonged illness like tuberculosis and malnutrition, and
   (*ii*) endocrine disorders—hyperthyroidism, severe diabetes and hypopituitarism.
3. *Idiopathic hyperlipemia:* This is an inborn error associated with high lipid levels of plasma. The fasting plasma is milk like and contains particularly high levels of triglyceride and fatty acids. Cholesterol and phospholipids also may be increased.
4. *Gaucher's disease:* Cerebrosides are increased in brain, liver and spleen. There is an imbalance between the synthesis and the breakdown of this lipid.
5. *Niemann-Pick's disease:* Sphingomyelins accumulate in liver, spleen, bone marrow, lung and lymph nodes. Gangliosides of brain show degeneration. Several variants of the disease occur. The abnormality here also is an imbalance between synthesis and breakdown of the sphingomyelins.
6. *Tay-sacks disease:* Abnormal glycosides accumulate in brain. There is retarded development, paralysis, dementia and blindness. The child does not usually survive beyond 2 to 4 years.
7. *Fabry's disease:* An abnormal galactosyl-sphingolipid accumulates in many tissues. Death occurs due to cardiac or renal failure.
8. *Hyper and Hypolipoproteinemias:* Inborn errors of metabolism leading to an increase or decrease or absence of the lipoprotein components of plasma occur. The fraction most often involved is the β-lipoprotein.

UNIT 7

Amino Acid Metabolism

BIO-DEGRADATION OF AMINO ACIDS

Most of the metabolic energy generated in the tissues comes from the oxidation of carbohydrates and try acyl glycerols, which together furnish up to 90 percent of the energy requirement of the adult human male. The remainder, from 10 to 15 percent, depending on the diet, is furnished by the oxidation of amino acids.

They can undergo oxidative degradation in three different metabolic circumstances.

(*i*) During the normal dynamic turnover of body proteins the amino acids released, if not needed for synthesis of new body proteins, may undergo oxidative degradation.

(*ii*) When amino acids are ingested in excess of the body's needs for protein synthesis, the surplus may be catabolized, since amino acids cannot be stored.

(*iii*) During fasting or in diabetes mellitus, when carbohydrates are either unavailable or not properly utilized, body proteins are called upon as fuel.

The α-amino groups of the 20-L-amino acids commonly found in proteins are ultimately removed at some stage in their oxidative degradation. If not reused for synthesis of new amino acids or other nitrogenous products, these amino groups are collected and ultimately converted into a single excretory end product, which in human beings and most other terrestrial vertebrates is Urea.

The removal of the α-amino groups of most of the α-amino acids is promoted by enzymes called transaminases or amino transferases. In these reactions, called transamination, the α-amino group is enzymatically transferred from the amino acid to the α-carbon atom of α-keto glutarate, leaving behind the corresponding α-ketoacid analog of the incoming amino acid and causing the amination of the α-ketoglutarate to form α-glutamate.

$$\text{L-}\alpha\text{-amino acid} + \alpha\text{-keto glutarate} \rightleftharpoons \alpha\text{-keto acid} + \text{L-glutamate}$$

Most transaminases are specific for α-ketoglutarate. Some of the most important transaminases, which are named for the amino group donor, are designated by the following equation:

$$\text{L-alanine} + \alpha\text{-ketoglutarate} \underset{\text{transaminase}}{\overset{\text{Alanine}}{\rightleftharpoons}} \text{pyruvate} + \text{L-glutamate}$$

$$\text{L-asparate} + \alpha\text{-ketoglutarate} \underset{\text{transaminase}}{\overset{\text{Asparate}}{\rightleftharpoons}} \text{oxaloacetate} + \text{L-glutamate}$$

$$\text{L-leucine} + \alpha\text{-ketoglutarate} \underset{\text{transaminase}}{\overset{\text{Leucine}}{\rightleftharpoons}} \alpha\text{-keto isocapoate} + \text{L-glutamate}$$

$$\text{L-tyrosine} + \alpha\text{-ketoglutarate} \underset{\text{transaminase}}{\overset{\text{Tyrosine}}{\rightleftharpoons}} \text{p-hydroxy phenyl pyruvate} + \text{L-glutamate}$$

Thus α-ketoglutarate is the common acceptor of amino groups from most of the other amino acids. The L-glutamate so formed serves to channel amino groups into certain biosynthetic pathway.

All transaminases have a tightly bound prosthetic groups and a share common reaction mechanism. The prosthetic group is pyridoxal phosphate, a derivation of pyridoxine or vitamin B_6 pyridoxal phosphate functions as an intermediate carrier of amino groups on the active site of transaminases. During the catalytic cycle it undergoes reversible transitions between its aldehyde form, pyridoxal phosphate which can accept amino groups, and its aminated form, pyridoxamine phosphate, which can donate its amino group to α-ketoglutarate. In this way the prosthetic group acts as a reversible transient carrier of amino groups from an α-amino acid to α-ketoglutarate.

Formation of Ammonia

Glutamate undergoes oxidative deamination by the action of L-glutamate dehydrogenase, which requires NAD^+ as the acceptor of the reducing equivalents.

$$\text{L-glutamate}^- + NAD^+ + H_2O \rightleftharpoons \alpha\text{-ketoglutarate} + NH_4^+ + NADH + H^+$$

This enzyme is present only in the mitochondria, where it is localized in the matrix. Glutamate dehydrogenase is responsible for most of the ammonia formed in animal tissue, since glutamate is the only amino acid whose α-amino group can be directly removed at a high rate in this manner.

Glutamate dehydrogenase is a complex allosteric enzyme. Its molecular weight is about 3,00,000 and it consists of six identical subunits, each containing a single polypeptide chain of 500 residues. It is strongly activated by the positive modulator ADP but inhibited by GTP, the product of the succinyl-CoA synthetase reaction in the citric acid cycle. Whenever the liver cell needs fuel for the citric acid cycle to form more ATP, glutamate dehydrogenase activity is increased, making α-ketoglutarate available for the citric acid cycle and releasing NH_3 to be excreted. On the other hand, whenever GTP accumulates in the mitochondria as a result of high citric acid cycle activity, oxidative deamination is glutamate is inhibited.

Ammonia can be salvaged and reused in the synthesis of amino acids. In this case glutamate dehydrogenase acts in a reverse, reducing ammonia and α-kitoglutarate to form glutamate. Instead of being the simple reverse of the NAD-linked reaction. Written above, however this reaction is promoted by the NADP-linked reaction.

$$NADPH + H^+ + NH_4^+ + \alpha\text{-kitoglutarate}^{2-} \rightarrow NADP^+ + \text{glutamate}^- + H_2O$$

The use of two different co-enzymes by glutamate dehydrogenase for the release and uptake of NH_3 makes possible independent regulation of the deamination of glutamate and the amination of α-kitoglutarate, even though both are catalyzed by the same enzyme.

The Carbon Skeletons of Amino Acids are Degraded by 20 Different Pathways

There are 20 standard amino acids in proteins, all having different carbon skeletons and there are 20 different catabolic carbon pathways for their degradation.

Figure 7.1 shows that the carbon skeletons of 10 amino acids are ultimately broken down to yield acetyl-CoA.

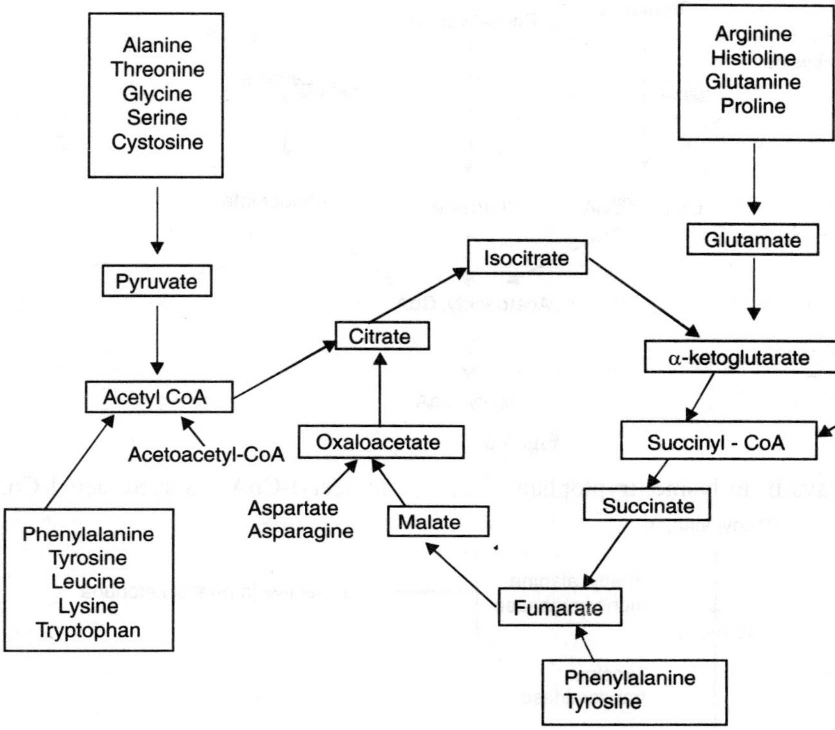

Fig. 7.1

Entry of the carbon skeletons of the common amino acids into the citric acid cycle.

Five amino acids are converted into α-kitoglutarate, three into succinyl CoA, two into oxaloacetate, and two yield fumarate. The individual pathways for the 20 amino acids are summarized by means of flow sheets, each leading to a specific point of entry into citric acid cycle. Some of the enzymatic reactions in these pathways that are particularly noteworthy for their mechanisms or their medical significance are singled out for special discussion.

The carbon skeletons of 10 amino acids yield acetyl-CoA, which enters the citric acid cycle directly. Five of the ten are degraded to acetyl-CoA via pyruvate. The other five are converted into aceto acetyl-CoA, which is then cleaved to form acetyl-CoA.

The five amino acids entering via pyruvate are alanine, cysteine, glycine, serine and threonine. Alanine yields pyruvate directly on transamination with α-kitoglutarate. The 4-carbon amino acid threonine is degraded into the 2-carbon glycine. Glycine has two pathways.

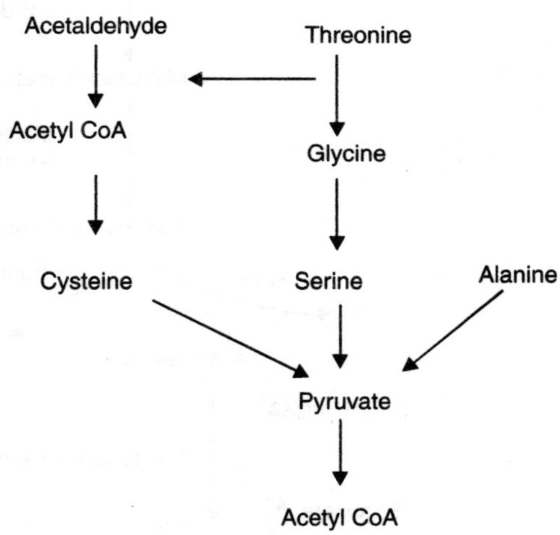

Fig. 7.2

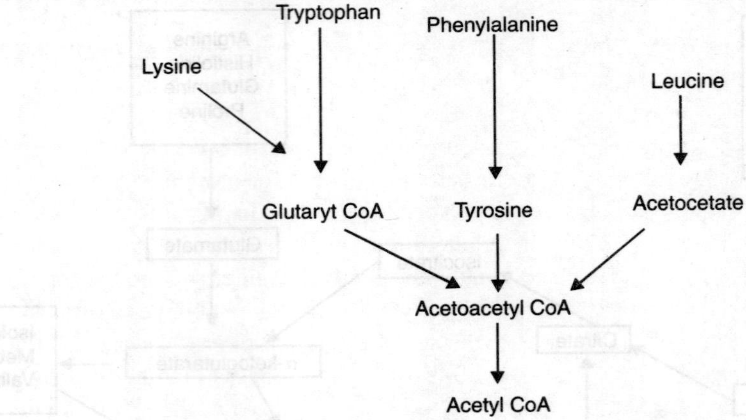

Fig. 7.3

Outline of the pathways from lysine, tryptophan ....to acetyl-CoA via aceto aceyl-CoA.

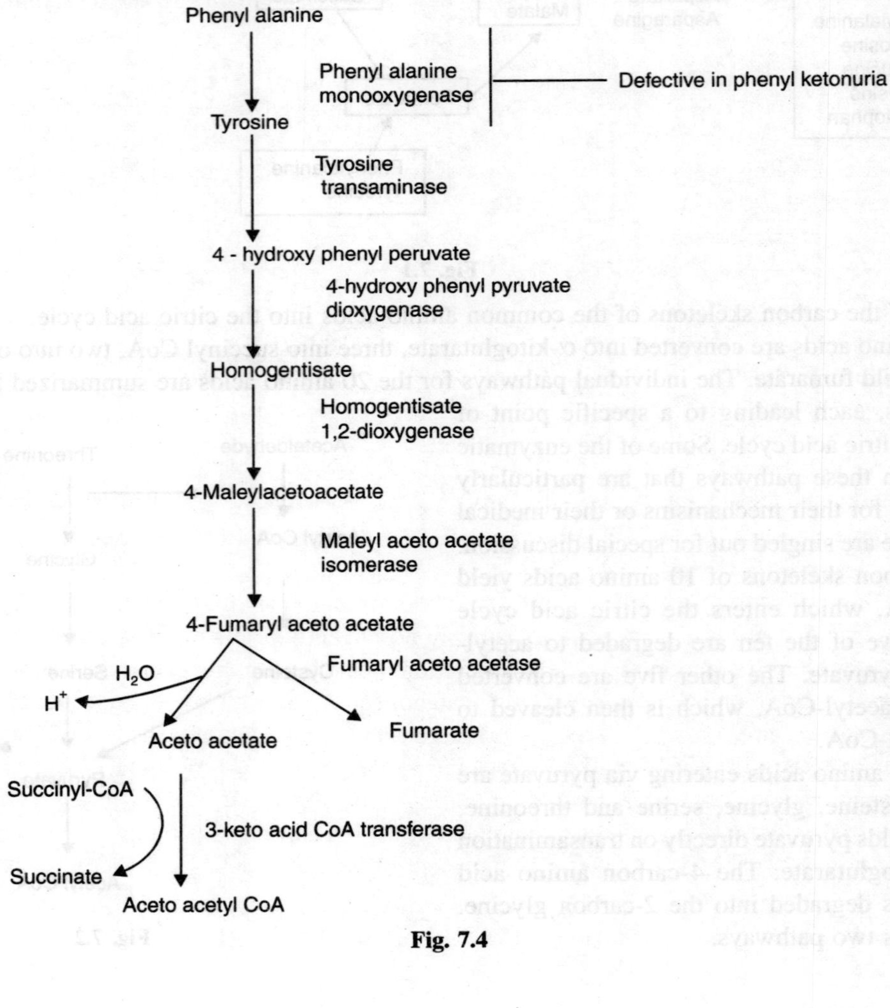

Fig. 7.4

It can be converted into serine, a 3-carbon amino acid, by enzymatic addition of a hydroxy methyl group carried by the co-enzyme tetrahydrofolate.

Portions of the carbon back bone of the amino acids phenyl alanine, tyrosine, lysine, tryptophan and leucine yield aceto acetyl-CoA which is then converted into acetyl-CoA.

Some of the intermediates in tryptophan catabolism are required precursors for biosynthesis of other important biomolecules, including serotonin, a neurohormone and nicotinic acid, a vitamin.

Many different genetic defects in amino acid metabolism have been identified in human beings.

The phenyl alanine tyrosine pathway requires special mention in this respect, since three of its enzymatic steps are particularly prone to genetic alteration, leading to three different types of heritable human metabolic defects. The enzyme in this pathway phenyl alanine 4-monooxygenase also called phenyl alanine hydroxylase which catalyzes the hydroxylation of phenyl alanine to tyrosine, is defective in some people. This defect is responsible for the genetic disease phenyl ketonuria. Phenyl alanine monooxygenase inserts one of the two oxygen atoms of O_2 into phenyl alanine to form the hydroxyl group of tyrosine. The other oxygen atom is reduced to H_2O by the NADH also required in the reaction.

$$\text{L-phenyl alanine} + \text{NADH} + \text{H}^+ + \text{O}_2 \rightarrow \text{L-tyrosine} + \text{NAD}^+ + \text{H}_2\text{O}$$

When phenyl alanine 4-mono oxygenase is genetically defective, a secondary pathway of phenyl alanine metabolism, normally little used, comes into play. In this minor pathway phenyl alanine undergoes transamination with α-kitoglutarate to yield phenyl pyruvate.

$$\text{Phenyl alanine} + \alpha\text{-kitoglutarate} \rightarrow \text{phenyl pyruvate} + \text{glutamate}$$

However, phenyl pyruvate is not metabolized further and thus is a dead end, since it (and also phenyl alanine) accumulates in the blood and tissues and is excreted in the urine. Excess phenyl pyruvate in the blood in early life impairs the normal development of the brain and caused severe mental retardation phenyl ketonuria (PKU).

Some human genetic disorder affecting amino acid metabolism.

Name	*Defective enzymes or process*
Albinism	Tyrosine 3-monooxygenase
Alkaptonuria	Homogentisate 1, 2-dioxygenase
Arginino succinic acidemia	Arginino succinate lyase
Homocystinuria	Cystathionine β-synthase
Maple syrup urine disease	Branched-chain α-keto acid dehydrogenase
Phenyl Ketonuria	Phenyl alanine 4-monooxygenase
Hyper Valinemia	Valine transaminase

The fourth enzyme in the phenyl alanine pathway homogentisate dioxygenase, also is defective in some people as the result of a heritable genetic mutation. Persons with this defect are unable to degrade homogentisate, an intermediate in phenyl alanine catabolism. As a result homogentisate accumulates in the body fluids and is excreted in the urine. Allowed to stand exposed to air. The urine of such a person will turn black. As the urine becomes alkaline due to decomposition of some of the urea to form ammonia, homogentisate is spontaneously oxidized by atmospheric O_2 to a black pigment, related to that present in the skin of black people. This genetic defect is known as alkaptonuria people with this condition usually suffer no significant impairment of health, apart from concern over their black urine.

Five Amino Acids Converted into α-kitoglutarate

The carbon skeletons of five amino acids are converted into α-kitoglutarate amino acids (argininie, histidine, glutamic acid, glutamine and proline) enter the citric acid cycle via α-kitoglutarate.

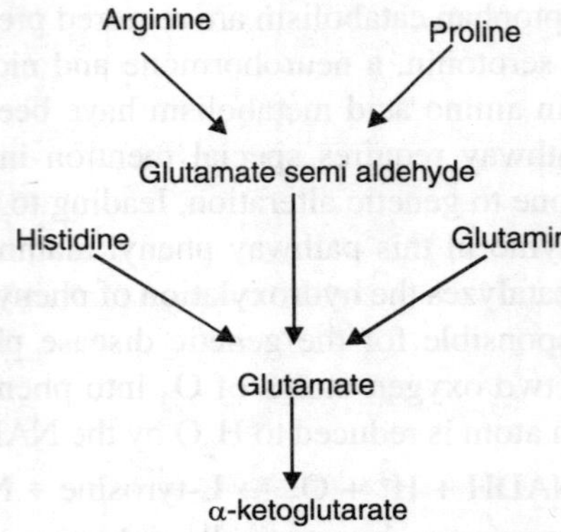

Fig. 7.5

Three Amino Acids are Converted into Succinyl CoA

The carbon skeletons of *methionine, isoleucine* and *valine* are ultimately degraded by pathways which yield *succinyl CoA*, and intermediate of the citric acid cycle.

The three α-keto acids derived by deamination of valine, isoleucine and leucine undergo oxidative decarboxylation by the same enzyme complex. The *α-keto acids dehydrogenase* system, which is genetically defective in some people, thus these α-keto acids accumulate in the blood and 'spill over' into the urine. This relatively rare conditions, which results in abnormal development of the brain and early death of infants unless treated, is called maple syrup urine disease because of the characteristic odor imparted to the urine by the keto acids. It is treated by rigid control over the diet to exclude, as far as possible, the common amino acids valine, isoleucine and leucine.

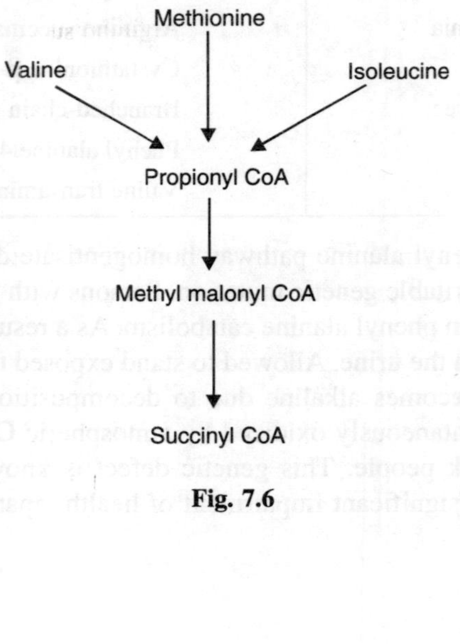

Fig. 7.6

Phenyl Alanine and Tyrosine Yield Fumerate

Phenyl alanine and tyrosine each yields two 4-carbon products, aceto acetate and fumarate. The aceto acetate enters the citric acid cycle in the form of acetyl CoA, whereas fumarate is of course an intermediate of the cycle.

The Oxalo Acetate Pathways

The carbon skeletons of asparagine and apsartic acid ultimately enter the citric acid cycle via oxaloacetate. The enzyme asparaginase catalyzes the hydrolysis of asparagine to yield aspartate.

$$\text{Asparagine} + H_2O \rightarrow \text{Aspartate} + NH_4^+$$

The aspartate then donates its amino group to α-kitoglutamate in a transamination reaction to yield glutamate.

$$\text{Aspartate} + \alpha\text{-kitoglutarate} \rightleftharpoons \text{Oxalo acetate} + \text{Glutamate}$$

The remaining carbon skeleton of aspartate, namely oxalo acetate enters the citric acid cycle.

Glucogenic and Ketogenic Amino Acids

Glucogenic

Alanine	Glycine
Arginine	Histidine
Asparagine	Methionine
Aspartic acid	Proline
Cysteine	Serine
Glutamic acid	Threonine
Glutamine	Tryptophan

Keto genic:

Leucine
Lysine
Tryptophan

Keto genic and Glucogenic:

Phenyl alanine
Tyrosine

Ammonia is Toxic to Animals

Ammonia resulting from the oxidative deamination of glutamate by glutamate dehydrogenase, a process that occurs in nearly all tissues. Ammonia is a very toxic substance, particularly to the brain. Ammonia is so toxic that injection of even very dilute solutions into the blood stream can render animals comatase.

The toxidity of ammonia to the brain is not completely understood, but two major factors can be identified. (1) The pK′ of ammonia is quite high, so that at the pH of the blood it occurs almost entirely as ammonium ion (NH_4^+). NH_4^+ ions are not readily permeant through the plasma membrane or mitochondrial membranes. However, free ammonia (NH_3), a neutral molecule, is freely permeant. Although only about 1 percent of the total ammonia in the blood occurs in the form of free NH_3 at pH 7.4, this small amount can penetrate membranes and gain entry into brain cells and their mitochondria.

The entry of ammonia into brain mitochondria leads to the formation of glutamate from ammonia and α-kitoglutarate, through the reverse action of glutamate dehydrogenase.

$$NH_4^+ + \alpha\text{-ketoglutamate}^{2-} + NADPH + H^+ \rightarrow \text{Glutamate} + NADP^+ + H_2O$$

Then the net result is that α-kitoglutarate is withdrawn from the pool of citric acid cycle intermediates in brain mitochondria, lowering the rate of oxidation of glucose, the major fuel of the brain.

Glutamine Carries Ammonia from many Peripheral Tissues to the Liver

The way taken to transport ammonia from the peripheral tissues to the liver or kidneys in most animals is to convert it into a nontoxic compound before exporting it via the blood. In many tissues, including the brain, ammonia is enzymatically combined with glutamate to yield glutamine by the action of glutamine synthetase, which promotes the reaction.

$$ATP + NH_4^+ + \text{Glutamate} \rightarrow ADP + Pi + \text{Glutamine} + H^+$$

In this reaction glutamate an acyl phosphate resulting from the phosphorylation of the 5-carboxyl group of glutamate by ATP. The bound glutamyl 5-phosphate combines with ammonia on the active site to form glutamine and release phosphate. The glutamine so formed is a neutral nontoxic compound which can readily pass through cell membranes, whereas glutamate, which bears a net negative charge, cannot. The glutamine is carried via the blood to the liver in most land animals, where it can be acted upon by glutaminase to yield glutamate and ammonia.

$$\text{Glutamine} + H_2O \rightarrow \text{Glutamate}^- + NH_4^+$$

The ammonia so formed is converted by the liver into urea. Glutamine is a major transport form of ammonia, it is present in normal blood in much higher concentration than other amino acids.

Alanine also plays a special role in transporting ammonia to the liver in a nontoxic form. Muscles, like other tissues, produce ammonia during the degradation of amino acids. In addition, ammonia also arises from the deamination of adenylate (AMP), a rather prominent process in very active skeletal muscles. Ammonia formed from these two sources is carried from the muscles to the liver by the amino acid alanine, through the action of the glucosealanine cycle. In this cycle ammonia is converted into the amino group of glutamate by the action of glutamate dehydrogenase.

$$NH_4^+ + \alpha\text{-ketoglutamate}^{2-} + NADH + H^+ \rightarrow \text{Glutamate}^- + NADP^+ + H_2O$$

The glutamate so formed now transfers its α-amino group to pyruvate, a readily available product of muscle glycolysis, by the action of alanine transaminase.

$$\text{Glutamate} + \text{Pyruvate} \rightleftharpoons \alpha\text{-kitoglutarate} + \text{Alanine}$$

The alanine, a neutral amino acid with no net charge at pH near 7, escapes into the blood and is carried to the liver. Here the alanine transfers its amino group to α-kitoglutarate by the action of alanine transaminase to yield glutamate, which then undergoes deamination to yield α-kitoglutarate and ammonia through the action of glutamate dehydrogenase. The ammonia so yielded is converted by the liver into urea.

The choice of alanine to transport ammonia from hardworking skeletal muscles to the liver is another example of the intrinsic economy of living organisms.

The glucose alanine cycle functions in a dual manner.

1. To carry amino groups from skeletal muscle to the liver to be converted into urea, and
2. To provide the working muscle with blood glucose made by the liver form the carbon backbone of alanine.

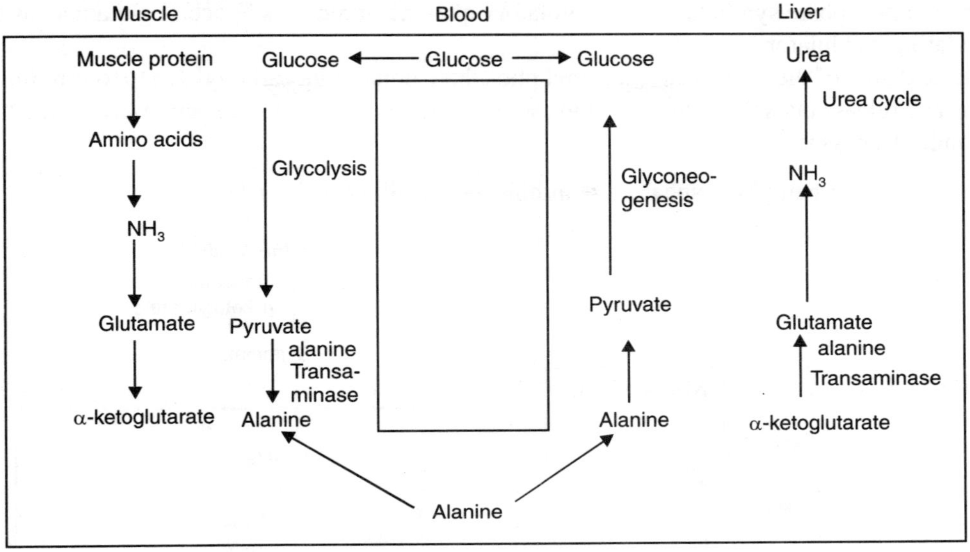

Fig. 7.7

Blood

Ammonia is converted into urea for excretion and pyruvate is rebuilt into blood glucose, which returns to the muscles. Animals have learned to solve two problems with one cycle. They combine the ammonia with the pyruvate to form alanine, a nontoxic neutral amino acid, which is passed via the blood to the liver for further processing.

Urea Cycle

In ureotelic animals the ammonia resulting from the deamination of amino acids is converted into urea in the liver by a cyclic mechanism, the urea cycle, was first discovered by Hans Krebs and Kuet Henseleit in 1932.

The first step of the urea cycle is, the amino group entry which arises in the form of free ammonia by the oxidative deamination of glutamate inside the mitochondria of the liver cell, catalyzed by glutamate dehydrogenase, which requires NAD^+.

$$\text{Glutamate}^- + NAD^+ + H_2O \rightleftharpoons \alpha\text{-kitoglutarate}^{2-} + NH_4^+ + NADH + H^+$$

The free ammonia so formed is immediately used, together with carbon dioxide generated in the mitochondria by respiration to form carbamoyl phosphate in the matrix in an ATP dependent reaction catalyzed by the enzyme.

The Roman numeral designates the mitochondrial form of this enzymes to distinguish it from the cytosolic form-II.

The latter has a different function, being required in nucleotide biosynthesis. The mitochondrial reaction is

$$HCO_3^- + NH_4^- + 2ATP^{4-} \rightarrow H_2N\underset{\underset{O}{\|}}{C}—O—PO_3^2 + 2ADP^{3-} + H^+ + Pi^-$$

Carbamyl phosphate synthetase I is a regulatory enzyme; it requires N-acetyl glutamate as a positive or stimulating modulator.

In the next step of the urea cycle carbamyl phosphate donates its carbonyl group to ornithine to form citrulline and release phosphate, in a reaction catalyzed by *ornithine transcarbamoylase*, a Mg^{2+} requiring mitochondrial enzyme.

$$\text{Carbamyl phosphate} + \text{Ornithine} \rightarrow \text{Citrullline} + Pi^{-} + H^{+}$$

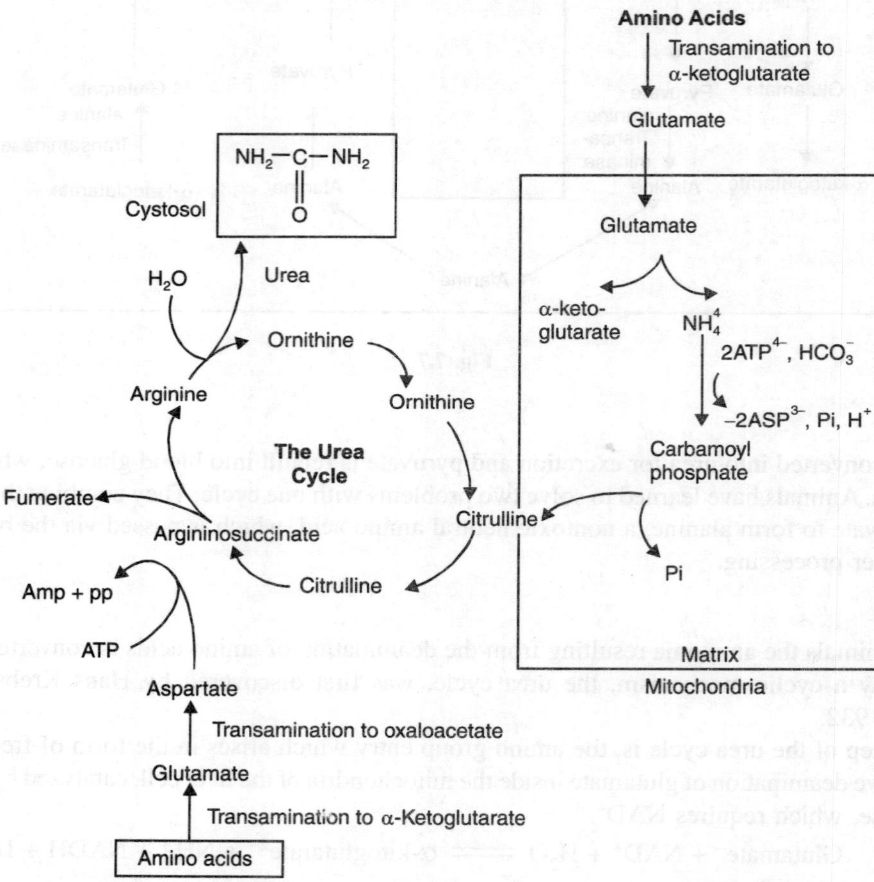

Fig. 7.8

The citrulline so formed now leaves the mitochondria and passes into the cytosol of the liver cell. The second amino group now is introduced in the form of L-aspartate which in turn acquired it from L-glutamate by the action of aspartate transaminase.

$$\text{Oxaloacetate} + \alpha\text{-glutamate} \rightleftharpoons \alpha\text{-aspartate} + \alpha\text{-kitoglutarate}$$

L-glutamate of course receives its amino group from most of the other common amino acids by transaminations to α-kitoglutarate. The transfer of the second amino group the citrulline occurs by a condensation reaction between the amino group of aspartate and the carbonyl carbon of citrulline in the presence of ATP to form arginino succinate. This reaction is catalyzed by arginino succinate synthetase of the liver cytosol, a Mg^{2+}—dependent enzyme.

$$\text{Citrulline + Aspartate + ATP} \rightleftharpoons \text{Argino succinate + AMP + PPi + H}^+$$

In the next step arginino succinate is reversibly cleaved by arginino succinate lyase to form free arginine and fumerate.

$$\text{Arginino succinate} \rightleftharpoons \text{Arginine + Fumarate}$$

The fumarate so formed returns to the pool of citric acid cycle intermediates.

In the last reaction of urea cycle the liver enzyme the liver enzyme arginase cleaves arginine to yield urea and ornithine.

$$\text{Arginine} + H_2O \rightarrow \text{Ornithine + Urea}$$

Ornithine is thus regenerated and can enter the mitochondria again to initiate another round of the urea cycle.

The overall equation of the urea cycle is

$$2NH_4^+ + HCO_3^- + 3ATP^{4-} + H_2O \rightarrow \text{Urea} + 2ADP^{3-} + 2Pi^- + AMP^- + PPi^{3-} + H^+$$

Note that for every molecule of urea formed a molecule of HCO_3^- must be consumed. The urea cycle thus disposes of two waste products, ammonia and bicarbonate. Moreover, this fact suggest that the urea cycle participates in the regulation of blood pH, which depends upon the ratio of dissolved CO_2 to HCO_3^-.

Disorders of Amino Acid Metabolism

Cystinuria: This is genetically inherited metabolic disease where 20–30 times than normal cystine is excreted in the urine. Cystinuria is a manifestation of renal transport defect not only for cystine but also for lysine, arginine and ornithine. Milne (1961) observed that cystinuria is due to intestinal transport defect for all these four amino acids. Cystinosis is also known as cystine-storage disease; in this congenital disease, cystine crystals are stored in several tissues including organs.

Homocystinuria: In this metabolic inborn error, plasma homocystine is excreted in large quantities through urine with methionine. It is due to decreased activity of hepotic cystathionine synthetase. In addition to these, the patients also suffer from mental retardation palpable liver, and bilateral posterior dislocation of lenses etc.

Phenylalanine and Tyrosine

Phenylalanine is an essential amino acid but tyrosine is not, because the latter is synthesized readily from the former. Tyrosine and phenylalanine have got immense metabolic role in the body.

Tyrosine is synthesized in the body from phenylalanine by a complex enzymatic processes requiring hydroxylation of phenylalanine. The reaction required enzyme phenylalanine hydroxylase and coenzyme NADP and folic acid coenzyme-tetrahydropteridine.

$$\text{Phenylalanine} + O_2 \xrightarrow[\text{NADP+H}^+]{\text{Hydroxylase}} \text{Tyrosine} + H_2O + \text{NADP}$$

tetrahydropteridine

Hereditary defects in phenylalanine and tyrosine metabolism: Four hereditary defective metabolic abnormalities of phenylalanine and tyrosine have been observed. They are considered as inborn error of metabolism as follows:

(*a*) **Phenylketonuria:** In this metabolism of phenylalanine is interfered with and as a result tyrosine cannot be formed from it in the normal way. The defect is due to absence of an enzyme phenylalanine oxidase responsible for this conversion. Children suffering from syndrome have defective mental development and a high plasma phenylalanine and its increased excretion through urine.

(*b*) **Tyrosinosis:** In this metabolic abnormality, p-hydroxyphenyl pyruvic acid cannot be oxidised and so excreted through urine in considerable amounts.

(*c*) **Alcaptonuria (ochrosiosis):** Due to deficiency of the enzyme homogenitisic acid oxidase, persons suffering with this disease are unable to oxidise homogenitisic acid which is excreted in urine. On standing homogenitisic acid (alcopton) present in urine becomes darker and finally black.

(*d*) **Abinism:** The defect in this congenital disease is due to lack of enzyme tyrosinase in the melanocyte cells and so the pigment melanin is not formed. Localized failure of pigment formation might also occur, *e.g.*, the eyes.

UNIT 8

Nucleic Acid Metabolism

Nucleic acids: Nucleoproteins contain phosphoric acid and other prosthetic group. Nucleoproteins are formed by the union of nucleic acid with basic protein. The nucleic acids viz., deoxyribonucleic acid (DNA) and ribonucleic acid (RNA) are found both in animal and plant cells. In the animal cells DNA is the chief nucleic acid—specifically found in chromosomes, whereas RNA is found in smaller quantity in the nucleus. RNA is also found in smaller quantity in the nucleus. RNA is also found in ribosomes inside the cytoplasm where they help in the synthesis of proteins. The breakdown and synthesis of RNA are continuously go in on within the cell with the help of intercellular enzymes. DNA is a stable compound and controls the hereditary characteristics of the cell. The nucleic acid is made up of many nucleotides (mononucleotides) making a double helical structure of the polynucleotide chains. Each mononucleotide is again composed of nucleoside and phosphoric acid. Each nucleoside is composed of a pentose and a base as follows:

(*i*) *Pentose*: The pentose, ribose is present in ribonucleic acid (RNA), whereas deoxyribose is present in deoxyribonucleic acid (DNA).

(*ii*) *Bases*: The bases are of two types:

(*a*) Purine bases—adenine and guanine are present in both the nucleic acids.

(*b*) Pyrimidine bases—cytosine and uracil are present in RNA whereas cylosine along with thymine is present in DNA. In addition to DNA and RNA, nucleosides are also component of a number of co-enzymes, like NAD and NADP and metabolically important compounds, such as ATP, UDPC etc.

Since nucleoprotein is a composite product, its complete metabolism will mean the life history of all the component parts in it. The protein parts of the nucleoprotein molecules contain a high proportion of diaminoacids and undergo the same fate as the other proteins in the body. The phosphoric acid part and the pentose molecule are treated by the body in the same way as the phosphates and carbohydrates derived from other sources. The special importance of the metabolism of nucleoproteins lies in the life history of its characteristic ingredients—the purine and pyrimidine bases.

Nucleic acid metabolism may be of two types—Exogenous and endogenous.

The exogenous nucleoprotein metabolism is undergone by the end products of nucleoprotein digestion, after they are absorbed. These end products are phosphoric acid, Pentose-carbohydrate. The Pyrimidine nucleosides which are not further digested, the two purine bases viz., adenine and guanine, and probably some nucleotide.

The endogenous nucleoprotein metabolism starts in a different way.

There are two binds of tissue nucleotidases. The phosphonucleotidase splits off phosphoric acid and forms and nucleosides. But the purine nucleotidase takes away the purine bases, leaving phosphoric acid molecule combined with carbohydrate.

Obtaining everything together it will be seen that metabolism of purine occurs under two conditions—first, when the purines remain combined as nucleotides and secondly, when they remain free as adenine and guanine.

The end product of purine metabolism is chiefly uric acid and slightly hypoxanthine and xanthine. But this is not all; all purine bases undergoing metabolic changes in the body, and not excreted in these forms. A good deal of it is lost in some unknown way.

The intermediate metabolism of pyrimidine bases, *e.g.*, cytosine, uracil and thymine is not clear. The catabolism of pyrimidines occurs mainly in the liver.

Nucleic Acids (DNA and RNA)

It is a well established fact that genetic information from one cell to the daughter cell is carried by DNA and different forms of RNA coordinate in protein synthesis.

Chromosomes of cell nucleus is composed of DNA, which is responsible for the continuity of genetic code from one cell to the other. Just before mitosis, the double helix of DNA separate from each other and with the help of an enzyme, DNA polymerase, a second chain exactly similar to the parent one is formed, thus giving it the original double helical structure. The two similar double helixes enter the two daughter cells formed as a result of cell division giving them the same amount of DNA as in the parent cell. In the process of meiosis (reduction division) such duplication of DNA in the parent cell does not occur. Therefore, the daughter cell contains half the number of original chromosomes.

Two types of RNA are also formed in the nucleus. One of them is single stranded messender RNA (mRNA) which has a smaller molecule than DNA. Its synthesis is catalyzed by the enzyme RNA polymerase. On being synthesized the mRNA leaves out of the nucleus and becomes incorporated into the ribosomes of the cytoplasm. The mRNA carries the genetic details to determine the sequence in which the amino acids are to be lined up in the polypeptide chain. In this process of synthesis another type of RNA—soluble RNA (sRNA) or transfer RNA (tRNA) is also formed in the nucleus to participate in the process of synthesis. The tRNA attaches itself to the activated amino acids forming sRNA amino acid and carries them to the mRNA at the ribosomes. The sRNA finds out the definite code word or codon on the mRNA, attaches to the specific amino acids and is set itself free to be used again. The long mRNA thread has a number of ribosomal structures known as polysomes (ergosomes) which are responsible for the synthesis of several amino acids forming a polypeptide. Finally the mRNA is set to free from the polypeptide chain of the completed protein and renews its help in protein synthesis.

The RNA codes or codons for different amino acids are as follows:

	Amino acids	*RNA code words*	
1.	Alanine (α-aminopropionic acid)	CCG	
2.	Arginine (α-amino, δ-guanidovaleric acid)	CGC	AGA
3.	Asparagine (γ-amide of α-aminosuccinic acid or α-amino-γ-succinamide)	ACA	AUA
4.	Aspartic acid (α-aminosuccinic acid)	GUA	
5.	Cyteine (α-aminoglutaric acid)	UUG	
6.	Glutamic acid (α-aminoglutaric acid)	GAA	

7.	Glutamine (δ-amide of α-aminoglutaric acid or α-amino-δ-glutaramide)	AAG	AGA
8.	Glycine (α-aminoacetic acid)	UGG	ACG CGG
9.	Histidine (α-amino β-methylvaleric acid)	ACC	
10.	Isoleucine (α-amino-β-methylvaleric acid)	UAU	UAA
11.	Leucine (α-aminoisocuproic acid)	UUG UUC	UCG UUA
12.	Lysine (α-ε-diaminocuproic acid)	AAA	AAU
13.	Methionine (α-amino-γ-methylthiobutyric acid)	UGA	
14.	Phenylalanine (α-amino-β-phenylpropionic acid)	UUU	CUU
15.	Proline (pyrrodiline-α carboxylic acid)	CCC	CCU CCA
16.	Serine (α-amino-β-hydroxypropionic acid)	UCU	UCC ACG
17.	Threonine (α-amino-β-hydroxy-*n*-butyric acid)	CAC	CAA
18.	Tryptophan (α-amino-β-indole propionic acid)	CGU	
19.	Tyrosine (α-amino-β-(parahydroxyphenyl) propionic acid	AUU	
20.	Valine (α-aminoisovaleric acid)	UGU	

Biosynthesis of Nucleic Acids

In human infants the amount of nucleoprotein rapidly increases as growth advances, although the chief food is milk which is almost free from nucleoproteins. This indicates that the nucleoproteins can be synthesized in the human body. Liver is the most probable site of nucleoprotein synthesis.

Nucleoproteins are compounds of simple basic proteins, protamines and histones with nucleic acid. Nucleic acids are composed of

(*a*) phosphoric acid,

(*b*) a pentose and,

(*c*) a nitrogenous base—purine and pyrimidine.

Both the nitrogenous bases are synthesised in the body not as such but as their corresponding nucleotides.

Purine nucleotides are adenylic acid and guanylic acid (guanine-5′-phosphate). These nucleotides are not formed directly. At first the nucleotide inosinic acid (hypoxanthine-ribose-5′-phosphate) is synthesized from the metabolic products of carbohydrate and protein. Different compounds of the purine ring are derived from formic acid, CO_2, glutamine, aspartic acid, glycine. The inosinic acid nucleotide, after being formed, is then converted into adenylic acid and guanilic acid.

Pyrimidine is also synthesized as nucleotides. Pyrimidine nucleotides are cytidylic acid, thymidylic acid and uridylic acid. The first step in the synthesis of pyrimidine nucleotide is the formation of orotic acid. It has been found by both invitro and invivo experiments that orotic acid is utilised for the synthesis of pyrimidine nucleotides. Carbamylphosphate and aspartic acid take the initial role in the synthesis of pyrimidine nucleotide. In the pyrimidine ring, nitrogen at position 1 and carbon at position 2 are donated by carbamylphosphate. Carbons at positions 4, 5, 6 and nitrogen at position 3 of the pyrimidine ring are donated by aspartic acid.

Part C
Microbiology

UNIT 1

Introduction

THE SCOPE OF MICROBIOLOGY

Microbiology is a branch of Biological Sciences, which deals with the study of microscopic living organisms. The term "Microbiology" is derived from Greek literature–*mikros* small; *bios*–life; and *logos*–study. Thus, microbiology deals with study of microorganisms. Microorganisms can be looked into and studied only with the help of microscope, as they are very tiny and invisible to the naked eye.

Microbiology deals with the form, structure, reproduction, physiology, metabolism and classification of microorganisms. This also includes the ability of microorganisms to cause physical and chemical variations in the environment and their reactions to different reagents both physical and chemical in nature.

Microbiology has its share in contributing as a model in the studies of genetics.

Molecular Biology, in addition to suggesting diagnostic methods suggests prevention and control of different diseases of microbial origin. With the techniques of genetic manipulations, genetic engineering is another development through microbes.

Microbiology has been diversified into the following branches:

(*a*) *Virology:* Deals with the study of viruses. Viruses taxonomically lie on the border of line between living and non-living matters. They cause different diseases among plants, animals and human beings.

(*b*) *Mycology:* Deals with the study of fungi *e.g.*, yeasts, molds and mushrooms.

(*c*) *Phycology:* Deals with the study of autotrophic eukaryotic microorganisms *e.g.*, Algae.

(*d*) *Bacteriology:* Deals with the study of bacteria. Bacteria are the largest group among microbes; it has more economic importance. Bacteria have tremendous influence on human beings with reference to health, agriculture and industry, etc.

(*e*) *Aquatic microbiology:* Microbiological examination of waters, water purification, biological degradation of wastes are studied in this branch.

(*f*) *Aero microbiology:* Deals with dispersal of disease causing microbes through air. Microbial population in air, their quality and number in air are studied.

(*g*) *Agricultural microbiology:* Deals with the study of role of microbes in agriculture, both useful and harmful; including importance of microbes in soil fertility.

(h) *Industrial microbiology:* Deals with the role of microbes in industrial production. Medicinal products like antibiotics and vaccines, fermented beverages, industrial chemicals are extracted from many microorganisms. From genetically induced microorganisms proteins and hormones are extracted.

(i) *Food microbiology:* Deals with food processing, food preservation, canning, pasteurization of milk, study of food-borne microbial diseases and preventive measures.

(j) *Medicinal microbiology:* Deals with the pathogenic microbes, their life cycle, physiology, genetics and reproduction, etc. Diagnostic procedures for causative microbes identification and preventive measures are also studied in medical microbiology.

(k) *Geochemical microbiology:* Deals with the role of microbes in coal, mineral and gas formation; oil and gas and recovery of minerals from low grade ores using microbes including prospecting for deposits of coal etc., are also studied.

(l) *Environmental microbiology:* Deals with the role of microbes in maintaining the quality of the environment, microbial influence in recycling and decaying of natural, domestic and industrial wastes; their role in biochemical cycles are also studied.

(m) *Exomicrobiology:* Deals with the study of microorganismal life in outer space.

(n) *Protozoology:* Deals with the study of protozoans in all their aspects. Harmful protozoans are known to cause several diseases *viz.*, Malaria, Amoebiasis, Sleeping sickness, etc.

Human life is depending on microorganisms as they are closely involved with the health and welfare of human beings, however, certain harmful microorganisms too are involved with human life. For example, beneficial activities of microorganisms involve the making of wine, cheese and yoghurt, production of penicillin, interferon and alcohol. Further, microorganisms are also involved in the processing of industrial and domestic wastes. Several harmful effects of microorganisms on human beings include causing diseases, spoiling of food and deterioration of materials like wooden articles, glass lenses and iron pipes.

Microorganisms are generally used for conducting research, for studying fundamental life processes in laboratories, as

(i) they grow as single living unit and can therefore be easily cultured in laboratory conditions.

(ii) they are highly adaptable to different organic substances which are utilized as carbon or energy source.

(iii) due to their microscopic size compared to plant and animal cell, they can be cultured within a limited space *i.e.*, in culture tubes and petriplates.

(iv) microorganisms' cell division rate is very high in laboratory conditions which encompasses to study many generations within short time.

Microbiology—Historical Events

Prior to the discovery of microorganisms certain processes caused by their life-metabolic activities, such as fermentation of wine, milk, curd preparation with yeast etc., were known to mankind.

With the works of Hippocrates (460–377 BC), Verona (2nd century BC), Lucretins (95–55 BC), Pliny (23–79 AD), Galen (131–211 AD) and other scientists of that time, the hypotheses of living nature of contagious diseases was described.

Columella and Varo (first century BC) suggested that diseases were caused by invisible organisms inhaled or ingested.

Kircher (1659) could find microscopic worms in the blood of a plague patient. Antony Van Leeuwenhoek (1683) could give description of various types of bacteria. He invented the simple microscope. He suggested that population of increase of animalcules were progeny of few parental organisms and described many different living and parasitic protozoa, globular yeast cells and filamentous fungi.

Robert Hook, a contemporary of Leeuwenhoek, developed compound microscope in 1678 and confirmed Leeuwenhoek's observations.

In 1762 Von Plenciz suggested that separate agent causes each disease.

Pasteur in 1857 established that fermentation was the result of microbial activity.

Robert Koch perfected bacteriological techniques in 1876.

Fraenkel (1886) described pneumococcus. Loeffler and Frosch (1898) observed that foot and mouth disease of cattle was caused by a microbe.

Walter Reed (1902) observed that yellow fever was caused by a virus which is transmitted through the mosquito bite.

Twort (1951) and Herelle (1917) discovered lytic phenomenon in bacterial culture. He also observed that, viruses that attack bacteria, are the agents responsible and termed them as bacteriophage.

Fleming (1925) made the accidental discovery that Penicillium fungus destroys *Staphylococci* bacteria.

Ruska (1934) invented electron microscope which enables detailed study of virus morphology.

Jerne proposed natural selection theory of antibody synthesis in 1955. Burnet in 1957 concluded clonal selection theory. Again Burnet in 1967 introduced concept of immunological surveillance.

Several new agents of infectious diseases have emerged *viz.*, HIV (identified in 1980), Outbreaks of plague in 1994, Cholera in 1995 and Dengue (Hemorrhage fever/Brain fever) in 1996.

Latest methods of several infectious diseases and vaccine production have been revolutionized. Recombinant DNA technology, PCR, Nuclear anaprobes, Radioimmunoassay, ELISA techniques etc., are some of the revolutionary diagnostic techniques developed.

MICROBIAL DIVERSITY AND TAXONOMY

Classification, identification and nomenclature comprise taxonomy of microorganisms. An organism can be classified only after its morphological, cultural, physiological and pathogenic characters have been studied in detail. Based on the common characters of properties, a set of organisms is considered into groups (taxa). As there are no formal rules to define the taxa, nomenclature is the name given for defined taxa as also governed by bacteriological code of nomenclature. One of the important aspects of taxonomy is classification *i.e.*, ordering of organisms into taxonomic groups.

(a) Criteria of Classification

This criteria does not help sometimes in characterization of genera or species which are not characterised by physiological tests or traditional biochemical tests.

Serological tests like agglutination, fluorescent antibody techniques etc., have limited role in classification but have enormous value in identification of microbes. However, genetic tools are the modern one for identification of bacteria by detecting a specific portion of an organism's genetic material.

(b) General Methods of Classifying Bacteria

(i) *Intuitive method:* Several microbiologists study the characteristics of bacteria like morphological, physiological, biochemical, genetical and molecular. Sometimes, it is difficult to assign an organism based on all the characters because a character important to a particular microbiologist may not be important to another, hence different taxonomists might arrive at different groupings. This approach *i.e.*, intuitive method is found useful sometimes.

(*ii*) *Numerical taxonomy:* This is based on different characters for each strain and each character is given equal weightage. The percentage of similarity in each strain can be determined by using the following formula:

$$\% \text{ S} = \text{NS}/(\text{NS} + \text{ND})$$

where

NS = Number of characteristics for each strain which are similar.

ND = Number of characteristics that are dissimilar or different.

S = Similarity – if it is high to each other, placed into groups larger and so on.

(c) Genetic Relationship

This classification depends on genetic relationship among microorganisms.

The % G + C determines the organism whether it is the same or different species. If two microorganisms have quite different mol % G + C, then the species are different and seem to be not related to each other. Further, characterization is established based on the two criteria:

(*i*) *DNA-DNA Hybridization or DNA-Homology:* After separating two strands and converting them into single strand, the later one is mixed with those obtained from other organism and will form heteroduplexes otherwise not.

(*ii*) *16s rRNA sequencing:* Due to its highly conservative nature of rRNA genes, the 16s rRNA nucleotide sequence gives better informations than DNA molecule. If there is no DNA-DNA homology between two organisms, still nucleotide sequence of their rRNA cistrons may be studied to observe the similarity. Therefore, the degree of similarity may be used as a measure of relatedness among organisms, but at a level beyond that of species.

Bacterial Nomenclature

A bacterium must be assigned with a name after identification and classification. The assigned name should be accepted internationally.

Carl Linnaeus gave hierarchial classification system similar to that of plants and animals. According to this, orders of bacteria divided into families, and in each family there are number of genera, each of which comprises one or more species.

In this process, each bacterium was described according to binomial nomenclature (genus and species). The genus name is always capitalized, often to just the first letter in capital letter and species name is written with a lower initial letter. The care must be taken while abbreviating the common genera which will share the same initial letters and then the first few letters are written, *e.g., Staphylococcus aureus* is abbreviated as *Staph aureus, Streptococcus pyogenes* is abbreviated as *Strepto pyogenes.* Generally bacterial names should be written in italic letters but if it is not possible then bacterial names are to be underlined. The unnamed species, if it is one written as sp. or if many, the spp. should be used. It (sp. or spp.) should also not be underlined or italicised. Most of the families end with aceae as in *Enterobacteriaceae* and order end in ales, *e.g., Spirochaetales*. Latinized names may be of either male (*Staphylococcus*) or female (*Salmonella*) or may be neuter gender (*Clostridium*). The plural forms of these genera are *Staphylococci, Salmonellae* and *Clostridia*, following strictly linguistic rules.

The bacterial nomenclature is currently regulated by a committee on systematic nomenclature. List of approved names are published in International Journal of Systematic Bacteriology. It is kept that most of the bacterial names should often be descriptive *e.g.*, *Mycobacterium tuberculosis* causes tuberculosis.

Evolutionary models: Recently evolutionary relationships of living organisms have been defined on the basis of ribosomal RNA sequences and other data.

Woese *et al.* (1990) noted that bacteria are distant from plants and animals, and, by contrast, plants and animals are not so far from each other. Therefore, they established a new superior concept of domains over the kingdom, and proposed three domains, Archaea, Bacteria and Eukarya in 1991.

In a modern sense, bacteria, cyanobacteria, actinomycetes, etc., are distributed in the domain—Bacteria; methanogens, extremely thermophilic, organisms, extremely halophilic organisms, etc., are in the domain—Archae, and molds, yeasts, basidiomycetes, algae and protozoa, etc., in the domain—Eukarya.

On the basis of common characters or properties, a set of organism is considered into groups (taxa). There are no formal rules to define the taxa. The features that are used to differentiate various organisms often have little to do with the fundamental basis for arranging the organisms into taxonomic groups.

Identification of bacteria must always be based on a phenotypic character that is group to correlate well with the generic information.

In general, the characters chosen for an identification scheme should be easily determined.

Further in identification, characters should also be few in numbers.

Classification of bacteria should be based on a DNA/DNA hybridization study.

Bacterial classification is quite difficult to determine.

However, bacterial classification involves large numbers of characters.

Classification of Microorganisms

Biologists divided living organisms into plants and animals in the past; the same was later accepted as two broad-living organism kingdoms. During the course of further studies, it came to light that microorganisms show similarities with plants as some are green and do not eat other organisms, on the other hand some bacteria which are motile, express similarities with animals. Thus bacteria were included in plant kingdom, slime molds were regarded as intermediary between plants and animals because they were considered as plants by botanists and animals by zoologists.

A German biologist Earnst Haeckel, in 1866 proposed a new kingdom as Protista, which included all microorganisms which doesn't possess any extensive development of tissues. Based on this character biologists included all the unicellular microscopic organisms in this kingdom, although few multicellular organisms have also been included (Fig. 1.1). Thus, the kingdom protista includes bacteria, algae, fungi and protozoans. This kingdom protista is further divided into two groups based on the cellular differentiation, namely lower protista which includes bacteria and cyanobacteria, and higher protista, which includes algae, fungi and protozoans.

The lower protista members are prokaryotic in their cellular organization, as the nuclear material is devoid of nuclear membrane while the higher protista members are eukaryotic, as the nucleus of the cell is enveloped by nuclear membrane.

Lower protista: Lower protista include bacteria and cyanobacteria which are prokaryotic in nature. "Bergey's Manual of Systematic Bacteriology"— contains a description of all established genera and species; it also provides a practical arrangement for differentiating the bacteia. As per this classification the bacteria are grouped into 19 (nineteen) sections on the basis of structure, genetic information, biochemical, nutritional, staining and ecological characters (Fig. 1.2).

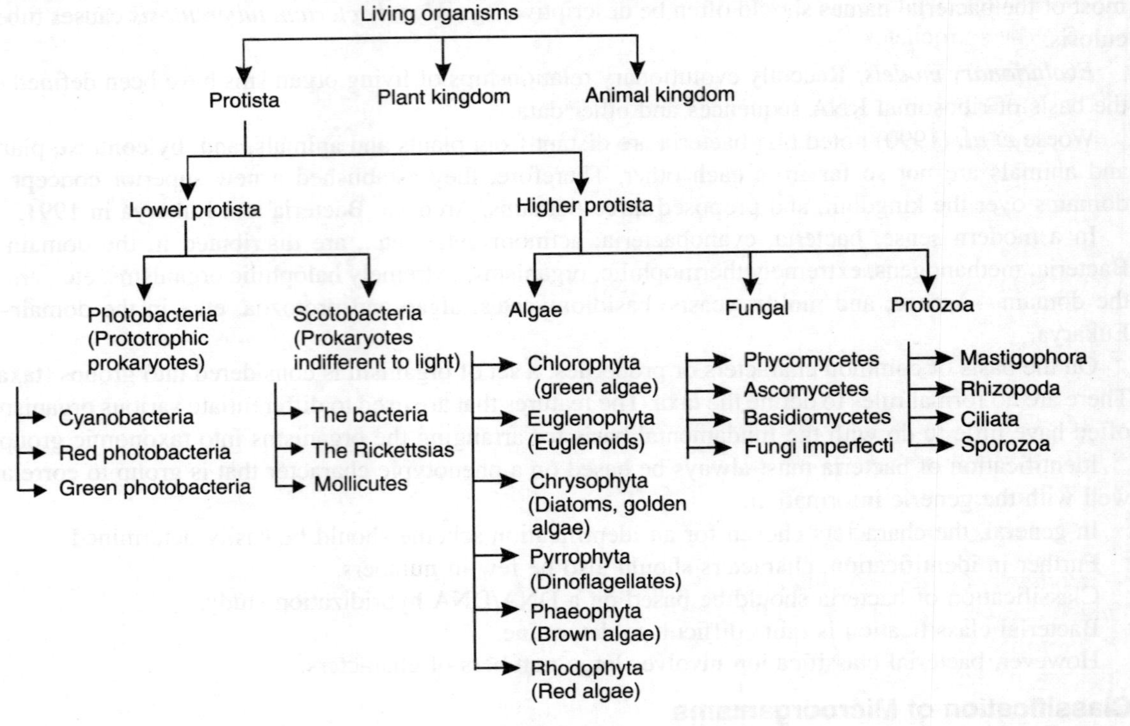

Fig. 1.1 Broad classification of microorganisms

19 (Nineteen) sections of bacteria:

1. The phototrophic bacteria
2. The gliding bacteria
3. The sheathed bacteria

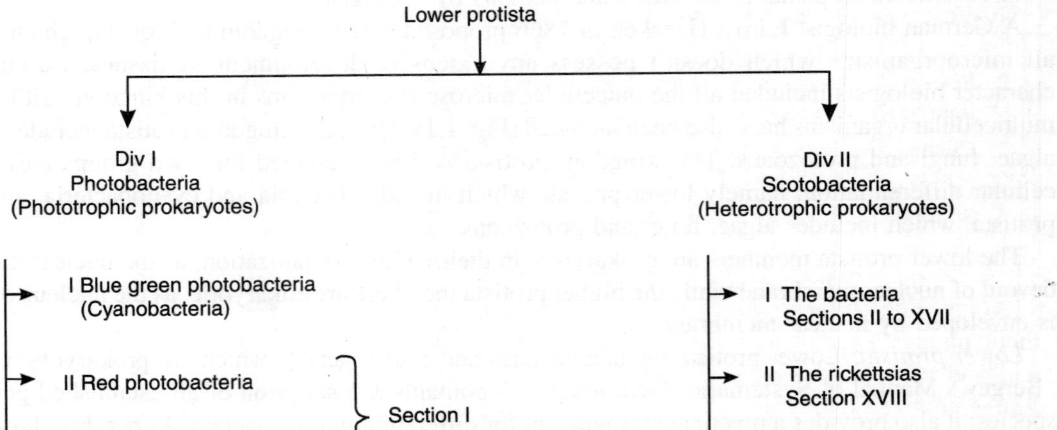

Fig. 1.2 Brief outline of bacterial classification

4. Budding and, or appendaged bacteria
5. The spirochetes
6. Spiral and curved bacteria
7. Gram negative aerobic rods and cocci
8. Gram negative facultative anaerobic rods
9. Gram negative anaerobic bacteria
10. Gram negative cocci and coccobacilli
11. Gram negative anaerobic cocci
12. Gram negative chemolithotrophic bacteria
13. Methane producing bacteria
14. Gram positive cocci
15. Endospore forming rods and cocci
16. Gram positive asporogenous rod shaped bacteria
17. Actinomycetes and related microorganisms
18. The rickettsias
19. The mycoplasmas.

Higher protista: The eukaryotic or higher protista include a highly divergent group of organisms which can be sub-divided into three major groups namely—Protozoa, Algae and Fungi.

Protozoa: The protozoa are unicellular, colorless, non-photosynthetic, motile organisms with complex life cycles. Protozoans are found in a variety of habitats and require sufficient moisture for growth and activity. Protozoan species are found in fresh water, marine water, sewage and terrestrial environments. Many protozoans are parasitic on other organisms. Protozoa members are classified into four major classes (Table 1.1).

Table 1.1: Major classes of Protozoa and their characteristics

Class	*Major characteristics*
1. *Mastigophora* (Flagellate protozoa)	Movement by one or more flagella. Cells divide by longitudinal fission.
2. *Rhizopoda* (Amoeboid protozoa) Reproduce by binary fission.	Motility generally by pseudopodia, but some form flagella.
3. Ciliata (Ciliate protozoa)	Motility by numerous cilia. Divide by transverse fission. Cell contains two nuclei.
4. Sporozoa	Generally non-motile, some show creeping or gliding motility. Multiple by multiple fission Parasitic.

Algae: The algae are aerobic photosynthetic organisms characterize by the presence of photosynthetic pigments and their ability to synthesize their food through photosynthesis. Algae inhabit in water, soil and other terrestrial environments. Algae occur in a variety of forms such as single thalloid, filamentous, rhizoidal and mucilagenus colonies. Algal species secrete various kinds of mucilages—they are polysaccharides in composition. The mucilage helps in motility or provide anchorage to the thalloid organism depending on the thallus structure, cell wall composition, pigments, motility, reproduction and reserve food materials, algae are classified into six major classes (Table 1.2).

Table 1.2: Major classes of algae and their characteristics

Class	*Major characteristics*
1. Chlorophyta (Green algae)	Found in fresh and salt water, soil, etc. Chlorophyll a, b, carotene and xanthophyll present. Stores starch as reserve material. It divides vegetatively or asexually to form flagellated zoospores.
2. Euglenophyta (Euglenoids)	Contain both fresh and marine water forms. Unicellular flagellates. Chlorophyll a, b, carotene present. Stores fat and starch like carbohydrates. It divides by longitudinal binary fission and rarely by sexual method.
3. Chrysophyta (Diatoms, golden algae)	Consists of a group of several diverse non-flagellar, unicellular algae. Some have calcium containing cell walls and others contain silica in their cell wall (diatoms). Stores oils and Lecucosin. Reproduce by asexual or sexual method.
4. Pyrrophyta (Dinoflagellates)	Primarily marine organisms. Two flagella with different structure and arrangement. Most are unicellular and contain carotenes and xanthophylls. Asexual reproduction and cell stores oils and starch.
5. Phaeophyta (Brown algae)	Multicellular and contain xanthophyll and carotene pigments. Found in sea water and structurally complex. Reproduce asexually by zoospores and sexually by gametes.
6. Rhodophyta (Red algae)	Contains phycobilins and chlorophyll a. Non-flagellate, unicellular or multi-cellular with complicated modes of sexual reproduction. Stores starch. Grows in salt water. Polysaccharides agar-agar, carrageenan and alginic acid are obtained from red algae.

Fungi: Fungi include the yeast and molds. Presence of a typical eukaryotic cell is their characteristic feature, a thread like mycelial thallus and fungi are non-photosynthetic. Majority of fungi are saprophytes but some are parasitic on plants and animals. Fungi are aerobic, while some yeasts are facultatively anaerobic. Many fungi are mesophilic but some are thermotolerant too. In fungi reproduction occurs both sexually and asexually.

Fungi are classified based on morphology, physiology, the nature of reproductive structures and mode of reproduction. Fungi are classified into four major classes (Table 1.3).

Table 1.3: Major classes of fungi and their characteristics

Class	*Major characteristics*
1. *Phycomycetes*	Mycelium nonseptate, aquatic or terrestrial, sporangiospores (asexual spores) motile or non-motile. Sexual spores are oospores or zygospores.
2. *Ascomycetes*	Mycelium well developed and septate. Sexual spores are borne in asci. Parasitic and saprophytic.
3. *Basidiomycetes*	Characteristics spores borne on basidia (club shaped structures)—basidiospores.
4. *Fungi imperfecti*	Reproduction by asexual process only. Mostly found on decaying wood.

Viruses: Viruses are very simple and tiny. Viruses are an intermediary stage between living and non-living. Viruses are acellular—without cellular organization. DNA or RNA is the genetic material in viruses but never both. Viruses are obligate parasites and they cannot be cultured on media. However, they can be cultured on cell media viz., bacteria, chick embryo cells, etc. Viruses are made up of proteins and nucleic acids; as they lack the enzymes for their synthesis, they depend on the host enzymes. Viruses can be crystallized. Viruses can be transmitted from one host to another. Based on their hosts, they are classified into seven types as follows:

1. Plant viruses — Infect plants
2. Bacteriophages — Infect bacteria
3. Cyanophages — Infect cyanobacteria
4. Myco viruses — Infect fungi
5. Mycoplasma viruses — Infect mycoplasma
6. Phyco viruses — Infect algae
7. Animal viruses — Infect animals.

Nutritional Classification of Microorganisms

Various chemicals by which a microbe is built are called nutrients. Environment contributes the chemicals. Nutrients that are taken up into the cell are converted into microbial cell constituents. This process is called anabolism. It is also called biosynthesis, since anabolism results in the biochemical synthesis of new material.

Biosynthesis is energy building process and each microbial cell should have a source of obtaining energy. As this energy is obtained from the environment, three kinds of energy sources are utilised, viz., light, organic chemicals and inorganic chemicals. Based on nutritional requirements and source of energy utilized, the microorganisms are divided into two classes—Autotrophs and heterotrophs.

1. ***Autotrophs:*** Autotrophs are self-feeding microorganisms. Autotrophic microorganisms use inorganic chemical or light as energy source and are able to grow in the absence of organic materials. They use environmental carbon dioxide as their sole source of carbon and convert this carbon dioxide into carbohydrates by the addition of hydrogen obtained from water.

Autotrophs are subdivided into photoautotrophs and chemoautotrophs based on the source of energy.

(*a*) ***Photoautotrophs:*** Photoautotrophs are microorganisms which use sunlight as an energy source and carbon dioxide as the sole carbon source. Pigmented bacteria and microalgae belong to this group. Varietal pigments present in these organisms trap the solar energy. Based on pigmentation and other features photoautotrophic bacteria can be divided as purple sulfur bacteria and Green sulfur bacteria, while the former contains bacteriochlorophyll and yellow carotenoid, the latter contains chlorobian, chlorophyll or bacteriviridin.

The source of hydrogen in all photoautotrophic microorganisms is water, but in photoautotrophic bacteria it is hydrogen sulphide. Photoautotrophic bacteria usually live in sulfur springs.

(*b*) ***Chemoautotrophs:*** These microorganisms obtain their energy from inorganic chemical sources whose oxidation provides the necessary energy. Ammonia, nitrates and non-metallic materials etc., which are easily available in nature are the sources of inorganic chemicals. Bacteria absorb inorganic molecules and on breaking them inside their body, the released energy is utilized. Based on the inorganic compound utilized in oxidation, chemoautotrophic bacteria are classified into hydrogen bacteria, nitrifying bacteria, iron bacteria and sulfur bacteria etc., chemoautotrophic microorganisms are represented by pigmentless bacteria.

2. ***Heterotrophs:*** Heterotrophs are microorganisms feeding on sources other than one self. They obtain readymade food directly from external sources like organic compounds. Heterotrophs include bacteria, fungi, protozoa, etc. They cannot grow with carbon dioxide as the sole source of carbon. The carbon must be supplied in organic form; *i.e.*, the reason they depend on organic compounds formed by

plants or animals. Based on their source of energy they are subdivided into photoheterotrophs and chemoheterotrophs.

(*a*) *Photoheterotrophs:* Besides using solar energy photoheterotrophs utilise organic materials as carbon source. They are considered as the intermediate forms between phototrophs and heterotrophs. These bacteria also possess bacteriochlorophyll and yellow carotenoid, like purple sulfur bacteria. *Example:* Non-sulphur purple bacteria.

Purple non-sulfur bacteria absorbs organic raw materials from the environment, however, they cannot be directly utilized as food. This organic source only serve as hydrogen resource and this released hydrogen is then utilized for reducing the carbon dioxide. These bacteria can synthesize their food in absence of oxygen and in the presence of light.

(*b*) *Chemoheterotrophs:* These microorganisms use principal carbon source and organic compounds as an energy source. It is the characteristic feature of this category that both energy and carbon requirement can be obtained by organic compound. This group is represented by many bacteria, protozoans and fungi.

As heterotrophs are pigmentless organisms, they obtain food either by parasitic method or by saprophytic method.

(*a*) *Parasitic microorganisms:* These microorganisms obtain their nutritional requirements by parasiting plants and animals. Parasitic bacteria may be either pathogenic or non-pathogenic. The pathogenic bacteria cause disease due to secretion of toxins. Non-pathogenic bacteria which survives in host are infact helpful. For example, human beings utilize the Vitamin B Complex synthesized by intestinal bacteria.

(*b*) *Saprophytic organisms:* Saprophytic microorganisms obtain food by oxidizing organic matter anaerobically. They grow on dead organic matter. Based on the nutritional source they may be fermenting microorganisms or putrefying bacteria. Fermenting microorganisms use carbohydrates as the substrate. They anaerobically respire and incompletely oxidise the carbohydrates, thus producing alcohols. Putrefying bacteria make use of protein or nitrogenous matter as their nutritional source and liberate methane gas, ammonia, etc.

Classification of Microorganisms Based on the Forms

Bacteria being the simplest prokaryotic microorganism, based on their forms, are classified into four main groups viz., spherical, spiral, rod shaped and pleomorphic.

(a) Spherical bacteria: They are the simplest form of bacteria in which bacteria appear like a minute sphere. It is also called as cocci (singular coccus). According to Greek literature 'Kokhos' means grain or kernel. Cocci appear in various arrangements and depending on the arrangements they are divided as follows (Fig. 1.3):

(*i*) Coccus — single spherical bacteria
(*ii*) Diplococci — two spherical bacteria
(*iii*) Tetracocci — cluster of four spherical bacteria
(*iv*) Staphylococci — bunch of spherical bacteria
(*v*) Streptococci — chain of spherical bacteria
(*vi*) Sarcina — cubical arrangement of eight or more spherical bacteria.

(b) Rod shaped bacteria: Rod shaped bacteria are also called bacilli (singular bacillus) and are the commonest in microbial world. According to Latin literature 'bacillum' means stick or rod. They are often straight and like rod but sometimes they are also slightly curved. They generally occur single and are called bacillus. Occasionally they occur in pairs called diplobacilli or they may occur in chain called streptobacilli (Fig. 1.3).

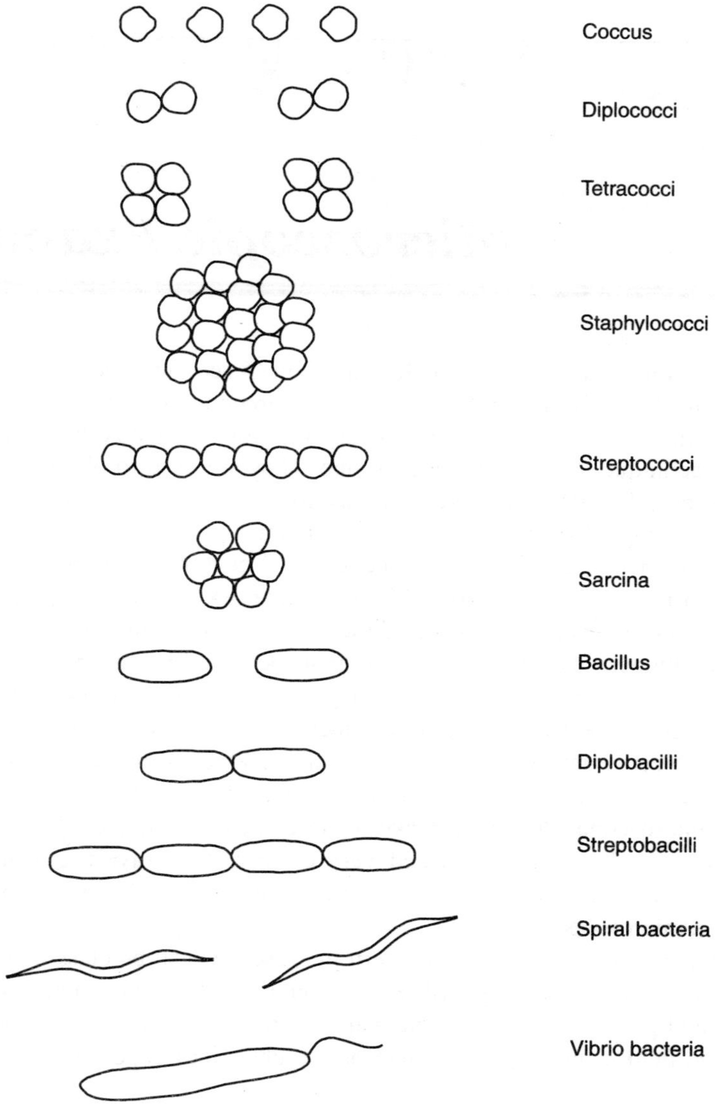

Fig. 1.3 Various forms of bacteria

(c) Spiral bacteria: These bacteria are helicoidal shaped or generally spiral. According to Greek literature "spira" means coil. They resemble like a cork-screw. A spirillum has more than one turn of helix whereas vibrio has less than one complete turn. Vibrio appears superficially like 'comma' shaped.

(d) Pleomorphic bacteria: Certain bacteria exhibit different shapes or forms, called pleomorphic bacteria.

Example: Acetobacter, Azotobacter, Rhizobium, etc.

UNIT 2

Microscopic Examination

Microbes are smaller; yet infinitely complex. It is difficult to see the structure of microbes as well as cell structure with naked eye. To discover the molecular components of microbes as well as cells, and also to find out how their individual components function, microscope is the only technical tool.

Microscopy is a technique of using the microscope. They are used to observe structures below the range of magnified lens. The detailed study and development of microbiology could not have been possible without microscope. Various microscopic techniques generate different magnifying views of microbes and cells. The invention of light microscope in 16th century is the most important contribution to the advancement of microbiology. In 1600 AD, Leeuwenhoek invented a simple, single lens microscope. Robert Hooke, an Englishman constructed the first compound microscope which has become as the forerunner of the modern day microscope. Microscope is an optical instrument with a lens or a combination of lenses for obtaining magnified images, as well as to show contrast and resolution. Magnification by a microscope is the collective product of the ocular and objective lenses; if the ocular lens is 10X and the objective lens is 10X, definitely the observed specimen magnifies 100 times.

In a microscope besides lenses, wavelength of the illumination is also important as it decides the resolving power of microscope. Resolving power of microscope is defined as the ability of lenses to distinguish between 2 smaller objects, located very closely. The wavelength of the light used for illumination and the numerical aperture (NA) of the objective are the two factors that decide the resolving power of a microscope.

Based on the principle of magnification, modern microscopes can be classified into two types viz., light microscope and electron microscope. Magnification in a light microscope is obtained by a system of optical lenses using light waves, whereas, the magnification in an electron microscope is obtained by a system of magnetic field using a beam of electrons. Light microscopes are of six types:

(*i*) Bright field microscope
(*ii*) Dark field microscope
(*iii*) Phase contrast microscope
(*iv*) Fluorescence microscope
(*v*) Ultraviolet microscope
(*vi*) Interference microscope.

Electron microscopes are of two types:

(*i*) Transmission Electron Microscope (TEM)

(*ii*) Scanning Electron Microscope (SEM)

Table 2.1: Showing units used for microscopic measurements

Unit name	*Fraction of a meter*	*Abbreviation*
Centimeter	10^{-2}	cm
Milimeter	10^{-3}	mm
Micron	10^{-6}	μ
Milimicron/Nanometer	10^{-9}	mμ/nm
Angstrom	10^{-10}	Å

Bright Field (Light) Microscope

The light or bright field microscope is the most common instrument used in microbiology laboratory. In this, the microscopic field is brightly lighted and microorganisms appear darker as they absorb some light. Though microorganisms do not absorb much light, process of staining them with a stain usually increases their light absorbing ability resulting in more contrast and color differentiation. Light microscopes produce a magnification of about 1000X to 2000X. Basic limitation of light microscope is not magnification but the resolving power.

Simple or dissecting microscope and compound microscope are the most common bright field microscopes used in biology and microbiology laboratories. Compound microscopes generally are equiped with three different magnifying objectives; they are oil immersion, high and low power objectives.

Living organisms cannot be viewed by bright field microscope because staining of the object/ organism makes it lifeless.

Dark Field Microscope

The concept of dark field microscope technique is bright object against a dark background. This objective is obtained by attaching the light microscope with a special condenser *i.e.*, dark field condenser.

In dark field condenser, the center of the top lens is opaque which prevents light rays passage; thus the specimen on the stage is illuminated only by the peripheral oblique rays. In such an arrangement, the light rays focussed, traverse across the specimen horizontally, almost at right angle to the objective lens. Subsequently the surrounding appears dark but the specimen appears as bright refractile object.

Dark field microscope is useful for the observation of live, unstained microorganisms. Since the microbes positioned on the stage are in moist, living condition, motility of microbes also could be observed.

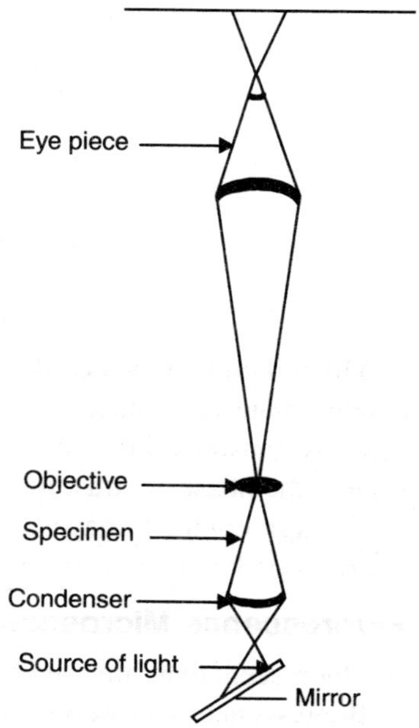

Fig. 2.1 Path of light in bright field microscope

Phase Contrast Microscope

It is a special microscope which enables the viewer to see transparent protoplasmic components without staining thus retaining the object organism alive. Frederick Zernike developed this microscope in 1933; hence it is also called Zernikemicroscope.

Phase contrast microscope is a modified bright field microscope attached with a phase contrast objective and phase contrast condenser. This modification makes it possible to distinguish unstained organelles within a cell by virtue of differing slightly in their thickness or refractive index.

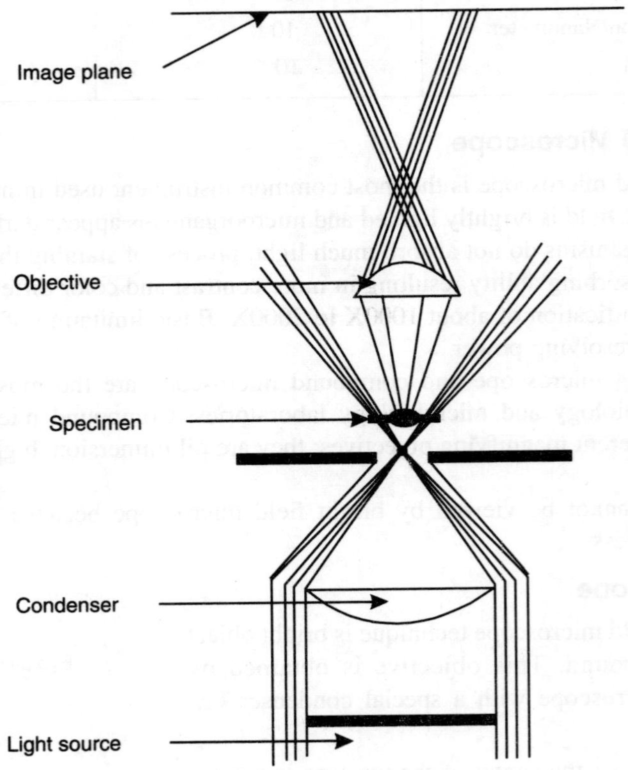

Fig. 2.2 Path of light through dark field microscope

The principle on which this microscope is devised is that the light passing through one material into another material of a slight difference in thickness, will undergo a change in phase. This difference in phase is translated into variation in brightness of the structure and, therefore, is detected by the viewer. Like in the phase contrast microscope, little phase changes results in interference of light waves resulting in high contrast images; certain structures which are not possible to stain and observe under light microscope become visible Fig. 2.3.

Fluorescence Microscope

As fluorescent dyes are used for staining of the specimen to view under this microscope, it is considered as fluorescence microscope. This differs from an ordinary bright field microscope in three ways:

(*i*) A dark-field condenser is used instead of bright field condenser.

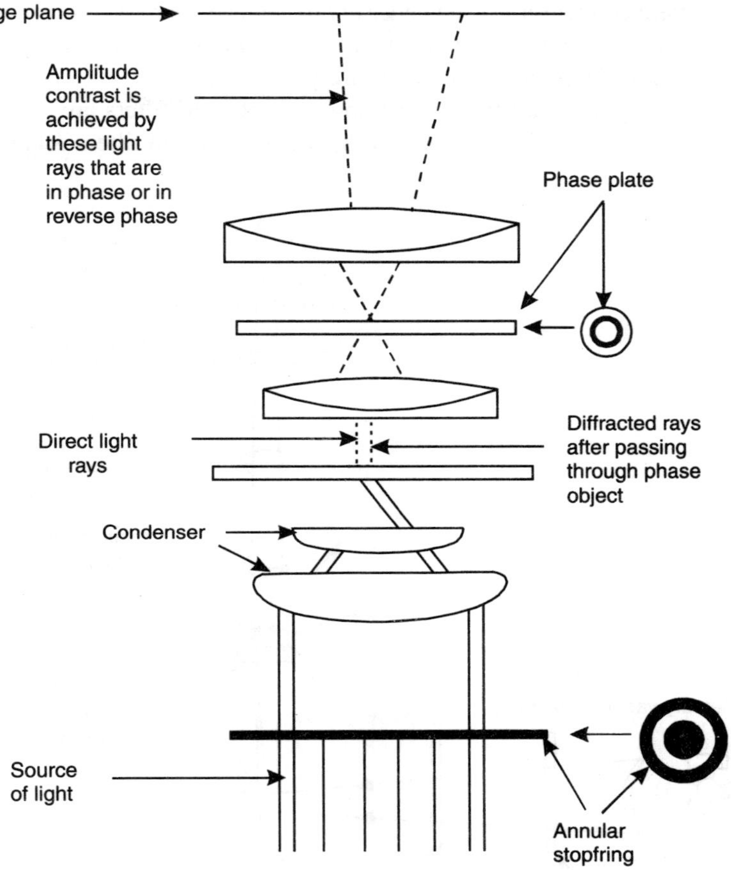

Fig. 2.3 Path of light through phase contrast microscope

(*ii*) It employs 3-filters to alter light rays which pass through the microscope and reach the eye of viewer.

(*iii*) Mercury vapor arc lamp is utilized in this microscope as an illuminating source.

The object appears brighter in this microscope, when fluorescent and the background appears lighter.

Components of Fluorescence Microscope

Magnification and resolution principle of a fluorescence microscope is similar to that of bright field microscope. Important components of fluorescence microscope are light source, heat filter, exciter filter, condenser and barrier filter.

Heat filter: Considerable amount of heat is produced by Infrared rays generated by the mercury vapor arc lamp. These rays do not contribute in fluorescence and cause stress on the filters within the system. Therefore, to eliminate these rays, a heat filter is used, which, however, does not interfere the transmission of UV rays.

Exciter filter: After light is cooled down by the heat filter, it passes through the exciter filter which absorbs undesired wavelength except the short wavelength, which excites the fluorochromes on the slide. Exciter filter is very dark and allows only the green, blue and violet or UV rays.

Condenser: Dark field condenser is used generally to achieve the best contrast of a fluorescent object in the microscopic field. Added advantage of dark field condenser is that the majority of UV rays are deflected by the condenser, thus observer's eyes are protected.

Precaution: One must make sure that all the above mentioned filters are in their respective places before viewing into the microscope because missing of any one or all filters is harmful to viewers eyes.

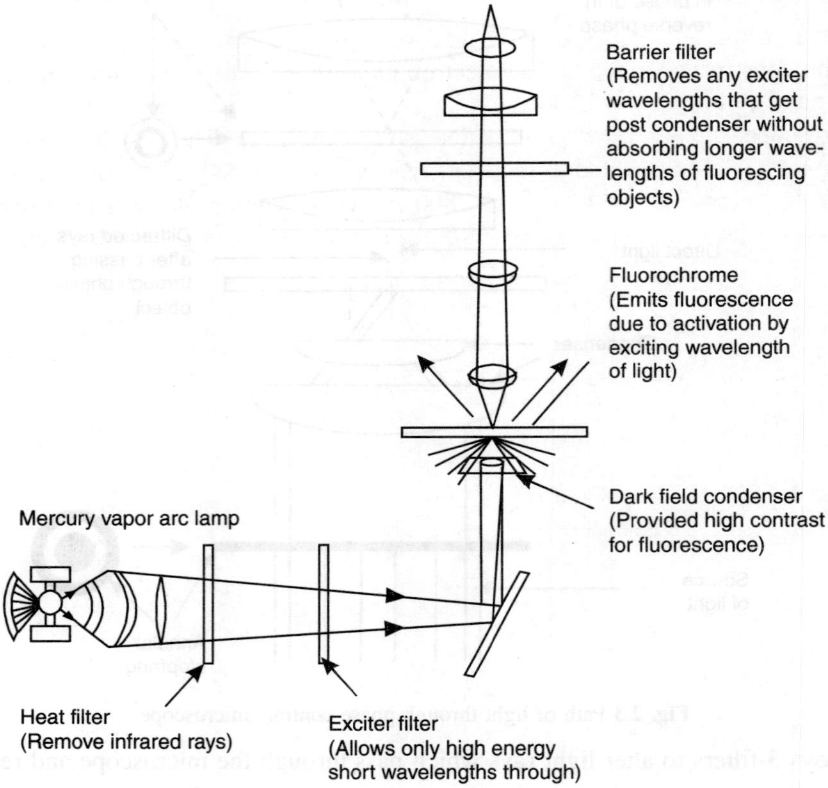

Fig. 2.4 Path of light through the fluorescence microscope

The Fluorescence Principle

To describe the reaction of fluorespar to UV rays, Stokes in 1852 used the word fluorescent. The atom absorbs a quantum of light and an electron is boosted to a higher energy level. When this displaced electron comes back to its original ground state, it emits a quantum of light. If this light is emitted only during the time of exposure or for a fraction of a second afterwards, it is known as *fluorescence*. If the emission of light continues for some seconds after the exciting light is cut off, it is called *phosphorescence*.

UV-light is used in fluorescence microscopy. The exciting light and resulting fluorescence rays are always of longer wavelength than exciting rays—this exactly coincides with Stakes' phenomena. As per this phenomenon, the light always loses energy when it passes through liquid so that the released light has to be of a longer wavelength. When the fluorescent dye fluorescein isothiocyanate is illuminated with blue light, green light is emitted. This happens due to the fact that light rays of shorter wavelength have more energy than those of longer wavelength.

Microbiological objects to be observed under a fluorescence microscope must be coated with certain special compounds which possess this quality of fluorescence. Such special compounds are called *fluorochromes*, *e.g.*, Auramine O, Acridine orange, Fluorescein, etc.

Electron Microscope

German scientists Knoll and Ruska in 1931 discovered electron microscopy. Von Borries and Ruska (1938) in Berlin constructed first practical electron microscope. The commercial instrument was first available in around 1940.

In this microscope, the source of illumination is electron beam. The principle and construction of electron microscope is easily related to those of light microscope. The range of wavelength of visible light used in light microscope is 4000 Å to 7800 Å, while with an electron microscope employing 60–80 KV electron, the wavelength is only 0.05 Å. The electron gun generates electron beam. These

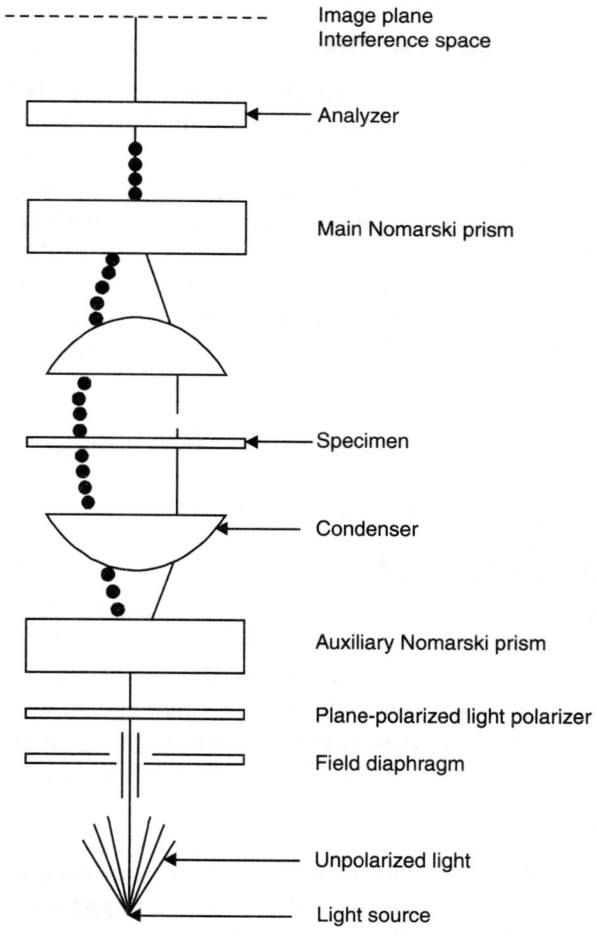

Fig. 2.5 Path of light through interference microscope

electrons are concentrated by other components of electron gun producing a fast moving narrow beam of electrons. Electrons are focussed by electromagnet layers. Electromagnet lens consists of wire encased in soft iron. When electric current is passed through the coil it generates an electromagnetic field through which electrons are focussed.

In total, there are three types of electromagnetic lenses. One is placed between the source of illumination and the specimen. This focusses the beam of electron on specimen which functions in a similar manner as that of light microscope. The other two lenses are on the opposite side of specimen, which magnify the image in similar fashion as objective and ocular in light microscope.

The final image in this microscope cannot be viewed directly as the electron rays are harmful to viewer's eyes. Therefore, they are either projected on the fluorescent screen through the airtight window or recorded on a photographic plate by a camera built into the instrument. Further, the entire electron microscope must be built in a vacuum or otherwise the electrons get scattered by virtue of colliding with air molecules and thus fail to get focussed.

Transmission Electron Microscope (TEM) and Scanning Electron Microscope (SEM) are two main types of electron microscope.

Transmission Electron Microscope (TEM)

In this, the electron source is commonly a tungsten filament at 30–150 kV potential. The electron beam passes through the center of ring like magnetic condenser and becomes converged on the specimen. After being transmitted through the specimen (hence the name "Transmission Electron Microscope" (TEM), the magnetic objective focuses the electron into a first (real) image of the object which is enlarged (2000 times). The magnetic projector lens then magnifies a portion of the first image producing magnification up to 2,40,000 or more.

The final enlarged image can be reviewed by striking a fluorescent screen which makes it visible. The image can also be thrown upon a photographic plate for permanent record. Portions of the photographs may be enlarged four to six times giving the picture in the range of two million times as large as the object.

$$\frac{\text{Magnifies 2 million times}}{\text{2 mm diameter}} \rightarrow \text{4 million max or 2.5 miles in diameter.}$$

It is also necessary to have specimen ultrathin.

Scanning Electron Microscope (SEM)

This microscope combines the mechanism of electron microscopy and television. It was invented by Knoll, Von Ardenne and Zworytein, and available commercially in early 1960s.

The specimen may be opaque and of any manageable thickness and size. In SEM, electrons are not transmitted through the very thin specimen from below but impinge on its surface from above. If the specimen is an electron conductor, it needs only to be held on an appropriate support. If it is non-conductor, it is allowed to dry, but if moist, freeze dried in liquid nitrogen; the specimen is then coated with metal vapour (gold) in vacuum.

The electrons originate at high energy (20,000 V) from a hot tungsten or lanthanum hexoboride cathode "gun". These electrons are sharply focussed, adjusted and narrowed by an arrangement of magnetic fields. Instead of forming abroad inverted cone rays, in SEM a needle sharp probe of about 5–10 mm in diameter is made. This primary probe in the form of a beam acts only as an exciter of image forming secondary electrons emerging from the surface of the specimen.

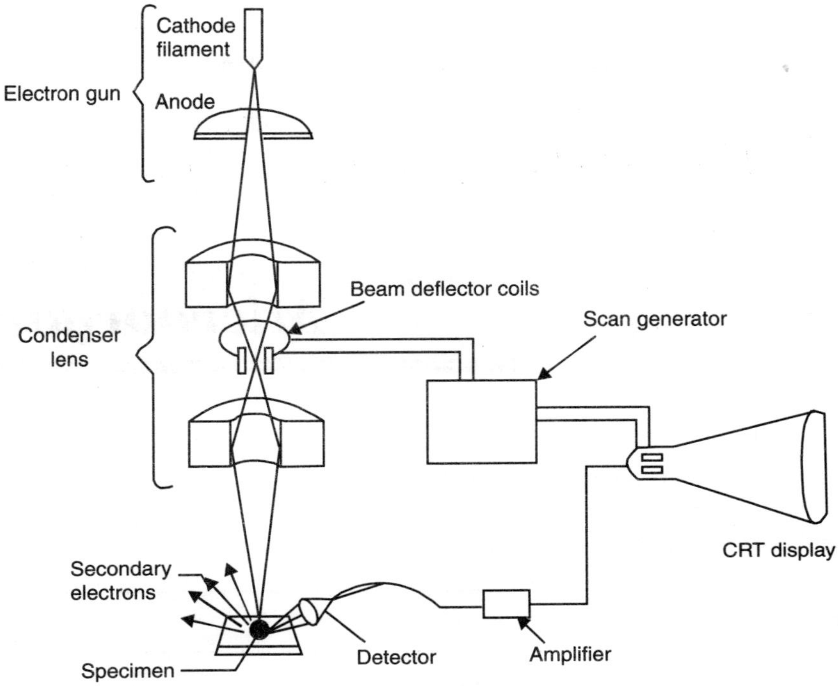

Fig. 2.6 The scanning electron microscope

The probe scans the specimen like that on a blank TV screen. The probe can impinge on depth and heights with equal speed and accuracy giving great depth of field and producing images with three dimensions. Images are elicited from wherever the probe strikes the metal coated areas of the specimen. Magnification is the ratio of final image to the diameter of area scanned.

Any of the secondary electrons with sufficient energy can emerge from the surface. Those that emerge not too far from the point of impact of the probe can be used to form an image. The useful secondary electrons are magnetically deflected to a collector or detector. Here they produce a signal that represents at any single moment, only 5–10 mm area or spot of impingement of the probe on the specimen. The successive signals from the collector are amplified and transmitted to a cathode ray *i.e.*, TV tube. The scanning beam and cathode rays are synchronized. The image seen by viewers eyes on TV screen is thus the sequence of signals representing in a raster pattern, the successive areas traversed by the primary probe beam. Exposure may range from a few seconds to one and half hours or more. The TV image may be photographed, video-taped or processed in motion on a computer.

UNIT 3 Microorganisms

Bacteria and viruses are grouped under prokaryotes. Viruses do not behave as living matter outside the host and do not have any specific cellular organization. Bacteria are typical prokaryotes and the structure and morphology of this prokaryote is discussed below.

BACTERIA

A bacterial cell consists of a compound membrane enclosing protoplasm. The protoplasm is a thick viscous fluid or almost jelly like colorless, transparent material. It contains high percentage of water and holds fine granules in suspension. The protoplasm contains various structures *i.e.*, nucleoid, vacuoles, mesosomes, ribosomes, polysaccharides, lipids, etc. Externally, bacterial cells may show Fimbriae and organs of locomotion like flagella.

Cell Membranes

The cell membrane consists of a thin cytoplasmic membrane, the cell wall proper and an outer slime layer or capsule.

Cytoplasmic membrane

This membrane appears as an interfacial thin film in young cells but becomes thicker in growing cells. This membrane comprises of about 10% of the total dry weight of the cell and it has 75% protein in it, 20 to 30% lipid and 2% carbohydrate. This protein consists of phosphoproteins as well as a number of enzymes and carriers. It is semipermeable and is responsible for gram and acid fast staining reactions. It possesses the property of selective permeability. In addition to this cell membrane, bacteria have internal cell membrane folding called *mesosomes*.

Cell wall

The cell wall is a rigid structure and is responsible for the form of the bacterial body. It behaves as a selectively permeable membrane and also serves as a protection to the cell. This does not stain easily and gets only lightly stained by certain basic stains. It comprises up to 20% of the dry weight of bacteria and represents the major cell component. Bacteria do not change shape of the protoplasm even when cells are kept in hypertonic solutions. The cell wall is composed of peptidoglycan which is composed of components, namely, N-acetyl glutamic acid and either L-lysine or mesodiamino pimelic acid. The rigid peptidoglycan layer is of 2 to 3 nm in thickness. It is located between the cytoplasmic membrane and the cell wall.

Detailed Structure of Cell Envelope

The cell envelope of bacteria is divided into two 75 Å unit membranes separated by 100 Å space called the *periplasmic space*. This space consists of a peptidoglycan layer. The outer membrane is called the *cell wall* and the inner membrane is called the *cytoplasmic membrane*.

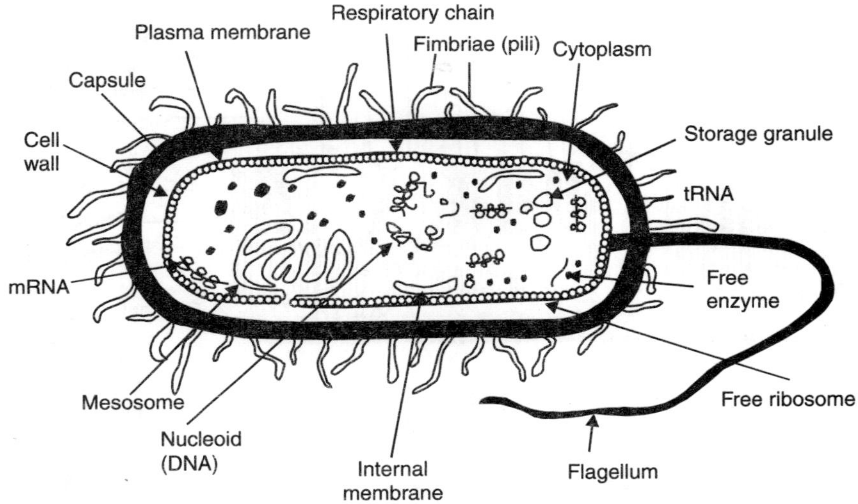

Fig. 3.1 A typical prokaryote

1. Inner membrane

It is also called cytoplasmic membrane. It is made up of proteins and lipids. Its major features are explained by the Fluid-Mosaic model of Singer and Nicholson (1972). According to them the

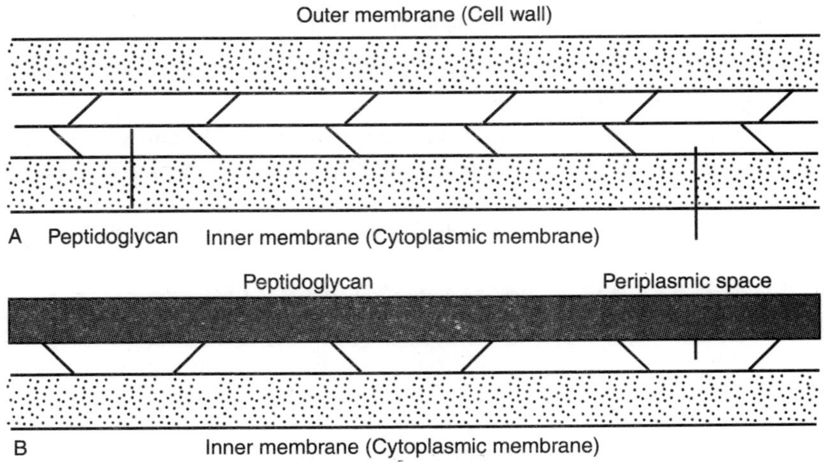

Fig. 3.2 (A) Gram-negative bacteria, (B) Gram-positive bacteria

cytoplasmic membrane is composed of continuous bilayer of phospholipid molecules in which globular proteins are embedded.

Biological membranes are considered to be quasifluid structures in which lipids and integral proteins are arranged in a mosaic manner. They proposed that the lipid protein interactions are hydrophobic.

The fluidity of the membrane is the result of this hydrophobic interaction. Phospholipids and most intrinsic proteins are amphipathic molecules, *i.e.*, both hydrophobic and hydrophilic groups can occur in the same molecule.

The globular proteins of the membrane are of 2 types, the extrinsic and intrinsic.

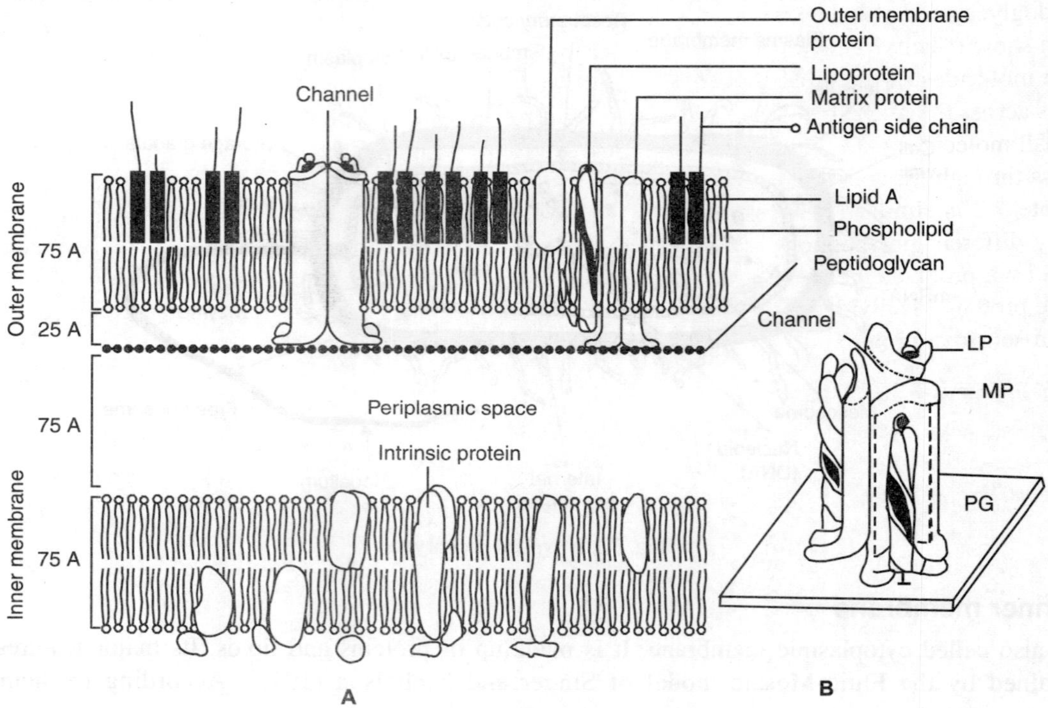

Fig. 3.3 Fluid-mosaic model of cytoplasmic membrane

The extrinsic or peripheral proteins are soluble and readily dissociate from the membrane.

They are entirely outside the lipid bilayer.

The internal or intrinsic (integral) proteins are relatively insoluble and dissociate with difficulty. Some intrinsic proteins may partially penetrate the lipid bilayer while others may penetrate right through the bilayer.

In such molecules the hydrophilic polar heads are exposed outside while the non-polar regions are embedded within.

The integral proteins are capable of diffusion through the lipid bilayer.

The phospholipids, when dispersed in water form a lipid bilayer wherein the polar head project into the aqueous phase and hydrophobic chains aggregate together to form a continuous bilayer.

2. Cell wall

The outer membrane or cell wall of gram negative forms is morphologically similar to the cytoplasmic membrane but contains less phospholipid and fewer proteins.

It contains a unique component called Lipopolysaccharide (LPS).

The cell wall consists of at least four different major protein components plus lipoprotein.

They are:

1. Matrix protein 1a and 1b which are strongly associated with peptidoglycan.
2. Protein 2.
3. Tol G protein (3a) and protein (3b) and the lipoprotein.

Matrix protein consists of two distinct peptides 1a and 1b. They are closely associated with the peptidoglycan and also the lipoproteins. They markedly show β-structure unlike many intrinsic proteins which show α-helix. They form hexagonal lattice structures covering 60% of the peptidoglycan layer. Three molecules of the protein form one hexagonal structure, through which passes a channel which passes across the outer membrane. They also form passive diffusion pores which allow rapid diffusion of small molecules like sugar, amino acids, etc. Nutrients with high molecular weight require receptors to pass through the membrane. Matrix proteins also function as receptors for phages.

Protein 2 is similar to matrix proteins and is associated with peptidoglycan. It represents a completely different polypeptide from the matrix proteins. It has been suggested that protein 2 may be coded by a prophage gene, and functions in surface exclusion of super-infecting phage.

The protein 3 is divided into 3a and 3b. 3a is identical to Tol G protein. This is heat modifiable and trypsin sensitive protein.

Lipopolysaccharides (LPS)

In gram negative bacteria, the outer membrane is covered by lipopolysaccharides. The LPS consists of a polysaccharide linked to lipid A. Lipid A has endotoxin properties and polysaccharide has antigenic property.

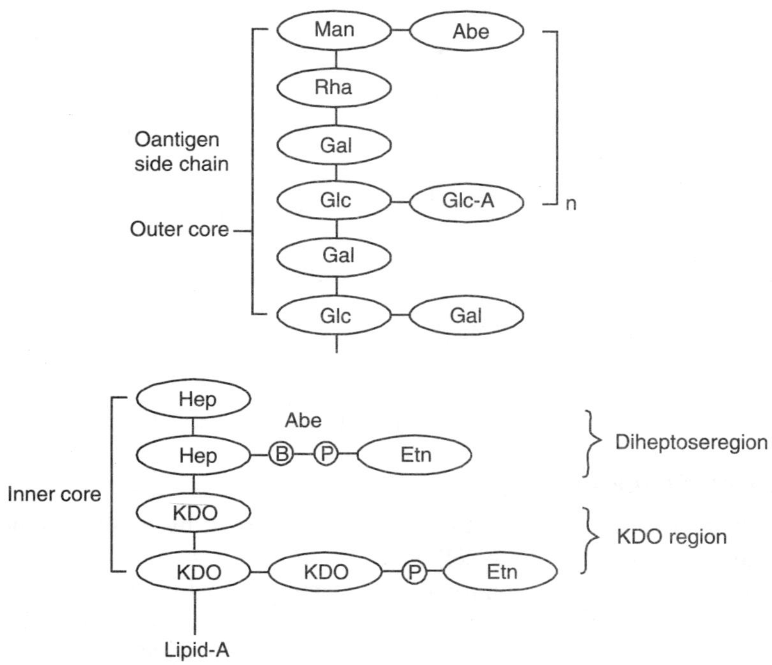

Fig. 3.4 Lipopolysaccharide molecule

Lipid A along with the first sugar molecule KDO of the polysaccharide plays an essential role in cell envelope.

Lipid A contains glucosamine, phosphate and fatty acid. The lipid A has a carbohydrate part which constitutes β-1, 6-D glucosamine disaccharide units. All the amino and hydroxyl groups of the disaccharide are substituted by polysaccharide chains or phosphates or pyrophosphates and fatty acids, which give lipid A into hydrophobic character. There are six long fatty acid chains, *e.g.*, lauric acids, myristic acid, palmitic acid etc. These chains give the fever inducing property to the LPS. The polysaccharide consist of 3 regions—the inner core, outer core and the O Antigen side chains. The rough strains of gram –ve bacteria completely lack these O Antigen side chains. Their LPS have only inner core and outer core. The inner core consists of the KDO region. These KDO, which is a 8-carbon L-ketosugar acid and heptose, which is 7-carbon heptose sugar caused L-glycero-D-mamoheptose are unique to bacterial cell wall. Smooth strains have O side chains which have antigenic properties and these side chains can be up to 40 in number.

Peptidoglycan

(Micro peptides, glycopeptides, mureins) are the structural elements of almost all bacterial cell walls. Halobacterium (bacteria which live in high saline concentration) do not have this layer. The cell walls of gram –ve bacteria have relatively low peptidoglycan content than gram +ve forms. There is a low degree of cross-linkage between the glycan strands of the peptidoglycan. In many gram negative bacteria the peptidoglycan is a monolayer or a bilayer. In *E. coli* the peptidoglycan consists of alternating residues of amino sugars attached to linear chains of polysaccharides. The amino sugars are namely N-acetyl glucosamine (NAG) and N-acetyl muramic acid (NAM) linked by β-1, 4 glycosidic bonds. Free tetrapeptide side chains are attached to the NAM residues. Some adjacent NAM residues are cross-linked by these chains.

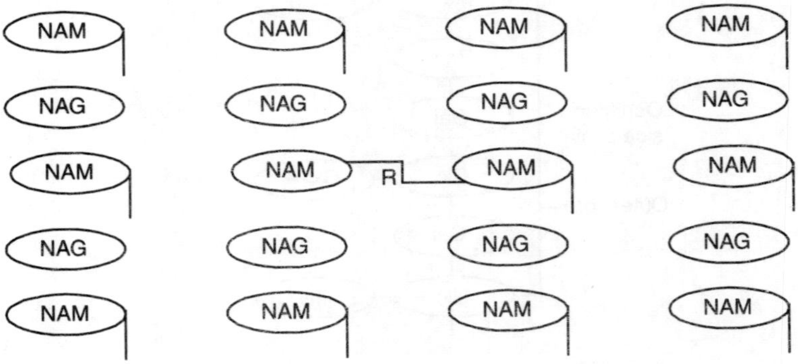

Fig. 3.5 Organization of peptidoglycan in *E. coli*

Gram Positive Cell Envelope

The cell envelope here consists of a peptidoglycan layer followed by a cytoplasmic membrane. The peptidoglycan make up to 40–90% of the cell wall.

Associated to the peptidoglycan are phosphorous containing acid polymers called *techoic acids* or *acidic polysaccharides* such as tuchuronic acids. In *Staphylococcus aureus*, the peptidoglycan is made up of linear glycan chains made up of alternating residues of NAM and NAG linked by β-1, 4 linkages. The NAM residue carries a short peptide chain of 4 amino acids namely L-alanine, D-glutamine,

L-lysine and D-alanine neighboring tetrapeptides and connected by a pentaglycine bridge. Because of the extensive cross-linking the cell wall has a rigid structural framework. The rest of the composition *i.e.*, of the cytoplasmic membrane is similar to gram –ve forms.

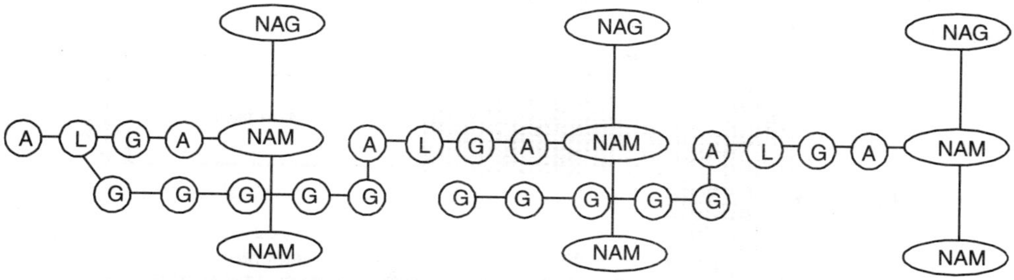

Fig. 3.6 Peptidoglycan arrangement in *Staphylococcus aureus*

Mesosome

A mesosome is formed by the invagination of the cytoplasmic membrane. It is more frequent in gram +ve bacteria.

The mesosome membranes may be coiled up *i.e.*, the lamellar form or as vesicles. The lamellar form becomes a part of the cell wall when cell wall is removed, whereas the vesicular type does not become a part but is released when cell wall is removed.

The exact function of mesosome is not clearly known. Mature mesosomes appear to be a link between the cytoplasmic membrane and nucleus.

Slime Layer and Capsule

Bacteria producing mucoid growths produce extracellular slimy or gelatinous polysaccharide material called the *slime layer* or *capsule.*

When this mucoid material remains adherent as a discrete convening layer of each cell, it is called capsule and when the substance is released freely as and when formed it is called the *slime* or *gum.*

The capsule is a part of the cell and slime is a secretion. Capsules are of definite shape, more or less of the same density and of a definite outline. Slime envelopes are amorphous and can be drawn out into manifold structures which are denser in the immediate vicinity of the cell.

In most cases, these gums are pure polysaccharides of glucose, galactose or levulose, whereas capsules are complex and show the presence of nitrogenous compounds and polysaccharides. It generally consists of (1) glycoproteins (2) protein polysaccharide complexes (3) a framework of polysaccharides with places filled with polypeptides.

Flagella

Prokaryotes have flagella in most cases as their locomotory organs. A flagella does not necessarily indicate locomotion but indicates the potential to move.

Flagella are long slender undulating organs. The thickness of flagella varies from species to species.

Flagella generally are composed of large amount of protein, little carbohydrates and sometimes lipids and are surrounded by a sheath of unknown chemical origin. When heated/acid treatment flagella decompose into individual protein molecules called *flagellins* (15,000–40,000 mw).

Flagella have H antigens and bacterial surfaces have O antigens. Flagella are seen to arise from interior of the cell.

The number and arrangement of flagella vary in different bacteria. They are constant in each species. They may be polar or lateral. The basal end of the flagellum shows a complex structure with a basal body, a hook and a filament. The basal body has 4 rings — M-ring, S-ring, P-ring and L-ring and a rod is present passing through them. The basal body is followed by the hook region which is curved or 'L' shaped followed by a filament.

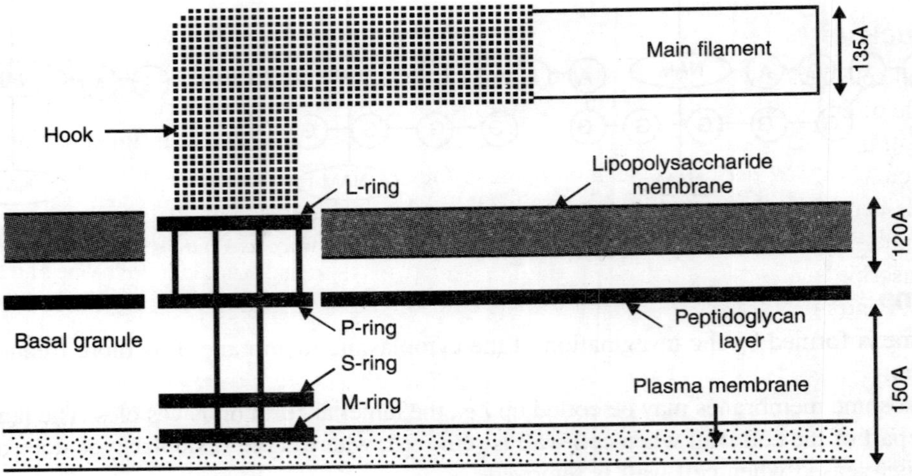

Fig. 3.7 Base of flagellum of a gram negative bacterium

Fimbriae or Pili

The term 'Fimbrial' was proposed by Duguid (1955) to designate certain structures present on the external surface of bacteria as external appendages on some bacteria. They are also called *pili*.

They measure from 0.1 to 4 m. They are not organs of locomotion.

Depending upon the dimensions, number, distribution, adhesive property and genetic control they have been divided into 1, 2, 3, 4, 5 and 6 fimbriae.

Type-I adhesive property to plants, molds, animal cells, etc. sex pili/fimbriae are involved in conjugation and are also called *F pilus*.

Mitochondria

A prokaryotic cell does not show any such organelle but they are able to perform the same chemical reactions with similar enzymes arranged on the respiratory membranes. The concentration of these enzymes will increase with decrease in cell size.

Vacuoles

They are cavities in the cytoplasm that contain fluid cell sap. Gas vacuoles have also been identified. They seem to help in storing O_2 in O_2 short conditions.

Cytosol

The cytosol consists of soluble materials and other components of the cell namely lipids, glycogen granules, metachromatic granules, pigments, free ribosomes, enzymes, tRNA, mRNA and the nucleoid.

Lipids are found as fat bodies or globules in the cells. They may be present in one single large globule or may be scattered and small. They store carbon atoms.

Glycogen granules function as reserve food stores when cells are inactive and are reused when the cells are rapidly growing.

Metachromatic/volutin granules are present only in growing cells and later get incorporated into vacuoles. They are acidic in nature as they get stained by basic dyes. They serve as reserve food material—phosphate storage structures. They are linked with energy metabolism and also involved in nucleic acid synthesis and cell division.

The Nucleus

Bacterial cells lack a well organized nucleus. A prokaryotic nuclear material contains a single very long molecule of DNA/RNA. Some viruses contain RNA as their genetic material. Both single-stranded and double-stranded DNA and RNA have been isolated in prokaryotes. Certain independent, self-replicating, circular molecules of DNA have been found in prokaryotes. They are called plasmids. These plasmids usually contain certain markers of drug resistance, etc.

The nuclear material does not form any chromosomes but there is the formation of chromatin bodies. The transcription and translation occur simultaneously (as there is no nuclear envelope and the ribosomes, the tRNA and mRNA are freely present in the cytosol) with the help of 70s ribosomes which are typical for prokaryotes.

Spore Formation

Prokaryotes, especially bacteria produce endospores in unfavorable conditions. These spores are very strong and are more resistant to extreme environmental conditions like cold, heat, desiccation, etc. Spore formation is limited to entirely two genera of rod-shaped bacteria namely *Bacillus* and *Clostridium*.

Spores may be spherical, ellipsoidal or cylindrical in shape. They may be either terminal, sub-terminal or central in their positions. The spores contain a thick surface. The outer surface is called *exine* and the inner surface is called *intine*. Following these outer coats, there are 3 layers namely, two outer membranes, the cortex and the inner membrane. All of them surround a central core or protoplast.

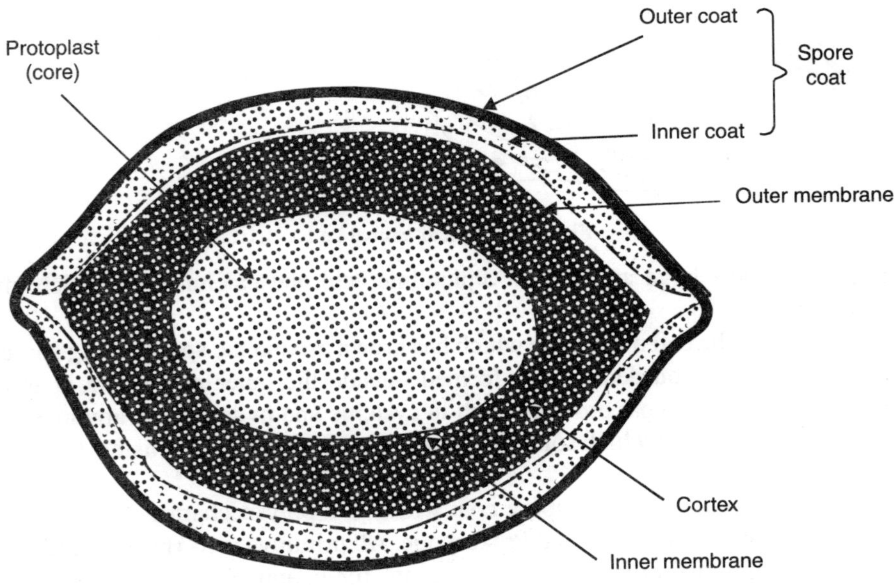

Fig. 3.8 Section through spore of *Bacillus megatherium*

The inner membrane is the unit membrane which encloses the spore cytoplasm and nuclear material.

The cortex is multilayered and contains many layers of peptidoglycan. The outer layer of peptidoglycan is loosely cross-linked and is hydrolyzed during germination. The inner layer is tightly cross-linked and remains during germination. The loose cortical peptidoglycan consists of 3 sub-units:

(*a*) Alanine sub-unit
(*b*) Tetrapeptide sub-unit
(*c*) A muramic lactum sub-unit.

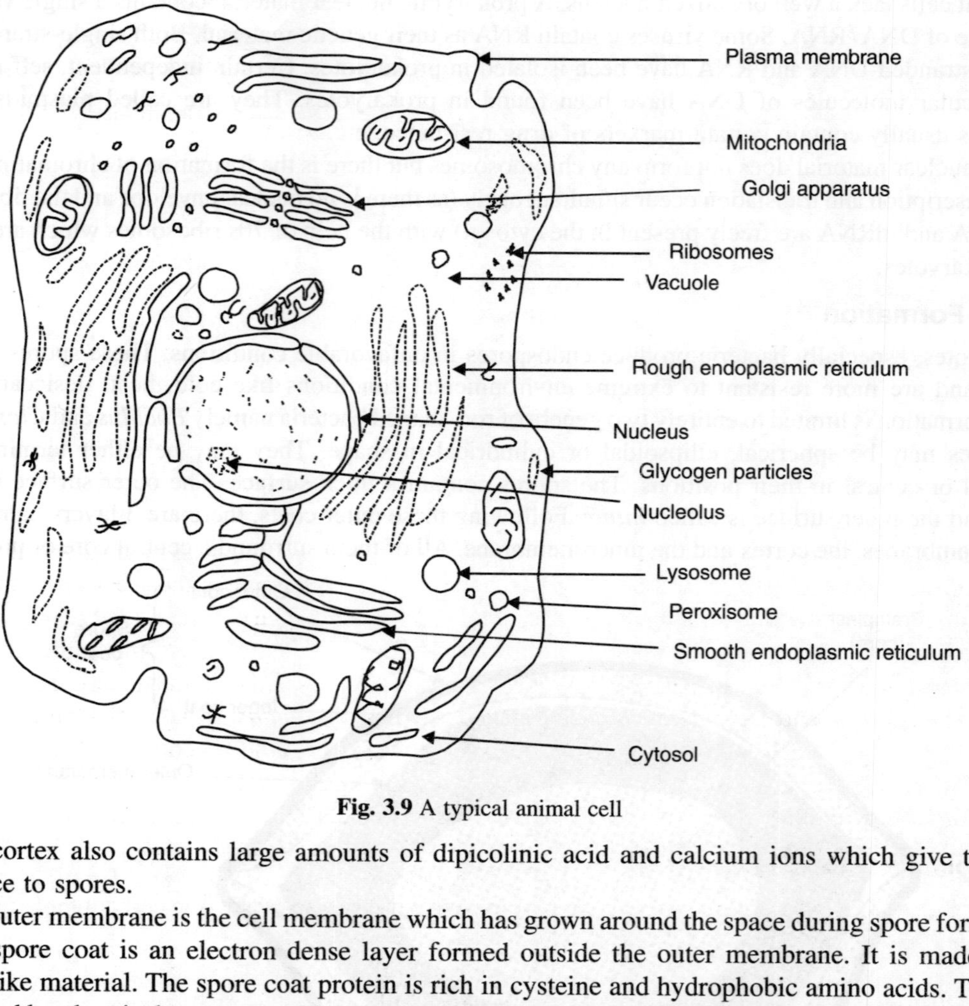

Fig. 3.9 A typical animal cell

The cortex also contains large amounts of dipicolinic acid and calcium ions which give the heat resistance to spores.

The outer membrane is the cell membrane which has grown around the space during spore formation.

The spore coat is an electron dense layer formed outside the outer membrane. It is made up of keratin-like material. The spore coat protein is rich in cysteine and hydrophobic amino acids. They are unaffected by chemical treatment which solubilize most proteins and are responsible for their resistance against chemicals.

An exosporium is formed around the spore coat which is a polysaccharide protein membrane.

Cells may be classified as Eukaryotic and Prokaryotic. All primitive cells (bacteria etc.,) are included into prokaryotes. Cells of higher organisms (yeasts, plants, fungi, animals) are grouped into eukaryotes. The prokaryotes and eukaryotes differ from each other in various aspects like size, morphology, cell

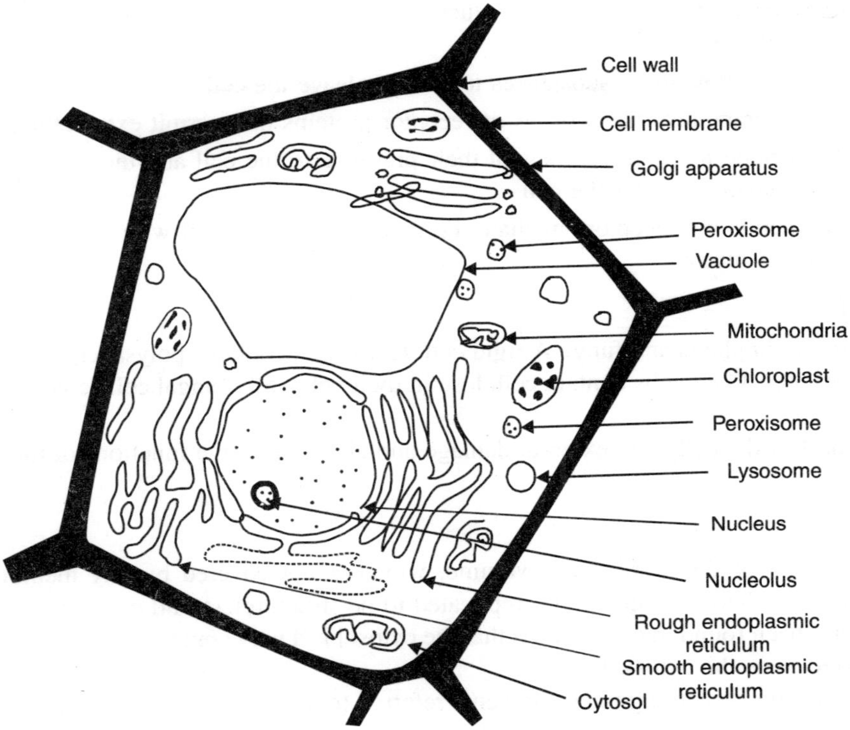

Fig. 3.10 A plant cell

wall, nucleic acid organization etc. To have a better view of the structure and functions of eukaryotic and prokaryotic cells and their organelles, we have to study the fine structure of these cells in detail. The above picture shows "the ultra structure of eukaryotic animal cell". Eukaryotic cell is basically marked by the presence of a clearly defined nucleus within an envelope. This is the most important difference between prokaryotes and eukaryotes. The only difference between a plant cell and an animal cell is the presence of a cell wall and also photosynthetic organelles called *chloroplasts*. Plant cells have rigid cell walls made up of cellulose and also have chloroplasts which perform the function of photosynthesis.

The eukaryotic cells are 100 to 10,000 times larger than prokaryotes.

The morphology and functions of various organelles of a eukaryotic cell are discussed below:

1. The plasma membrane

The plasma membrane is a semipermeable membrane and serves to sequester the cell contents from the environment.

It plays an important role in the interaction of the cell with its surroundings.

It acts as a selective barrier to the outside small, lipid soluble molecules like oxygen, carbon dioxide, ethanol and urea which are able to diffuse into or out of the cells.

Large molecules like glucose cannot simply move across the cell membrane. Hence, there are certain transport proteins which help in this process of transport.

The membrane is composed of lipids and proteins. The lipids are arranged in 2 layers with proteins intercalating in between them. These proteins may move and change position in the membrane thus making its structure a transient one.

There are transport proteins available for glucose like molecules to move in and out of the cell and ion-channels for both cations and anions.

For example, Ca^{2+} allow these substances to enter or leave the cell.

The plasma membrane also has a series of receptor proteins that permit extracellular molecules.

For example, peptide hormones, to signal their presence to the cell and thereby alter intracellular chemistry without having to enter the cell.

Other plasma membrane receptors mediate the interaction of the cell with its neighboring cells or with the extracellular matrix.

2. Cell wall

Cell wall is present in plants and fungi. A rigid cell wall which contains polysaccharides is the characteristic of eukaryotes. Plant cells contain cellulose in their cell wall. Fungal cells contain chitin in their cell walls.

Cell walls protect the cell from external damage and also help in the functions of the plasma membrane.

3. Cytosol

Approximately 50% of the intracellular volume, which is not covered by any membrane is called cytosol/the cell sap. Cytosol is generally implicated to be structureless and contain only soluble molecules. But it has been found in certain cells that the cytosol is a very complex network of protein fibres which is referred to as the cytoskeleton.

This includes actin-containing microfilaments referred to as "stress-fibres" and the tubulin containing microtubules. These fibers play an important role in the maintenance of cell shape and mobility. They interact with specialized proteins located in the plasma membrane, for example, and thus serve as true 'skeletal elements'. Actin fibres are clearly visible in microvilli of the intestinal epithelium.

4. The endoplasmic reticulum

Eukaryotic cells contain a dense interlacing membrane called *endoplasmic reticulum.* They seem to be continuous in parts of the cell to be continuous with the nuclear membrane. The ER have been classified into smooth ER and rough ER based upon their morphology. The ER which have 'ribosomes' attached to their surface are called *rough endoplasmic reticulum* and those which do not have ribosomes are called *smooth* ER.

The ribosome attached to the ER are involved in the synthesis of specialized proteins which are formed in the cytoplasm and are attached with certain peptides (signal peptides) on the ER.

It is important to note that all proteins are synthesized on free ribosomes.

Once initiation occurs, the ribosomes get attached to the ER to produce the proteins.

The proteins include secretory proteins and proteins destined to take up residence in other organelles (except nucleus, peroxisomes, mitochondria) and the plasma membrane.

Thus, cells with actively secreting proteins have high RER content.

The specialized proteins often have carbohydrate molecules covalently attached to them (glycosylation) for targeting. This process starts in the RER. This regulation or modification after translation is called *post translational modification.*

The SER has several roles. It is involved in steroid hormone biosynthesis, phospholipid biosynthesis and foreign particle metabolism and also drug metabolism.

5. The golgi apparatus

The Golgi apparatus is structurally and functionally very similar to the ER.

The Golgi consists of a serves of plate-like membranes which are interconnected. It also has certain membrane vesicles. The functional relationship between these two cytoplasmic membranous structures is that protein sequestered in the inner of the RER are packaged into membranous vesicles. These membranous vesicles bud off from the RER and then attach and fuse to the Golgi apparatus. Here the glycosylated proteins undergo further glycosylation and other types of post transcriptional modifications. In addition, the proteins are all sorted. These sorted proteins may remain in the ER, or Golgi or included into the lumen of vesicles that join with plasma membrane to release their contents. Some vesicles budding from Golgi may form lysosomes.

Thus, the specialized function of ER and Golgi is to jointly pack and target the proteins.

6. Lysosomes

Lysosomes are membrane structures that contain, in their lumen, a range of hydrolytic enzymes. All these enzymes are most efficient as catalysts at acid pH values, and hence are called Acid Hydrolases.

The lysosome membrane contains a proton pumping ATPase, which keeps the lumen in acidic condition (pH 5.5). The lysosome hydrolyses external particles taken up by endocytosis or phagocytosis. They are involved in the breakdown of endogenous structures. They are thus often called as 'suicide bags' of the cell.

7. Peroxisomes

They resemble the lysosomes in their appearance but have different function and biogenesis.

They do not arise from ER but they arise from preexisting peroxisomes like mitochondria. Peroxisomes are found in all eukaryotic cells and range from 0.5 to 1.5 mm in size and are bound by a single membrane. They contain catalases and peroxidases. Peroxisomes also carry out B-oxidation of fatty acids.

Ribosomes structure

In prokaryotes the ribosomes are smaller in size and are named 70s ribosomes and in eukaryotes they are called 80s based on their sedimentation coefficient.

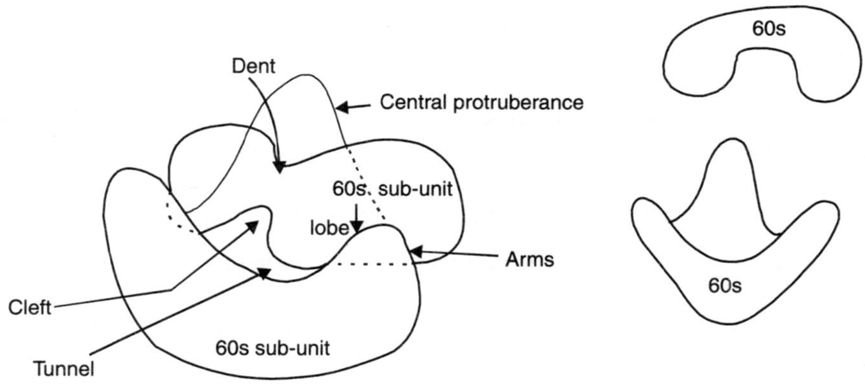

Fig. 3.11

Ribosomes are involved in protein synthesis by translating the mRNA. The basic structure of ribosomes is given below. A ribosome consists of 2 sub-units 70s–30s and 50s and 80s–40s and 60s. According to Stoffler and Wittmamis model, the ribosome structure is as follows:

The 30s sub-unit is a bipartite structure. A transverse hollow or left divides the 30s sub-unit into two parts, a smaller 'head' and a larger body. The two lobes protenule unequally and to different extents. A shallow dent may be present on the dorsal surface.

The 50s sub-unit is associated to 30s sub-unit to form 70s ribosome. A central protruberance is most prominent, and lateral protruberance are also seen. It is similar in structure to an armchair.

When the two sub-units assemble, the 30s sub-unit approaches the larger 50s sub-unit with its dent facing the vaulted seat of the 50s ribosome forming a tunnel through which mRNA passes.

8. Mitochondria

In common with the nucleus, mitochondria have a double membrane. They are tubular structures of about 0.1–0.5 mm × 1.2 mm. The inner membrane is extensively invaginated and these infoldings are referred to as existal.

Mitochondria replicates independently of the nucleus. They also contain their own DNA and ribosomes. The DNA encodes some of the proteins found in the mitochondria and also tRNA and mRNA of the mitochondria.

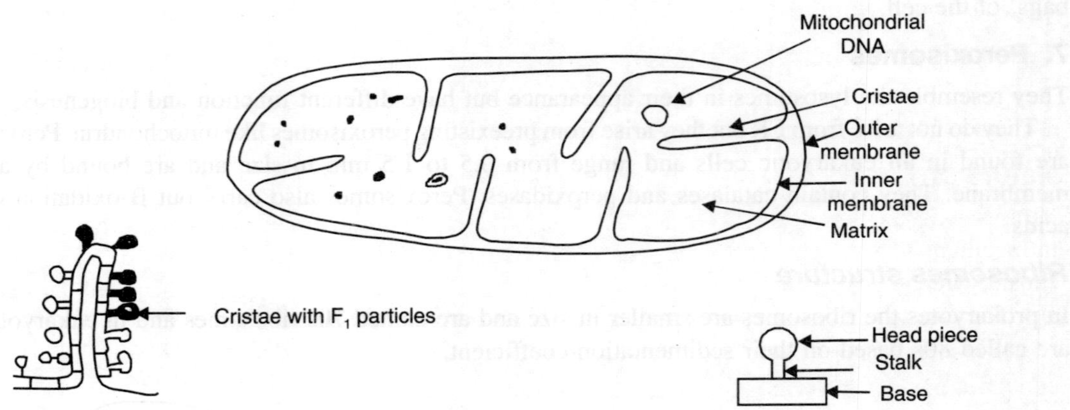

Fig. 3.12

Mitochondria acts as the energy producing, oxidative machines of the cells. The enzymes of an important metabolic pathway of the cell, the TCA, are all found in the mitochondrial matrix.

Electron transport system is also located here. B-oxidation of fatty acids is also done here.

Eukaryotic cells have definite advantage in possessing the above organelles.

(*a*) Various metabolic pathways are efficiently conducted by interacting enzyme systems present in the cell.

(*b*) Organelles separate enzymes which perform opposing reactions in different compartments thus controlling the reactions.

(*c*) Substances can be stored at single concentrations and can be released when required.

(*d*) Organelles can be maintained by indisequilibrium with the cell.

The Nucleus

There is clear distinctly marked enveloped nucleus present in eukaryotes. The nucleus is enveloped by a nuclear envelope which is continuous with the E.R.

The nucleus contains the genetic material in the form of chromatin. The chromatin, during cell division organizes itself into structures called *chromosomes*. All eukaryotes have DNA as their genetic material.

The transcription step of protein synthesis occurs in the DNA and so the mRNA informed in the nucleus out of which it is transported and then translated in the cytoplasm by 80s ribosomes (typical for eukaryotes).

VIRUSES

They are defined on the basis of genetic material and the symmetry or shape of the virus. Based upon genetic material viruses are divided into DNA viruses and RNA viruses. The DNA viruses are placed in the sub-phy—Deoxyvira and RNA viruses are placed in the sub-phy—Ribouia. Based upon the symmetry they are either helical, icosahedral or complex.

Some examples of DNA viruses are: (*a*) Pox virus, (*b*) Adeno virus, (*c*) Parvo virus and (*d*) Herpes virus etc.

Some examples of RNA viruses are: (*a*) Retrovirus, (*b*) Picorna virus, (*c*) Rhabdovirus, etc.

Viruses have either DNA or RNA covered by a capsid (dipoprotein). Certain virus infect bacteria and are called bacteriophages.

Nature of Virus

Viruses exist in two different stages, (*a*) the extracellular infectious stage called the *virion* and the intracellular viral stage which consists of the nucleic acids. The virion consists of a protein coat or capsid, which encloses a genome of either RNA or DNA. The entire structure is called the *nucleocapsid*. The capsid may be a polyhedron, a helix or a combination of both.

Their size ranges from 20 to 300 nm. Viruses have very simple structures. The simplest viruses contain a genetic material surrounded by a protein capsid.

More complex viruses contain lipids and carbohydrates in addition to proteins and nucleic acids and are called *enveloped viruses* as they are covered by a membranous envelope derived from the host cell. Enveloped viruses may be made up of general capsomeres.

The envelope protects the virus and also serves for transmission from one cell to another. The envelope consists of a lipid bilayer and proteins with special functions.

Glycoproteins and matrix proteins are the main proteins present. Glycoproteins have hydrophobic ends fixed in the lipid bilayer and hydrophilic glycosylated ends towards the surface.

The outer surface of virions often contain spikes which are made up of glycoproteins.

The envelope and capsid proteins are specified by viral genes whereas the lipid and carbohydrates are derived from the host.

Viruses do not have any cytoplasm and hence any cytoplasmic organelles. They utilize the host machinery (ribosomes) for protein synthesis. Antigens are present in the spike like capsomere. Viruses do not have any independent metabolism and growth or division.

1. Polyhedral/Icosahedral symmetry

Viruses with polyhedral symmetry may be with naked capsid, *e.g.*, Herpes simplex virus or Enveloped, *e.g.*, Polio virus or adenovirus.

Watson and Crick have shown that polyhedral capsids can exist in tetrahedral, octahedral and icosahedral shape but the icosahedron is the most stable symmetry to pack and bond the subunits.

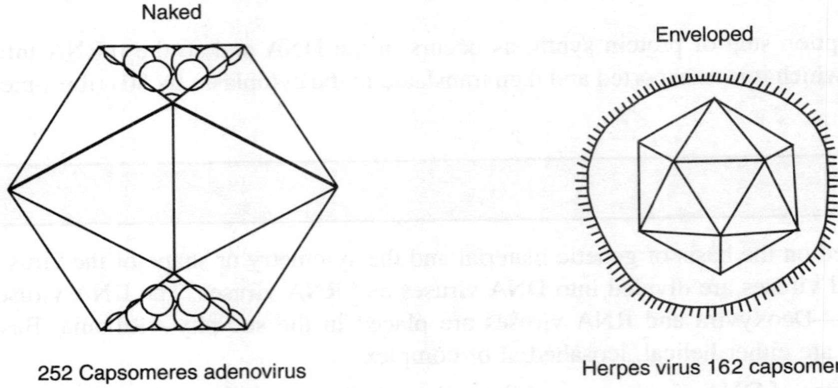

Fig. 3.13

Adenovirus is a typical icosahedral virus. But it does not have an envelope. It is approximately 80 nm in size with dsDNA as the genetic material. It contains 252 capsomeres which are roughly spherical. From 12 points on the capsid surface arise spike-like structures made up of fibre polypeptides. The 3 surface antigens of the virus are distributed in the icosahedral capsid and the spike capsomere. The antigen-A forms the faces and the edges of the icosahedron, the B and C antigens are present in the spike like capsomere—B antigen on the spherical and C on the spike part.

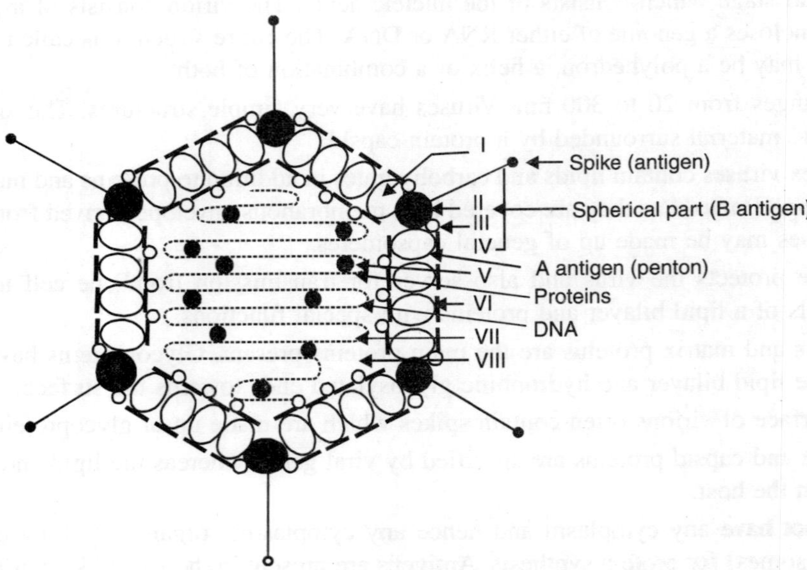

Fig. 3.14 Structure of adenovirus

The virus particle consists of at least 14 protein species. The dsDNA is normally linear but sometimes may circularise. The circular form is more infectious. The precursor mRNA molecules formed have sequences complementary to the noncontagious segments of DNA.

Therefore, splicing/ligation occurs during mRNA formation with elimination of intervening precursor segments. This virus attaches to susceptible cells by apical fibres (spikes) and is then taken into the cells, losing both pentons and fibres in the process. It now reaches the nucleus losing other proteins at the nuclear membrane. Inside the nucleus, the replication as described above occurs.

2. Helical symmetry

The helical symmetry is present is viruses like tobacco mosaic virus (TMV) and bacteriophage (MB). TMV is a typical helical symmetry virus. In 1936, Stanley isolated TMV from the sap of infected tobacco plants in the crystalline form. Here the helical capsids are made up of monomers arranged around a single rotational axis. The monomers curve into a helix because they are thicker at one end than the other. The helical capsids may be naked (TMV) or enveloped (influenza virus).

The TMV is a rod-shaped virus about 300 nm long and 15 to 18 nm in diameter. Studies have shown that the virus consists of a protein tube with a lumen of 20 Å enclosing a single strand of helically coiled RNA. The tube is made up of a number of identical sub-units (monomers) of protein arranged in a helical manner.

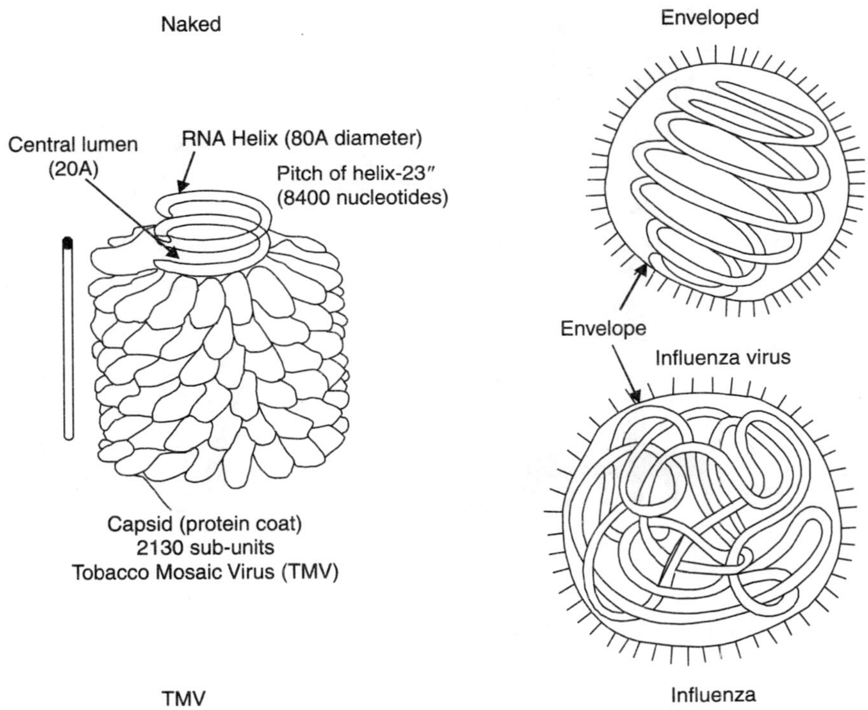

Fig. 3.15

The RNA is a single-stranded molecule coiled into a helix 80 Å in diameter. It follows the pitch of the protein helix. Each turn of RNA helix contains 49 nucleotides. RNA is infective by itself but being susceptible to the action of enzymes is less infective without the protein coat which serves as a protection.

3. Complex symmetry

Complex viruses are divided into 2 groups: those without identifiable capsids, and those with capsids to which are attached additional structures. Vaccinia viruses are examples of virus without definite capsid. Here, the nucleic acid is covered by several coats.

Bacteriophages of the T-even series (T2, T4, T6) are examples of complex viruses with capsid and attached structures.

The T4 phage infects *E.coli.* It is a tadpole-shaped virus with head and tail regions.

The head capsid is made up of about 2000 sub-units. It encloses or packages a 500 μm, long double stranded DNA. The head capsid consists of two 10-faceted equatorial bands with a pyramidal vertex at their end and the tail is helical in symmetry and hence the phage shows a combination of both icosahedral and helical symmetry (binal symmetry).

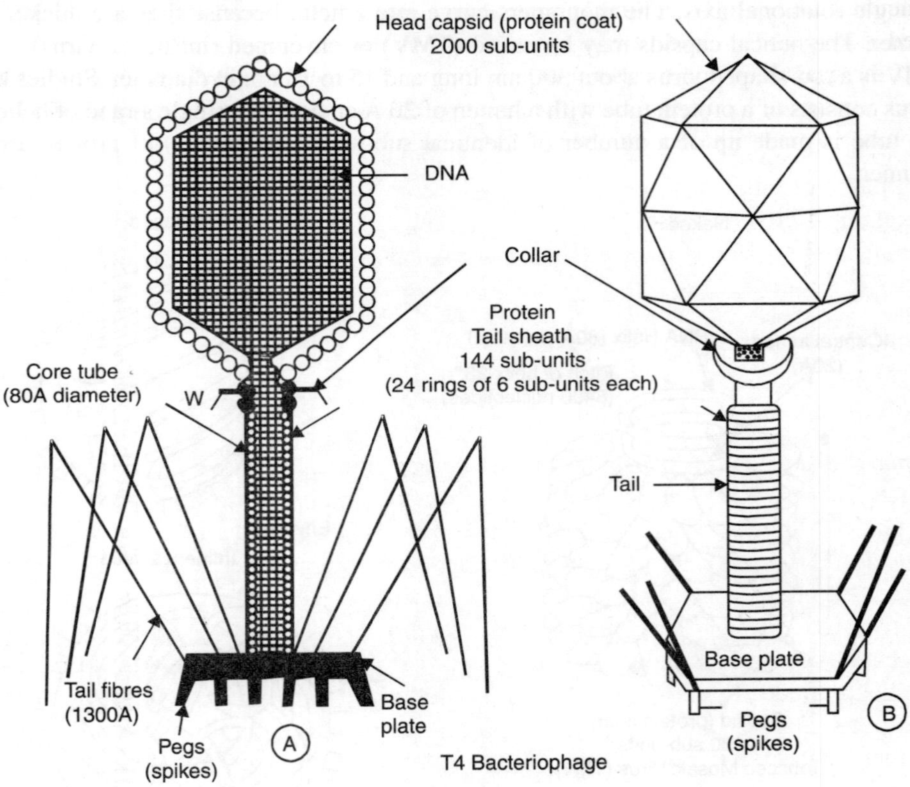

(A) Sectional view, (B) External view, Only the basal parts of the tail fibres have been shown, W = Whiskers.

Fig. 3.16 Structure of the T4 bacteriophage

The tail consists of a core tube 80 Å in diameter, through which the DNA passes out, surrounded by a protein tail sheath. This sheath consists of 144 sub-units arranged in 24 rings of 6 sub-units each. The sheath is connected to a thin disc, called the *collar*, at the upper end and to a basal plate at the lower end.

The base plate is hexagonal and has a pin or spike at each corner. The six corner also contains a tail fibre which is 1300 Å long. This serves for attachment of the phage to the host cell. Complex virus may show some valuation in their structure. In lambda coliphage the head is a regular icosahedron unlike T4 where it is a prolate icosahedron. In T3 and T7, the tail is short and stubby and ends in a baseplate like structure T1 and T5 have no tail sheath and their fibres are rudimentary. Pox viruses are brick shaped, void or flattened cylindrical, Rabies virus is bullet shaped.

MICROBIAL GROWTH

Population growth is studied by analysing the growth curve of a microbial culture. When microorganisms are cultivated in liquid medium, they are usually grown in a batch culture or closed system *i.e.*, they are incubated in a closed culture vessel with a single batch of medium. Because no fresh medium is provided during the course of incubation, nutrient levels decline and concentrations of wastes increase. If the growth of microorganisms reproducing by binary fission is plotted as the logarithm of cell number versus the incubation time the resulting curve is composed of four distinct phases. These different phases of microbial growth curve are as follows:

1. Lag phase or initial slow phase
2. Accelerated growth phase
3. Exponential phase or log phase
4. Phase of decrease in growth rate
5. Stationary phase
6. Death phase
7. The declining phase
8. The survival phase.

1. *Lag phase or initial slow phase:* When microorganisms are introduced into fresh culture medium, no immediate increase in cell numbers or mass is usually seen, and therefore this period is referred to as the lag phase. Although cell division does not occur right away and there is no net increase in cell mass, new cell mass components are being synthesized before growth can begin. The medium may be different from the one in which the organism was growing previously. In this case, new enzymes would be needed to utilize different nutrients. Possibly the organisms have been injured and require time to recover. Whatever the causes, eventually the cells recover, replicate their DNA, begin to increase in mass and finally divide. The lag phase varies considerably in length with the condition of the microorganisms and the nature of the medium. This phase may be quite long if the inoculum is from an old culture or that has been refrigerated. Inoculation of a culture into a chemically different medium also results in a long lag phase. On the other hand, when a young, vigorously growing exponential phase culture is transferred to fresh medium of the same composition, the lag phase will be short or absent.

2. *Accelerated growth phase:* It is an intermediary phase between lag phase and log phase. As the cells mature physiologically at the end of lag phase, the cells start dividing. As all the cells don't complete lag phase at same time, all these cells don't divide simultaneously. Due to this reason, the number of cells dividing will increase gradually in the accelerated growth phase.

3. *Exponential phase or log phase:* The cell number increases logarithmically and the cells divide at the maximum rate in the log phase. As a function of the exponent (2^1, 2^2, 2^3, 2^4, ...) the cell number during this period increases. So, the log of the number of cells plotted against time will show a straight line at this phase.

Cells are smaller in size as they are constantly dividing during this phase. This phase is the most suitable period for studying biochemical and physiological properties of cells because the cells are metabolically more active during this phase. This phase is also extremely sensitive to the change in environmental factors like pH, temperature, aeration, concentration of nutrients, etc.

4. *Phase of decrease in growth rate:* As the bacterial cells increase during log phase, it results in gradual decrease in nutrients quantity and also there is accumulation of toxic waste produced by the bacterial metabolism. However, these factors will reduce the cell division and the cells do not multiply with same rate as in the log phase. In between log phase and stationary phase this phase is transition phase.

5. *Stationary phase:* After the due course of time the growth of the organism stops and the growth curve becomes horizontal. This stationary phase is usually attained by bacteria at a population around 10^9 cells per ml.

Other microorganisms normally do not reach such high populations; protozoans and algal cultures often have a maximum concentration of 10^6 cells/ml. The final population size depends upon the type of organisms, growth and the nutrients available. In this phase the total number of viable organisms remain constant. This may result due to a balance between cell division and cell death or the population may just stop dividing but may be metabolically active.

Microbial populations may enter into a stationary phase due to various reasons. One important factor is nutrient limitation. If a nutrient is severely depleted, the microbial growth will slow-down. Aerobes are often limited due to oxygen availability because it is not very soluble and may be used up quickly and only surface of the culture will have sufficient oxygen concentration to support growth. The cells beneath the surface will not grow unless the culture is shaken or aerated in some way. Population growth may also cease due to accumulation of toxic waste products. For example, streptococci can produce so much of lactic acid that growth is inhibited. They can also enter into stationary phase due to depletion of sugar in the medium.

6. *Death phase:* Harmful environmental changes like nutrient depletion and build up of toxic wastes lead to the decline in the number of viable cells, characteristic of the death phase. When death of a microbial population, live its growth during the exponential phase, is usually logarithmic, *i.e.*, a constant proportion of cells die every hour. But the turbidity of the culture remains the same as the cells fail to lyse after dying. The only way to determine that a bacterial cell is dead is by incubating it in a fresh culture medium and, if it does not grow and reproduce it is assumed to be dead.

Although most of the microbial population usually dies in a logarithmic fashion, the death rate may reduce after the population has been drastically reduced. This is due to the extended survival of particularly resistant cells.

7. *The declining phase:* The declining phase is characterized by an exponential decline in the number of viable cells; a cell during this phase is considered to have died or to have become inviable when it is no longer capable of multiplying. Bacterial death rate is not uniform like that of log phase where the increase in number is uniformly exponential, because few species of bacteria like *Neisseria gonorrhoeae*, die rapidly as the cells are very susceptible to autolysis.

8. *The survival phase:* Rate of the cell death is balanced by the rate of cell division during survival phase. If the nutrient levels or the environmental conditions are changed, unbalanced growth results because the rates of synthesis of cell components vary in relation to one another until a new balanced state is reached. Thus the survival of the cell depends on the type of organism and the environmental conditions, some microbes perish within a few days while some microbes might remain viable for months and even for years. The cells first construct new ribosomes in order to enhance the protein

synthesis. This is then followed by an increase in protein and DNA synthesis. Finally there is a rise in the reproductive rate. Unbalanced growth also results when a bacterial population is transferred to a poor medium previously, in the rich medium the organism may have been able to obtain many cell components directly from the medium but when shifted to a poor medium they need time to make the enzymes required for the biosynthesis of unavailable nutrients. Consequently cell division and DNA replication continue after the shift-down, but net protein and RNA synthesis slows down, cells become smaller and reorganize themselves metabolically until they are able to grow once more and balanced growth is resumed.

Synchronous Growth

Bacterial populations growth patterns do not conclude about growth patterns of individual cell as the distribution of cell size and cell age in most bacterial cultures is at random. Information about the growth behavior of individual bacteria can be obtained by the study of the synchronous cultures. These cultures are composed of cells *i.e.*, are all at the same stage of cell cycle. Any measurement studies made in such cultures are equivalent to studies made on individual cells.

Synchronous cultures are obtained by various techniques. Synchrony can be induced by manipulations of environmental conditions, usually cyclic. It is accomplished by repetitive shifts of temperature or by furnishing fresh nutrients to cultures that have first entered stationary phase. Synchronous population can be obtained from a random population by physical separation of cells that are at the same stage of the cell cycle. This can be achieved by differential consideration or by centrifugation. This is the selection method. The selection for obtaining synchronous cultures is Helmsteller-cummings variance. This is based on the principle that some bacteria have the capacity to adhere tightly to the cellulose nitrate membrane used for filtration. The unsynchronous culture is filtered through the membrane filter. Inverting the filter and flushing it with fresh medium will wash off the loosely attached cells. After this, the only cells that are found in the effluent stream of the medium are those that arise through fusion of the attached cells, hence all are at the same stage of the cell cycle. For a certain initial period of time the newly formed cells grow in size, then abruptly they all double in number. Synchrony in the culture is rapidly lost because various cells of the culture do not all divide at the same time, age or size following the previous division. Synchrony is almost lost in the third division cycle.

Synchronous cultures rapidly loose synchrony because various cells of a population do not all divide at the same time, age or size. From the rate of loss of synchrony one can calculate the distribution of age of cells at division. Using electronic counting devices one can determine the distribution of cell size of an exponentially growing culture. These two studies allow the calculation of growth rate as a function of cell size.

Growth Yields and Effects of a Limited Nutrient

When microbial growth is limited by low concentration of required nutrient the final number increases with that of initial.

For example, a sample of washed cells collected from a known volume of medium can be analysed for total protein or nitrogen. An increase in the microbial proportion will result in high protein or nitrogen concentration. Chlorophyll determinations can be used to measure algal populations, and the quantity of ATP is an indication of the living microbial mass present.

The *y*-value is often expressed in terms of grams of cells formed per gram of substrate utilized and is an index of the efficiency of conversion of the nutrients into cell mass. Aerobic microorganisms in rich media may assimilate from 20–50% of the sugar carbon taken up. In dilute media, some bacteria seem to be able to increase this efficiency and assimilate up to 60% of the sugar carbon taken up.

During fermentations when bacteria grow in nutritionally rich media, all the sugar that is taken up is used to generate energy in the form of ATP and other substances like amino acids are used as raw materials for macro molecular biosynthesis.

When microbial growth is limited by the low concentration of a required nutrient, the final number of cells or yield of cells increases with the initial amount of limiting nutrient present. The rate of growth reflects the rate of nutrient uptake by microbial transport proteins. At sufficiently high nutrient levels the transport proteins are saturated and the growth rate does not rise with a further increase in the nutrient level.

The amount of microbial mass produced from a nutrient can be expressed as growth yield,

$$y = \frac{\text{mass of microorganisms formed}}{\text{mass of substrate used}}.$$

Continuous Cultures

These are batch cultures in which nutrients are not renewed nor wastes removed. Exponential growth tests for few generations and soon the stationary phase is reached. Cultures can also be grown in an open system, a system where constant environmental conditions are maintained through a provision for continual addition of nutrients and removal of wastes, and the microbial population can be maintained in the exponential growth phase and at a constant biomass concentration for extended period of time.

There are two major types of continuous culture systems commonly in use. The chemostats and the turbidostats.

1. *Chemostats:* A chemostat is constructed in such a way that sterile medium is fed into the culture vessel at the same rate as media containing organism is removed. The culture medium for the chemostat possesses an essential nutrient in limiting quantities, for example, an amino acid. Because of this limited nutrient, the growth rate is determined by rate at which new media is fed into the growth chamber and the final cell density depends on the limiting nutrient. The rate of nutrient exchange is expressed as the dilution rate (D), the rate at which medium flows through the culture vessel relative to the vessel volume,

where 'f' is the flow rate (ml/hr)

V is the vessel volume (ml)

$D = f/V$

For example, f = 50 ml/hr, and V is 100 ml, the dilution rate is 50/100 = 0.5 hr.

Both the microbial population and the nutrient concentration is kept constant, the microbial density remains unchanged and the generation time shortens *i.e.*, the growth rate increases. If the rate of dilution becomes too high, the organisms can actually be washed out of the vessel even before they can multiply.

2. *Turbidostat:* The second type of continuous culture system is equipped with a photo cell that measures the absorbance or turbidity of the culture in a growth vessel. The flow rate of medium through the vessel is automatically regulated in order to maintain a predetermined turbidity. The turbidostat is different from a chemostat in many ways. The dilution rate in a turbidostat varies rather than remaining constant. Its medium also does not have a limiting nutrient.

Continuous culture systems are very useful because they provide a constant supply of cells in exponential phase growing at a known rate. They make possible the study of microbial growth at low nutrient levels, close to those present in natural environment such conditions are essential in research.

Mathematics of Growth

In any biological system, growth can be defined as an orderly increase of all chemical components. Increase of mass does not really reflect growth because the cells could simply increase their content of storage products. In a well balanced medium to which they have become fully adopted, bacteria show a balanced state of growth. During this period, an increase in the biomass is accompanied by an increase in all other measurable properties.

Example: Proteins, RNA, DNA and intracellular water content.

Since the rate of increase of all components of the population is the same measurement of any of the component is sufficient to determine the growth rate. During the exponential phase, each microorganism is dividing at constant intervals. Thus, the population will double in number during a specific length of time called the *generation time* or *doubling time*. This can be illustrated with a simple example. Assume that a culture tube is inoculated with one cell that divides every 20 minutes, *i.e.*, the population will be 2 cells after 20 minutes, 4 cells after 40 minutes and so forth. Because the population is doubling at the end of every generation time, the increase of population is always 2^n, where n is the number of generations. The resulting population is exponential.

Let N_0 be the initial population.

N_t be the population at time t.

n be the number of generations in time t. Hence, population at a given time n_t is

$$N_t = N_0 \times 2^n$$

Calculating n, the number of generations, where all logarithms are to the base 10,

$$\text{Log } N_t = \log N_0 + n.\ \log 2$$

$$n = (\log N_t - \log N_0)/\log 2 \text{ when } \log 2 = 0.301$$

$$n = (\log N_t - \log N_0)/0.301 \quad ...(1)$$

The rate of growth in a batch culture is expressed as K *i.e.*, the growth rate constant which is the number of generations per unit time, which is often expressed as generations per hour *i.e.*,

$$K = n/t \text{ substituting } n \text{ from 1}$$

$$K = (\log N_t - \log N_0)/0.301\ t \quad ...(2)$$

The time, it takes for a population to double in size *i.e.*, the mean generation time, g, can be calculated. If the population doubles *i.e.*, $t = g$ then

$$N_t = 2N_0 \text{ substituting } 2N_0 \text{ in 2}$$

$$K = (\log (2N_0) - \log N_0)/0.301\ g$$

$$= (\log 2 + \log N_0 - \log N_0)/0.301\ g$$

$$= (0.301 + 1)/0.301\ g$$

$$K = 1/g$$

$$g = 1/K$$

For example, assume that a bacterial population increases from 10^3 cells to 10^9 cells in 10 hours.

then $K = (\log N_t - \log N_0)/0.301t$

$K = (\log 10^9 - \log 10^3)/0.301 \times 10$

$= (9 - 3)/3.01 = 2 \text{ generations/hr}$

$g = 1/K$

$g = 1/2$ generations/hr

= 0.5 hour or the generation time of that particular cell is 0.5 hr or 30 minutes.

Generation time varies markedly with the species of the organisms as well as with the environmental conditions. Generation time in nature is much longer than in culture.

UNIT 4 Microbial Techniques

When certain species of microbe is present in a very less number in comparison to the total number of microorganisms, such culture is called mixed culture whereas, a microbial culture containing exclusively one species of microbe is called pure culture. The method of screening a pure culture by separating one type of microbe from a mixed population, is termed isolation.

PURE CULTURE TECHNIQUES

A microbial culture containing only one species of microbe is called pure culture. Different techniques are established for the isolation of pure culture from mixed culture as follows:

1. Micromanipulator utilization

One single viable cell could be transferred on the culture medium to develop axenic pure culture by using micromanipulator which is used with a microscope for picking up a single colony of microbes from a mixed population.

2. Isolation by streak plate technique

In this method the tip of a fine structure, a wire loop called inoculation needle with a glass or wooden handle; the needle is of nichrome wire which is bend to form a loop, is used to transfer microbes from culture broth. Such inoculation needle is used to transfer culture in colony formed on solid culture medium. In such cases, the colony from solid medium develops due to microbial sample, which is streaked on the surface of nutrient agar medium in a sterile petridish. This technique involves following steps:

(*a*) Hold the broth culture containing tube in left hand and shake it.

(*b*) Sterilize the wire loop of the inoculation needle on burner flame.

(*c*) Remove the cotton plug of the broth culture tube by little finger of right hand.

(*d*) Flame the mouth of the test tube immediately.

(*e*) Inside the test tube insert the wire loop to form a thin film and replace the cotton plug.

(*f*) The thin film in the loop is streaked in either a zigzag manner by removing the loop backwards and forwards firmly. Care should be taken that loop should not be firmly pressed against the agar surface.

(*g*) Incubate the petridish in incubator at a required temperature.

(*h*) Growth of the bacteria will be visible after an over night incubation on the streaked marks.

3. Isolation by spread plate method

Some media are selective as they possess chemicals or dyes which enrich the media. They have growth inhibitory effect on some microorganisms and separate the dominant species. Chemical dyes like crystal violet and malachite green are used to inhibit the growth of bacteria and yeast.

Enrichment medium is preferred by adding many number of growth factors. This enhances the recovery and growth of the desired microorganism the differential media by virtue of their distinguish organism growing together in EMB (methylene blue complex) # of eosin methylene blue complex while other enteric bacteria generally do not show metallic sheen. Similarly, on Mac Conkey agar medium, *E. coli* colonies develop with brick red in colour due to fermentation of lactose. On the other hand, *salmonella typhi* does not give this appearance due to non-fermenter of lactose. The other media are selective media such as those which of their special composition promote the growth of one organism and inhibit the growth of others. The minimal medium lack certain growth factors. The medium supports the growth of those microorganisms whose nutritional requirements do not extend those of the corresponding wild type of strain. Such media are helpful when anxotrophs are required.

4. Isolation by pourplate method

In this method, the sample is mixed with known quantity of distilled water. The suspension is poured *i.e.*, plated (1 ml) on a presterilised petridish; spread the sample with the help of loop, and then pour the melted agar medium after cooling it, over the sample containing plate. Incubate the petridish at an optimum temperature (27°C ± 1°C) in an incubator. Sample dilutions are to be made as demonstrated in Fig. 4.

Staining of Microorganisms

The coloring agents impart a color to the colorless microorganisms. Due to this coloration the microorganisms become visible, so as to observe its cell shape and structures.

1. **Simple staining:** The aqueous or alcoholic solution of a single basic dye is called simple staining for example methylene blue, crystal violet, carbol fuschin and saframine. Sometimes a chemical helps to intensify the stain, such chemical is termed mordant. The mordant is closely associated with stain and shows its affinity. In case, stain is applied to the fixed smear, then washed off, slide kept for drying and then it is to be examined under microscope. Such procedure helps in studying the cellular shape and structure details of microorganisms.
2. **Differential staining:** The most important differential stain used in bacteriology is gram stain which was discovered by Danish Bacteriologist Hanschristian Gram. It is the only procedure based on staining technique which divides bacteria into two large groups. Gram positive and Gram negative. The procedure of staining is diagrammatically represented in Fig. 4.6.

As the crystal violet imparts purple color to all cells, it is called primary stain. The purple dye after washing off from the smear, is to be covered with iodine (mordant), both types of bacteria appear dark violet or purple in color. The smear is then washed off with ethanol or ethanolacetone solution which acts as decolorizing agent, removes the purple color from the cells of some species. Thus the ethanol is rinsed off and again stained with safranine, a basic dye, which acts as counter stain, washed again, bottled dry and examined under microscope.

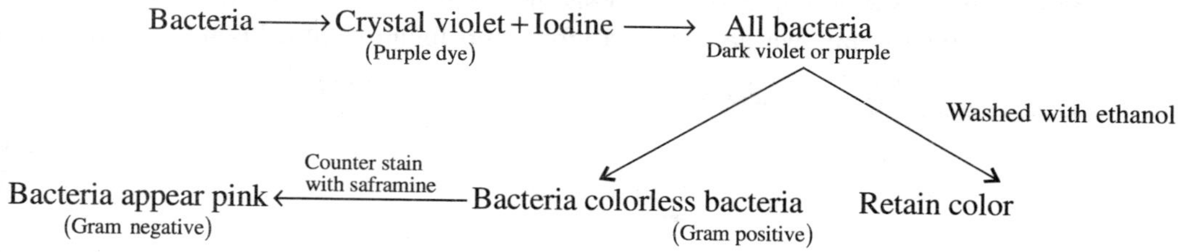

Theories (Mechanism) of staining: The mechanism of staining depends upon the structural differences in their cell walls of Gram-negative and Gram-positive bacteria. It could be confirmed on the basis that if cell wall of Gram-positive bacteria is removed and treated with Gram stain the bacteria turned Gram-negative. The cell wall of Gram-positive bacteria consists of a single 20–80 nm homogeneous peptidoglycan or murein layer lying outside the membrane, while the Gram-negative cell wall has a 1–3 nm thick peptidoglycan layer surrounded by a 7–8 nm thick outer membrane. Added to this during the Gram staining technique the bacterial cell size appear to be about two third than the cells before staining, hence it causes a degree of shrinkage of bacterial cells. On the other hand, some bacteria may appear Gram-positive and the remainder Gram-negative. Such bacteria are termed Gram-variable.

Enumeration of Microorganisms

Microorganisms grow in a variety of physical and chemical environments. To differentiate orderly growth with that of inorderly growth, the term "balanced growth" is used. "Balanced growth" as defined by Campbell, is doubling of every biochemical unit of the cell within the time duration of a single division without a change in the rate of growth.

Several methods are available for measuring microbial growth. The choice depends upon the measurement, objectives and on available techniques usefulness. However, no matter which method is used, considerable care is required in interpreting and concluding the results.

(*i*) *Direct microscopic count*: Bacteria can be accurately counted with the help of Petroff-Hausser counting chamber. This includes a glass slide, a coverslip which is framed and kept 1/50 mm above the slide so that bacterial suspension is present in each ruled square of the slide. The area of square is 1/400 mm^2; glass cover slip rests 1/50 mm above the slide hence volume over a square is 1/20,000 mm^3 or 1/20,000,000 cm^3, for example—if in one square, an average of five bacteria are present, then it can be considered $5 \times 20{,}000{,}000$ or 10^8 bacteria per ml.

(*ii*) *Pour plate technique*: Pour plate method is based on the fact that one viable cell gives rise to one colony. Therefore, a colony count on an agar plate reveals the viable microbial population. For carrying out this, a measured amount of the sample of bacterial suspension is mixed with agar medium (in liquid form at 40–45 ±°C) and incubated. It is plated after thoroughly mixing. Each bacterium grows, reproduces and forms a viable-visible mass in the form of colony. These are counted either by using colony counter or with the help of large magnifying lens. In case several colonies are appearing and overlapping each other, the sample is diluted so that the colonies are accurately counted. This method is called pour plate method.

(*iii*) *Streak plate method*: In this method 0.1 ml sample containing bacteria is spread over the surface of an agar plate using a sterile glass spreader or inoculation needle. The plate containing bacterial suspension has to be incubated until the colonies appear and the colonies are counted.

(*iv*) *Membrane filtration*: This method is based on the principle *i.e.*, cell mass is directly proportional to cell number. This can be obtained after centrifugation of a known volume of culture and weighing the pellet obtained. This is called fresh weight but dry weight of cells is obtained by drying of the pelleted cells at 90–110°C overnight.

(*v*) Electronic method.

UNIT 5

Microbiology: Control of Microorganisms

Sterilization is freeing of an article from all the living organisms, including viable spores. This can be achieved by exposure to lethal physical or chemical agents or in special cases of heat liable situations, by filtration.

Disinfection

To dislodge the infection or the infecting source is the process of freeing an article from harmful or infectious microorganisms. As the resistant spores are rarely encountered, satisfactory disinfection can be achieved by the destruction of all the vegative organisms.

To understand the basis of sterilization by lethal agents, it is necessary to understand the kinetics of death in a microbial population. Death in an organism is the irreversible loss of the capacity to reproduce, which is detected by quantitative plating method. Number of cells which die on exposure to a lethal agent always increase with the increase in exposure period *i.e.*, the kinetics of death is exponential and the number of survivors decrease with time.

In the practice of sterilization the microbial population to be destroyed is always a mixed one, the method of sterilization is, always meant and is capable of killing the most resistant of the organisms, *i.e.*, the sporulating form. Hence the spore suspensions of known resistance are always used to access the reliability of sterilization methods or the indicators of sterilization.

Indicators Used in Sterilization

1. Spores of a toxigenic strain of clostridium tetani are used as the microbial test of heating efficiency. A paper strip impregnated with 1 million (10^6) spores is placed in envelopes and inserted into suitable packs. These are placed along with the articles to be sterilized in the most inaccessible part of the sterilizer. After sterilization these spores are inoculated into suitable media and incubated at 37°C for 5 days under strictly anaerobic conditions.
2. A Browne's tube may be available for dry heating checking. A satisfactory cycle produces a colour change after 1 hr at 160°C.
3. An adhesive tape stuck on to the articles before autoclaving show a change in color after a satisfactory cycle of autoclaving.
4. A Bowie Dick autoclave test strip indicates the degree of steam penetration into a pack.
5. Thermocouples provide a very satisfactory way of controlling sterilization cycles.

Sterilization can be carried out by various methods which are classified as follows:

1. Physical methods
2. Chemical methods
3. Radiation.

Physical methods of sterilization are divided as follows:

Dry Heat

Burning (or) incineration

Discarded materials in the hospitals like soiled bandages, used syringes and cotton scrubs used disposable petridishes and contaminated material from a laboratory and other laboratory waste may be easily burnt off or may be discarded by incineration.

Incinerator should be designed for controlled, fuel powered, high temperature destruction of large volumes of waste with minimal emission of harmful gaseous elements. For efficient combustion two stage process is used involving a primary and a secondary combustion chamber with the later operating at a higher temperature. Temperature controllers and recorders should be fitted to both combustion chambers to ensure that they operated at specified temperatures. Burners for the oxidation of gases in the effluent and filters for the removal of acid gases form effluent, should be included.

Primary chamber operates at a temperature above 800°C while a temperature of at least 400°C for a minimum period of 1 sec is recommended for combustion in the secondary chamber. The danger from spattering can be avoided by using a *Bassi bo*.

Burning

Red heat has its main application in the sterilization of inoculating wires and loops which should be held almost vertically in a bunsen flame until red hot along their whole length. This is also known as heating to in candescence.

Flaming

Instruments like scissors, scapels, blades and slide coverslips etc. are sterilized by dipping into alcohol and passing through the flame. This is also known as surface sterilization.

Hot Air Sterilizers

Heating by exposure to hot air is the accepted method for sterilization articles that cannot be penetrated by steam. These articles should be capable of withstanding the high sterilizing temperatures required.

Example: 160–180°C. This method is particularly used to sterilize glass ware, syringes and metallic instruments.

A hot air oven is a double jacketed chamber with a copper plating inside and an asbestos wall on the outside. This is filled with several adjustable racks and a thermometer is inserted inside. Material should be so packed that it allows free circulation of the hot air inside the chamber so that uniform temperature is maintained inside the chamber. It should also have a temperature indicator, a thermostat and a tuner to ensure a proper sterilization cycle.

All the articles to be sterilized should be packed in aluminium foil or craft paper and stoppers of cotton wool may be used. At the temperatures used for sterilization the paper or cotton wool might charr and give off volatile substances which might be antibacterial. A sterilization cycle maintains the following temperatures for complete sterilization of the articles:

160°C for 1 hr
170°C for 40 mins
180°C for 20 mins

The sterilization hold time does not start till the temperature of the chamber has reached the chosen temperature and a further period is allowed for raising the temperature of all the articles there into that temperature. (The length of this period should be established during the commissioning of the instrument in the lab.)

Cooling may take a long time because of the protective insulation of the oven. Cooling should be gradually done so that the glass ware does not crack. It can be hastened by a controlled cooling fan to circulate cool air which is drawn through a sterilizing filter. Do not open the door of the chamber before the temperature reaches below 60°C. Rubber and plastic made articles and liquids cannot be sterilized in this way as they cannot withstand those high temperatures.

Moist Heat

Microorganisms and their spores are killed at lower temperatures by moist heat than by dry heat for which they must be in contact with water such as the condensing steam—hence it can be used only to sterilize culture media and other liquids which are required to retain their water content.

Boiling in Water Bath at 100°C

Heating at this temperature for 10 minutes kills all the vegetative bacteria, some viruses and some but not all spores. A sterilizer of enamelled or nickeled copper is used, filled with sufficient water and allowed to boil. Hold time starts after the water starts boiling. Medical or surgical instruments are still sterilized by this method. Use a sterile forceps to pick up these articles so that bacteria do not contaminate them.

Steaming at 100°C

Arnolds or Koch's steam sterilizer is used where water boils and the steam generated surrounds the articles to be sterilized. It is also called as free steaming. Steaming for 30–60 minutes at 100°C is employed to disinfect or sterilize some heat labile media like DCA, TCBS, Selective F broth etc., because they are selective media and do not support the growth of all the organisms a common use of the steamer is to dissolve agar and distribute the ingredients during the initial preparation of culture media.

Pasteurization

Heat labile fluids may be disinfected by this procedure, by heating for 1 hr at 56°C or 10 minutes at 65–75°C, which is sufficient to

1. Thermal death time
2. Mode of action
3. Decimal reduction.

Thermal death time

It is the shortest period of time taken to kill a suspension of bacteria or spores at a specified temperature and under specified conditions.

Decimal reduction time

It is the time in minutes to reduce the population by 90% or the time required for the thermal death time curve to pass through one log phase.

The two above terms explain the time temperature relationship to killing in the thermal death time. The temperature is selected as a fixed point and the time is varied.

The decimal reduction time measures 90% death than total call killing. In the determination of the above values experimental conditions should be rigidly controlled since all these have a bearing on the susceptibility of organisms to heat.

Example: Nature of medium ppt. number of organisms etc. kill the mesophilic vegatative bacteria. If it is serum, do not allow the temperature to rise above 59°C above which it coagulates. It is best carried out in water baths with well controlled temperature.

Pasteurization is of two types:

1. *Hold process*: The temperature here is held at 63°C for 30 minutes.
2. *Flash process*: The temperature is held at 72°C for 15–20 seconds.

Sterilization by this method is not complete as spores are not killed. Some vegetative cells are not killed.

Tyndallization

Media may be sterilized by exposure to free steam at 100°C for 20–45 minutes, depending on the volume of the medium, for 3 consecutive days. Spores that survive the first heating will germinate in the medium and these will be killed when reheated. Those organisms will not be killed whose spores cannot germinate in the medium or under the conditions maintained. It is therefore not applicable to non-nutritive media.

Autoclaving

Autoclaving provides moist heat at temperature higher than 100°C, which is the boiling point of water under normal atmospheric pressure. This steam is biocidal. As the steam condenses on the cooler articles, it releases its latent heat and gives moisture both of which together cause denaturation of microbial proteins.

The greater the pressure of steam, the higher the temperature reached. A temperature of 121°C is reached only when the 'P' is 15 lbs/in^2 only if the atmosphere consists pure steam. If some air remains inside, the 'P' of steam is lower than that indicated on the pressure gauge and the temperature will be correspondingly lower for which an autoclave should have a temperature and pressure gauge.

The autoclave has a cylindrical chamber with thick walls, an electrical coil to heat the water at the bottom, a pressure gauge, a steam outlet and a safety valve on the lid, butterfly screws to tighten the lid.

Hold Time

The sterilization hold time is the time for which the entire load needs to be exposed to pure, saturated stream, to be sterilized. Where the steam makes contact with the coolee surfaces it condenses and releases the latent heat of condensation and also makes place for fresh stream and helps in sterilization. Efficient condensation takes place only if the steam is pure which is why all the air should be expelled.

After the required pressure and temperature have been attained and the sterilization has been carried out for the given time, the cooling of the material specially aqueous media is very important. If the steam is released immediately, the media in containers will boil over violently. Hence cooling should occur slowly but should not be very slow which reduces the nutritive value of the medium.

Filtration

Aqueous liquids, including solutions of heat liable substances, may be sterilized by forced passage through a filter of porosity small enough to retain any microorganisms contained in them.

Five main types of filters have been employed for bacteriological purposes:

1. Earthern filters

Berkfield are made from kieselguhrs, a fossil diatomaceous earth, which after use can be sterilized by autoclaving. They can be scrubbed clean with a brush and boiled in distilled water chambers, and filters are made up of unglazed porcelain are produced in various grades of porosity. These earthenware filters are made in the form of hollow candles through which the fluid is forced in or out. When the poles get clogged. The candles are heated to redness and allowed to cool.

2. Asbestos filters

These filters called the SEITZ filters consists of a disk of asbestos supported on a perforated metal disk within a metal funnel. The whole assembly is loosely assembled, sterilized by autoclaving and tightened by screwing down the funnel and used. The disk is discarded after use.

3. Scintered glass filters

These are made from finally ground glass fused to make the glass particles adhere. They may be supplied in the form of a disk fused into a glass funnel. After use filters are washed with running water in the reverse direction and cleaned with warm strong sulphuric acid.

4. Membrane filters or collodion or millipore filters

These membranes are manufactured from a variety of polymeric materials such as cellulose nitrate, cellulose diacetate, polycarborate and polyester. Being less absorptive than other filters they have a faster rate of filteration. Membranes for filtration are chosen based on organism to be filtered, the volume and type of medium to be filtered. These membranes are sterilized by autoclaving.

STERILIZATION BY RADIATION

Non-ionising Radiation

These are electromagnetic or non-particulate radiations which are rays with wavelength longer than those of visible light and are absorbed as heat. They are:

1. Infrared radiation

These bring about sterilization by hot air in which the organisms are killed as a result of oxidation due to the generated heat. These radiations are a convenient method of sterilization of a large number of syringes in a relatively shorter time reaching up to 190°C in 10 mts.

2. Ultraviolet radiation

Ultraviolet portion of spectrum includes all radiations from 150 Å to 3900 Å. Wavelengths around

2650 Å have the highest bactericidal efficiency. Although the radiant energy of sunlight is partly composed of U.V. light, most of the shorter wavelengths of this type are filtered out by the earth's atmosphere (Ozone, Clouds and smoke). Consequently the U.V. radiation at the surface of the earth is restricted to the span from about 2670 Å to 3900 Å from this we may conclude that sunlight under certain conditions has microbicidal capacity but to a limited degree.

Mercury vapour lamps are available which emit a high concentration of U.V. light in the most effective region, 2600 Å to 2700 Å. Germicidal lamps which liberate U.V. radiations are widely used to reduce microbial populations.

Example: They are used extensively in hospital operating rooms in *aseptic filling rooms* in pharmaceutical industry. These are bactericidal and to a lesser extent sporicidal. But their energy levels are low and penetrating power is less. Hence they should be used only for the disinfection of pre-cleaned surfaces such as interior of a safety cabinet.

Exposure of eyes and skin to U.V. rays should be avoided.

Mode of Action

Ultraviolet light is absorbed by many cellular materials but most significantly by the nucleic acids, where it does the most damage. The absorption and subsequent reactions are predominantly in the pyrimidines of nucleic acids. One important alteration is the formation of pyrimidine dimer in which two adjacent pyrimidines become bonded. Unless dimers are removed by specific intracellular enzymes, DNA replication can be inhibited and mutations can result.

Ionizing Radiations

These are particulate and electromagnetic radiations. The most important of these are the γ radiations which are emitted from radioactive elements usually Cobalt 60. As a result of the major research programs with atomic energy, large quantities of radioisotopes have become available as by products of atomic fission.

γ-rays are similar to X-rays but are of shorter 'λ' and higher energy. They are capable of great penetration into matter and lethal to all life, including microorganisms.

Because of their great penetrating power and their microbial effect, γ-rays are attractive for use in commercial sterilization of materials of considerable thickness or volume. Development of radiation sources for large scale use and the design of equipment to eliminate any possible hazards to the operators is under study. Since there is no appreciable rise in temperature it is called as cold sterilization. Radiant energy particles makes a direct hit on some essential substances such as DNA within the bacterial cell, causing ionization which results in the death of the cell.

Cathode Rays

When a high voltage potential is established between a cathode and an anode in an evacuated tube, the cathode emits beams of electrons, called Cathode rays (or) Electron beams. Special type of equipment has been designed which produces electrons of very high intensities and these e^{-1}s are accelerated to extremely high velocities. These intense accelerated e^{-1}s are microbicidal as well as having other effects on biological and nonbiological materials.

The e^- accelerator produces the high voltage e^- beam, which is used for sterilization of surgical supplies; the material can be sterilized after it has been packaged and at room temperature.

STERILIZATION BY CHEMICALS

Aldehydes

Formaldehyde is highly lethal to all kinds of microbes and spores, killing bacterial spores as easily as vegetative forms. It is applied as an aqueous solution or gaseous form. It is non-injurious to fabrics, wood, leather, metals, etc.

Glutaraldehyde in a 2% aqueous solution is even more efficient and less irritant. It is stable in acidic solution and very active in alkaline solution which is marketed as Cidex.

Cidex is a bactericidal and sporicidal and effective against viruses also. It is particularly useful for the sterilization of articles which cannot be subjected to sterilizing temperatures such as cystoscopes, anesthetic equipment, plastic materials and thermometers.

These solvents are two of those few which are sporicidal.

Alcohols and Volatile Solvents

70–75% aqueous solution—(V/V) of isopropyl alcohol is optimally bactericidal but has a very little bactericidal effect outside this range *i.e.*, where absolute *i.e.*, not diluted with water as when diluted too much. This is used in skin disinfection before hypodermic injections or reinpuncture etc. Isopropyl alcohol is not subject to excise regulations and is much cheaper than ethyl alcohol.

Chlorohexidine or iodine (0.5%–2%) in 70% alcohol is also used for the same purposes and are superior to alcohol alone.

Acetone and ether are only weakly antibacterial and are not effective alternatives to alcohols. Chloroform has a limited use as a bactericidal agent. It rapidly kills bacteria in vegetative forms and is removed from solution by evaporation.

Metallic Salts and Metallic Organic Compounds

Mercuric Chloride is sometimes used as a disinfectant in a 1:1000 solution. Mercuric salts are highly bacteriostatic but are not effectively bactericidal.

Silver nitrate is a bactericidal agent. A drop of 1% solution placed in the eye of a new born baby was widely and successfully used as a preventive measure against gonococcal conjuctivitis.

Halogens

Chlorine and Iodine are bactericidal and sporicidal. Chlorine has a special place in the treatment of water supplies, and combinations of hypochlorites and detergents are useful for cleansing of disinfection in the food and dairy industry. Various hypochlorite preparations are viricidal. Hypochlorite solutions are also used for general disinfection and disposal of glassware such as microscopic slides, pipettes etc, disinfecting surfaces contaminated with blood. Hypochlorites corrode metals and damage clothes.

Mode of Action

Chlorine

When free chlorine is added to water it forms hypochlorous acid

$$Cl_2 + H_2O \rightarrow HCl + HCl(O)$$

Hypochlorites and Chloramines also undergo hydrolysis with the formation of hypochlorous acid. The hypochlorous acid thus formed undergoes further decomposition to

$$HClO \rightarrow HCl + (O)$$

The oxygen released is called nascent O_2 and is a strong oxidizing agent and through its action, organisms are destroyed. This is partly also due to chlorine and its compounds combining directly with proteins of the cell membranes and enzymes.

Iodine is used as a rapid skin disinfectant and is valuable for preparation of the skin for surgery. Few individuals are hypersensitive to I_2. Betadine, a water soluble complex of Iodine and polyirnyl pyrolidone can be used as a bactericidal antiseptic for intact skin and for disinfection of superficial wounds.

Mode of Action

It is an oxidising agent and this fact may account for its antimicrobial action. Oxidizing agents can irreversibly oxidize and thus inactivate essential metabolic compounds such as proteins with sulfhydroxyl group. It may also involve the halogenation of tyrosine units of the enzyme and other cellular components which require tyrosine activity.

Dyes

Aniline dyes and acridine dyco are active against gram positive organism but less active against gram negative organism.

Acridine dye acts by binding to nucleic acids in the inactive or denatured state. As these dyes intercalate with nucleic acids the organisms do not remain viable for long after staining.

Phenols

Carboxylic acid is powerfully microbicidal and is used for decontamination of injective discharges, bathrooms, bedpans and hospital floors. They are too toxic and irritant to be applied and left on objects. They are not inactivated by the presence of organic matter.

Chlorophenols and chloroxylenols (dettol) are less toxic and less irritant and are also less active and inactivated by organic matter.

Mode of Action

Depending on the concentration of phenol it produces different effects on the cells. It causes disruption of cells, precipitation of cell proteins, inactivation of enzymes and leakage of cell contents. The lethal effects of phenol are presumed to be due to the physical damage to the membrane structure which initiates further deterioration.

Surface Active Agents

These possess wetting and cleansing properties as well as some disinfectant properties.

The cationic detergents, the quarternary ammonium compounds have antibacterial as well as detergent activity. They are relatively non-toxic. They are bacteriostatic and more active against gram positive bacteria.

The anionic agents are active against gram positive bacteria.

Mode of Action

Heavy metals

Heavy metals and their compounds act antimicrobially by combining with cellular proteins and inactivating them.

Example: In the case of mercuric chloride the inhibition is directed at enzymes which contain sulfhydreyl group. High concentration of salts of heavy metals coagulate cytoplasmic proteins resulting in damage or cell death. Salts of heavy metals are also precipitants and in high concentration such as salts could cause cell death.

UNIT 6 Microbial Metabolism

Energy and nutrients are two basic requirements for every biological process. Nutrition supplies the basic cell building requirements. Nutrition not only provides energy but also acts as precursor for growth of microorganism. Biochemical composition of a microbial cell decides the nutritional requirement of it.

Apart from energy, the microorganisms also require a source of carbon, nitrogen, sulphur, phosphorus, oxygen and other metal ions for growth. Microorganisms get these from the environment. Based on the source of energy they depend upon, microorganisms are classified as phototrophs, and chemotrophs. Phototrophs utilize the energy trapped in the solar radiations (solar energy) for their metabolic requirements and convert it into chemical energy while chemotrophs use the energy trapped in the chemical bonds (chemical energy) for their metabolism.

Carbon is the prime requirement of any organism. Based on the source of carbon, microorganisms can be either autotrophs or heterotrophs. Autotrophs (self feeders) use carbon dioxide of the atmosphere as their source of carbon. These microorganisms fix the atmospheric carbon into carbohydrates required for growth and metabolism. As far as carbon source is concerned, these microbes are independent. Heterotrophs (feeders on others) depend upon other organisms for carbon *i.e.*, they depend upon carbon already fixed by autotrophs or present in other heterotrophs.

1. *Phototrophs:* They are photosynthetic bacteria. They fix carbon, utilizing the energy of solar radiations through a process called *photosynthesis*. For example, green non-sulphur bacteria and purple non-sulphur bacteria.

2. *Chemotrophs:* They utilize the energy present in chemical compounds and carbon in the form of carbon dioxide. Example, *Beggiatoa* (utilises hydrogen sulphide as energy source), *Nitrosomonas* (utilizes ammonia as energy source), *Thiobacillus* (utilizes iron), *nitrobacter* (utilizes nitrites) etc. Generally same organic compound can act both as a source of energy as well as carbon. All the parasites of man, animals and plants are chemotrophs and all saprophytes are also chemotrophs.

Microbial metabolism is of two types namely anabolism and catabolism.

Anabolism is a process in which essential biomolecules required for growth are generated by the utilization of energy. In other words, anabolism is the synthesis of complex molecules into simple molecules. It is also known as biosynthesis.

For the biosynthesis (anabolism) of different essential biomolecules, following anabolic processes take place in organisms:

(*a*) Synthesis of carbohydrate like glucose, sucrose, cellulose, etc.

(*b*) Synthesis of lipids, glycolipids, phospholipids, etc.
(*c*) Synthesis of amino acids and protein.
(*d*) Synthesis of nucleic acids.
(*e*) Synthesis of other growth requirements like vitamins, hormones, etc.

Catabolism: All processes in which the nutrients absorbed in the form of biomolecules as food are broken down or digested to release energy are called catabolism. Catabolic processes also convert complex organic compounds stored in the cells of microbes like glycogen granules, polyphosphate, etc., to simpler forms. Organic compounds like carbohydrates, protein or fat can be catabolized as per the requirement of the organism. The major catabolic processes involved in microbial cell metabolism are:

(*a*) Glycolysis
(*b*) Pentose phosphate pathway
(*c*) Entner doudoroff pathway (EDP)
(*d*) Tricarboxylic acid cycle (TCA)
(*e*) Fermentations
(*f*) Glyoxylate hydrolysis.

Apart from these ATP and NADP, H also acts as a link between anabolism and catabolism.

ATP: Adenosine triphosphate is the main energy carrier molecule in all living organisms. ATP is needed for the anabolic generative process of an organism by which organic macromolecules required for the growth are synthesized. ATP is formed by the catabolic degradative process in which macromolecules are broken down or digested and energy is generated.

In 1930, the structure of ATP was first given by Lohman and confirmed by Alexander Todd *et al.*, in 1948. ATP consists of an adenosine and three phosphate groups. These three phosphate groups are bounded by high energy bonds to adenosine unit and during one or two of these phosphate groups are detached from the ATP, large quantity of energy is released (Fig. 6.1).

$$\underset{\text{Adenosine triphosphate}}{\text{ATP}} \longrightarrow \underset{\text{Adenosine diphosphate}}{\text{ADP}} + \underset{\text{Inorganic phosphate}}{\text{Pi}} + \text{energy}$$

$$\text{ADP} \longrightarrow \underset{\text{Adenosine monophosphate}}{\text{AMP}} + \text{Pi} + \text{energy}$$

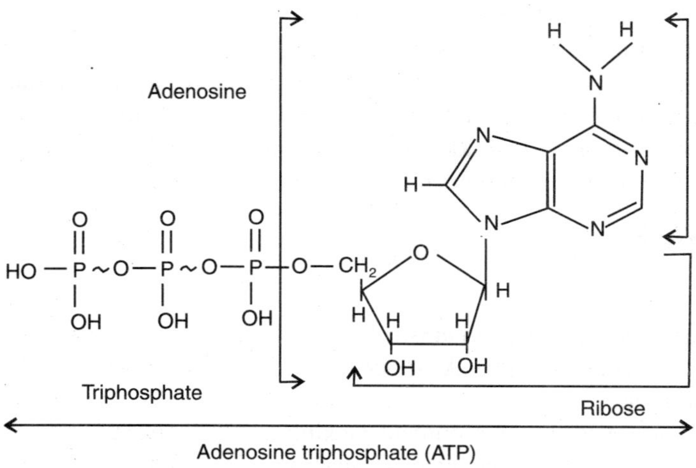

Fig. 6.1 ATP molecule's chemical structure

The energy thus released is the instant energy utilized in anabolic processes. The ATP used in anabolic process is replenished by catabolic process by reversal reactions. The trapping of chemical energy, released by the oxidative reactions of the cell in the form of ATP, is called *Phosphorylation*.

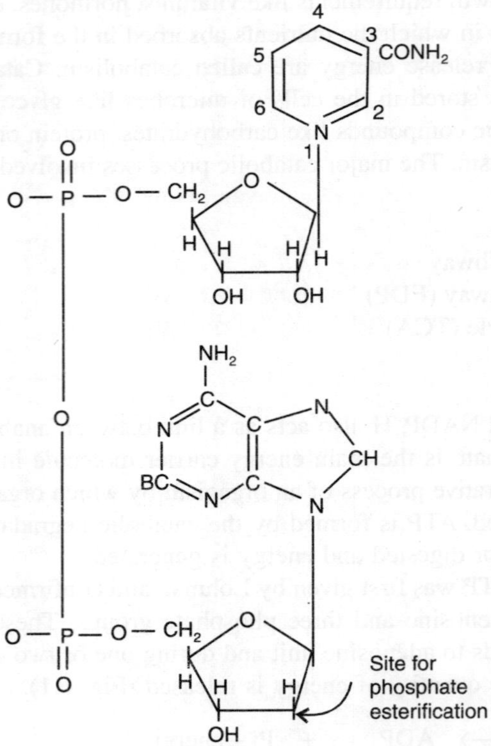

Fig. 6.2 Chemical structure of nicotinamide adenine dinucleotide (NAD)

NADP: Nicotinamide adenine dinucleotide phosphate (NADP) and nicotinamide adenine dinucleotide (NAD) are the carriers of electrons in the cells. Therefore, NADP and NAD serve as the reducing power of the cells in the form of NADP/NADPH$_2$ or NAD/NADH$_2$.

NADP or NAD thus functions as coenzymes of a large number of oxidoreductase enzyme. They act as electron acceptors during enzymatic removal of hydrogen atoms from specific substrate molecules. Ultimately, reduced NADP or NAD *i.e.*, NADPH$_2$ or NADH$_2$ release energy, and ATP is generated.

NAD structurally consists of a nicotinamide moiety, two adenine nucleotides, whereas, in NADP, a phosphate group is esterified to the second adenine nucleotide at 2-hydroxyl group (Fig. 6.2).

METABOLIC PATHWAYS

1. Glycolysis

Glycolysis is the most ancient type of mechanism by which organisms obtain energy from organic compounds in absence of molecular oxygen. As it occurs in the absence of oxygen, therefore, it is also called *anaerobic fermentation*.

Since living organisms arose in the environment lacking oxygen, anaerobic fermentation was the only method to obtain energy. However, glycolysis or anaerobic fermentation is present in both aerobic and anaerobic organisms. Most higher organisms have retained the glycolytic pathway of degradation *i.e.*, glucose to pyruvic acid as a preparatory pathway for complete aerobic catabolism of glucose. Glycolysis also serves as an emergency mechanism in anaerobic organisms to produce energy in the absence of oxygen.

The complete pathway of glycolysis from glucose to pyruvate (Fig. 6.3) was elucidated by Gustav Embden (who gave the manner of cleavage of fructose 1,6-diphosphate and pattern of subsequent steps) and Otto Meyerhof (who confirmed Embdens work and studied the energetics of glycolysis), in late 1953s. Therefore, the sequence reaction from glucose to pyruvate is also called Embden Meyerhof pathway or glycolysis (EMP).

The overall balance sheet of glycolysis is given below:

$$\text{Glucose} + 2\text{ADP} + 2\text{Pi} + 2\text{NAD} \longrightarrow \text{Pyruvate} + 2\text{ATP} + 2\text{NADH} + 2\text{H}^+$$

In anaerobic organisms, pyruvate is further converted to lactate or other organic compounds like alcohol, etc., after using $NADH^+ + H^+$ formed during glycolysis:

$$\text{Pyruvate} + \text{NADH} + \text{H}^+ \longrightarrow \text{Lactate} + \text{NAD}$$

In aerobes, the pyruvate is converted to acetyl CoA as a preparatory step for entrance into tricarboxylic acid cycle for complete oxidation of glucose.

$$\text{Pyruvate} + \text{NAD}^+ + \text{CoA} \longrightarrow \text{Acetyl CoA} + \text{NADH} + \text{H}^+ + \text{CO}_2$$

Glycolysis is carried out by the help of ten enzymes for ten reactions of the glycolytic pathway. These enzymes are present in the soluble portion of the cytoplasm. All the intermediates of the glycolytic pathway are phosphorylated compounds. The most important use of phosphate groups is in the production of ATP from ADP and phosphate.

The complete reactions of glycolytic pathway can be divided into two stages. In the first stage, ATP is utilized and glucose is converted into two molecules of three carbon compounds, glyceraldehyde 3-phosphate. In the second stage ATP is produced and glyceraldehyde 3-phosphate is converted into pyruvic acid resulting in a net synthesis of two molecules of ATP. The complete reaction with respective enzymes is shown in Fig. 6.3.

Apart from glucose, other types of sugar (monosaccharides, disaccharides, polysaccharides) can also enter the glycolytic pathway by following pathways:

(a) Polysaccharides *e.g.*, Glycogen

$$(\text{Glucose})_n + \text{HPO}_4^{2-} \xrightarrow[\text{Glycogen}]{\text{Glycogenphosphorylase}} (\text{Glucose})_{n-1} + \text{Glucose-1-phosphate}$$

$$\text{Glucose-1-phosphate} \xrightarrow{\text{Phosphoglucomutase}} \text{Glucose-6-phosphate}$$

Glucose-6-phosphate can enter as an intermediate of glycolysis.

(b) Disaccharides *e.g.*, Sucrose

$$\text{Sucrose} + \text{H}_2\text{O} \xrightarrow{\text{Fructuranosidase}} \text{Glucose} + \text{Fructose}$$

$$\text{Maltose} + \text{H}_2\text{O} \xrightarrow{\text{Glucosidase}} 2\text{Glucose}$$

$$\text{Lactose} + \text{H}_2\text{O} \xrightarrow{\text{Galactosidase}} \text{Glucose} + \text{Galactose}$$

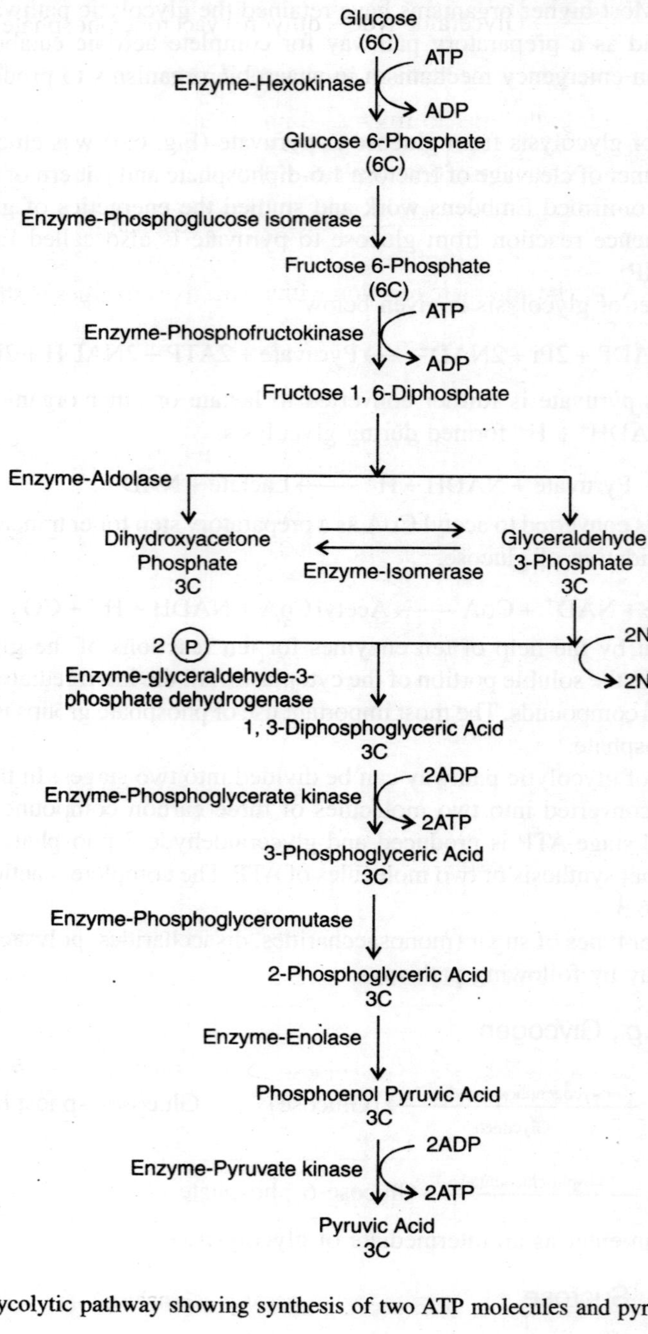

Fig. 6.3 Glycolytic pathway showing synthesis of two ATP molecules and pyruvic acid (glycolysis)

(c) Homosaccharides *e.g.*, Fructose

Fructose can enter the glycolysis by changing to glyceraldehyde 3-phosphate.

$$\text{Fructose} + \text{ATP} \xrightarrow{\text{Fructokinase}} \text{Fructose-1-phosphate} + \text{ADP}$$

$$\text{Fructose-1-phosphate} \xrightarrow{\text{Aldolase}} \text{Glyceraldehyde + dihydroxyacetone phosphate}$$

$$\text{Glyceraldehyde + ATP} \longrightarrow \text{Glyceraldehyde-3-phosphate + ADP}$$

Mannose enters glycolysis after converting to fructose 6-phosphate

$$\text{Mannose + ATP} \xrightarrow{\text{Hexokinase}} \text{Mannose-6-phosphate + ADP}$$

$$\text{Mannose-6-phosphate} \xrightarrow{\text{Mannose phosphate isomerase}} \text{Fructose 6-phosphate}$$

Glycerol can also enter the glycolysis. Phosphate is converted into pyruvic acid resulting in a net synthesis of two molecules of ATP. The complete reaction with respective enzymes is shown in Fig. 6.4.

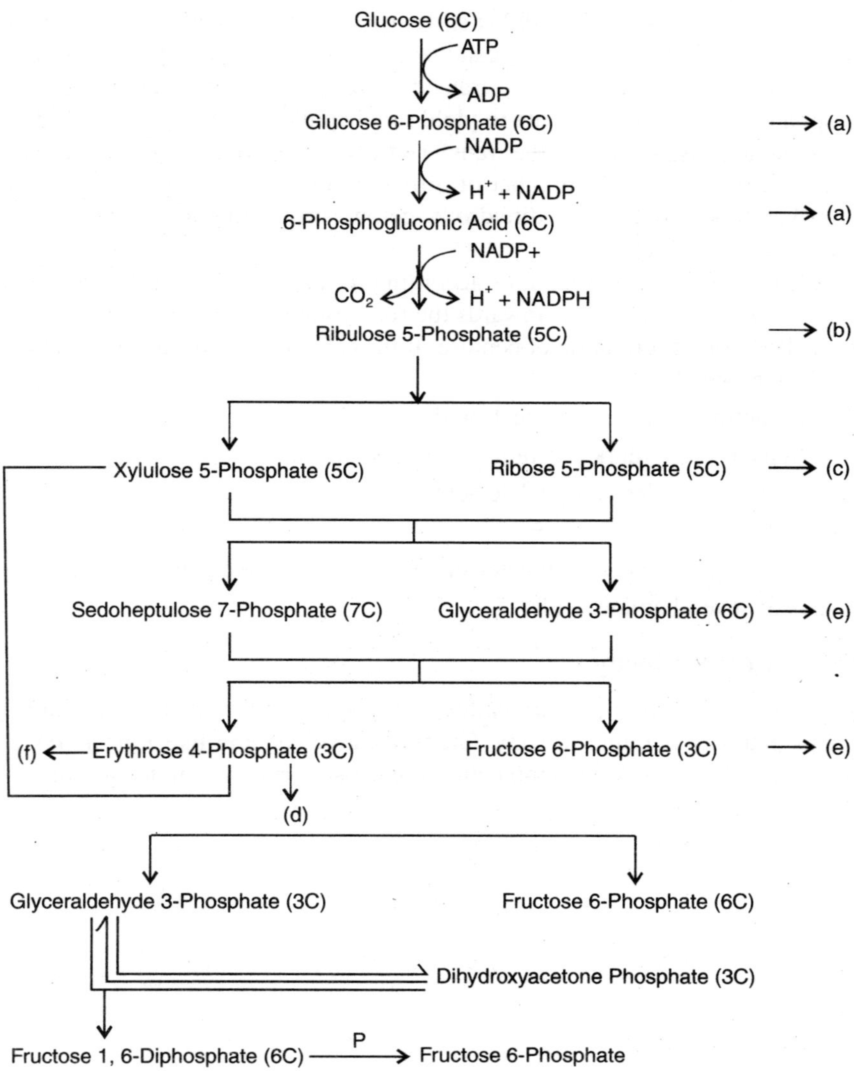

Fig. 6.4 The pentose phosphate pathway

2. Pentose Phosphate Pathway (PPP)

Pentose phosphate pathway is an alternative of glucose degradation. This pathway, also called hexose monophosphate shunt (HMP) or phosphogluconate pathway is not the major pathway, but is a multi-purpose pathway. Its main function is to generate reducing power in the extramitochondrial cytoplasm in the form of NADH. Its second function is to convert hexoses into pentoses, required in the synthesis of nucleic acid. Its third function is complete oxidation of pentose. The reactions of phosphogluconate pathway takes place in the soluble portion of extramitochondrial cytoplasm of cells. The bacteria which show PPP are *Bacillus subtilis, E. coli, Streptococcus faecalis* and *Leuconostoc mesenteroides*. Apart from microorganisms the prominent tissues which show PPP are liver, mammary gland and adrenal cortex. The complete PPP is given in Fig. 6.4.

There are three enzymes involved in PPP *i.e.*, transketolase, transaldolase and ribulose phosphate epimerase. Ribulose phosphate 3-epimerase catalyzes the conversion of ribulose 5-phosphate into the epimer xylulose 5-phosphate. Transkelotase transfers the glycoaldehyde group (CH_2OH-CO-) from xylulose 5-phosphate to ribose 5-phosphate to yield sedoheptulose 7-phosphate and glyceraldehyde-3-phosphate. Transkelotase also catalyzes the transfer of glycoaldehyde group from a number of 2-keto sugar phosphate to carbon atom on a number of different aldose phosphate. Transaldolase acts on the products of transketolase and transfer dihydroxyacetone group to form fructose 6-phosphate and erythrose 4-phosphate (Fig. 6.4).

Pentose phosphate pathway thus functions according to the needs of the cell. If the requirement of reducing power is more than it proceeds towards the formation of NADPH but if pentoses are required it functions in the direction of formation of pentose. If the cell requires instant energy, the PPP stops and glycolysis and TCA proceed:

(*a*) to anabolic reactions that require electron donors
(*b*) to Calvin-Benson cycle (dark reactions of photosynthesis)
(*c*) to synthesis of nucleotides and nucleic acids
(*d*) to step *e* of glycolysis
(*e*) to glucose 6-phosphate which can enter the pentose phosphate pathway or glycolysis
(*f*) to synthesis of several amino acids.

3. Entner-Doudoroff Pathway

Apart from glycolysis, Entner-Doudoroff pathway is another pathway for oxidation of glucose to pyruvic acid. This pathway is found in some Gram-negative bacteria like *Rhizobium, Agrobacterium* and *Pseudomonas*. In this pathway each molecule of glucose forms two molecules of NADPH and one molecule of ATP. The complete pathway is shown in the Fig. 6.5.

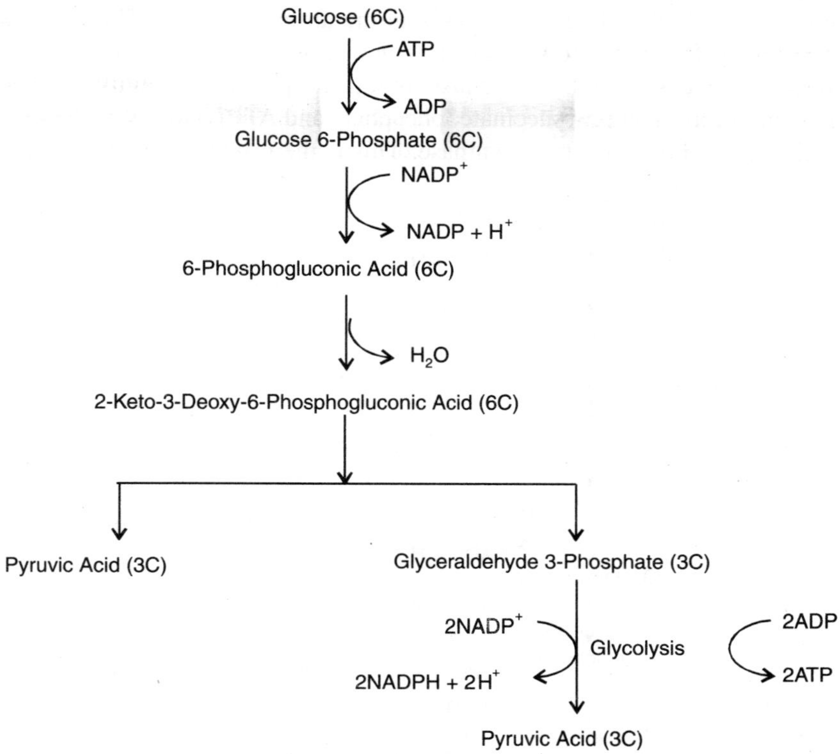

Fig. 6.5 Entner-Doudoroff pathway

4. Tricarboxylic Acid Cycle

The tricarboxylic acid cycle was first given by H.A. Krebs in 1937. H.A. Krebs then gave the name citric acid cycle. The cycle is of universal occurrence in all the aerobic organisms and leads to complete oxidation of glucose to CO_2 and H_2O, while glycolysis leads to incomplete oxidation of glucose to pyruvate. Tricarboxylic acid cycle completely oxidises it to release large amount of energy in the form of $NADH^+ + H^+$ mainly and GTP. $NADH^+ + H^+$ enters the respiratory chain where each $NADH^+ + H^+$ produces three ATP molecules. GTP is converted to ATP by substrate level oxidation. Another form of energy is in the form of substrate of $FADH_2$, which also enters the respiratory chain to form two molecules of ATP.

All the reactions of tricarboxylic acid cycle take place in the inner compartment of mitochondrion. Some of these enzymes occur in the matrix of inner compartment, while rest of them occur on the inner mitochondrial membrane. For the start of the cycle the pyruvate formed in the glycolysis is first converted to acetyl CoA by preparatory reaction.

The reaction is irreversible and is not itself a part of the tricarboxylic acid cycle. It is carried out with the help of the enzyme pyruvate dehydrogenase. Acetyl CoA then enters the cycle after combining with oxaloacetate to form citrate, after which a cycle of reactions occurs (Fig. 6.6) leading to the formation of six CO_2, NADH, eight $NADH^+ + H^+$, one $FADH_2$ and from one molecule of glucose.

There are few key steps in the tricarboxylic acid cycle which control the cycle as per need of the cell. The first of these controls is the preparatory reaction. The activity of pyruvate dehydrogenase is reduced in the presence of excess ATP and again increases in the absence of ATP. There are two more steps which can control the cycle. These are the isocitrate dehydrogenase reaction (which requires ADP as positive regulation), succinate dehydrogenase reaction (promoted regulation) and succinate dehydrogenase reaction (promoted by succinate, phosphate and ATP). However, the key control of the cycle is the reaction carried out by citrate synthase. This is the primary control of the cycle.

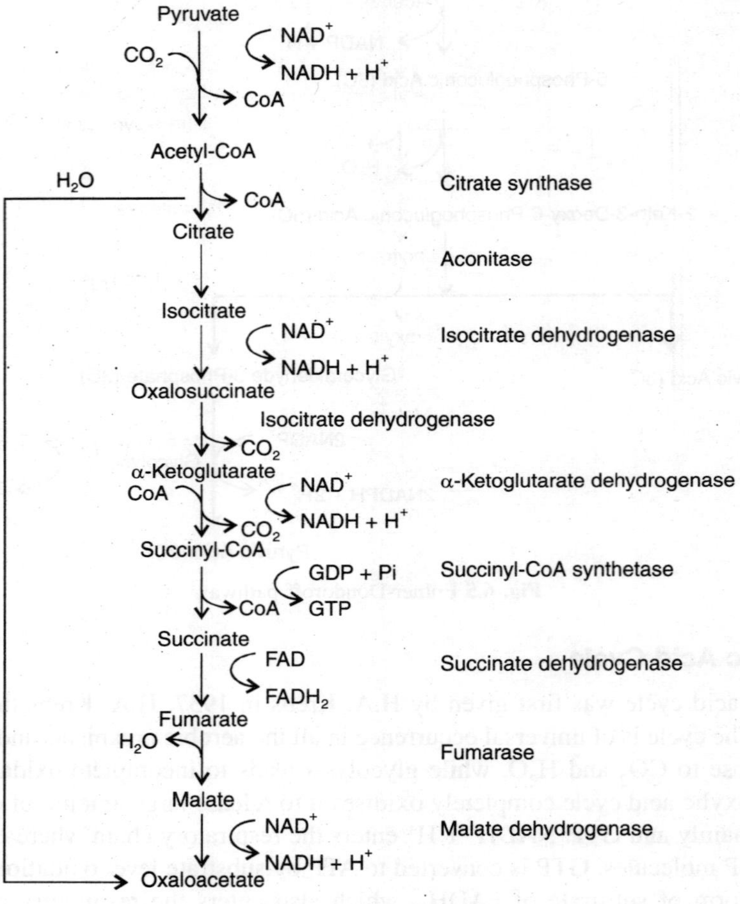

Fig. 6.6 Tricarboxylic acid cycle

5. Glyoxylate Cycle

Glyoxylate cycle was given first by H.A. Kreb's and H.R. Kornberg. This cycle is a modified form of tricarboxylic acid cycle found in plants and those microorganisms which utilize fatty acids as the source of energy in the form of acetyl CoA. In this cycle, the CO_2 evolving steps of tricarboxylic acid cycle were by-passed and instead a second molecule of acetyl CoA is utilized which condenses with glyoxylate to form malate. Succinate is a byproduct, used for biosynthesis, particularly in gluconeogenesis. The overall reaction of glyoxylate cycle is given below:

$$2\,\text{Acetyl CoA} + NAD^+ + 2H_2O \longrightarrow \text{Succinate} + 2\text{CoA} + NADH + H^+$$

The two key enzymes, isocitrate lyase and malate synthase of glyoxylate cycle are localized in cytoplasmic organelles called glyoxysomes. Glyoxylate cycle goes on simultaneously with the tricarboxylic acid cycle, while tricarboxylic acid provides energy, glyoxylate cycle provides succinate for the formation of new carbohydrate from fats as shown in Fig. 6.7.

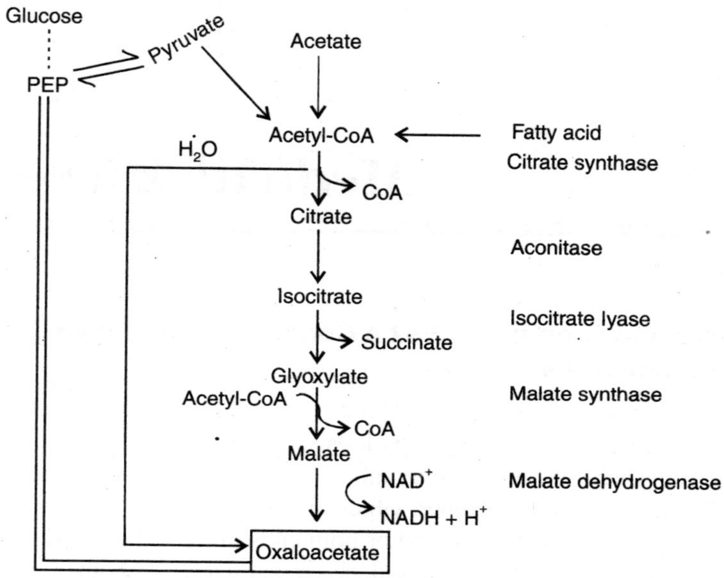

Fig. 6.7 Glyoxylate cycle

UNIT 7

Medical Microbiology

Progenote (Nyberg and Cronhjort) 1992 a Common Ancestor Cell for *Prokaryotes* and *Eukaryotes*

Under electron microscope the ultra structure of bacteria (typical prokaryotic cell) can be studied. It is observed that many bacteria can be surrounded by loose sling layer, called capsule. Beneath the capsule is a cell wall which gives shape to the cell. Internal to cell wall is cell membrane. The folded imagination of the cell membrane is called mesosomes, which are present in many bacteria. Some bacteria have one or more flagella, which may arise from one or both ends of the cell or may arise from all over the cell. The nucleoid is in the form of highly folded circular DNA without a nuclear envelope. The cytoplasm contains 70s ribosome storage granules, vacuoles, inclusion bodies and in photosynthetic bacteria chromatophores.

Capsule: Many bacterial species like diplococcus pneumococcus etc. are surrounded by a capsule. It is a viscous or gelatinous loose aggregates of substance secreted surrounding the cell wall. Chemically the capsule is made up of proteins, polysaccharides and lipids but more particularly polysaccharides. The capsule provides protection in unfavorable conditions.

The presence of a capsule is important in the disease causing ability, *i.e.*, virulence of some bacteria.

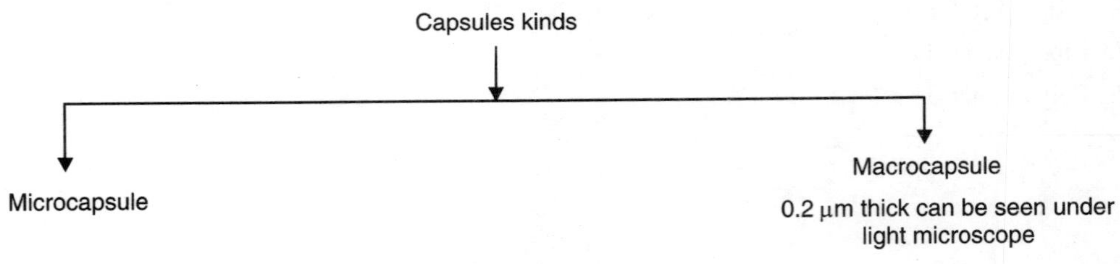

Fig. 7.1

Capsule provides buoyancy to the cell; which permits it to float, where aerobic conditions are greater. The capsule may also function in the storage of food substances and a site for disposal of waste material.

Bacterial capsules are species specific and therefore can be used for immunological distinction of closely related species.

Cell wall: It is the outermost boundary in a majority of bacteria. The cell wall is thin, about 10–25 nm thick and provides rigidity to the cell. It is visible only under electron microscope. It accounts for 20–30% total dry weight of the cell.

Chemically cell wall is made up of strong fibres of heteropolymers called mucopeptides or peptidoglycans.

Depending upon the cell wall composition, bacteria are divided into two types viz., gram +ve bacteria and gram –ve bacteria.

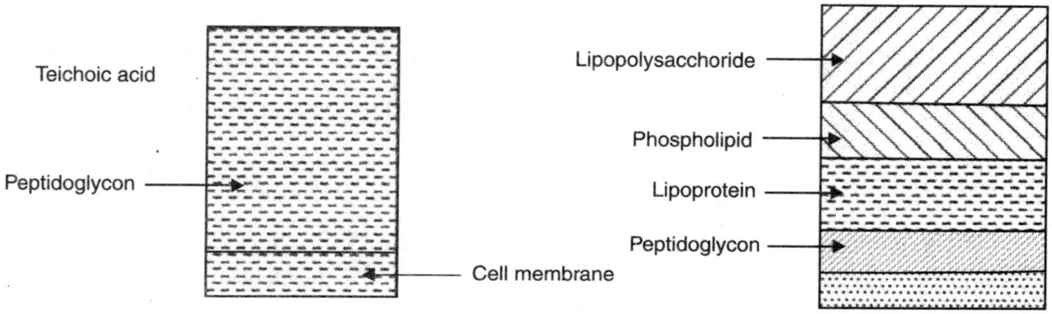

Fig. 7.2

A schematic representation of gram –ve and gram –ve cell wall. Comparison between gram +ve and –ve cell wall

Gram +ve cell wall	*Gram –ve cell wall*
1. Cell wall is thick	1. Cell wall is thin
2. Few amino acids present	2. Many amino acids present
3. Sulfer containing amino acids are absent	3. Sulphur containing amino acids present
4. Teichoic acid present	4. Teichoic acid absent
5. Peptidoglycon present	5. Peptidoglycon present in less quantity
6. Lipproteiry absent	6. Lipproteiry absent
7. Lipopolysaccharide absent	7. Lipopolysaccharide present
8. Polysaccharide present	8. Polysaccharide absent.

Fungi

The fungi (singular fungus; Latin fungus = mushroom) are eukaryotic, spore-bearing; microorganisms of great practical importance. The study of fungi is termed as *Mycology* (Gr. mykes = mushroom): They occur in water, soil, air and upon living and dead plants and animals. They are also geographically well distributed occurring even in *deserts* and *icy polar* regions.

Fungi are quite *distinct from plants* as they *lack chlorophyll* and are placed in a separate kingdom called Fungi (Myceteae). They are heterotrophic in nutrition and they live as:

Saprophytes (decomposing plants and animals remains)

Parasites (living on other organism)

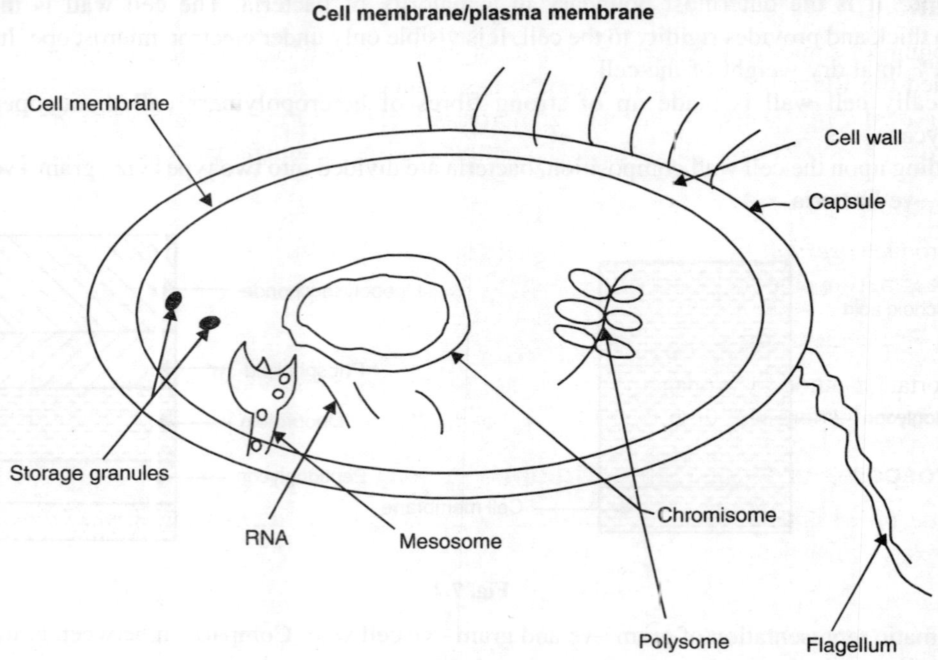

Fig. 7.3 Ultra structure of bacteria

Symbionts (living with other org. with mutual benefit)

Parasitism can be *obligate* when they do not grow or reproduce away from the host, or *facultative* when they can grow or reproduce on living hosts and also on dead remains of the host.

General Features

A typical fungus consists of tubular filaments enclosed by a rigid cell wall. The filaments, called *hyphae* (sing. hypha), branch repeatedly into a complicated, radially-expanding network called the *mycelium* (Pl. mycelia) which makes up the *thallus*, or undifferentiated body.

Fungi may be a *single cell* as in *yeasts* or a group of cells generally referred to as moulds (molds).

The hyphae may be septate (with cross walls of regular intervals) or aseptate (tubular). The septa (sing. septum) may be complete membranous barriers allowing limited exchange of ions or may be with central pores as in some ascomycetous fungi (sac fungi). The septal pore allows protoplasmic streaming. The cells of the hyphae may be uninucleate, binucleate or multinucleate. When the cells are binucleate with genetically different nuclei, the hypha is called dikaryotic, and when the cells are multinucleate, hypha is called heterokaryotic.

Fungal Cell

Fungal *cell wall* is composed of *cellulose* and *chitin*, and differs from those of plant and bacterial cell walls which lack chitin. The *fungal protoplast* has the *same* general structure as the protoplast of other *eukaryotes*. The nucleus is bounded by a nuclear membrane that is two layered with nuclear pores. The cytoplasm consists of *mitochondria, ribosomes, endoplasmic reticulum, microbodies,* microtubules and vacuoles. Golgi bodies are not always present in fungi.

Modification of Mycelia

The mycelium of some fungi have long *strands* of *hyphae* which are *cemented* together to form as a tissue called the Rhizomorph. The rhizomorph has a thick hard cortex and a growing tip somewhat *resembling a root tip* (hence the name rhizomorph (Gr. rhiza = root). Rhizomorphs are common in Basidiomycetes such as Armillaria mellea, a common soil-borne fungal pathogen of trees.

Reproduction

Fungi reproduce *asexually* and *sexually*. Asexual reproduction also called somatic or vegetative reproduction does not involve the union of gametes.

Asexual Reproduction

Very important method of propagation of the species. *Spore* formation is a *characteristic feature* of fungi and asexual spores fall into the following categories based on their method of formation:

1. Arthrospores or Oidia (Sing. Oidium)

The *hyphae break up* into their *component cells* that behave as a chain of spores called oidia or arthrospores *e.g.*, powdery mildews.

2. Chlamydospores

Spore transformed from cells of *vegetative hyphae* and surrounded by *thick walls* are chlamydospores *e.g.*, fusarium.

3. Blastospores

Spores formed by *budding*.

4. Sporangiospores

Spores formed *inside sac-like* structures called *sporangia* (sing. sporangium) are called sporangiospores. Sporangia are produced at the tips of special hyphae called sporangiosphores. Sporangiosphores bearing one or more flagella are motile in the +ce of water and are called zoospores. If non-motile, these spores are called aplanospores.

5. Conidia (Conidiospores)

Conidia (Sing. conidium) are single-celled or multicelled spores produced at the *tips* or *sides* of *special hyphae* called *conidiophores*. Conidia are quite varied in size, shape, color, septation and ontogeny (developmental history of an individual organism).

Sexual Reproduction

Essentially involves the union of two nuclei of different genetic make-up.

Steps:
1. Plasmogamy—Union of two protoplasts, bringing the two compatible nuclei close together.
2. Karyogamy—Fusion of two nuclei
3. Meiosis—Restoration of haploidy.

The sex organs of fungi are called *gametangia*. The male gametangium is called *antheridium* and the female gametangium is called *Oogonium*. When the male and female gametangium cannot be distinguished morphologically they are called *isogametangia*.

The various means by which compatible nuclei are brought together by plasmogamy are:

1. Planogametic copulation

Fusion of *naked gametes* when one or both of them are motile (*e.g*., Allomyces sp.).

2. Gametangial contact

Two gametangia come in contact but *do not fuse*, the *male nucleus* migrates through a pore or the *fertilization tube* into the female gametangium (*e.g*., Saprolegnia, Pythium).

3. Gametangial copulation

Two gametangia or their protoplasts fuse and give rise to a *zygote* that develops into a resting spore or zygospore (*e.g*., Rhizopus sp.).

4. Somatogamy

Two somatic cells with little differentiation fuse bringing the nuclei together (*e.g*., Basidiomycetes).

5. Spermatization

A *special male* gamete called *spermatium* (Pl. spermatia) *fuses* with *female receptive hyphase.* The spermatium empties its contents into the receptive cell during plasmogamy.

Sexual Spores

Sexual spores are the characteristic spores formed after the fusion of two nuclei. Less varied and less frequent compared to asexual spores.

Types:

1. *Oospores:* The *male nucleus* from an theridium comes in *contact* with the *oogonium* and *fuses* with the female nucleus or *Oosphere* resulting in the *formation* of *Oospores* after fertilization inside oogonium.
2. *Zygospores:* Produced by gametangial copulation and *fusion* of two similar gametangia or *isogametangia*. These are *thick walled*, smooth.
3. *Ascospores:* Produced in a *special sac-like* structure called the ascus (Pl. asci.). These are then may be multiple of eight in some species.
4. *Basidiospores:* Single celled sexual spores borne on a club shaped structure called basidium (Pl. basidia).

Fructification

The *fungal spores* (both asexual and sexual) may be *surrounded* by *Protective tissues*, together forming structures called Fructifications or *fruiting bodies*.

Asexual fruiting bodies	—	*Acervulus* (Pl. acervuli)
		Pycnidium (Pl. Pycnidia)
Sexual fruiting bodies	—	*Cleistothecium.*
		Perithecium.
		Apothecium.

Mycobacterium-Tuberculosis

Introduction: Mycobacteria are slender bacilli and sometimes exhibit filamentous forms resembling fungal mycelium and hence they are so named.

Hausen (1868) discovered the first (Lepra bacillus), Robert Kock (1882) isolated mammalian tubercle bacillus.

Mycobacterium tuberculosis: Mycobacterium tuberculosis cause pulmonary tuberculosis. Robert Koch named the organism Mycobacterium tuberculosis in 1886.

Morphology: Mycobacterium tuberculosis is nonmotile, non-sporing and non-capsulated bacillus arranged singly or in groups. They are acid fast due to presence of mycolic acid in cell wall and weakly gram +ve. With Ziel-Neelsen stain, Mycobacterium tuberculosis look slender, straight or slightly curved rod.

Identification

Staining: Mycobacteria are acid-fast (*i.e.*, they stain red or pink) because of the presence of N. glycolyl-muranic acid in their cell walls. Ziehl-Neelson stain is most commonly employed acid fast technique.

Pathogenesis and Clinical Disease

1. *Primary infection:* Primary infection may occur in any organ such as lung, tonsils, intestine or skin but in children the usual site is in the lungs and is known as "primary complex". This consists of a small peripheral or subpleural focus of tuberculous pneumonia in the lung parenchyma "Glon focus". The outcome of Mycobacterium tuberculosis infection depends on nutritional status and immunocompetancy. Infected patients are at most risk for developing active tuberculosis during the first year following exposure.
2. *Active tuberculosis:* Active tuberculosis may develop as a progression of primary infection or by reactivation of primary infection or by reactivation of a quiescent primary infection.
   (*a*) When the Mycobacterium Tuberculosis is not controlled at the site of primary infection, the phagocytosed bacilli will diffuse through the lymphatic vessels and blood stream.
      (*i*) The lesions may occur in any organ (lungs, kidneys, bone, meninges and intestines); but often of lungs or other sites of high oxygen tension are frequently the sites of infection.
      (*ii*) Medical examination shows chronic caseous granuloma.
   (*b*) *Reactivation of a latent infection:* Reactivation may occur even after 20 years or more of primary infection following lowering of body resistance and most commonly found in western countries.
3. *Miliary (disseminated) tuberculosis:* Occasionally, the primary infection may disseminate by lymphatics or blood stream leading to numerous small miliary tubercles in different organs.

Tuberculosis in immunosuppressed persons:

Immunosuppressed persons are at high risk from tuberculosis which include:

(*a*) transplant recipients.
(*b*) those with malignant disease.
(*c*) persons with AIDS in whom diagnosis of tuberculosis is made before the other features of AIDS are apparent.

Immunoprophylaxis: BCG Vaccine

Dose and administration: Administration of vaccine is done intradermally in a dose of 0.1 ml reconstituted BCG vaccine. BCG vaccine should be administered soon after birth.

Leprosy

Leprosy is endemic and widespread in the underdeveloped areas of the Tropics of the Southern United States, and in some mediterranian countries of Europe. Of the 15–20 million world wide cases, one third are in India.

Mycobacterium leprae: Mycobacterium leprae is the only pathogenic bacteria of man that has not yet been convincingly cultivated in vitro.

This bacterium causes a dreadful disease––Leprosy in human beings; this is also called as Hansen's disease, first described by Hansen in 1874 in Norway.

Morphology: When scrapings from skin or mucous membranes of nasal septum of a patient with lepromatous leprosy is stained by Ziehl-Neelsen staining, the acid fast lepra bacilli are visible in globularmasses. The bacilli are slender, slightly curved or straight rods. They are gram +ve.

Cultivation: Mycobacterium leprae is found only in cases of human infection. They have not yet been grown on artificial media or tissue culture although conflicting claims have been made.

The generation time of lepra bacillus from animal experiments is found to be 12–13 days on the average.

1. *Mouse:* When ground tissue or nasal scrappings from lepromatous leprosy containing lepra bacilli are inoculated intradermally into foot pad of mouse and kept at low temperature (20°C), a granulomatous lesion develop at the site of infection in 1–6 months time with limited multiplication of bacilli.

   European hedge hoylbred in captivity is suitable for propagation of this virus.
2. *Armadillo:* When a banded Armadillo is inoculated with lepra bacilli, generalized infection develops with extensive multiplication of the bacilli and the lesions produced.

   Resistance: Mycobacterium leprae can survive in warm humid environment for 9–16 days, in moist soil for 46 days and in ultraviolet light for 30 minutes.

   Pathogenesis: Leprosy is a chromic granulomatous disease that usually involves skin, peripheral nerves and nasal mucosa. Incubation period is very long and varies from 3–15 years.

Sources of infection are nasal discharge and skin lesions. Method/root of entry is through inhalation of infectious aerosols or through skin contact (damaged skin). The principal target of lepra bacilli is the Schwann cell and the resulting nerve damage causes manifestations of leprosy which include anaesthesia and muscle paralysis. Most likely to occur in children.

Clinical types: There are two main forms of leprosy.

(*i*) Patients with a high degree of cell-mediated immunity develop the tuberculoid form.

(*ii*) *Lepromatous leprosy:* In this form of disease cell-mediated immunity is either absent or deficient.

Some patients occupy an intermediate position, which is termed as Borderline tuberculoid form and this may turn into any of the two main forms.

Borderline type: In this type there is an early unstable tissue reaction with mild transient tissue lesion, often resembling maculoanaesthetic patches.

Laboratory diagnosis of leprosy: Diagnosis of leprosy is based primarily on clinical signs and symptoms.

Specimens: Skin biopsy and scrapings from lesions and nasal mucosa.

Control: So far there has not been prepared any living attenuated vaccine for leprosy.

Treatment: WHO recommended multidrug therapy (MDT) for all leprosy cases based on dapsone, rifampicin and clofazimine.

Table showing:

WHO study group's recommendations for multidrug therapy

Type of leprosy	*Drug*	*Dose*	*Frequency*	*Total duration*
1. Tuberculosis Borderline (Paucibacillary)	Rifampicin	600 mg	Monthly supervised	6 months
	Dapsone	100 mg	Daily unsupervised	
2. Lepromatous Multibacillary	Rifampicin	600 mg	Monthly supervised	2 years or more
	Dapsone	100 mg	Daily supervised	
	Clofazamine	300 mg	Monthly supervised	
		+ 50 mg	Daily unsupervised	

"Typhoid"

Typhoid fever in human beings is caused by *Salmonella typhi*. Infection due to S. typhi is prevalent in India and other Asian countries. The incubation period ranges from 10 to 14 days. Eberth (1880) first observed the typhoid bacillus in mesenteric lymph nodes. Salmon and Smith in 1885 isolated the American-hog-cholera bacillus (S. Choleraesuis) and from the name of the first author the term "Salmonella" derived.

Morphology: The Salmonellae are motile—facultatively anaerobic gram –ve nonsporing bacilli, and primarily intestinal parasites of vertebrates. Most strains are motile due to presence of peritrichous flagella.

Cultural characters: Salmonella grow on ordinary culture media and in Mac Conkey's agar and DCA media optimum temperature for growth is 37°C.

Pathogenesis: S. typhi causes infections in human beings.

Infection occurs almost via oral route, usually with water or food contaminated by sewage or via the hands of carrier. All virulent strains of salmonellae can survive in gastric acidity and penetrate the intestinal mucosa and submucosa.

Typhoid fever: The bacilli attach themselves on the surface of epithelial cells of the villi in small intestine and pass through them to submucous coal, where they are phagocytosed by neutrophils and macrophages. The virulent bacilli resist intracellular killings and multiply within these cells. These cells enter the mesenteric lymph nodes, where after a period of multiplications, the bacilli invade the blood stream via thoracic duct and a transient bacteraemia follows. Liver, gallbladder, lymph nodes, spleen, kidney, lungs and bone marrow are infected during primary bacteraemia in first 7–10 days.

(*a*) *Septicaemia stage:* By about 10th day, the parasitized cells undergo necrosis and the bacilli pass into blood leading to a secondary and heavier bacteria, which causes clinical illness at about 14th day after infection. Some bacteria undergo lysis liberating endotoxin into blood.

(*b*) *Stage of localization:* From blood stream some organisms localize in organs like—gallbladder, liver, spleen and bone etc. Some bacilli are discharged from the gallbladder into the intestine which cause inflammation of "Payer's patches" of intestine producing necrosis leading to typhoid urcers.

The typhoid fever causes severe mental clouding of consciousness (typus-cloud) and includes headache, anorexia, and congestion of mucosa membrane.

Brady cardia and leucopania skin rashes "rose spots" 1–5 mm may appear on chest and abdomen in 2nd and 3rd week.

The disease lasts for 4–5 weeks.

The organism appear in stool during second to third week and in urine during third to fourth week.

Relapse may occur in about 5–10% cases. Incidence of relapse is higher in patients treated with **chloramphenicol**.

Ampicillin Trimetho primsulfamethoseazole Ciprofloseacin and Norflosiacin.

Diarrhoea

Diarrhoea is the most common presenting symptom of lower intestinal tract infection. When there is more than three bowel movements daily and the stool takes on a form that is softer or more liquid form than usual, it is called diarrhoea.

There are two basic types of (Tenesmus: Painful straining while passing the stools).

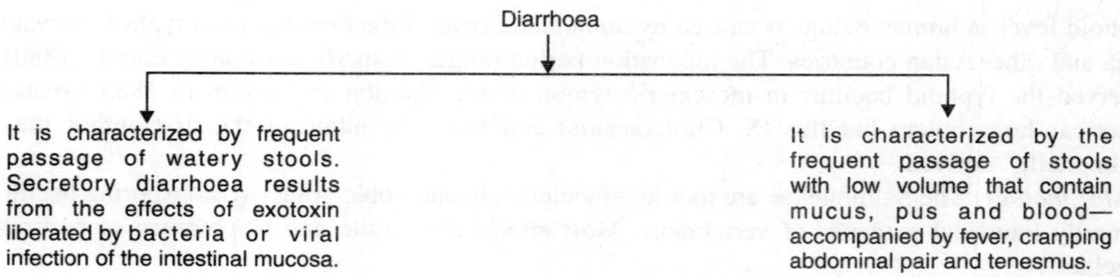

Fig. 7.4

Dysentery is caused by cytotoxins released by bacteria or by the invasion of mucous membranes by bacteria or protozoan parasite.

Epidemiology: There are *two modes* of transmission of infectious agents responsible for diarrhoea.

1. **Direct person to person transmission:** This type of transmission is usually seen in overcrowded places and with poor personal hygiene, *e.g.*, in foils, mental institutions, overcrowded residential areas.
   - Contamination of food during or after cooling. (Flies)
   - In underdeveloped countries, unhygienic environment may lead to contamination of pond water by faeces of infected persons.
2. **Food borne transmission:** It involves distribution of infected food products (*e.g.*, meat, poultry products, seafood or vegetables and fruits) which got contaminated before processing (by zoonotic bacteria).

Aetiology of infective diarrhoea

Bacterial diarrhoea can be divided into two groups, those caused by invasive and those caused by noninvasive microorganisms.

A. *Noninvasive bacterial pathogens*

Food poisoning leading to diarrhoea is caused by the following toxin mediated bacteria:

(*i*) Vibrio cholerae
(*ii*) Enteropathogenic E. coli (EPEC)
(*iii*) Verotoxin producing E. coli (VTEC)
(*iv*) Clostridium
(*v*) Staphylococcus aurens
(*vi*) Bacillus cereus
(*vii*) Clostridium difficile
(*viii*) Shigella dysenteriae
(*ix*) Vibrio parahaemolytions.

B. *Invasive bacterial pathogens*

These bacteria cause diarrhoea or food poisoning by invasive mechanism and not by toxin (non-toxin type)

(*i*) Salmonella species,
(*ii*) Campylobacter species,
(*iii*) Shigella species,
(*iv*) Entero invasive E. coli (EIEC)
(*v*) Enterohaemorrhagic E. coli (EHEC)
(*vi*) Yersinia enterocolitica
(*vii*) Yersinia pseudotuberculosis.

Pathogenesis: A. colonisation and proliferation:

(*i*) **Multiplication:** Most diarrheal infections results from multiplication of the bacteria in the gastro intestinal tract.
(*ii*) **Adherence:** Some diarrhoea causing bacteria attach to intestinal mucosal cells. Adherence is mediated by surface proteins of bacteria.
(*iii*) **Pathogenic mechanisms:** Once established, the aetiologic agents cause infection by *three* ways.
  (*a*) By releasing toxins that affects fluid secretion, cell function or neurologic function.
  (*b*) By invading the intestinal mucosa, causing cellular destruction.
  (*c*) By adhering to mucosal epithelium thus preventing normal functions of absorption and secretion.

B. *Noninvasive bacteria*

Toxin producing bacteria may release enterotoxins or cytotoxins.

(*i*) *Enterotoxins:* Such as cholera toxin and enterotoxin produced by enterotoxigenic E. coli (ETEC). Enterotoxins are responsible for biochemical lesion and there is no histopathological effects on intestinal mucosa.
  1. Sulfonamines Ex. Bactrim D.S.
  2. Quinolones Ex. Ciprofloxacin
  3. Aminoglycocies Ex. Garamycin and Amikacin.

Syphilis

Syphilis is an important form of treponematoses because of its severity and long term effects on untreated cases. There are two types of syphilis.

Types of syphilis

Acquired

Congenital

The acquired syphilis may occur in two forms, venereal and nonvenereal (endemic) syphilis.

Acquired (adult) veneral syphitis is acquired by sexual contact. This disease is caused by *"spirochaetes"* (Speria: Coil; Chaite: hair) are large motile flexible, slender and helical organisms.

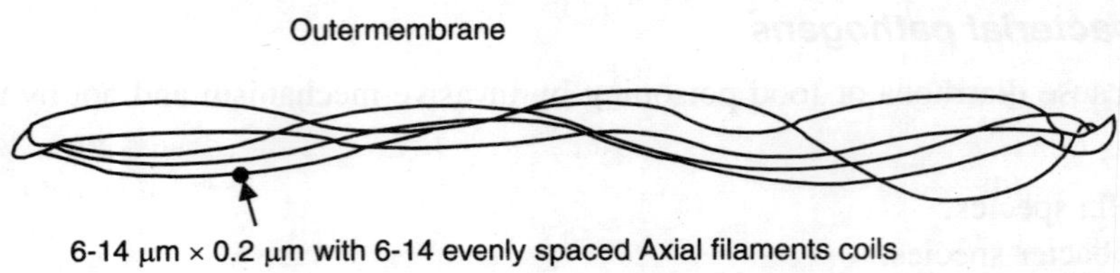

Fig. 7.5 Axial coils

Morphology: These are slender cork screw like filamentous helices with tapering ends. It is actively motile rotating steadily about its long axis.

Pathogenesis: Veneral syphilis is acquired by sexual contact. The trepanomes penetrate mucous surfaces or abraded skin and travels from the site of infection (usually the skin) to the draining lymph nodes in approximately 30 minutes, where they multiply. During incubation period, the bacteria not only invade the lymphatics but also disseminates to the blood stream. After an incubation period of 2–4 weeks (10–90 days) the clinical disease sets in. The principal lesion is endearteris, which can lead to endothelial scarring, and over the long term, can lead to intense inflammatory reaction which is followed by extensive tissue necrosis and replacement fibrosis.

Clinical disease: Clinically, syphilis is divided into early syphilis (primary and secondary stages) and late syphilis (tertiary and quarternary stages). Late syphilis develops usually two (2) years after the onset.

A. **Primary stage:** Usually a papule appears on the genital area, 2 to 4-weeks after initial exposure to T. pallidum that ulcerates forming a classical chancre of primary syphilis characterized by a flat, dull, red and indurated ulcer excluding serous fluid cell hard chancre. The chancre containing abundant spirochaetes.

   When syphilis is contracted by doctors and nurses non-venereally as occupational hazards, the usual site of primary lesion is finger.

B. **Secondary stage:** It starts to manifest itself 3 to 8 weeks after healing of primary chancre. Secondary lesions result from multiplication of the treponomes and their dissemination through the blood.

   Clinical signs include low grade fever, lymphadenopathy. Rashes develop on skin including on the palms and soles.

C. **Latent stage:** In this stage, the infection remains dormant for variable period (2 years is usual benchmark) without any clinical symptoms but with +ve serology.

D. **Tertiary or late stage:** Represented by tertiary lesions appearing any where in the body, characterized by degenerative necrotic lesions that do not contain spirochaetes.

Clinical manifestations:

Include severe cortitis and neurological manifestations—

- blindness,
- mental incapacitation and
- musculoskeletal problems.

Congenital syphilis: T. pallidum can cross the placental barrier from about 16th week of gestation until late in pregnancy. Congenital syphilis tends to occur when the mothers are afflicted with early syphilis during the course of pregnancy, which results in an infection of new born.

1. *Latent infection:* Includes sero +ve.
2. *Early infection:* Widespread maculopapular rash.
3. *Late infection:* Manifestations occur after 2 years. Late Bony destruction, with craniofacial alterations, saddle nase. Hutchinson's triad—notched teeth, intestinal herotitis, and 8th nerve failure—deafness.

Administration of Penicillin

Ex: Benzathine penicillins.

(a) Early syphilis

(*i*) Benzathine benzyl penicillin—a single dose (24 lacs 1 M)

(*ii*) Alternatively, doxcycline 100 mg orally twice a day for 15 days.

(b) Late Syphilis

Late syphilis of greater than 2 years, duration—Benzathine benzyl penicillin 24 lacs 1 M once weekly for 3 weeks.

Congenital syphilis: Treatment is almost similar to that of adult syphilis.

Control: Protective vaccines are not available. Sex hygiene. Prophylactic use of sex barriers such as condoms and antibiotics.

Pneumonia

Pneumonia, a respiratory symptom is caused by *Pneumocystis Carinii*, a unicellular organism with a tropism for growth on respiratory surfaces of mammals. P. carinii, although classified as protozoan, recent findings proved phylogenetically, it is related to fungi. Because P. carinii and fungi have similar cyst wall ultrastructures. Cysts resemble those of Ascomycetes. The 16S rRNA subunit of P. carinii is most homologous with that of Ascomycetes.

P. carinii pneumonia is a major cause of morbidity and mortality in AIDS.

Epidemiology: The organism is very common in the environment. By the age of 5 years, about 75% of children-develop antibodies to P. carinii.

Transmission: Apparently occurs from host to host by droplet inhalation and close contact.

Morphology: It is cystic, oval to fusiform, spores are (1–3 μm dia) sporocyst 4–7 μm dia.

Amature spore case–5 μm dia containing 8 spherical spores. After the nature spore case ruptures to release the intracystic bodies, the cyst wall remains and appears as empty oval or collapsed structure.

Clinical syndrome

The pathogenecity of P. Carinii is closely related to the immune status of the host.

1. In immunocompetant persons the infection is asymptomatic.
2. In immuno compromised persons (Particularly those with (AIDS), or other severe immune disorders the organism produces interstitial pneumonia.

Fewer non-productive cough and respiratory distress are main complaints. Infections of eye, ear, liver and bone marrow are reported among AIDS patients.

Lab. diagnosis

Sputum, bronchial washings.

Bronchio alveolar lavage samples.

Smear of specimen and impression smears of being tissue are stained "Giemsastains".

Nucleus appear purple. Cytoplasm—light blue. Cell wall appears as unstained.

Treatment

Trimethoprim, sulfamethoxazole or a pentamidine isothionate.

"AIDS". Acquired immune deficiency syndrome was first discovered in USA in 1981 **amongst homosexuals, haitians, heroin, addicts and haemophiliacs,** in whom the incidence of Kaposis sarcoma (KS) and Pneumocystis pneumonia were alarmingly high.

The cousative virus was isolated first by **Luc Montagnier** of Pesteur institute, **Paris in 1983** which he called **Lymph Adenopathy Associated virus (LAV). Robert Gallo** and colleagues isolated a retrovirus from AIDS patients of **USA in 1984** and named it **Human-T-cell lymphotropic virus III (HTLU-III)**. Since the virus has been named differently by different workers, the **international committee on virus nomenclature,** named it human immuno deficiency virus (HIV).

Transmission

Known routes

1. Inoculation in blood
   (a) Transfusion of blood and blood products.
   (b) Sharing needles of intravenous drug abusers.
   (c) Tattoo needles.
2. Sexual contact: Anal and vaginal intercourse (unhygenically).
3. Transplacental or perinatal transmission.
   (a) Transplacental transmission.
   (b) Transmission through breast milk.

Routes not involved

Close personal contact with house hold members and health care personnel not exposed to blood.

HIV infects leading to a number of **immunological deficiencies.**

Morphology: HI virus is in icosahedral in shape; 90–120 nm in dia. Its outer envelope is liprotein with spikes present on the surface.

Epidemiology: It has been suggested that the virus originated in Africa. The virus spread to USA through **Haiti** among male homosexuals and drug **addicts in 1970s**, which came out into the open in 1981. The virus spread to Europe from USA and then rest of the world. The first AIDs patient in India was reported from Mumbai in 1986.

The number of seropositive cases in India was estimated **7 millions in 1998.** HIV infection incidence is very high in Manipur (**178 per 1000;** mostly among injected drug abusers), and next in **Maharashtra (108)** followed by **Tamil Nadu (78)**. AIDS is not worldwide in distribution.

Transmission: HIV is transmitted through blood, semen, vaginal fluid and from an infected mother to her child.

HIV infected person remains asymptomatic for a variable periods of time, often 10–15 years. Virus replicates during the period but levels in the blood decrease and is kept checked by humoral and cellular immune response to HIV.

Clinical syndromes: Acute illness associated with seroconversion incubation period ranges between 2–4 weeks. Viraemia occurs during that period and the infection can be transmitted. Clinical features include flu, fever, sore throat, maculopapular rash and lymphadenopathy.

HIV infected patients have unexplained fever, fatigue, night sweats, weight loss and chronic diarrhoea other respiratory infection like lymphocytic interstitial pneumonia in infants and children.

Serological tests: Enzyme linked immuno sorbent assay (ELISA) or agglutination procedures are employed for routine screening.

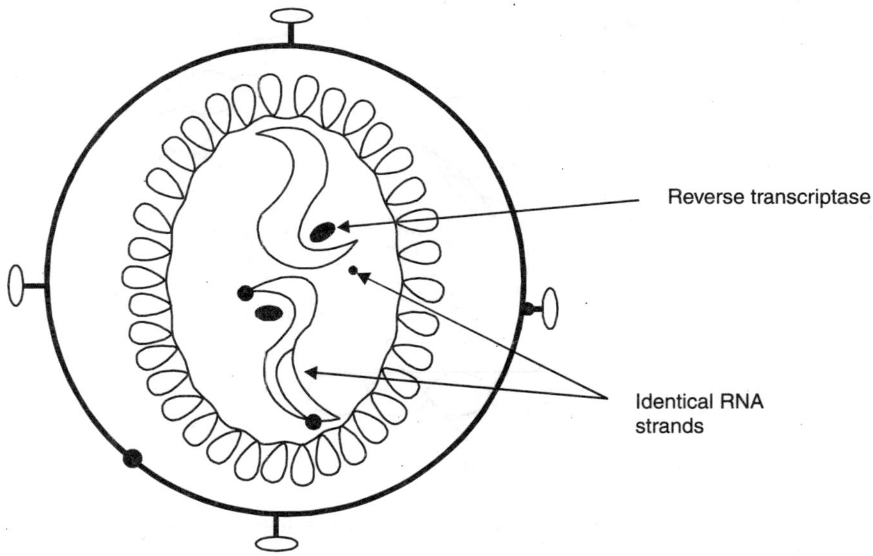

Fig. 7.6 HIV Enveloped Virion

Prevention: WHO has recommended the following preventive measures for which health education is necessary.

1. *Sexual contact:* Safer sex should be practiced by the usage of condoms.

2. Contaminated needles, syringes or other skin piercing equipments should not be shared with others.
3. All blood and blood products are to be screened. This also applies to donation of semen, cornea, bone marrow, kidney and other organs.
4. Isolation of AIDS and patients and treatments against viruses are to be done.

Risk factors: High risk group includes female sex workers and intravenous drug abusers.

Vaccines: A overall effect of recombinant vaccine has not been established in human beings.

Drugs: Azidodideoxythymidine also known as Zydovidine is till now the drug of choice.

Hepatitis Virus

The hepatitis viruses, a heterogenous group of viruses (A to E and G) with tropism for the liver, are the major courses of viral hepatitis. Hepatitis B and C viruses are also responsible for many cases of hepatocellular carcinoma. Except hepatitis B, others A, C, D, E and G contain RNA genome.

Hepatitis A virus (HAV)

HAV causes an acute highly contageous form of hepatitis. It is a subacute disease that occurs in children and young adults.

General properties

1. HAV is enterovirus. It is unenveloped, 27 nm in diameter with an icosahedral symmetry.

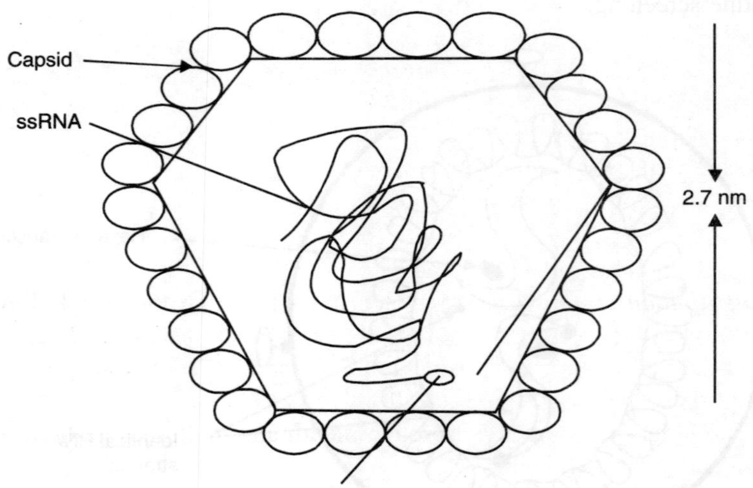

Fig. 7.7 Hepatitis A virus morphology

2. HAV is stable to treatment with 20% ether, at 2.4 pH and a temperature of 60°C for one hour.

Epidemiology

Transmission: HAV causes hepatitis and is excreted in the stools of patients during and after the onset of jaundice. The virion spreads rapidly in a community.

(*a*) *Faecal-oral route:* Most infections occur via spread by the faecal-oral route from contaminated water, food or dirty hands.
(*b*) Person to person transmission may result in out breaks, where personal hygiene is poor.
(*c*) HAV in shell fish occurs from sewage contaminated water.
(*d*) Virus can be transmitted by food handlers and children.

2. HAV is found worldwide especially in areas with poor sanitation.

Pathogenesis: Following ingestion of contaminated food and drink HAV infects cells in gut and spreads the target, the parenchymal cells of the liver via blood. Virus is produced in the these cells and excreted via the bile into the stool approximately 10 days before symptoms appear.

Clinical syndrome: Symptoms produced by HAV result immune-mediated damage to the liver.

Clinical infection: The incubation period ranges from 2–6 weeks (median 28 days).

(*a*) The preicteric stage is characterized by fever, fatique, nausea, vomiting, and sometimes abdominal pain.
(*b*) The icteric phase is characterized by jaundice, hepatomegaly and tenderness in right upper quadrant of abdomen.

Biochemical feature: Liver function tests such as ALT and bilirubin supplement the clinical and pathological diagnosis.

Immunity: Immunity after HAV infection is usually life long. Relapses have been observed, but chromic carriers of HAV are not known.

Prophylaxis

Prevention includes improved sanitation.

Vaccine: Formalin inactivated cell culture derived HAV vaccine with 90% protection are now available but not universally accepted. Attenuated strains obtained by cell cultures are under trial for preparation of vaccine and should be available shortly. *Example:* Live-52 tablets.

Structure: Picorme virus, Pico: small rna = RNA.

Polio

Polio is the causative of poliomyelitis (1, 2 and 3)

Incidence and distribution: Enteroviruses are exclusively human pathogens. Poliovirus is worldwide in distribution. Warm weather favors spread of viral. Faeces of the patient and carrier is the important source of dissemination in the community, usually due to sewage contamination of drinking water.

Transmission occurs by the faecal-oral route.

Pathiogenesis: The virus enters the body through mouth and multiplies in the intestinal epithelium. Incubation period ranges from 9–12 days.

Initial Viraemia

From the intestinal epithelium the viruses spread to the adjacent **lymphatic tissues of oropharynx** and **intestine**. Then they enter the blood stream producing initial viraemia. The patient may form antibody at this stage, preventing the infection from spreading (Abortive infection).

Second viraemia: In patients who fail to control the initial viraemia, there is a second major viraemia from where they are disseminated in spinal cord and brain by passage across the blood-CNS barrier. Children are most susceptible to poliomyelitis.

Clinical Disease

The earlier phase of viraemia includes fever, headache, drowsiness, constipation and sore throat which lasts for 1 to 5 days. Then one of the following types of responses may occur:

Abortive infections: With fewer, headache and sore throat within 3–4 days of exposure occurs among 4–8% of the infected people.

Nonparalytic poliomyelitis: (*Aseptic meningitis*). Approximately 1% of patients develop stiffness and pain in back and neck. This illness losts for 2–10 days. Patient recovers rapidly and recovery is complete.

Paralytic poliomyelitis: It is the major illness, occurs in 0.1–2% of infected patients. It appears 3–4 days after the minor illness has subsided. Flaccid paralysis occurs due to damage of lower motor neurons. Paralysis is classified as spinal, bulbar and bulbospinal depending on nerve involvement.

Maximal recovery takes place within 6 months but residual paralysis lasts much longer.

Post-polio syndrome: It may occur in 20–30% of the original victims 25–30 years later. These patients suffer a deterioration of the originally affected muscles.

Pathological diagnosis: Virus can be recovered from pharyngeal aspiratious, (1–3 days) and faeces (up to 5 weeks).

Prophylaxis: Medical recommendations prescribe the use of attenuated vaccines.

(*i*) Inactivated polio vaccine.
(*ii*) Oral attenuated polio vaccine.

Size: 20–30 nm in dia,
Icosahedral symmetry Naked.
Possessing mRNA.

Rabies

Rabies is a zoonotic disease

- Sylvatic rabies, occur in wild animals *e.g.*, jackals, foxes, wolves, mongoose and bats.
- *Urban rabies* is transmitted by domestic animals *e.g.*, dogs, cats, horses and cattle.

Transmission: The principal reservoir for urban rabies is infected, unvaccinated dog; the virus responsible for this disease is Rhabdovirus; and is transmitted to man due to dog's bite.

Virus is shed in saliva of infected animals. Along with saliva it enters the bite-wound. Occasionally infection may be transmitted via a scratch or, even more rarely inhalation of contaminated aerosolized materials of animal.

Pathogenesis: The virus multiplies in muscles locally and invades peripheral nerve fibres at the neuromuscular function and spreads along Isen peripheral nerves and ascends to CNS (spinal cord and brain). It multiplies in the gray matter in brain and propagates via afferent neurons to peripheral tissues (including salivary glands), from which the virus is excreted.

Incubation period is long and variable which depends on the portal entry and dose of the virus. On an average incubation period is 1–3 months.

Changes: Degeneration of nerve cells, occur in cortex, mid-brain, basal ganglia and medulla.

Clinical disease: After a prodromal period of headache, *malaise* and low grade fever and anxiety of 1–10 days infected person develops characteristic fear of water (Hydrophobia). Though patient is thirsty attempts at drinking provokes violent contraction of diaphragm and inspiratory muscles.

Approximately 10% to 20% to patients exhibit flaccid paralaysis. Death results from neurological and pulmonary failure.

Vaccine: HDCV vaccine 1.0 ml is given intra muscularly on days 0, 3, 7, 14, 30 and 90.

India: Human diploid cell strain vaccine (HDCV) costlier. BPL vaccine.

Diagnosis in animal: The brain of rabied animal is removed carefully and preserved in 50% glycerol saline for isolation of virus.

Virus Structure

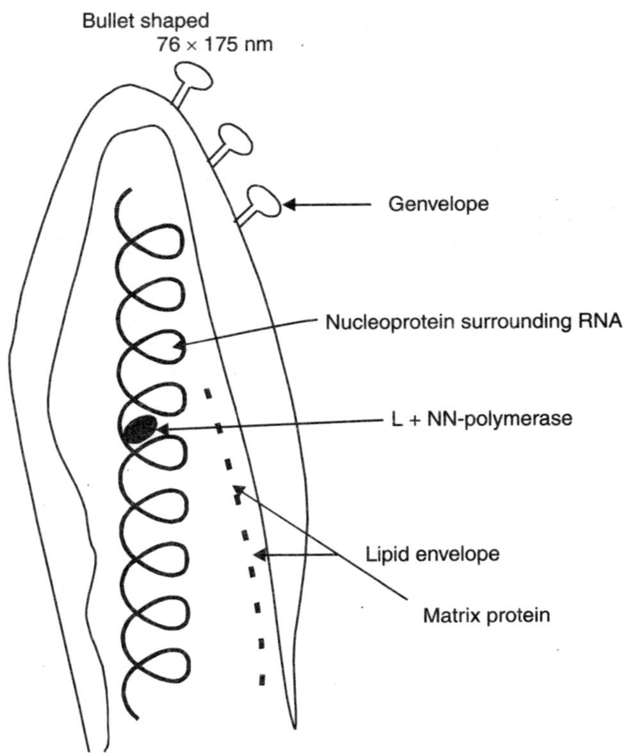

Fig. 7.8

Candidacies

Some saprophytic fungi of environment that usually do not produce disease may cause infection among immunocompromised patients of AIDs, and with wide use of antibiotics, steroid and immuno suppressive drugs of the medically significant yeast species candida and cryptococcus are encountered more frequently among human infections.

Candida albicans, C. Glabrata, C. parapsiolosis, C. Viswannathii, C. Kruse, C. tropicals, C. stellatoidea, C. guilliermondii. C. Albicans is responsible for most (90%) of the human infections.

Morphology: C. Albicans is Gram +ve. Spherical or ovoid budding cells, these are yeast form. At times appear as elongated filaments due to joined end to end, resembling hyphae (pseudohyphae).

Epidemiology: Candida species are worldwide in distribution.

Habitat: C. albicans is present as normal flora of intestine in 40–80% healthy people. They can also be isolated from the oral cavity, vagina, and rectal area of healthy people.

Pathogenecity: Candidiasis is an endogenous infection but cross-infection may occur mother to baby or from baby to baby in nursery.

Superficial lesions: Infection of the mouth occurs in infancy and old age; it is characterized by white patches on the mucosal surface (mouth thrush).

Vaginal thrush: Characterized by typical white lesions on epithelial surfaces of vulva, vagina and cervix. Following sexual contact male sex organ gets infected:

Skin: Infection of the skin usually occurs in sites that may become abnormally moist and warm such as axillae, groin, submammary folds and sometimes too clefts.

Nails: Due to frequent immersion of hands and feet in water infection of finger webs, nail folds occur reddened swellings resembling pyogenic paronychia.

Candidal endocarditis follows surgery of heart valves. C. albicans is involved.

Internal organs like lungs, kidneys, and other organs may be secondarily invaded by C. albicans, when the patient is suffering with J.B. and cancer.

Laboratory diagnosis: Candiolosis is diagnosed on clinical grounds and by a gram +ve strain or by examining a KOH preparation of specimen and confirmed by culture.

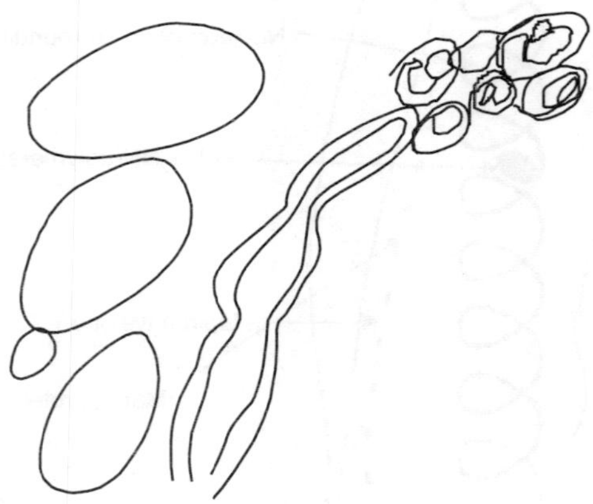

Fig. 7.9

Scrappings from the lesions of skin, nails or mucous membranes are examined in a wet film of KOH or gram stained smear. The fungus appears as budding yeast cells.

Treatment: For superficial infections tropical applications of Polyene (Nystatin) or imidazole (Miconazole, clotrimazole) is effective for candidiasis.

Systemic infection: Amphotericin B is administered along with 5 fluorocytosine, combined therapy.

Amoebiasis

Amoebiasis is defined as the presence of Entamoeba histolytica or *E. coli*—in human body with or without clinical manifestations of the disease. It may manifest as diarrhoea, dysentery or in any of its extra intestinal complications.

Geographical distribution: W.H.O. has estimated that 50% of inhabitants of some developing countries, 10% of world population as a whole and even 1% of Americans are infected with E. histolytica. Amoebiasis produces an estimated 40 million disabling infections and 40,000 deaths annually.

Morphology: Entamoeba histolytica causes amoebiasis, an infection in the large intestine in man. The morphology of the parasite can be studied in both stained and unstained preparations 3 phases are described in the morphology *i.e.*, trophozoite, precyst and cyst. Livetrophozoite of E. histolytica vary is size 10–60 μm.

The trophozoite do not have a fixed shape because of constantly changing position. They move by pseudopodia.

Epidemiology: These amoebae are transmitted by oral-faecal route. Factors such as poverty, ignorance, mental retardation and others that impair personal hygiene facilitate the spread of disease.

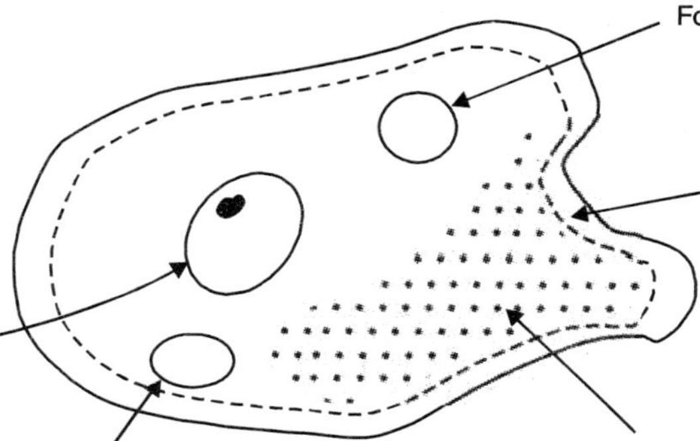

Fig. 7.10

During encystinent this protozoan parasite becomes rounded and is surrounded by a smooth refractile non-staining wall. The size of the cyst varies between 12–15 μm. The cyst is initially uninuclear but by binary fission soon, develops into a binuclear and quadrinuclear body. The cyst containing 1–4 nuclei may be passed in the faeces.

Source of infection: Cysts reach healthy lumen beings through water and vegetables contaminated with infective faeces, through food contaminated by flies or the hands of infected food handlers. It is easy to remember 5 Fs in mode of transmission of E. histolytica; these are—faeces, fingers, food, flies, and fluids.

Pathogenesis: E. histolytica can live in the human of the colon without tissue invasion and by feeding on bacteria and other bowel contents. The trophozoite penetrate the colonic epithelium by pseudopodial motility. They burrow their way deeper by continuous lysis of tissue cells till they reach submucous coat. Amoebae rapidly multiply here, increase in number, form colonies destroy the tissue in their viscinity and use the cytolysed material as their food. Thus amoebae—destroyed area of submucosa leads to abscess formation which breaks down to form ulcer.

Incubation period is 1 to 4 weeks. The intestinal lesions can lead to various other organs of the body. The commonest form is hepatic amoebiasis. The other forms could be pulmonary, and spheric. If uncontrolled-leads to cerebral amoebiasis and even antaneous amoebiasis.

Clinical features of amoebiasis: 6–8 amoebic dysenteries per day; consisting of faeces with blood and mucus. Acidic in nature. Pain and tenderness in the right abdomen. Fever is present in majority cases which is high continuous chills, rigors and sweating may be present.

Prevention: Since man is the principal reservoir of this parasite, all infections should be treated. Hygiene of the environment and provision of safe drinking water must be ensured because chlorination is not wholly effective. Boiling water is a safe and effective method of getting rid of Entamoeba cysts. Health education regarding use of properly washed and disinfectant treated fruit and raw vegetables must be imparted widely. Aqueous solution of iodine or 5% acetic acid are good disinfectants. Food exposed to dust, flies and cockroaches should not be consumed.

Drug treatment: Emetine hydrochloride 1 mg/kg body weight/day till acute symptoms last—followed by chloroquine 250 mg twice a day for 14–21 days plus Metronidazole 750 mg thrice daily for 10 days.

Malaria

Parasite: **Plasmodium vivax.** Transmitted through Amopheles mosquito.

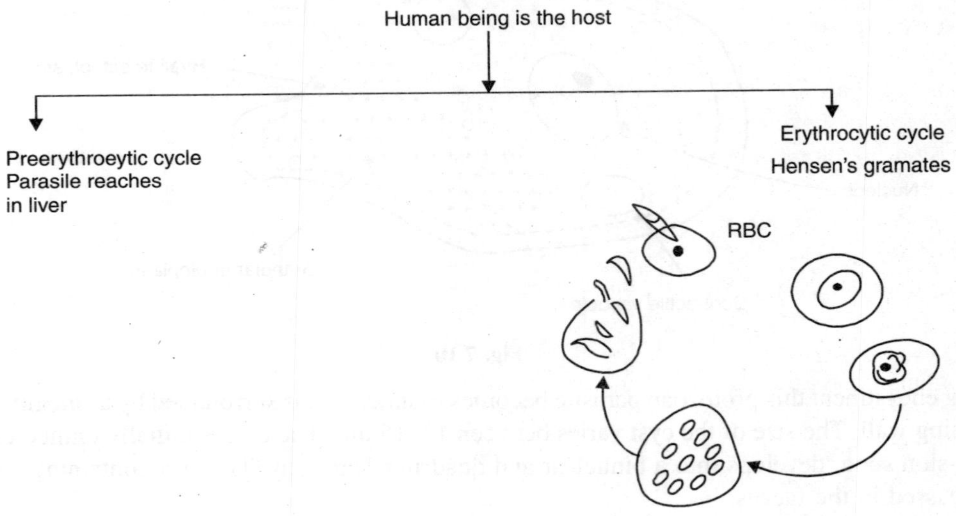

Fig. 7.11

Symptoms of Malaria:

Chillness
High temperature
Burning of eyes
Body pains—Indigestion
Prevention: Prevent mosquito breeding.
Drugs: Aspirin—Antibiotics and pain relievers.

Malaria

Malaria is caused by parasites of genus Plasmodium. They complete their life cycle in two hosts. The asexual cycle (Schizogony) occurs inside the red blood cells of the man (intermediate host), whereas the sexual cycle (Sporogony) occurs in the invertebrate host—a mosquito. Thus there is alternation of generations.

Geographical distribution: Malaria is seen in all countries.

Habitat: After passing through the hepatic parenchymal cells, the parasites reside within the RBCs and are carried by blood to various organs.

Human cycle: When an infected female anopheles mosquito bites man, sporozoites are infected together with saliva and circulate in the blood stream for less than an hour, by which time some of them have invaded the hepatocytes (liver cells) where in pre-erythrocytic cycle—pre-erythrocytic forms develop. These normally rupture in 6–15 days and release thousands of merozoites. Some merozoites enter erythrocytes and erythrocytic phase or blood.

The erythrocytic phase continues until drugs or immunity of the host persists and remains dormant in the hepatocytes for considerable periods and undergo pre-erythrocytic cycle thus liberating merozoites into the blood stream causing relapse of these infections.

Having undergone erythrocytic schizogony for sometime, some of the merozoites instead of developing into trophozoites and schizonts, give rise to gametocytes which settle in the peripheral blood stream—which is the infective to the vector mosquito species.

Mosquito cycle: This starts about 10 days after the malarial infection has occurred in man. The male and female gametocyctes are called microgametocyte and macrogametocyte respectively. The insect vector is female anopheles mosquito. The mosquito during her blood meal from an infected person or a carrier such as blood containing sexual or both sexual and asexual malaria parasites. The asexual parasites taken with the blood meal are destroyed in the stomach of mosquito. Several microgametocytes surround one macrogametocyte. However, only one microgametocyte could enter into the macrogametocyte to form a zygote. The zygote acquires active power of movement and it is called ookinet with a pointed end. It penetrates through the epithelial living and the muscular wall of the stomach wall forming oocyst. Oocyst increases in size and with successive nuclear divisions containing several hundreds of sickle shaped bodies called sporozoites. The sporozoites have special predilection for the salivary glands, pass through the gland and ultimately reach the lumen of alimentary duct. The mosquito is now infective and is capable of transmitting the infection to human beings.

Clinical features: Fever peaks followed by anaemia and splenomegaly. The disease tends to be particularly severe in children and pregnant woman.

Prevention and control: Control measures include

(*i*) prevent mosquitoes from feeding on human beings; destroy adult mosquitoes and its larvae. Eliminate mosquito breeding sites and to treat cases of malaria.

(*ii*) For individual protection, use of mosquito repellents, bed nets, antimosquito fumigants can be used.

Drugs chloroquine, resochin are recommended.

Leishmaniasis

Leishmania donovani—Kala-azar.

Leishmania topica—Oriental sore.

Leishmania braziliensis—Cutaneous leishmaniasis.

The genus of this pathogenic protozoan has been named after Sir William Leishman who discovered the flagellate parasite causing Kala-azar. 3 different species cause 3 different diseases.

In 1903, L. donovani was reported by Sir William Leishmanin the spleen smears of a soldier in England who died from fever contracted at Dum Dum, Calcutta.

Geographical distribution: It is prevalent in India, China, Tropical Africa and South America. In India kala-azar is common in Bengal, Bihar, and Eastern part of U.P. It is referred to by various names such as *Burdwan fever, Jwarvikar, Sarkari Bemari, Kala Dukh* (Blackmisery), etc.

Habitat: The amastigote form is found in the cells of retiaclo endothelial system throughout the body especially in liver, spleen and bone-marrow. It is present in the peripheral blood in few numbers. The promastigote form has been found in the gut of the sand flies of the species Phlebotomus argentipes in India.

Morphology: In the course of development the parasite passes through two stages. Amastigote forms occurs in the gut of the insect vector.

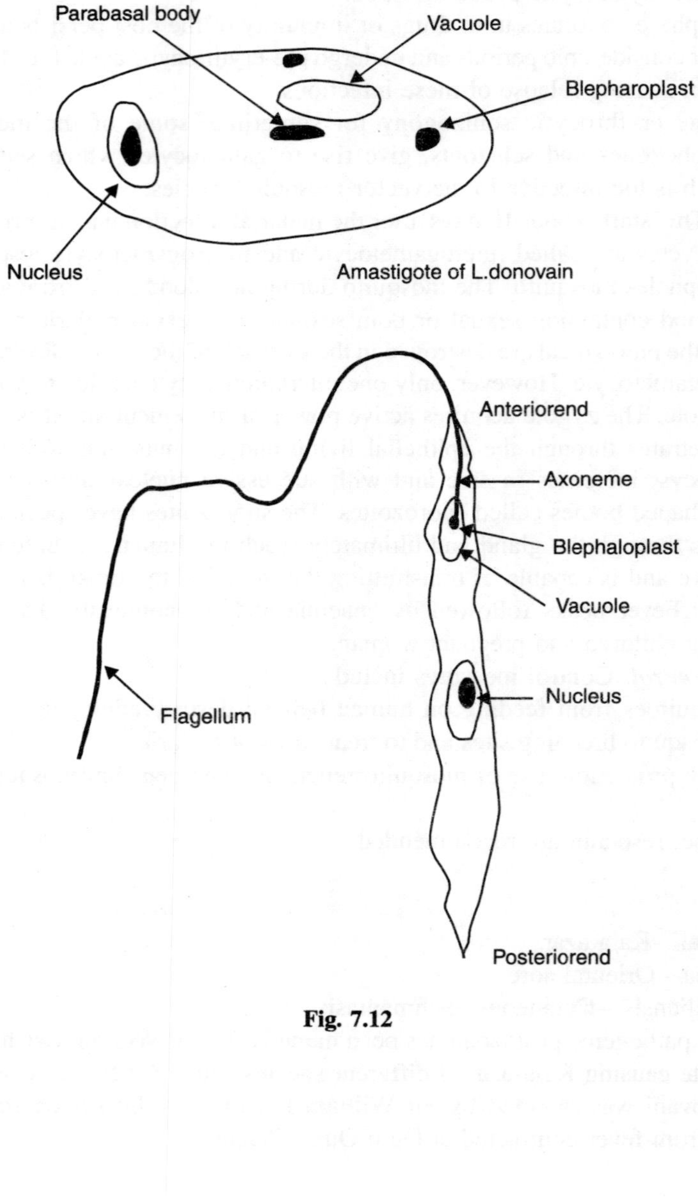

Fig. 7.12

Clinical Features

Incubation period is 3–6 months, many exceed 1–2 years.

Phlebotomus is the chief vector for the transmission of Leishmaniasis in human beings.

- Irregular fever,
- Progressive enlargement of spleen and liver,
- Loss of weight,
- Voracious or perverted appetite,
- Complications like diarrhoea, and dysentery.
- Commonly affects children

Laboratory diagnosis: Depends upon demonstration of Leishmania antibodies in the serum.

Vector control: Consists of reducing the sand fly population by insecticidal spraying using DDT.

Measures to reducing man-fly contact such as mosquito nets, and repellants.

Health education: Making the general public aware about how the disease is caused and how it can be controlled.

Treatment:

Two widely used drugs are:

(*i*) *Meghimine antimoniate* 20 mg/kg up to a maximum of 850 mg/day for a minimum period of 20 days.

(*ii*) Sodium stibogluconate 20 mg/kg body weight up to a maximum of 850 mg/day for a minimum period of 20 days.

Industrially Important Microbial Enzymes

Production of microbial enzymes: A large number of commercially useful enzymes are manufactured from microorganisms. Examples are:

- Proteases,
- Amylases,
- Glucose isomerase,
- Glucose oxidase,
- Renin,
- Pectinases, and
- Lipases.

1. Starch processing
2. Textile industry
3. Detergents
4. Cheese processing

Though many of these enzymes can be obtained from animal and plant sources, production from microorganisms is easier and economical.

Production of proteases: Bacterial alkaline proteoses are largely used in detergents. These enzymes in detergents make them act at lower T., and improved cleaning performance. Proteases are also stain removers in laundry. Proteases for detergents are commercially derived from Bacillus licheniformis.

Recombinant DNA technology adopted Bacillus strains yield an enzyme "KERAZYME" which dissolves hair and useful in cleaning drains clogged with hair.

Fungal proteases of Aspergillus are used in bread-making; they are also used as digestive aids as they help to digest meat products. Proteases are also used in textile industry for removing proteinaceous matter and in the leather industry for softening hides.

Amylases have various commercial applications. They are used in textile and the paper industry mainly to dissolve and remove starch or to coal starch to paper. In laundry they are used in removal of spots from clothes in confunction with proteases. Amylases are also used in brewing industry in the initial breakdown of starchy substances.

Amylase is extracted from fungi—Aspergillus niger, A. oryzae, and bacteria—**Bacillus subtitis, and B. diastaticus.**

Amylases include α-amylase, β-amylase, and glucamylase in addition to other categories.

Glycerol: It has a number of commercial applications (including the production of explosives and propellants). Mainly it is used as a solvent in food colouring agents, as a lubricant in tooth pastes, candles, cake icings and in cosmetics and pharmaceutical industry.

It is produced by using yeast-saccharomyces cerevisiae or bacteria such as Bacillus subtitis.

Methane: Methanogenic bacteria produce methane.

Archaea bacteria—Extremophilic bacteria live in springs at the T° of boiling H_2O and also near volcanic eruptions in the deep sea at temperatures well over 100°C. Some Archea thrive inside the digestive tract of cows, termites, and marine animals where they produce Methane.

Some Archea (Halobacterium halobrium) can live in saturated salt solutions. It is interesting to note that these microorganisms contain a light sensitive red pigment halorhodopsin, which is found in the retina of many vertebrate animals.

Cheese processing: All cheeses result from the tactic acid fermentation of milk. Milk is pasteurized and inoculated with an appropriate starter culture for lactic acid fermentation. This leads to formation of lactic acid and curd. Cheese is the solid product obtained by separating casein and butter from "whey", the liquid portion that is discarded.

Streptococcus thermophilus is employed for acid production in cooked curds since it is more heat tolerant or a combination of S. thermophilus and S. lactis for curds. The curd is shrunk and pressed, followed by salting and cheeses are the product of metabolic activities of lactic acid bacteria.

Swiss cheese is made with a culture of Lactobacillus bulgaricus and streptococcus thermophilus.

In India the most common indigenous cheese made is *Panir*. This is a semihard cheese which is not ripened.

The watery fluid the whey, formed during separation of curd is very good drink for convalescing patients who cannot digest fat. This contains 5% lactose and small amounts of minerals, vitamins, and soluble proteins.

Food products: Microbial spoilage of foot (Putrefaction, Rancidity and Fermentation). Types of food spoilage and control.

Various food products serve as ideal sources of nutrition for microorganisms. Food deterioration and spoilage may occur when the microorganisms naturally contaminating food articles are allowed to grow beyond certain limits. The awareness of food spoilage and the need to take measures to prevent it arose even before the establishment of microbiology as a branch of science. The first man to suggest the role of microorganisms in spoiling foods was **A. Kricher**, a monk who as early as 1658 examined decaying meat, milk and other substances and saw what he referred to as "worms invisible to naked eye". The first man to use heat for food preservation was **Louis Pasteur (1860)** while dealing with wine and beer. Today more refined methods of food preservations have been evolved. The microorganisms growing in food substances may sometimes produce toxic substances which are lethal to consumers. The less lethal "aflatoxins" have been shown to be produced by fungi in various food stuffs and stored grains (**seed pathology**). Another aspect of food contamination and spoilage is the possibility of spread of food borne pathogenic microorganisms that cause various human ailments such as gastroenteritis.

Uncontrolled growth of specific microorganisms in food stuffs is dangerous, however, **the controlled growth of specific microorganisms in certain kinds of food is highly desirable and is most**

accepted. The use of microorganisms in the production of **wine, beer and fermented foods such as Soy sauce, tempeh, idli, pickles, vinegar,** etc. is widely practiced.

Important concern of the food microbiologists are:

(*i*) **the preservation of food by preventing microorganisms** responsible for spoilage,
(*ii*) **the detection of food poisoning** by microorganisms and preventing it, and
(*iii*) **the use of desirable microorganisms** in fermented foods and beverages.

Food spoilage: It is defined as any change in a food that renders it unsafe for human consumption. Foods are classified according to their susceptibility to microbial spoilage as:

(*a*) *perishable,*
(*b*) *nonperishable,* and
(*c*) *semi-perishable.*

(*a*) Perishable foods are those having short storage lives such as meat, fish, poultry, fruits, vegetables, eggs and milk.
(*b*) Nonperishable foods are those that have long shelf life such as sugar, cereals, flour and dry products. However, if the storage conditions are not proper, even these food substances can get spoiled.
(*c*) Semi-perishable foods include vegetables such as potatoes, sweet potatoes, cassava, apples, etc., which generally remain unspoiled for a reasonably long period.

If stored under warm and humid conditions, fungi and bacteria grow on, almost all types of foods. In addition to the environmental conditions, the kinds of organisms that colonise the food article- determine the nature of spoilage. For the colonization of foods the initial inoculum comes from various sources.

Primary Sources of Contamination of Foods

Soil and water: Mic. orgs. from soil and H_2O can be naturally expected in food stuffs and the common bacterial genera are:

- Acinetobacter
- Alcaligenes
- Bacillus,
- Citrobacter
- Clostridium
- Corynebacterium
- Enterobacter
- Micrococcus
- Proteins
- Pseudomonas
- Serratia and streptomyces.

Common fungi from soil are—Aspergillus, Rhizopus, Penicillium, Trichothecium, Botrytis, Fusarium. A large number of yeasts also come from soil.

Yogurt: The yogurt starter is a mixed culture of Lactobacillus bulgaricus and streptococcus thermophilus in 1:1 ratio. The coccus grows faster than the rod and is mainly responsible for acid production while the rod adds flavour and aroma. The associative growth of 2 organisms result in acid production at a rate greater than that produced by them individually. The water content of milk should be first reduced at least by on fourth to get thick yogurt. The milk is heated to 82–90°C for 30–60 min., and cooled to around 45°C before the stertor culture is added. If the fermentation is allowed to go on at 45°C the product forms fastest and should be brought to low temperature when the acidity reaches around 0.65–0.70%. This may require only around 4 hours. At lower temperatures fermentation takes longer time. However, temperature below 25°C will not yield a good product.

Single cell protein—Spirulina.

Food borne infection Salmonellosis and Cl. Botulinum.

Spirulina occurs more commonly in brackish water than in fresh water. It forms phytoplanleton covering the surface of water. The filaments are spirally twisted hence the name.

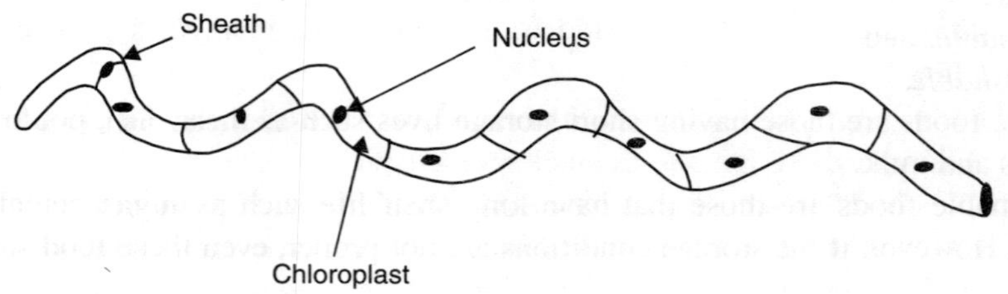

Fig. 7.13

Cyano bacteria (blue green Alga) are oxygenic phototrophic bacteria that use water as an electron donor in photosynthesis giving out oxygen. They may grow as mats on the surface of bare soil, as primary colonizers.

Cell structure: The wall of the cyanobacterial cell corresponds to that of –ve bacteria. The peptiologly can accounts for 22–25% of wt. of the walls and contains N-acetyl muranic acid.

Food preservation: Methods of food preservation have been in used for centuries. Modern methods of food preservation are much more refined. Two principal goals of food preservation methods are:

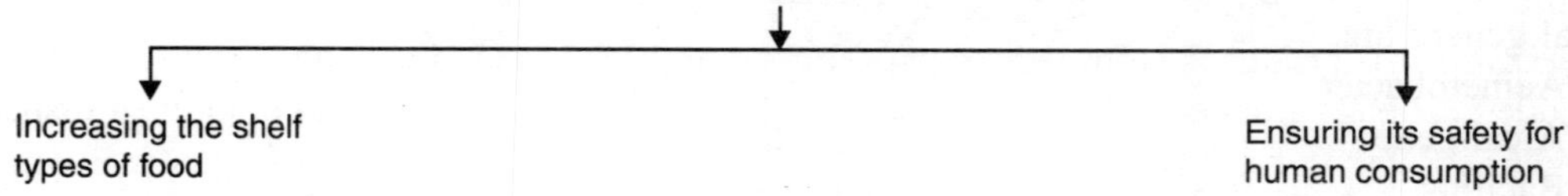

The principles of modern methods of food preservation are:

(*i*) elimination of sources of contamination,
(*ii*) inhibition of the growth of unwanted microorganisms, and
(*iii*) destruction and removal of microorganisms already present in food.

Types of Food Preservation

1. *Asepsis:* To prevent primary contamination use of sanitary methods like washing utensils and anything which comes in contact with food including hands, work surfaces, etc. is important in minimizing contamination of food. Chlorinated water. High T° is deployed to minimize prequadicate

Destructive effect of microorganisms

Pasteurization and destruction of disease producing organisms (Milk) 62.8°C Mycobacterium tuberculosis, coxiella burnetix yeasts moulds, gram +ve, –ve bacteria.

Sterilization destruction of all viable organisms. Canning of food for example uses heat for sterilizing. Heating above the boiling point 115°C for 15 mints. storage of milk increases by this process.

2. Low temperature refrigeration deep freezing 5°C – 20°C prevents microbial growth quick freezing
3. *Drying for preservation:* Ancient process—25% moisture storage ability of dried foods depends on humidity and temperature of storage environment under high humidity moulds tend to grow quickly. Rehydration of dried foods requires long time.
4. *Wood smoke contains:* Tribals volatile compounds possessing. Formaldehyde, phenol, cresols and to molecular not fatty acids.
5. *Irradiation:* γ-irradiation is the most effective and is used for the preservation of sea foods, poultry, pork, vegetable, fruits.

In addition to billing spoiling organisms, radiation can inactivate enzymes involved in autocatalytic spoilage.

Fermented foods: Several types of foods with characteristic texture flavour are produced by microbial fermentation. Many of these include ripened cheeses, pickles, sauerkraut and fermented sausages are preserved products as their shelf-life is extended considerably over the raw materials from which they are made. The vitamin content of fermented food is increased alongwith increased aroma, flavour and digestibility. The microorganisms that produce the desirable changes may be the natural flora of the raw material or may be natural flora of the raw material or away added as involved.

Lactic acid bacteria: The lactic acid bacteria consists of lactobacillus, leucnostoe, pediococans and streptococcus. Those organisms are widespread in nature. All members of this group produce lactic acid during fermentation from hexoses.

Lactic acid bacteria: The lactic acid bacteria consists of the species of lactobacillus, leuconostock, pediococcus and streptococcus. They are widespread in nature. All members of this group produce lactic acid during fermentation.

Other organisms involved in fermentation: In other fermentation processes fungi such as yeasts, Aspergillus spp. Pennicillium spp. and the bacterial species such as Bacillus subtitis are involved.

Sauerkraut: It is produced from lactic acid fermentation of shredded cabbage. Salt (2.5%) is added to shredded cabbage to help extract the juices, control microbiota and maintain an even dispersal of bacteria. Anaerobic conditions develop as a result of continued respiration of plant cells and also because of bacterial metabolism.

The production of sauerkraut involves a succession of bacterial populations. Coliform bacteria perform the initial fermentation.

Food born infections

Food poisoning due to contaminating microorganisms has been known since the time of Louis Pasteur. Food poisoning organisms such as salmonellae, clostridia, and vibriocholeree, may not show sign of spoilage but still contain toxins thus causing epidemics of food poisoning.

Salmonellosis: The salmonellae are gram –ve nonsporing rods widely distributed in nature. All species of salmonella may be presumed to be pathogenic for man. Typhoid fever caused by S. typhi is the most severe of all the diseases caused by this germs. Paratyphoid fever is caused by S. paratyphi A and B. The third disease syndrome caused by salmonellae is gastroenteritis. The symptoms develop 12–14 hours after ingestion of food and persist for 2–3 days. The mortality r e is about 4%.

Salmonellosis involves two toxins—an enterotoxin and a cytotoxin. The main carrier foods of salmonellosis are meat, poultry and eggs. Some of the inoculum comes from the intestinal tract of slaughtered animals. Boiling while cooking can kill the pathogen.

Cholera: Vibrio cholerae causes cholera. The symptoms of this gastroenteritis include diarrhoea, cramps, weakness, chills, headache and vomiting. The foods through which food poisoning occurs are—oysters, shrimps, crabs, lobsters, clams and related shell fish. The prevention involves heat treatment of sea foods at pasteurization temperatures.

Botulism: Clostridium botulinum, the Gram +ve anaerobic, spore forming rod produces a highly toxic, soluble exotoxin while growing in foods. The toxic substance is a neurotoxin formed within the organism and released upon autolysis. The botulism toxin is the most toxic substance known. There are 7 types of toxins A, B, C, D, E, F and G produced by different strains. A toxin is the most potent. This toxin is a polypeptide that can attach to as neural tissue and causes neuroparalysis 60–70% cases are fatal. The foods frequently contaminated with the organism and smoked, pickled and canned—where anaerobic conditions prevail. Whereas cooking removes the toxin as the toxin is sensitive to heat.

The symptoms of botulism include nausea, vomiting, dizziness, dryness of mouth and throat, paralysis, double vision and finally respiratory failure and death.

Prevention of botulism involves proper heat sterilization of the food before canning. As the bacterium is spore forming, high temperatures above 121°C are required to kill endospores. Home canned foods should be heated at boiling point for a few minutes before use as heat destroys the toxin if formed.

Microbial insecticides

Biotechnology can be defined as the commercial application to technological principles of life sciences. Thus in broader sense several industrial techniques pertaining to plants, animals or microorganisms fall under the purview of biotechnology.

Insect resistant plants: Bacillus thuringiensis is a common soil bacterium known to produce a toxin against.

Lepidopteran insect such as the cotton bollworm, maize black cutworm etc. The bacterium produces during sporulation an **intracellular protein toxin,** called the **parasporal body,** that acts as insecticide to specific groups of insects. The parasporal crystal is not the direct toxin but after exposure to **alkaline conditions** in the hindgut of the insect, undergoes **solubilization** and releases the **protoxin** (precursor of the toxin). After the **protoxin** reacts with the **protease enzyme** of **the insect,** the active toxin is released. The toxin binds to the plasma membrane thus creating **osmotic imbalance,** loss of **ATP** and finally **cell lysis**. The toxin is, however, ineffective for humans because of the lack of alkaline conditions, and the specific **protease** in the intestine. The toxin is, however, has to be ingested by the insect; it is harmful to only a specific insects and harmless to other biota.

Different B. thuringiensis strains carry different toxin producing genes-Bt-gene viz. cry genes—(crystal protein). The cry-I genes are toxic to Lepidoptera, cry-II to Lepidoptera and Diptera, cry-III to coleoptera and cry-IV to Diptera.

With the insertion of the Bt-genes into plants and by doing the genes express in plants to produce the toxin, scientists have succeeded in producing genetically engineered insect—resistant plants. Different range of transgenic Bt-crops *e.g.*, Bt-tobacco, Bt-maize and Bt-cotton. Lepidopteran bollworm is a serious pest in cotton; and the use of Bt-cotton drastically reduces the use of pesticides.

Virus resistant plants: RNA molecule which is complementary to the normal gene transcript RNA (mRNA) is called antisense RNA. Antisense RNA can decrease the synthesis of the gene product by forming a duplex molecule with the normal sense RNA. The antisense RNA-mRNA duplex is also easily degraded. It is therefore, possible to prevent the replication of a virus, by creating transgenic plants that synthesize antisense RNA that is complementary to virus coat protein mRNA.

The plants thus produced were completely protected against, low concentrations of the virus (*e.g.* cucumber mosaic virus); but at high concentrations the resistance brokedown. Some more improvements in this technology are essential.

Vaccines developed by recombinant

Biotechnology

Different viruses are used as vaccine in an inactivated form to protect human beings against active virulent virus infections. The safe virus used for protection is vaccinia, "**virus**" which was used by Edward Jennea in 1976 as a vaccine against smallpox. Presently new methods have been used to introduce foreign genes, into vaccinia virus. A plasmid is constructed with some vaccinia genes and the recombinant plasmid is inserted into cells as a calcium phosphate coprecipitate.

The virus recombines with its homologous sequences in the plasmid and in doing so it incorporates the foreign gene into its own DNA. The recombinant virus then replicates, matures and is released from the cell. Thus the vaccinated individual must develop antibodies against the foreign gene and must be protected when challenged with the pathogenic virus.

Vaccine: Vaccination is a process of preventive immunization. It involves infection of an antigen to create an antibody response that will protect the organism against future infections.

Vaccine came from vacca (Latin) = cow used exudates from pustules of cowpox and vaccinated an sysold boy James Phips who has shown protection from pathogen—small pox.

Mycorrhizas

The fine feeder roots of many vascular plants are invaded by specific non-pathogenic fungi, the root and the fungus forming a distinct morphological unit known as mycorrhiza (mybes: mushrooms or fungus, rhiza: roots)

Physiologically they represent a case of symbiosis and not parasitism.

Endomycorrhizas – Eclomycorrhizas

Mycorrhizal roots are thicker.

Ex: in Abies, Fagus, Eucalyptus and Pinus tree. Better intake of Nitrogen Phosphorus and Potassium.

Penicillium: Penicillium genus and its species occur as green molds and blue molds on citrus fruits and other food stuffs. The vegetative mycelium of Penicillium is septate and branched.

The top soil contains this fungi ranging from 100 to 400 thousands per gram of soil. Since fungi prefer aerobic conditions, their number dwindles with the depth of earth's crust. These fungi play a major role in decomposition of litter and other plant materials. Most fungi have enzymes to decompose **pectin**, the substance present in the middle lamella of plant tissues.

Their role in soil fertility includes their ability to ramify between soil particles and give the soil a crumble structure (binding together of fine soil particles) so important in agricultural soils.

In the living organisms phosphorus occurs in the organic form in the protoplasm. On death of the living organism, the organic phosphorous is converted into inorganic phosphoric acid. Phosphoric acid is converted into insoluble salts of calcium, iron, Mg and aluminium. Insoluble phosphorous is made soluble by various phosphate solubilizing fungi like P. digitatum.

UNIT 8

Environmental and Industrial Microbiology

Microbiology has diversified and extended its scope into various applied fields like environmental engineering and industrial microbiology on the productive side, several species of microbes are responsible for producing improved quality and variety of products for betterment of human life. On the other side microbes are capable degrading pesticides, wastes generated in several industries degrading lognocellulosic solid wastes into compost, degrade pesticides and some microorganisms are also capable of degrading petrol and petroleum products. Some microorganismal species are also capable of bioleaching ores in mineral sectors.

"Waste"—A resource: Solid and liquid wastes are the major source of pollution, when they are generated particularly in large quantities. These wastes can be utilized for biogas production and the slurry generated as by-product is used as compost for agricultural fields.

Organic compost: Compost is defined as the process of mixture of decayed organic matter added to soil for the better growth of plants.

Process of composting: The crop/plant residues are degraded in specially designed soak pits where cattle dung including cattle urine are allowed to remain for a desired period so as to conserve nutrients to that confined area. These contents are either exposed to environment or loosely covered with a mud pack so as to prevent water logging during rainy season. Incerbating in this way for 6–8 months, the material is ready to use in form as an organic fertilizer. Most composts generally consist of a mixture of pectin, hemicellulose, cellulose and lignin. Usually the process of compost formation is aerobic. This composted fertilizer can also be used as soil conditioner.

Factors Responsing Composting

(*a*) *Microorganisms:* The suitable microbes selection depends on the type of composting process *i.e.*, aerobic or anaerobic, type of raw material, etc. certain cellulolytic cultures viz., Aspergillus, Trichoderma, Penicillium and Trichurus accelerate composting for efficient recycling of dry crop wastes with high carbon: nitrogen ratio and reduce the composting period by about 30 days. Enrichment of partially composted crop wastes could be achieved by Azotobacter and Phosphate solubilizers to improve the nitrogen.

Anaerobic forms of microorganisms presence in dung or biogas enables compost making efficiently. Basically, the compost carries agriculturally useful microorganisms which help in the promotion of soil fertility.

(*b*) *Soil:* The soil can be described as a natural medium for plants growth consisting of minerals, living organisms and organic materials. The soil microbial metabolism and their biological activities in the soil contribute to soil fertility and texture.

(*c*) *Organic matter:* The quantity of organic matter present in any soil determines it's natural suitability for plant cultivation. The quality of compost is assessed not only by its NPK content but also in the substantial quantities of humus which is important to retain soil organic matter and fertility both in tropical and sub-tropical soils.

Role of compost: Due to compost application, soluble nitrogen contents of agricultural wastes reduce. Composting also results in phosphorus compounds becoming bound up in new microbial cells so that run-off could be avoided. Higher crop yields are claimed by composts against directly applied animal manures. The application of organic compost, farmyard manure, green manure and cereal residues influence favorably the physical, chemical, physicochemical and biological properties of soil.

Biofertilizers: So as to meet the food requirements of increasing population, the nitrogen fertilizer requirement for crop production is considered as the most important prerequisite. Biological nitrogen fixation can be the key to develop nitrogen content of soil—because of the high cost and several other demerits of chemical fertilizers. For the production of a good and efficient biofertilizer first of all an efficient nitrogen fixing strain is required, then its inoculum is produced. Packing, storing and maintenance are other aspects of biofertilizer production.

Criteria for strain selection. The efficient nitrogen fixing strain is evolved in laboratory, maintained and multiplied on nutritionally rich artificial medium before inoculating on seed or soil. In soil, the strain has to survive and multiply to compete for infection site on roots against hostile environment in soil steps for preparing biofertilizers. The isolated strain is inoculated in small flasks containing suitable medium for inoculum production. The volume of the starter culture should be a minimum of 1% to obtain atleast 1×10^9 cells/ml. Now the culture obtained is added to the carrier for inoculant (biofertilizer) preparation. Carriers carry the nitrogen fixing organisms to the fields. In some cases carrier is first sterilized and then inoculated, while in other cases it is first inoculated and then sterilized by UV radiation. The inoculum is now packed with 10^9 –10^{10} viable cells per gram. Ultimate moisture content should be around 40–60%. For large scale production of inoculum culture fermenters are used.

(*a*) *Seed pelleting:* Direct seed coating with sugary syrup or gum arabic and useful nitrogen fixing strains especially the coating of rhizobia over specific host legume seeds are another method for obtaining fruitful results. Immediate after the seed coating $CaCO_3$ is added to sticky seeds. The practice of seed inoculation dates back to 1896—as Voecher used this technique. In many soils the module bacteria are absent or are not adequate in either quantity or quality to meet the nitrogen requirements of the plants. Therefore, it is necessary to inoculate seeds or seedlings with most effective rhizobia.

(*b*) *Inoculant carriers:* Most of the inoculants are mixture of broth culture and a finely milled, neutralized carrier material, carrier is a substance with properties such as, nontoxicity, moisture, absorption capacity, non-lump forming material, easy to sterilize, inexpensive, easily available inexpensive with good buffer capacity, so that it can elongate and maintain the growth of nitrogen fixing microorganisms which it is carrying. The most frequently used carrier for inoculant production is peat. However, where peat is not available alternative carriers like vegetable oils, mineral soils, polyacrylamide, vermiculite, filter mud, bagasse, coal, charcoal and lignite are used. Carriers have to be neutralized by adding precipitated calcium carbonate (pH 6.5–7.0).

(*c*) *Standards of quality for inoculants:* Biofertilizers should follow certain standards. The inoculant should be carrier based and contain a minimum of 10^8 viable cells per gram of carrier on dry mass basis within 15 days of manufacture and 10^7 within 15 days before the expiry date marked on the packet when the inoculant is stored at 25–30°C. The inoculant should have a maximum expiry period of 6 months from the date of manufacture. pH of inoculant should be in between pH 6.0 and 7.5.

The biofertilizer inoculant should be stored in a cool place away from direct heat preferably at a temperature of 15°C. The bioinoculant should be packed in 50–75 µ low density polyethylene packets.

Green manuring: Green manuring is defined as a "farming practice where a leguminous plant which has derived sufficient benefits from its association with appropriate species of Rhizobium is ploughed into the field soil and then a non-legume plant is sown and allowed to get benefitted from the already present nitrogen fixer. This practice of green manuring was in practise from time immemorial in India and China.

Some of the cultivated and annual legumes like sesbania rostrata, Glycine, whitii, Indigofera limifolia, crotalaria juncea, C. strianta, cassia mimosoides, cyamosis painas, Leucaemia leucocephala, etc. contribute nitrogen.

Algal and Other Biofertilizers

Biological nitrogen fertilizers play a vital role to enhance soil fertility, soil productivity and environmental quality. Anabaena azollae, a cyanobacterium lives in symbiotic association with the free floating water fern Azolla. Anabaena azollae can grow photoautotrophically and fixes atmospheric nitrogen. The nitrogen fixing cyanobacteria such as A. azollae and A. variability when immobilized in polyurethane foam and sugar cane waste have significantly increased the nitrogen fixing activity and ammonia secretion. The inoculation of cyanobacteria in rice crop significantly influenced the growth of rice crop by secretion of ammonia in flood water. The use of neem cake coupled with the inoculation of Azolla greatly increased the nitrogen utilization efficiency in rice crop.

Besides contributing Nitrogen, the cyanobacteria add organic matter, secrete growth promoting substances like auxins and vitamins mobilize insoluble phosphate and improve physical and chemical nature of the soil.

Biofertilizers Aiding Phosphorus Nutrition

Tropical soils are deficient in phosphorus. Further most of them solubilize P and thus make it available for plant growth. It is estimated that in most tropical soils, 75% super phosphate applied is fixed and only 25% is available for plant growth. There are certain fungi viz., Aspergillus awamori, Penicillium digitatum, etc. and bacteria polymyxea, Pseudomonas striata etc., which can solubilize unavailable form of P to available form.

Vesicular-arbuscular mycorrhizal (VAM) fungi colonize roots of several crop plants. They are zygomycetous fungi belonging to the genera Glomus, Gigaspora, Acaulospora, Sclercystis etc. These are obligate symbionts and cannot be cultured on synthetic media. They help plant growth through improved phosphorus nutrition and protect the roots against pathogens.

Aquatic microbiology: Water is most essential part of protoplasm and creates a state for metabolic activities to occur smoothly. Therefore, no life could exist without water; there are thousands of microorganisms which live in water and transported through water.

About three-fourth of earth surface is covered by water in the form of oceans, rivers, lakes, streams and ponds etc. However, water is in continuous circulation. Such phenomenon of movement of water is considered hydrological cycle or water cycle. Water is also present deposited below the land which is recovered by bore wells. Water is also present in the form of ice in North and South poles of the earth, and on the top of high mountains etc. From the earth surface water is lost through evaporation, transpiration and exhalation. Almost 93% of evaporated sea-water cools and falls back in the form of rain on the sea and about 900 cubic miles of water moves along with wind current over land surface and joins the clouds; double of this quantity of water is obtained through evapo-transpiration from land source.

Water available naturally is grouped among four types. They are (*i*) atmospheric water, (*ii*) surface water; (*iii*) stored water and (*iv*) ground water.

Atmospheric waters: Atmospheric water consists of rain water and snow. Several microorganisms in the form of dormant propagules, cells and dust particles remain suspended in water and snow. Many bacteria could be insolated from rain water. Dust particles and bacteria are washed from the atmosphere after heavy rain or snow fall.

Surface water: The water bodies, that are stagnant or flowing on earth surface are known as surface water. It is found in the form of several water bodies such as oceans, rivers, streams, lakes, reservoirs etc. Rain water and snow immediately falling on the earth gets contaminated by the micro-organisms present in soil. Total microbial number of water corresponds to the microbial population of soil, quantity of organic materials present in soil and mature of microorganisms present in soil including their activities on the whole water-microorganisms are governed by climatic, chemical and biological conditions of water. For example, microbial population in pools, ponds and lakes would be several times higher than that of flowing streams and rivers. Anthropogenic activities like frequent bathing of cattle, contamination by excreta and influence of effluents in water bodies are causative factors for water-microbial load.

Stored water: The stagnant water present in reservoirs, lakes and ponds etc. are considered stored water. In general during storage the microorganisms get reduced in number; thus it establishes to some extent the purity and stability. However, in stagnant water the microorganisms are affected by sedimentation, temperature, ultraviolet light, food supply, osmotic influence and other microorganisms activities too.

Ground water: Depending on different types of soil particles, their porosity varies. The percentage, space and pores size regulate the quantity of rain water that could be soaked and held in soil. The large pores allow a free percolation of water due to gravitational force. This is known as gravitational water. As water moves down through soil pores, bacteria and other microorganisms clubbed with suspended particles are filtered. Therefore deep bore wells contain negligible number of microorganisms.

Water microorganisms: A large number of saprophytic and pathogenic microorganisms are found in water which belong to the groups–bacteria, fungi, algae, protozoa and nematoda. Several animal viruses are also transmitted through water.

Many of the bacteria found in water belong to groups as follows:

(*i*) Fluorescent bacteria—Pseudomonas, Alginomonas, etc.
(*ii*) Chromogenic rods—Xanthomonos, etc.,
(*iii*) Coliform group—E. coli, Aerobacter, etc.,
(*iv*) Proteus group
(*v*) Spore formers—Bacillus sps.
(*vi*) Pigmented and non-pigmented cocci—Micrococcus sps.

Several types of saprophytic and pathogenic fungi, have been reported from a variety of water sources including potable water. Table 8.1. In addition, there are several water borne protozoan, bacterial and fungal pathogens, which are transmitted through water and cause serious diseases.

Marine Microbiology

It deals with microorganisms living in sea. About 97% of earth's water is present in sea. Ocean at it's greatest depth is slightly more than 11,000 meters deep. Therefore, pressure in the marine environment increases by about 1 atm/10 meters in depth. The bacteria growing in vertically differentiated marine environment can be categorised in three groups:

(*a*) *Barotolerant*—are the bacteria growing between 0 and 400 atm but best at normal atmospheric pressure.

(*b*) *Moderate barophiles*—are the bacteria growing optimally at 400 atm, but still grow at 1 atm.

(*c*) *Extreme barophiles*—are the bacteria growing only at higher pressure *i.e.*, 6000 to 11,000 meter depth.

The hydrostatic pressure can reach 600 to 1,100 atm in the deep sea, whereas the temperature is about 2 to 3°C. Enhanced pressure does not affect the barotolerants. Certain bacteria living in the gut of depose invertebrates like amphipoda and holothurians are truly basophilic. They play a significant role in nutrient cycling. In the marine aquatic system, the major source of organic matter is from phytoplanktons. A common planktonic alga is synecococcus which can reach to a density of 10^4 to 10^5 cells/ml at the sea surface. Picocyanobacteria which is very small cyanobacteria may represent 20–80% of total phytoplankton biomass. In turn they act as a source of food for marine fish and other creatures.

Recently Archeobacteria, a very interesting group of bacteria have been discovered from the marine ecosystem. About 1/3 of oceanic picoplanktons are archeobacteria associated with hostile environment *i.e.*, hot springs, thermo acidophilic regions and highly saline regions.

Microbiology of Domestic Water (Fresh Water Microbiology)

Lakes, reservoirs and rivers provide the major fresh water bodies which are used for potable or drinking water. Due to growth of phytoplanktons and decomposition of organic matter, however, nutrient status varies throughout the year. Further, animals activities including human beings also influence the nutrient level of lakes, reservoirs and rivers. The poor nutrient lakes are known as oligotrophic lakes, whereas the rich-nutrient lakes are called eutropic lakes. The nutrient-rich lakes contain high amount of bottom sediments containing organic matter. Eutrophication (enrichment of lake due to high concentration of nutrients) of lakes occur by multifarious ways. The eutrophic lakes support luxuriant growth of bacteria and algae. Some of the fast growing algae at optimum condition bloom well showing their maximum population. This phenomenon is known as water blooming, and microorganisms associated with it are called water blooms. The microorganism growing in lakes are the genera anabaena, microcystis, nostoc, oscillatoria, oedogonium, spirulina, diatoms and protozoans etc. The two most important lakes in India which are fully eutrophicated are Dal Lake (Srinagar J & K) and Naina Lake (Nainital) which are the only source of potable water.

Among rivers and streams, to process the added organic material, the capacity is very limited. If too much organic material is added, the water becomes anaerobic. This situation arises in those rivers and lakes which are present near the urban areas. In streams and rivers untreated or inadequately treated municipal wastes and other organic materials are discharged. These organic wastes cause changes in microbial community. Release of organic materials added is not too high, it support the growth of several algae. Algae produce O_2 during photosynthesis in day light and evolve CO_2 in day and night. This leads to diurnal oxygen shifts. In the meantime oxygen level completes the self-purification process. This demand of oxygen is expressed in terms of biological oxygen demand (BOD) or the chemical oxygen demand (COD).

Microbiology of Waste Water

There are several ways by which microorganisms enter in water supply, for example, broken sewer lines, congested centres, inappropriate treatment and lack of awareness among people also add in contamination of water. These contaminated waters are waste waters. Unhygenic environment at public places of water collection is also one of the reasons. People suffering from communicable diseases also discharge pathogenic microbe in water through their excreta, for example, amoebic dysentery, typhoid fever, bacillary, dysentery, polymyelitis, etc., Bacteria in water multiply at 20°C to 35°C.

The Faecal Coliforms

Based on concerned water-microbial examination, its suitability for drinking may be ascertained. Intestinal bacteria present in water generally do not survive in aquatic environment due to physiological stress, but if entered human system in the meanwhile, they cause serious problems. The characteristic group of intestinal bacteria are the coliforms. Coliforms are defined as facultative anaerobic. Gram-negative, nonsporing, rod shaped bacteria that ferment lactose with gas formation within 48 hours at 35°C. The coliform bacterial groups are present in water due to faecal contamination *i.e.*, discharge of human faeces, other animals' faeces in water.

Coliforms are the members of the family Enterobacteriaceae which includes E. coli, Enterobacter aerogenes, and Klebsiella pneumoniae. These bacteria make up, about 10% of intestinal microorganisms of human beings, and other animals and therefore, have found wide spread use as indicator organisms, as an index of possible water contamination. If such bacteria are not detectable in a sample of 100 ml, the water can be said as potable water.

Microbes in Bioremediation

Petroleum and its by-products are hydrocarbons. It is a rich source of organic matter and is oxidized if comes in contact with air and moisture. Bioremediation is an important process. In this process oil or other pollutants are utilized by microorganisms if added with inorganic nutrients. The significance of bioremediation in oil spill of marine environment is well studied.

Bacteria and fungi are the main agents which decompose oil and oil product. Besides cyanobacteria, yeast and algae have been shown to oxidise hydrocarbons. The simplest hydrocarbon pollutant is methane. It is degraded by a specialized group of bacteria called methanotrophic bacteria. It is well established fact that oil in insoluble in water and is less dense; it floats on the surface and forms sticks or oil films. Hydrocarbon-oxidizing microorganisms develop rapidly in such films. Oil is present both in the absence of O_2 as well as in the presence of O_2 which is evident by the presence of natural oil deposits.

Bacteria like pseudomonas, cyanobacteria, corynebacteria and mycobacteria are capable of degrading petroleum products. Traditional microbial processes using yeasts and bacteria: certain chemicals which do not exist in nature which are man-made are termed Xenobiotics. They are synthesized compounds such as pesticides. Pesticides are toxic chemicals which act by interfering with microbial reactions in the target organisms. Due to these pesticides microorganisms of soil which are important in maintaining soil fertility. Such microorganisms also detoxify pesticides in soil.

The microorganisms that are able to metabolize pesticides and herbicides enlisted in Table 8.1.

Table 8.1: Showing microorganisms and pesticide degradation products

Name of microbe	*Chemical*	*Degradation product*
Yeast	DDT	1, 1-dichloro 2-2 bis ethane (P-chlorophenyl)
Clostridium sporogenes	y-BHC	1, 2, 3, 5-tetra chlorobenzene
Pseudomonas putida	y-BHC	y-BTC & αBHC
Protens vulgaris	DDT	TDE
Aspergilus niger	Aldrin	Dieldrin
Penicillium Chrysogenum		
Trichoderma viridae	Malathion	Dyfoseon
Mucor plumbeus		

These are diverse group of microorganisms which are able to metabolize pesticides and herbicides, including genera of both fungi and bacteria.

Characteristics of Microbial Metabolism

Most of the metabolic activities in the microbial world are meant for production of energy. Most of the organic molecules can serve as a source of energy to at least some microbes. However a few group are foreign substances to microorganisms. Among insecticidal compounds, the halogen-containing chemicals must be regarded as foreign, or unstable material as such for microorganisms.

Microorganisms, if mutated may have adaptability towards chemicals that are initially toxic to them. In such cases the pesticide degrading metabolic activities become higher. In general microbes alter the pesticide degradation process by using several mechanism. They are as follows.

A. Enzymatic

Incidental metabolism of pesticides which cannot serve as energy sources.

1. (*a*) Wide-spectrum metabolism involving hydrolases, oxidases, etc.
   (*i*) Insecticides as substrates.
   (*ii*) Insecticides as electron acceptors or donors.
   (*b*) Co-metabolism of compounds structurally very similar to the natural substrate.
2. *Catabolism:* Insecticides serve as energy sources.
3. *Detoxification:* Serving as resistance mechanism.

B. Non-enzymatic

1. Photosynthetic breakdown.
2. Contribution via pH change.
3. Production of organic and inorganic reactants.
4. Production of cofactors.

A. Enzymes from microbes: Enzymatic process is most important to find out whether the microbe derives energy from the process or not. It is generally possible that incidental metabolism is more prevalent form of microbial metabolism when the amount of pesticide is high, coupled with favorable chemical structure of pesticide that allows it to be microbially degradable and utilizable as a carbon source.

B. Non-enzymatic process: Certain pesticides are photochemically altered in the environment. There are two ways in which microbial product can promote photochemical reactions. In the first case, microbial products can act as photosensitizers by absorbing the energy from light and transmitting it to insecticidal molecule. In another case, microbial products can facilitate photochemical reactions by serving as donors or acceptors. Scanty information is available on the importance of the microbial formation of organic products capable of reaching with pesticides. Such reactants can be postulated to include amino acids, peptides, alkylating agents, organic acids vitamins etc. Insecticidal chemicals are known to react with amino acid particularly with an—SH moiety.

Ultimately, microbial production of co-factors which are used in both enzymatic and non-enzymatic reactions should not be overlooked. Co-factors are those which promote the overall reactions involving an organic chemical without becoming a part of the chemical reaction product derived from that chemical.

Microbial Insecticides

In common there are three different processes pertaining to insecticidal metabolism viz., (*a*) Hydrolytic processes, (*b*) Reductive systems and (*c*) Oxidation.

(*a*) *Hydrolytic processes:* Majority of the microbial activities are based on hydrolytic processes. It does not occur in other biological group. For example, major metabolic product in Trichoderma viridae is the phenol of mexacarbate (*i.e.*, hydrolysic product) as compared with metabolism in animals which release various oxidation products. The reason is many of the microbes secrete hydrolytic enzymes exogenously as in the case of fungi. Almost all the exozymes secreted by microbial cells are related to the metabolism of large molecules.

(*b*) *Reductive systems:* The major conversion product of parathion is aminoparathion. This is due to microorganisms. On the other hand, products of oxidative reactions is such as para-oxon diethyl-thiophosphoric acid which occurs due to animal metabolism. Another important microbial reaction on insecticide is reductive dechlorination. The reaction proceeds by replacing a chlorine atom on a non-aromatic carbon with a hydrogen atom. The best known examples are the conversion of DDT to TDE and DDE.

(*c*) *Oxidation:* There are many oxidative reactions such as epoxidation of cyclodiens such as altrins and hepataculor to corresponding epoxides, oxidation of thioethers to sulphoxides and sulphones, oxidative dealkylation of alkylamines, aromatic ring opening, decarboxylation, etc.

The most important aspect is to identify one by one the key metabolic reactions and the stable endproducts so as to provide the necessary information to understand the processes, tendencies and role of microorganisms in altering the character of the important group of environment pollutants. DNA technology and useful products from microbes:

In twentieth century experiments through biotechnology were done for the production of potential microbial strains which, could degrade pollutants. Dr. Anand Mohan Chakraborty (Indian born American Scientist) has produced a genetically engineered strain of Pseudomonas putida that utilized complex chemical compounds and it is called commonly as super bug.

Genetic recombination is of fundamental importance to the living organisms for reassortment of genetic material.